ALGEBRA AND ELEMENTARY FUNCTIONS

H. S. Bear

University of Hawaii

Page-Ficklin Mathematics Series

BURGESS PUBLISHING COMPANY
Minneapolis, Minnesota

Copyright © 1976 by Page-Ficklin Publications.

All rights Reserved. No part of this book may be reproduced by any means, nor transmitted, nor translated into a machine language without the written permission of the publisher.

Printed in the United States of America
Library of Congress Catalog Number 76-184869
ISBN 0-8087-2855-5

Marketed by Burgess Publishing Company

PREFACE

This text contains the material for a standard college algebra course, as well as all those topics which are generally covered in a precalculus or elementary functions course. Since much of the difficulty students have with calculus stems from inadequate skill in algebraic manipulation, many schools want to include work in algebra as part of the student's general preparation for further work. This text has enough material to accommodate a wide range of such introductory courses.

Our presentation is deliberately informal, with none of the "definition—theorem—proof" format which has been so fashionable (and so unsuccessful) in recent years. Although formal rigor is missing, the material is nevertheless presented accurately, and there is nothing the student will have to unlearn in subsequent courses. Since some material contains review work, we have tried always to present some feature or point of view which will be new to the student. In particular, the geometric content of each new idea has been emphasized throughout the book.

A great many new examples have been added in this edition, and the exercises have been completely redone. Each example has a set of exercises keyed to it, and these exercises are grouped together in a carefully graduated sequence. This format has several real advantages, both for the students and the instructor. The instructor can select only certain topics from a given section, and still be sure of assigning problems only on the material covered. The student has a readily identified example in the text to consult as a model for each exercise.

Each even-numbered exercise is very like its odd-numbered predecessor. In an elementary functions course taught from this material, the author assigned odd-numbered exercises for homework, and selected quiz problems form the even-numbered exercises. The answers to the odd-numbered exercises are provided, and great care has been taken to make these as free from error as possible.

Each section ends with a set of review questions for the student. These constitute a modified programmed learning approach to the new material and nomenclature. The student should be encouraged to cover the answers given in the margin, and write the answer to each question before uncovering the answer. The commitment to an answer involved in writing it plays an important role in fixing the new material in the student's active vocabulary.

H. S. Bear
March, 1976

CONTENTS

CHAPTER 1.	THE REAL NUMBERS		
	1.1	Set Notation	1
	1.2	The Number Line	5
	1.3	Axioms for Addition	9
	1.4	Axioms for Multiplication	12
	1.5	The Distributive Law	17
	1.6	Axioms for the Order Relation	22
	1.7	Groups, Rings, and Fields	26
CHAPTER 2.	ALGEBRAIC EXPRESSIONS		
	2.1	Integer Exponents	34
	2.2	Fractional Exponents	40
	2.3	Radical Expressions	45
	2.4	Special Products	50
	2.5	Factoring	54
	2.6	Rational Expressions	59
	2.7	Review for Chapter	66
CHAPTER 3.	EQUATIONS AND INEQUALITIES		
	3.1	Linear Equations	69
	3.2	Higher Degree Polynomial Equations	76
	3.3	Special Techniques	81
	3.4	Linear Inequalities	87
	3.5	Further Inequalities	92
	3.6	Review for Chapter	96
CHAPTER 4.	ANALYTIC GEOMETRY		
	4.1	Intervals	101
	4.2	Coordinates in the Plane	108
	4.3	Circles	115
	4.4	Linear Equations	122
	4.5	Inequalities	130
	4.6	Parabolas	137
	4.7	Ellipse and Hyperbola	145

4.8	Review for Chapter	154

CHAPTER 5. FUNCTIONS

5.1	Relations	157
5.2	Functions	162
5.3	Sum, Product, and Composition of Functions	169
5.4	Polynomial Functions	175
5.5	Rational Functions	181
5.6	Implicit Functions	188
5.7	Inverse Functions	194
5.8	Review for Chapter	202

CHAPTER 6. EXPONENTIAL AND LOGARITHMIC FUNCTIONS

6.1	The Exponential Function a^x	205
6.2	The Logarithmic Function $\log_a x$	210
6.3	Tables of Common Logarithms	216
6.4	Computation with Logarithms	221
6.5	Exponential and Logarithmic Equations	224
6.6	Review for Chapter	230

CHAPTER 7. CIRCULAR FUNCTIONS

7.1	Sine and Cosine Functions	233
7.2	Graphs of Sin x and Cos x	241
7.3	Graphs of $y = A \sin (Bx + C), y = A \cos (Bx + C)$	247
7.4	The Other Circular Functions	254
7.5	Sum and Reduction Formulas for Sine and Cosine	263
7.6	Multiple Angle and Half Angle Formulas	271
7.7	Further Indentities	275
7.8	Inverse Circular Functions	280
7.9	Trigonometric Equations	286
7.10	Review for Chapter	289

CHAPTER 8. TRIGONOMETRY

8.1	Angles	293
8.2	Tables and Trigonometric Functions	300
8.3	Right Triangles	307
8.4	Law of Sines	312
8.5	Law of Cosines	321
8.6	Polar Coordinates	326

8.7	Rotation of Axes	333
8.8	Review for Chapter	340

CHAPTER 9. SYSTEMS OF EQUATIONS

9.1	Two Equations in Two Variables	343
9.2	Two Linear Equations	349
9.3	Linear Systems in Several Variables	356
9.4	Matrix Methods	362
9.5	Second and Third Order Determinants	368
9.6	General Properties of Determinants	375
9.7	Review for Chapter	385

CHAPTER 10. COMPLEX NUMBERS

10.1	Addition and Multiplication	389
10.2	The Algebra of Complex Numbers	396
10.3	Conjugate and Modulus	402
10.4	Complex Roots of Polynomial Equations	409
10.5	Polar Form of Complex Numbers	415
10.6	Review for Chapter	425

CHAPTER 11. PERMUTATIONS, COMBINATIONS, AND PROBABILITY

11.1	Permutations	427
11.2	Combinations	433
11.3	Distinguishable Permutations; Partitions	438
11.4	Probability	442
11.5	Review for Chapter	448

CHAPTER 12. SEQUENCES AND SERIES

12.1	Induction	450
12.2	Sequences	456
12.3	Arithmetic and Geometric Sequences	459
12.4	Series	464
12.5	Convergence; Infinite Series	471
12.6	Review for Chapter	478

CHAPTER 13. VECTORS

13.1	Vectors and Scalars	481
13.2	Components	491
13.3	Dot Product	497
13.4	Vectors and Straight Lines	502
13.5	Parametric Equations and Vector Functions	508
13.6	Review for Chapter	513

APPENDIX A. ANSWERS TO ODD NUMBERED
              PROBLEMS   A-1

APPENDIX B. TABLES   B-1
            Table I. Four-Place Common Logarithms of
            Numbers   B-2
            Table II. Natural Logarithms   B-4
            Table III. Four-Place Values of Trigonometric
            Functions   B-6
            Table IV. Four-Place Logarithms of
            Trigonometric Functions   B-13

INDEX

1

THE REAL NUMBERS

1-1 SET NOTATION

The student is no doubt familiar with the basic facts about sets and their operations. Our development in this text will not depend on any specific knowledge of sets, but it is convenient on occasion to use the terminology of set theory. Therefore we will review some of the basic ideas and notation.

A *set* is simply a collection of objects, which we call the *elements* of the set. One can speak of sets of students, sets of books, etc., but we are really interested in sets whose elements are mathematical objects. Thus we speak of sets of numbers, sets of points, sets of lines, and so on.

We use the notation $x \in S$ (read "x epsilon S", or "x belongs to S") to mean that x is one of the elements in the set S. We write $x \notin S$ for "x does not belong to S", or "x is not an element of S". If S and T are sets, we write $S = T$ to mean that S and T are just different names for the same collection. That is, every element of S is an element of T, and vice versa. It is convenient to have a name for the set with no elements, and we denote this *empty set* by $\emptyset$. Hence the empty set is the one set $\emptyset$ such that $x \notin \emptyset$ for all x.

To describe a set with only a few elements, we can simply display all the elements of the set within braces. Thus $\{2, 4, 6, 8\}$ is the set of positive even integers less than 10, and $\{2, 3, 5, 7, 11, 13, 17\}$ is the set of the first seven prime numbers. We also use braces to indicate large finite sets and infinite sets when a definite pattern can be established with a few elements of the set. For example, we write $\{1, 2, 3, \ldots, 99, 100\}$ to indicate the set of the first 100 positive integers, and $\{1, 3, 5, \ldots\}$ to indicate the set of all odd integers. The three dots inside the braces mean "and so on."

Another way of describing a set is to specify some condition that the elements must satisfy. We use the *set-builder notation*, $\{x: P(x)\}$, to mean "the set of all elements x such that $P(x)$ is true". Here $P(x)$ stands for some condition on x. For example, $\{x: x > 0\}$ is the set of positive numbers.

If S and T are any two sets, then we can form two new sets from the elements of S and T. The *union* of S and T, denoted $S \cup T$, is the set containing all the elements which are in either S or T. Hence

$$S \cup T = \{x: x \in S \text{ or } x \in T\}$$

The *intersection* of S and T, denoted $S \cap T$, is the set consisting of those elements which are in both S and T. Thus

$$S \cap T = \{x: x \in S \text{ and } x \in T\}$$

Example 1 Let $S = \{2, 4, 6, 8\}$, $T = \{2, 3, 5, 7, 11, 13, 17\}$, and $U = \{1, 3, 5, 7\}$. Indicate the following sets by displaying their elements in braces: $S \cup T$, $S \cap T$, $S \cup U$, and $S \cap U$.

Solution.

$$S \cup T = \{2, 4, 6, 8, 3, 5, 7, 11, 13, 17\}$$
$$S \cap T = \{2\}$$
$$S \cup U = \{1, 2, 3, 4, 5, 6, 7, 8\}$$
$$S \cap U = \emptyset$$

Note that it does not matter in what order the elements are listed. For example, $S \cup U$ could equally well be written

$$S \cup U = \{2, 4, 6, 8, 1, 3, 5, 7\}$$

We say that a set S is a *subset* of another set T provided every element of S is also an element of T. We write $S \subset T$ to indicate that S is a subset of T. The notation $\subset$ does not mean that T necessarily has any elements other than those of S. Thus $S \subset S$ for every set S. If both $S \subset T$ and $T \subset S$, then $S = T$.

Example 2 Let $S = \{p, d, q\}$. List all subsets of S.

Solution. We agree that the empty set, $\emptyset$, is a subset of *every* set, so $\emptyset$ is one of the subsets of S. The set S itself is also a subset of S. Hence the eight subsets of S are:

$$\emptyset, \{p\}, \{d\}, \{q\}, \{p, d\}, \{p, q\}, \{d, q\}, \{p, d, q\}.$$

We will use the word "equal" and the sign "=" in a very specific way in this text. We write $A = B$ to mean that A and B are both names for the same mathematical object. If A and B are numbers, then $A = B$ means that A and B are names for the same number. For example, when we write

$$1 + 3 = 8/2$$

we simply mean that "1 + 3" and "8/2" are just different ways of expressing the same number. If A and B are sets, then $A = B$ means that A and B both denote the same collection of elements.

An immediate consequence of our use of the equal sign is that if $A = B$, then we can replace A by B in any expression which makes sense. For example, if $A = B$ (and A, B, C are numbers), then it is clear that $A + C = B + C$, since $A + C$ and $B + C$ are just different ways of writing the same thing. Similarly, if $A = B$ (and A, B, C are sets), then $A \cup C = B \cup C$.

Since our primary concern in this course is with algebraic properties of numbers, we will generally use letters ($a, b, c, x, y, z, A, B, C, p, q, r$, etc.) to stand for numbers without explicitly saying so. When we use letters to stand for other mathematical objects, such as sets or functions, it will be clear from the context.

SECTION 1-1, REVIEW QUESTIONS

1. A set is a _____ of objects.   collection
2. The objects in a set are called its _____.   elements
3. We express the fact that x is an element of S with the notation x ___ S. $\in$

S	4. The expression $x \notin S$ means that x is not an element of _____.
∅	5. The empty set is denoted _____.
∉	6. For every element x, x _____ ∅.
braces	7. We can indicate a finite set by listing its elements within _____.
four	8. The set $\{a, b, c, d\}$ has _____ elements.
$x \in S$ and $x \in T$	9. $S \cup T = \{x : x \in S \text{ or } x \in T\}$, and $S \cap T = \{x : \text{_____}\}$.
union intersection	10. $S \cup T$ is the _____ of S and T, and $S \cap T$ is the _____ of S and T.
subset	11. If every element of S is an element of T, then S is a _____ of T.
⊂	12. The fact that S is a subset of T is indicated in symbols: S _____ T.

SECTION 1-1, EXERCISES

Let $A = \{3, 8, 27\}$, $B = \{1, 2, 3, 4, 5, 6, 7, 8\}$ and $C = \{1, 8, 27, 64, 125\}$. Find the sets indicated.

Example 1
1. $A \cup B$
2. $A \cup C$
3. $B \cup C$
4. $A \cap B$
5. $A \cap C$
6. $B \cap C$
7. (a) $(A \cap B) \cup (A \cap C)$
   (b) $A \cap (B \cup C)$
8. (a) $(A \cup B) \cap (A \cup C)$
   (b) $A \cup (B \cap C)$

Let $A = \{a, b, c\}$, $B = \{a, d, e, f, g, h\}$, and $C = \{a, b, i, j, k\}$. Find the sets indicated.

9. $A \cup B$
10. $A \cup C$
11. $B \cup C$
12. $A \cap B$
13. $A \cap C$
14. $B \cap C$
15. (a) $(A \cap B) \cup (A \cap C)$
    (b) $A \cap (B \cup C)$
16. (a) $(A \cup B) \cap (A \cup C)$
    (b) $A \cup (B \cap C)$
17. If $S \subset T$, then $S \cap T = $ _____.
18. If $S \subset T$, then $S \cup T = $ _____.
19. If $S = \emptyset$, then $S \cap T = $ _____.
20. If $S = \emptyset$, then $S \cup T = $ _____.
21. If $S \cup T = \emptyset$, then $S = $ _____.
22. If $(A \cap B) \cup C = A$, then which of the following must be true?
    (a) $C \subset A$
    (b) $A \subset C$
    (c) $A \subset B \cup C$
    (d) $B \subset A$
23. If $(A \cup B) \cap C = C$, then which of the following must be true?
    (a) $A \cap C = \emptyset$
    (b) $A \subset C$
    (c) $C \subset B$
    (d) $C \subset A \cup B$

24. If $A \cup (B \cap C) = B$, then which of the following must be true?
    (a) $A \subset B$ (b) $B \subset A$
    (c) $C \subset B$ (d) $B \subset A \cup C$
25. List all subsets of each of the following sets.   *Example 2*
    (a) $\{x\}$ (b) $\{x,y\}$ (c) $\{x,y,z\}$
    How many subsets do you think $\{x,y,z,w\}$ has?

1-2 THE NUMBER LINE

The first numbers of any interest to anyone were the *natural numbers*, or *counting numbers*:

$$1, 2, 3, 4, 5, 6, \ldots.$$

Over the centuries, larger systems of numbers have been developed from these, leading finally to the real number system.

The sum of two natural numbers is again a natural number, so we say that the natural numbers are closed under addition.* Since the difference of two natural numbers need not be a natural number (for example, $3 - 7 = -4$, and -4 is not a natural number), the natural numbers are not closed under subtraction. The smallest system of numbers containing the natural numbers which is closed under both addition and subtraction is the set of all *integers*:

$$\ldots, -3, -2, -1, 0, 1, 2, 3, \ldots$$

Now the set of all integers is closed under addition, subtraction, and multiplication. However, the quotient of two integers clearly need not be an integer. The smallest system of numbers which is closed under all four operations (addition, subtraction, multiplication, and division) is the set of all fractions, or ratios m/n of integers with $n \neq 0$. This is called the system of *rational numbers*. Since $m/1$ is the same as the integer m, the rational numbers include the integers, which include the natural numbers.

There are many numbers which are not rational, and these are called *irrational*. Thus an irrational number is one which cannot be expressed as the ratio of two integers. For example, it can be shown that the radicals $\sqrt{2}, \sqrt{3}, \sqrt{5}$, etc., are irrational. The number π, which is the ratio of the circumference of a circle to its diameter, is irrational. The number e (approximately 2.71828), which is used as a base for natural logarithms, is also irrational.

---

*In general, a set is *closed* under an operation if whenever the operation is applied to elements of the set, the result is again an element of the set.

The relations among the various subsystems of the real numbers are summarized in the following chart:

{natural numbers} ⊂ {integers} ⊂ {rational numbers} ⊂ {real numbers} ;

{rational numbers} ∪ {irrational numbers} = {real numbers}.

Every real number is either rational or irrational, and within the rational numbers we distinguish the smaller families consisting of the integers, and the natural numbers.

Example 1 Give the most exact classification of each of the following numbers (that is, the smallest of the above subsystems to which the number belongs): $-\sqrt{5}$; $3/2$; $-8/2$; $(-2)(-3)$; $\pi/2$; -5; $-10/3$.

Solution. $-\sqrt{5}$ and $\pi/2$ are irrational; $3/2$ and $-10/3$ are rational; $-8/2$ and -5 are integers (remember, $-8/2 = -4$); $(-2)(-3)$ is a natural number, 6.

A common way of picturing the real numbers is with a *number line* as shown in Fig. 1-1. We label some point on the line 0, and refer

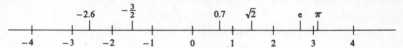

Figure 1-1

to this point as the *origin*. We then pick an arbitrary unit of length, and mark off the points 1, 2, 3 units to the right of the origin. We associate these points with the numbers 1, 2, 3, etc., and, in general, we associate the point x units to the right of the origin with the number x. The point x units to the left of the origin is associated with the number $-x$. Thus -2 denotes the point 2 units to the left of the origin, and $+2$ denotes the point 2 units to the right.

In addition to the two operations, addition and multiplication, the real number system has the order relation "greater than". We express the fact that x is greater than y with the notation $x > y$, and this corresponds to the geometric idea that x is to the right of y on the number line. For example, the statement $\pi > 2$ corresponds to the fact that π lies to the right of 2 on the number line. We can also express the order relation with the symbol $<$ (less than). Thus $x < y$ means the same as $y > x$. The symbols $\leq$ and $\geq$ mean "less than or equal", and "greater than or equal", respectively.

Example 2 Put the correct symbol, $<$ or $>$, between each pair of numbers: 2 ___ $-\pi$; -1 ___ 0; 3/2 ___ $\pi/2$; -7 ___ -8; -50 ___ 1.

Solution. $2 > -\pi$; $-1 < 0$; $3/2 < \pi/2$; $-7 > -8$; $-50 < 1$.

The distance from the point x on the number line to the origin is called the *absolute value* of x, and is denoted $|x|$. For any number x, $|x| = |-x|$; and both $|x|$ and $|-x|$ stand for the nonnegative one of the numbers x and $-x$. Thus the absolute value of a positive number is the number itself, and the absolute value of a negative number is its negative (since the negative of a negative number is positive). For example, $|-5| = 5$; $|2| = 2$; $|-\sqrt{3}| = \sqrt{3}$; $|0| = 0$; $|-\tfrac{1}{4}| = \tfrac{1}{4}$; and $|-\pi| = \pi$.

Example 3 Solve the equations: (a) $|x - 5| = 2$; (b) $|2x| = 8$.

Solution. (a) The two numbers whose absolute value is 2 are 2 and -2. Therefore if $|x - 5| = 2$, we have either

$$x - 5 = 2 \quad \text{or} \quad x - 5 = -2.$$

Hence $x = 7$ or $x = 3$.
(b) The two numbers whose absolute value is 8 are 8 and -8. Therefore $|2x| = 8$ means

$$2x = 8 \quad \text{or} \quad 2x = -8.$$

Hence $x = 4$ or $x = -4$.

SECTION 1-2, REVIEW QUESTIONS

1. The numbers 1, 2, 3, ... are called counting numbers, or _____ numbers.   **natural**

2. The natural numbers are closed under _____ but not subtraction.   **addition**

3. The numbers ..., $-3, -2, -1, 0, 1, 2, 3, \ldots$ are called the _____.   **integers**

4. The integers consist of the natural numbers, their negatives, and the number _____.   **zero**

5. The system of all quotients of integers is the system of _____ numbers.   **rational**

6. The rational numbers are _____ under all four operations.   **closed**

7. The numbers which are not rational are called _____.   **irrational**

8. Every number is either _____ or irrational.   **rational**

integers   9. The rational numbers include the _____, which include the the natural numbers.

greater than   10. The statement $x > y$ is read: "x is _____ than y".

11. The statement "x is greater than y" means the same as "y is less than x",

$<$ which is written y _____ x.

$|x|$ 12. The distance from x to 0 is denoted _____.

nonnegative   13. For any number x, $|x|$ is a _____ number.

$-x$ 14. $|x|$ is the nonnegative one of the numbers x and _____.

x 15. If $x < 0$, $|x| = -x$; and if $x \geq 0$, $|x| =$ _____.

SECTION 1-2, EXERCISES

Let N denote the set of natural numbers, I the set of integers, Q the set of rational numbers, P the set of irrational numbers, and R the set of real numbers.

1. N is closed under what operations $(+, -, x, \div)$?
2. I is closed under what operations?
3. Q is closed under what operations?
4. Which of the following is true?
    (a) If $x \in N$, then $x \in R$.
    (b) $I \subset P$.
    (c) $Q \subset P$.
    (d) Q is a subset of I.
    (e) If $x \in I$, then $x \in Q$.
    (f) N is a subset of Q.
    (g) Every element of N is an element of Q.
    (h) If $x \in I$, then $x \in N$.
    (i) $N \subset P$.

Give the most precise classification of the following numbers (natural number, integer, rational number, or irrational number):

Example 1   5. $-\frac{3}{2}$ 6. $\sqrt{4}$ 7. $\frac{\sqrt{7}}{3}$ 8. $-\frac{\pi}{2}$

9. -8 10. $\frac{4}{\sqrt{6}}$ 11. $-\sqrt{9}$ 12. $-\frac{3}{127}$

Put the correct symbol $<$, $>$, or $=$ for each pair of numbers.

Example 2   13. -5 _____ -6 14. $\sqrt{16}$ _____ 4

8 *The Real Numbers*

15. -3 _____ 0
16. $\dfrac{5}{8}$ _____ $\dfrac{5}{16}$
17. $\dfrac{4}{3}$ _____ $\dfrac{\pi}{3}$
18. $\sqrt{2}$ _____ $\sqrt{3}$
19. $\dfrac{1}{\sqrt{2}}$ _____ $\dfrac{1}{\sqrt{3}}$
20. $\dfrac{1}{3}$ _____ $\dfrac{2}{5}$
21. $\dfrac{1}{8}$ _____ $\dfrac{3}{16}$
22. $-\dfrac{1}{9}$ _____ $-\dfrac{4}{27}$
23. $-\dfrac{1}{3}$ _____ $-\dfrac{3}{10}$
24. -11.01 _____ -11.02

Solve the following equations.

25. $|x| = 3$
26. $|-x| = 3$
27. $|x| = -3$
28. $|2x| = 8$
29. $|-5x| = 0$
30. $|3x| = 27$
31. $|x + 2| = 5$
32. $|x - 1| = 6$
33. $|x - 3| = 4$
34. $|-2x + 1| = 0$

Example 3

1-3 AXIOMS FOR ADDITION

So far we have talked informally about certain set operations, and given names to certain classes of numbers. In the process we have become familiar with some notation, and introduced a way of visualizing the set of real numbers. Now we can start with our actual mathematical development.

In this section, and the three following sections, we state axioms for the real number system, and show how some of the basic facts (theorems) of algebra follow from these axioms. This section deals with properties of addition.

An operation such as addition is simply a rule which associates a third number with each pair of numbers. For example, addition associates the number 5 with the pair (3, 2), and this fact is indicated by the symbolism $3 + 2 = 5$. The following are the axioms for addition.

1. *Commutative Law* For all a and b, $a + b = b + a$.

2. *Associative Law* For all a, b, and c, $a + (b + c) = (a + b) + c$.

3. *Existence of an Identity* There is a number, 0, such that for all a, $0 + a = a$.

4. *Existence of Inverses* For every number a, there is a unique additive inverse, $-a$, such that $a + (-a) = 0$.

Now let us draw some simple conclusions from these axioms.

Axioms for Addition

First notice that the commutative and associative laws allows us to rearrange and regroup the terms in any sum. For example,
$$x + [y + (z + w)] = y + [x + (w + z)]$$
or
$$x + [y + (z + w)] = (x + y) + (z + w).$$
There are of course many other possible arrangements which are equal to these. Since grouping and arrangement does not change the value of a sum, we will omit parentheses in this kind of sum and write simply
$$x + y + z + w.$$
Since the additive inverse of a number is unique, axiom 4 implies that whatever $a + b = 0$, we know that $b = -a$ and $a = -b$. In particular, since $0 + 0 = 0$, and $a + (-a) = 0$, we can conclude that $-0 = 0$ and $-(-a) = a$.

We now simplify our notation and write

$\qquad a - b \qquad$ instead of $a + (-b)$

$\qquad -a - b \qquad$ instead of $(-a) + (-b)$

$\qquad -a + b - c + d \qquad$ instead of $(-a) + b + (-c) + d$

and so on. The operation of subtraction is just a combination of addition and taking the *additive inverse*. In other words, to subtract b means to add the additive inverse of b.

By rearranging the terms, we see that
$$(a + b) + (-a - b) = 0$$
and
$$(a - b) + (-a + b) = 0.$$
Consequently, the expressions in parentheses in each of these equations are inverses of each other; that is,
$$-(a + b) = -a - b$$
and
$$-(a - b) = -a + b.$$

You get the additive inverse of any expression in parentheses by changing the sign of each term.

We also refer to the additive inverse of a number or expression as its *negative*. Thus the negative of a is $-a$, the negative of $-a$ is a, the negative of $a - b$ is $-a + b$ (or $b - a$), and so on.

Example 1 Write the negatives of each of the following: (a) -8; (b) $x - y$; (c) $-(3 - x)$; (d) $-x + 1$; (e) $2 - x + y - z$.

Solution. (a) $\qquad -(-8) = 8$

(b) $\qquad -(x - y) = -x + y = y - x$

(c) $\qquad -[-(3 - x)] = 3 - x$

(d) $\qquad -(-x + 1) = x - 1$

(e) $\qquad -(2 - x + y - z) = -2 + x - y + z$

Example 2 Remove parentheses and simplify $a - [b - (c - d) + a]$.

Solution. We start with the inside parentheses and simplify as follows:

$$a - [b - (c - d) + a] = a - [b - c + d + a]$$
$$= a - b + c - d - a$$
$$= -b + c - d.$$

Example 3 Show that the following *cancellation law* holds: If $a + c = b + c$, then $a = b$.

Solution. We suppose that $a + c = b + c$. Because of our conventions for the equal sign, we get the same result if we add the same number to $a + c$ and to $b + c$. We add the number $-c$ to both of these expressions, and conclude

$$(a + c) + (-c) = (b + c) + (-c)$$
$$a + [c + (-c)] = b + [c + (-c)] \qquad \text{associative law}$$
$$a + 0 = b + 0 \qquad \text{definition of inverse}$$
$$a = b \qquad \text{definition of 0}$$

SECTION 1-3, REVIEW QUESTIONS

1. The operation of addition is a rule which associates a unique number with each _____ of numbers.   **pair**
2. The number that addition associates with the pair (3, 2) is _____.   **5**
3. The commutative law says that $a + b =$ _____.   **b + a**
4. The associative law says that $a + (b + c) =$ _____.   **(a + b) + c**

identity	5. The number 0 such that $0 + a = a$ for all a is called the additive _____.
−a	6. The additive inverse of a number a is denoted _____ .
negative	7. The inverse $-a$ is also called the _____ of a.
0 a	8. The negative of 0 is _____ , and the negative of $-a$ is _____ .

SECTION 1-3, EXERCISES

Write the negative (additive inverse) of each of the following.

Example 1

1. -3
2. $-\frac{5}{2}$
3. 2
4. $x - 2$
5. $-x + 2$
6. $-(x - 1)$
7. $3x - 5$
8. $7 - x$
9. $2 + x + y$
10. $r - 3 - s$
11. $4 - u - v$
12. $r - s + t$

Remove parentheses and simplify each of the following.

Example 2

13. $a - (b + a)$
14. $(a - b) + (b - a)$
15. $x - (y + z) + y$
16. $x - [x - (x - y)]$
17. $-[a - (a - b)]$
18. $-[-a + (b + a) + (a - b)]$

Give examples to support your answer to the nex two questions.

19. Is subtraction a commutative operation? That is, is $a - b = b - a$ for all a, b?
20. Is subtraction associative? That is, is $a - (b - c) = (a - b) - c$ for all a, b, c?

Example 3

21. Show that if $a - c = b - c$, then $a = b$. Justify each step.

1-4 AXIOMS FOR MULTIPLICATION

The axioms for multiplication are almost exactly parallel to those for addition. The only difference is that while *every* number has an additive inverse, one number (zero) does not have a multiplicative inverse.

1. *Commutative Law*   For all a and b, $a \cdot b = b \cdot a$.
2. *Associative Law*   For all a, b, and c, $a \cdot (b \cdot c) = (a \cdot b) \cdot c$.

3. *Existence of an Identity* There is a number, 1, such that for all a, $1 \cdot a = a$.

4. *Existence of Inverses* For every number a except 0, there is a unique multiplicative inverse, a^{-1}, such that $a \cdot a^{-1} = 1$.

We explicitly assume that 0 and 1 denote different numbers, so the additive and multiplicative identities are different. The multiplicative inverse, a^{-1}, of a number is also called its *reciprocal*.

In exactly the same way as for addition, we see that grouping and the order of the factors do not change the value of a product. Thus

$$x \cdot [y \cdot (z \cdot w)] = y \cdot [x \cdot (w \cdot z)]$$

or $\qquad x \cdot [y \cdot (z \cdot w)] = (x \cdot y) \cdot (z \cdot w).$

Hence we omit the parentheses and write $x \cdot y \cdot z \cdot w$ instead of one of the more complicated expressions above.

We saw in the last section that if $a + b = 0$, then necessarily $b = -a$ and $a = -b$. In the same way, from axiom 4 we know that if $ab = 1$, then $b = a^{-1}$ and $a = b^{-1}$. In particular, we have $1^{-1} = 1$ (since $1 \cdot 1 = 1$) and $(a^{-1})^{-1} = a$ (since $a \cdot a^{-1} = 1$). Notice that

$$-(-a) = a \quad \text{and} \quad (a^{-1})^{-1} = a$$

are completely parallel, and follow in exactly the same way from parallel assumptions.

We defined subtraction, $a - b$, as the sum of a and the inverse of b:

$$a - b = a + (-b).$$

In exactly the same way, we define division as the product of a and the *multiplicative* inverse of b:

$$a \div b = a \cdot b^{-1}.$$

We also use the notation a/b for division, so

$$\frac{a}{b} = a \div b = a \cdot b^{-1}.$$

It is important to remember that none of the expressions a/b, $a \div b$, $a \cdot b^{-1}$ make any sense unless $b \neq 0$, since 0 does not have a multiplicative inverse.

We list below some laws for reciprocals which are parallel to those developed in the last section for negatives. Each new law is obtained from a law for addition by substituting multiplication for addition, the multiplicative inverse a^{-1} for the additive inverse $-a$, and the multiplicative identity 1 for the additive identity 0. Since the axioms are the same for multiplication and addition, the proofs will be the same, with the single proviso that $a \neq 0$ when you write a^{-1}.

Multiplication	Addition	
$1^{-1} = 1$	$-0 = 0$	(1)
$(a^{-1})^{-1} = a$	$-(-a) = a$	(2)
$(ab)^{-1} = a^{-1}b^{-1}$	$-(a+b) = -a-b$	(3)
$(ab^{-1})^{-1} = ba^{-1}$	$-(a-b) = b-a$	(4)
$(ab^{-1})(cd^{-1}) = (ac)(bd)^{-1}$	$(a-b)+(c-d) = (a+c)-(b+d)$	(5)
$(ab^{-1})(cd^{-1})^{-1} = (ad)(bc)^{-1}$	$(a-b)-(c-d) = (a+d)-(b+c)$	(6)

We now restate the laws above for reciprocals in fractional notation, recalling that a/b means $a \cdot b^{-1}$, and hence $1/b$ means b^{-1}.

$$\frac{1}{1} = 1 \tag{1'}$$

$$\frac{1}{\frac{1}{a}} = a \tag{2'}$$

$$\frac{1}{ab} = \frac{1}{a} \cdot \frac{1}{b} \tag{3'}$$

$$\frac{1}{\frac{a}{b}} = \frac{b}{a} \tag{4'}$$

$$\frac{a}{b} \cdot \frac{c}{d} = \frac{ac}{bd} \tag{5'}$$

$$\frac{\frac{a}{b}}{\frac{c}{d}} = \frac{ad}{bc} \tag{6'}$$

Notice that (5') is just the familiar rule for multiplying fractions: *Multiply the numerators and multiply the denominators.* Similarly, (6') gives the rule for dividing fractions: *Invert the denominator and multiply.*

Example 1 Write as a simple fraction and tell which of the laws [(1'), (2'), etc.] is involved. Write the additive analogue to each statement: (a) $1/\frac{1}{8}$; (b) $\frac{1}{3} \cdot \frac{1}{7}$; (c) $1/\frac{2}{3}$; (d) $\frac{2}{3} \cdot \frac{4}{5}$; (e) $\frac{4}{7}/\frac{3}{5}$.

Solution. (a) From (2'),
$$\frac{1}{\frac{1}{8}} = 8$$
$$-(-8) = 8$$

(b) From (3'),
$$\frac{1}{3} \cdot \frac{1}{7} = \frac{1}{3 \cdot 7} = \frac{1}{21}$$
$$(-3) + (-7) = -(3 + 7)$$

(c) From (4')
$$\frac{1}{\frac{2}{5}} = \frac{5}{2}$$
$$-(2 - 5) = 5 - 2$$

(d) From (5'),
$$\frac{2}{3} \cdot \frac{4}{5} = \frac{2 \cdot 4}{3 \cdot 5} = \frac{8}{15}$$
$$(2 - 3) + (4 - 5) = (2 + 4) - (3 + 5) = 6 - 8$$

(e) From (6'),
$$\frac{\frac{4}{7}}{\frac{3}{5}} = \frac{4 \cdot 5}{7 \cdot 3} = \frac{20}{21}$$
$$(4 - 7) - (3 - 5) = (4 + 5) - (7 + 3) = 9 - 10$$

Example 2 (This is the multiplicative analogue of Example 2 in Sec. 1-3.) Remove parentheses and simplify $a \cdot [b \cdot (c \cdot d^{-1})^{-1}]^{-1}$.
Solution. Start with the inner parentheses, and use rules (3) and (2):
$$a \cdot [b \cdot (c \cdot d^{-1})^{-1} \cdot a]^{-1} = a \cdot [b \cdot c^{-1} \cdot d \cdot a]^{-1}$$
$$= a \cdot b^{-1} \cdot c \cdot d^{-1} \cdot a^{-1}$$
$$= b^{-1} \cdot c \cdot d^{-1}$$

Example 3 (This is the multiplicative analogue of Example 3 in Sec. 1-3). Show that the following *cancellation law for multiplication* holds: If $ac = bc$ and $c \neq 0$, then $a = b$.

Solution. Since $c \neq 0$, c^{-1} exists and we get the same number if we multiply ac and bc by c^{-1}:

$(ac)c^{-1} = (bc)c^{-1}$

$a(cc^{-1}) = b(cc^{-1})$ associative law.

$a \cdot 1 = b \cdot 1$ definition of multiplicative inverse.

$a = b$ definition of 1.

SECTION 1-4, REVIEW QUESTIONS

commutative 1. The law $ab = ba$ is the _____ law for multiplication.

associative 2. The law $a(bc) = (ab)c$ is the _____ law for multiplication.

1 3. The multiplicative identity is _____.

a^{-1} 4. The multiplicative inverse of a is denoted _____.

5. Every number has an additive inverse, and every number except 0 has a

multiplicative _____ inverse.

reciprocal 6. The multiplicative inverse, a^{-1}, is also called the _____ of a.

7. For every law for addition and subtraction there is a parallel law for multiplication and division, except that division by

zero _____ is impossible.

SECTION 1-4, EXERCISES

Write the reciprocal (multiplicative inverse) of each of the following. (These are the multiplicative analogues of Ex. 1-12 of Sec. 1-3.)

1. $\dfrac{1}{3}$ 2. $\left(\dfrac{5}{2}\right)^{-1}$ 3. 2 4. $\dfrac{x}{2}$

5. $\dfrac{2}{x}$ 6. $\dfrac{1}{x-1}$ 7. $\dfrac{3x}{5}$ 8. $\dfrac{7}{x}$

9. $2xy$ 10. $\dfrac{3x}{y}$ 11. $\dfrac{4}{uv}$ 12. $\dfrac{rt}{s}$

Write each of the following as a simple fraction, and justify your answer with one of the laws (1'), (2'), ... , (6'). Then write the additive analogue.

16 The Real Numbers

13. $\dfrac{3}{4} \cdot \dfrac{5}{8}$
14. $\dfrac{1}{\frac{3}{7}}$
15. $\dfrac{\frac{2}{3}}{\frac{7}{8}}$ Example 1

16. $\dfrac{1}{\frac{1}{9}}$
17. $\dfrac{1}{2} \cdot \dfrac{1}{5}$

Remove parentheses and simplify each of the following. Write the simplified version in fractional notation (i.e., using no negative exponents).

18. $a(b^{-1}a)^{-1}$
19. $(ab)^{-1}a$ Example 2
20. $[(ab)^{-1}c]^{-1}$
21. $[(a^{-1}b)^{-1}b]^{-1}$
22. $[(ac)(abc)^{-1}]^{-1}$
23. $(ac)(bc)^{-1}(a^{-1}b)$

Give examples to support your answer to the next two questions.

24. Is division a commutative operation? That is, is $a/b = b/a$ for all a and b different from zero?
25. Is division an associative operation? That is, is $(a/b)/c = a/(b/c)$ for all $a, b,$ and c, with b and c different from zero?
26. Show that if $a/c = b/c$, with $c \neq 0$, then $a = b$. Justify each step. Example 3

1-5 THE DISTRIBUTIVE LAW

In Section 1-3 we saw how the standard rules for addition and subtraction followed from the axioms for addition. We did the same kind of thing for multiplication and division in the last section. Now we develop the rules for treating the two operations together. That is, we now consider sums of products, differences of products, sums of fractions (quotients), and differences of fractions.

The axiom which relates the two operations of addition and multiplication is called the (left and right)

Distributive Law: For all $a, b,$ and c

$$a(b + c) = ab + ac$$

$$(b + c)a = ba + ca$$

This is the only axiom which involves both addition and multiplication.

The first consequence of the distributive law is the important fact that

$$a \cdot 0 = 0. \qquad (1)$$

To see this, we start with identity $0 + 0 = 0$, multiply both sides by a, and use the distributive law:

$0 + 0 = 0$	definition of 0
$a(0 + 0) = a \cdot 0$	definition of "="
$a \cdot 0 + a \cdot 0 = a \cdot 0$	distributive law
$a \cdot 0 = 0$	subtract $a \cdot 0$

The laws of signs for multiplication follow from (1) as follows:

$a + (-a) = 0$	definition of $-a$
$[a + (-a)] \cdot b = 0 \cdot b$	definition of "="
$ab + (-a)b = 0$	distributive law and (1)

Therefore $(-a)b$ is the additive inverse of ab, or

$$(-a)b = -(ab). \qquad (2)$$

If you change the sign of one factor, (2) says that you change the sign of the product. Therefore if you change the sign of both factors you change the sign of the product twice, and hence

$$(-a)(-b) = -[-(ab)] = ab. \qquad (3)$$

From the above laws of signs we can now verify the left and right distributive laws for subtraction:

$$a(b - c) = ab - ac; \; (b - c)a = ba - ca. \qquad (5)$$

The first equation above is checked as follows:

$$a(b - c) = a(b + (-c)) \quad \text{definition of subtraction}$$
$$= ab + a(-c) \quad \text{distributive law}$$
$$= ab - ac \quad \text{by (2).}$$

Example 1 Write out the following products.

(a) $3(x - y)$ (b) $-a(2c - 3b)$ (c) $-3x(2y + z)$

Solution (a) $3(x - y) = 3x - 3y$
(b) $-a(2c - 3b) = -2ac + 3ab$
(c) $-3x(2y + z) = -6xy - 3xz$

In the example above, we used the distributive law to expand a product. The next example illustrates how the same law can be used to simplify a sum by collecting like terms together.

Example 2 Simplify: (a) $2a + 3a$ (b) $3ab + 4x - 5ab + 6x$ (c) $xy - 3x(y + 2z) + xz$

Solution. (a) $2a + 3a = (2 + 3)a = 5a$
(b) $3ab + 4x - 5ab + 6x = 3ab - 5ab + 4x + 6x$
$$= (3 - 5)ab + (4 + 6)x$$
$$= -2ab + 10x$$
(c) $xy - 3x(y + 2z) + xz = xy - 3xy - 6xz + xz$
$$= -2xy - 5xz$$

The rules for addition and subtraction of fractions are also consequences of the distributive laws. For example, we have

$$\frac{a}{c} + \frac{b}{c} = ac^{-1} + bc^{-1} \quad \text{definition of quotient}$$
$$= (a + b)c^{-1} \quad \text{distributive law}$$
$$= \frac{a + b}{c} \quad \text{definition of quotient}$$

A similar argument works for subtraction, so we have the rules for addition or subtraction of fractions with the same denominator:

$$\frac{a}{c} + \frac{b}{c} = \frac{a+b}{c}; \quad \frac{a}{c} - \frac{b}{c} = \frac{a-b}{c}. \tag{4}$$

To add (or subtract) fractions with different denominators, we first rewrite the fractions so they have the same denominator. Since $\frac{a}{b} = \frac{ad}{bd}$ and $\frac{c}{d} = \frac{bc}{bd}$, we can use (4) to get

$$\frac{a}{b} + \frac{c}{d} = \frac{ad}{bd} + \frac{bc}{bd} = \frac{ad+bc}{bd}. \tag{5}$$

Similarly,

$$\frac{a}{b} - \frac{c}{d} = \frac{ad}{bd} - \frac{bc}{bd} = \frac{ad-bc}{bd}. \tag{6}$$

Example 3 Simplify

(a) $\frac{3x}{y} - \frac{2}{y}$ (b) $\frac{1}{a} + \frac{1}{b}$ (c) $\frac{1}{x} - a$ (d) $\frac{2y}{3x} + \frac{a}{b}$

Solution. (a) $\frac{3x}{y} - \frac{2}{y} = \frac{3x-2}{y}$

(b) $\frac{1}{a} + \frac{1}{b} = \frac{b}{ab} + \frac{a}{ab} = \frac{b+a}{ab}$

(c) $\frac{1}{x} - a = \frac{1}{x} - \frac{ax}{x} = \frac{1-ax}{x}$

(d) $\frac{2y}{3x} + \frac{a}{b} = \frac{2by}{3bx} + \frac{3ax}{3bx} = \frac{2by+3ax}{3bx}$

SECTION 1-5, REVIEW QUESTIONS

1. The axiom which relates addition and multiplication is called the

    distributive   _____ law.

2. The distributive law is what we must use to handle sums of products, sums of fractions, differences of products, and differences of

    fractions   _____.

    ab + ac   3. The distributive law states that $a(b+c) = $ _____.

    0   4. For any number a, $a \cdot 0 = $ _____.

5. From the fact that $a \cdot 0 = 0$ we get the rule of signs: $a(-b) =$ _____. — $-(ab)$
6. For any a and b, $a(-b) = -(ab)$, and $(-a)(-b) =$ _____. — ab
7. The distributive law also gives us the rule for adding fractions with the same denominator: $\dfrac{a}{c} + \dfrac{b}{c} =$ _____. — $\dfrac{a+b}{c}$
8. To add fractions with the same denominator, you just add the numerators; to add fractions with different denominators you must first rewrite them so they do have a common _____. — denominator
9. A common denominator for $\dfrac{a}{b}$ and $\dfrac{c}{d}$ is _____. — bd
10. Since $\dfrac{a}{b} = \dfrac{ad}{bd}$ and $\dfrac{c}{d} = \dfrac{bc}{bd}$, we have $\dfrac{a}{b} + \dfrac{c}{d} =$ _____. — $\dfrac{ad+bc}{bd}$

SECTION 1-5, EXERCISES

Write out the following products.

1. $a(x + y)$
2. $x(b - a)$ Example 1
3. $-2u(r + s)$
4. $-x(a - b)$
5. $3x(2u - v)$
6. $-2rs(-x + y)$

Simplify the following.

7. $3x - 2x$
8. $7x + 3b - 5x$ Example 2
9. $ab - 2t + 4ab - t$
10. $-uv + 2x - 10x - 4uv$
11. $3xy + 4ab - 4xy$
12. $-x - (3x - y) + 2y$
13. $ab + 5b(c - 2a) + 6bc$
14. $2rs - 3s(t + r)$

Add the following fractions.

15. $\dfrac{3a}{x} - \dfrac{2b}{x}$
16. $\dfrac{a}{bc} + \dfrac{2}{bc}$ Example 3
17. $\dfrac{x}{2a} - \dfrac{4x}{2a}$
18. $\dfrac{-2a}{3xy} + \dfrac{b}{3xy}$
19. $\dfrac{1}{x} + \dfrac{1}{y}$
20. $\dfrac{1}{x} - \dfrac{1}{y}$
21. $\dfrac{a}{b} + \dfrac{c}{x}$
22. $-\dfrac{u}{v} + \dfrac{r}{s}$
23. $\dfrac{a}{b} - \dfrac{b}{a}$
24. $\dfrac{2x}{y} + \dfrac{3y}{2x}$

*25. Show that $\dfrac{-1}{a} = -\left(\dfrac{1}{a}\right)$ if $a \neq 0$.

*26. Show that $\dfrac{1}{-a} = -\left(\dfrac{1}{a}\right)$ if $a \neq 0$; i.e.; that $(-a)^{-1} = -(a^{-1})$. Hint: check that $(-a)(-a^{-1}) = 1$.

I-6 AXIOMS FOR THE ORDER RELATION

In this section we look at the axioms for order, the axiom which relates order and addition, and the axiom which relates order and multiplication. Since $x < y$ means the same as $y > x$, we arbitrarily pick the sign $<$ to write our axioms. Each statement involving $<$ clearly implies a companion statement involving $>$. The first two axioms for order are the following:

1. *Trichotomy Law*: For every two numbers a and b, exactly *one* of the following statements is true:

$$a < b, \quad a = b, \quad b < a.$$

Another way of stating the trichotomy law is that if a and b are *distinct* numbers, then either $a < b$ or $b < a$. We will refer to any statement involving $<$ or $>$ as an *inequality*.

2. *Transitivity Law*: For any three numbers a, b, and c, if $a < b$ and $b < c$, then $a < c$.

We frequently use the transitivity property implicitly and write chains of inequalities like

$$-9 < -5 < -1 < 4 < 27.$$

We infer from the transitive law that any number in such a chain is less than any number to the right of it. For example, $-9 < 4$ and $-5 < 4$ can be read from the chain above.

A number a is *positive* if $a > 0$, and *negative* if $a < 0$. Hence every non-zero number is eigher positive or negative, and zero is the only number which is neither positive or negative.

The next axiom gives the relationships between order and addition.

3. *Addition preserves order*: For any three numbers a, b, c, if $a < b$, then $a + c < b + c$.

In other words, an inequality is preserved if you add the same quantity to both sides.

As a first simple consequence of this axiom we show that

$$\text{if } a > 0, \text{ then } -a < 0. \tag{1}$$

$a > 0$ assumption

$a - a > 0 - a$ Axiom 3 (add $-a$)

$0 > -a$ definition of 0 and $-a$.

The same type of argument also shows that

$$\text{if } a < 0, \text{ then } -a > 0. \tag{2}$$

Combining these results with the fact that $-(-a) = a$, we immediately get

$$\text{if } -a > 0, \text{ then } a < 0, \tag{3}$$

and

$$\text{if } -a < 0, \text{ then } a > 0. \tag{4}$$

Our final axiom for order relates order and multiplication.

4. *Multiplication by a positive number preserves order*: if $a > b$, and $c > 0$, then $ac > bc$.

The immediate consequence of Axiom 4 is the fact that the product of two positive numbers is positive: if $a > 0$, and $b > 0$, then $a \cdot b > 0 \cdot b$, or $ab > 0$. Combining this fact with the laws of signs of the last section, and (1), (2), (3), (4) above we see that the product of a positive number and a negative number is negative, and the product of two negative numbers is positive.

If a is any number other than zero, then a is positive or negative. In either case $a^2 > 0$, since the product of two positive numbers is positive, and the product of two negative numbers is also positive. Since $1 = 1 \cdot 1$, we conclude that $0 < 1$. Also, since $a \cdot \frac{1}{a} = 1$, and 1 is positive, both a and $\frac{1}{a}$ are positive, or both are negative. That is, if $a > 0$, then $\frac{1}{a} > 0$, and if $a < 0$, then $\frac{1}{a} < 0$.

Successively adding or subtracting 1 from both sides of the inequality $0 < 1$ gives us the ordering of the integers:

$$... < -4 < -3 < -2 < -1 < 0 < 1 < 2 < 3 < 4 < ....$$

Axioms for the Order Relation

Axiom 4 says that an inequality is preserved if both sides are multiplied by a positive number. If you multiply (or divide) both sides of an inequality by a *negative* number, the inequality is *reversed*. To see this, suppose $a > b$ and c is negative. Then $a - b$ is positive, and c is negative, so their product is negative:

$$(a - b)c = ac - bc < 0.$$

Therefore $ac < bc$, and the inequality sign is changed from $>$ to $<$ by multiplying by the negative number c.

One consequence of this last rule is that an inequality is reversed if you change the sign of both sides, since that is the same as multiplying both sides by -1:

$$\text{if} \quad a > b, \quad \text{then} \quad -a < -b. \tag{5}$$

Example 1 Fill in the correct sign, $>$ or $<$:

(a) if $2x + 3 < 8$, then $2x$ ____ 5.
(b) if $-3x < 12$, then x ____ -4.

Solution. (a) If $2x + 3 < 8$, then $2x < 5$, since the inequality is preserved by subtracting 3 from both sides.

(b) If $-3x < 12$, then $x > -4$, since the inequality is reversed when both sides are multiplied by a negative number.

Example 2 Fill in the correct sign, $>$ or $<$:

(a) if $4 < x$, then $\dfrac{1}{x}$ ____ $\dfrac{1}{4}$;
(b) if $a < b < 0$, then $\dfrac{1}{a}$ ____ $\dfrac{1}{b}$.

Solution. (a) Since 4 and x are both positive, so is $4x$, and hence also $\dfrac{1}{4x}$. Multiplying both sides of $4 < x$ by $\dfrac{1}{4x}$ gives

$$\frac{1}{4x} \cdot 4 < \frac{1}{4x} \cdot x,$$

$$\frac{1}{x} < \frac{1}{4}.$$

24 *The Real Numbers*

(b) If a and b are both negative, then $ab > 0$, and hence $\frac{1}{ab} > 0$. Multiplying both sides of $a < b$ by $\frac{1}{ab}$ preserves the inequality and gives

$$\frac{1}{ab} \cdot a < \frac{1}{ab} \cdot b,$$

$$\frac{1}{b} < \frac{1}{a}.$$

SECTION 1-6, REVIEW QUESTIONS

1. The trichotomy law states that $a < b$, or $a = b$, or _____. $b < a$
2. Any statement involving $<$ or $>$ is called an _____.   inequality
3. A number a is positive if a _____ 0, and negative if a _____ 0. $>$ $<$
4. If $a > 0$, then $-a$ _____ 0, and if $a < 0$, then $-a$ _____ 0. $<$ $>$
5. If you add the same number to both sides of an inequality, the inequality is _____.   preserved
6. The product of two positive numbers is positive, and the product of two negative numbers is _____.   positive
7. The product of a positive number and a negative number is _____.   negative
8. An inequality is preserved if you multiply both sides by a _____ number.   positive
9. An inequality is reversed if you multiply both sides by a _____ number.   negative

SECTION 1-6, EXERCISES

Are the following positive or negative?

1. $(-4)(-7)$ 2. $(3)(-1/2)$ 3. $(1/-2)(1/3)$
4. $(6)(1/-5)$ 5. $-3/-4$ 6. $(1/7)(0)$

Supply the appropriate sign, $>$ or $<$.

7. If $3x > x + 4$, then $2x$ _____ 4.   Example 1

8. If $3x > x + 4$, then 0 _____ $-2x + 4$.
9. If $x + 5 < 8$, then x _____ 3.
10. If $5 - x < 0$, then x _____ 5.
11. If $2x + 1 > x - 3$, then x _____ -4.
12. If $x - 3 < 4x + 2$, then $3x$ _____ -5.
13. If $2x < 6$, then x _____ 3.
14. If $3x > -9$, then x _____ -3.
15. If $-2x < -8$, then x _____ 4.
16. If $6 - 2x > 0$, then x _____ 3.
17. If $3x + 2 > x - 4$, then x _____ -3.
18. If $1 - 2x > 3x - 9$, then x _____ 2.

Example 2
19. If $1 < x$, then $\frac{1}{x}$ _____ 1.
20. If $2 < 3x$, then $\frac{1}{x}$ _____ $\frac{3}{2}$.
*21. Verify: if $0 < a < b$, then $\frac{1}{a} > \frac{1}{b}$.
*22. Verify: if $a < 0 < b$, then $\frac{1}{b} > a$.
*23. Verify: if $a - b > 0$, then $\frac{1}{b - a} < 0$.

*1-7 GROUPS, RINGS, AND FIELDS

We have introduced the real number system simply as a set of objects which satisfy certain axioms. There are many other structures studied in mathematics which satisfy some of these axioms, but not all of them. For example, we can consider sets of objects, which have only one operation defined, instead of two as in the real number system. In this section we will look at some of these other structures, and see how they differ from the real number system.

The axioms we have given so far for the real number system fall into the following groups:

Addition Axioms	Multiplication Axioms
A1 Commutative law	M1 Commutative law
$a + b = b + a$	$ab = ba$
A2 Associative law	M2 Associative law
$(a + b) + c = a + (b + c)$	$(ab)c = a(bc)$

26 The Real Numbers

A3 Existence of identity   M3 Existence of identity
$$a + 0 = a$$ $$a \cdot 1 = a$$
A4 Existence of inverses   M4 Existence of inverses
$$a + (-a) = 0$$ $$a \cdot a^{-1} = 1 \quad \text{if } a \neq 0$$

Order Axioms

O1 Trichotomy law
$$a = b \quad \text{or} \quad a < b \quad \text{or } a > b$$
O2 Transitivity
If $a < b$ and $b < c$ then $a < c$

In addition there are the axioms relating addition and multiplication, addition and order, and multiplication and order:

Mixed Axioms

A–M Distributive law
$$a \cdot (b + c) = a \cdot b + a \cdot c$$
A–O Addition preserves order
If $a < b$ then $a + c < b + c$
M–O The product of positive numbers is positive
If $a > 0, b > 0$ then $ab > 0$

There is one order axiom we have not yet given. This is the important *completeness axiom*. First we make the definition: a set S of numbers is *bounded above* if there is some number M such that $x \leq M$ for every $x \in S$. Any number M which works in this definition is called an *upper bound* for S. We can now state the final order axiom.

Completeness Axiom If S is a non empty set of numbers which is bounded above, then there is a smallest upper bound for S.

The completeness axiom is necessary to ensure that our number system has numbers other than rational numbers. The rational numbers (that is, numbers of the form m/n, where m and n are positive or negative integers) satisfy all the other axioms. It is not hard to show that there is no rational number r such that $r^2 = 2$. Since the rational number system satisfies all the other axioms, but $\sqrt{2}$ is not a rational number, it follows that there is no way to prove that there is a real number $\sqrt{2}$ without the completeness axiom.

Any set of objects with a single operation satisfying A1, A2, A3, and A4 is called a *commutative group*. Hence the set of all numbers is a commutative group under addition. The axioms M1, M2, M3, and M4 are exactly the same as the addition axioms, except for the proviso that $a \neq 0$ in M4. Hence the real numbers *other than zero* also form a commutative group under multiplication. Any subset of the real number system will clearly satisfy the commutative and associative laws of addition and multiplication. Such a subset will therefore be a group under addition or multiplication provided

1. The sum (or product) of any two numbers in the set is again in the set.
2. The additive identity 0 (or multiplicative identity 1) is in the set.
3. The additive (or multiplicative) inverse of each element of the set is in the set.

We refer to Condition 1 by saying that the set is *closed under addition*, or *closed under multiplication*.

Example 1 Which of the following sets of numbers form a group under the given operation: (a) The even integers under addition. (b) All integers under multiplication. (c) The integers $\{-3, -2, -1, 0, 1, 2, 3\}$ under addition.

Solution. (a) The even integers are closed under addition. The additive identity 0 is an even integer. The additive inverse (negative) of every even integer is again an even integer. Therefore the even integers do form a group under addition.

(b) The set of all integers is closed under multiplication, and contains the multiplicative identity 1. However, the multiplicative inverse (reciprocal) of an integer other than 1 is not an integer, so condition 3 is not satisfied.

(c) The set $\{-3, -2, -1, 0, 1, 2, 3\}$ contains the additive identity 0 and the inverse of each element. However, the set is not closed under addition (2 + 3 is not in the set), so it is not a group.

An algebraic structure which has both operations, addition and multiplication, is called a *commutative ring*, provided it satisfies A1, A2, A3, A4, M1, M2, and A-M. That is, the addition and multiplication must satisfy the usual axioms, except that a ring need not have a multiplicative identity or multiplicative inverses. A subset of the real numbers is a ring provided

1. It is closed under addition and multiplication.
2. It contains zero and the negative of each of its elements.

Example 2 Which of the following sets of numbers is a ring: (a) The set of integers which are divisible by 4. (b) The set consisting of zero and the odd integers.

Solution. (a) The sum of two integers divisible by 4 is also divisible by 4, as is the product. The identity 0 is divisible by 4, and the negative of any number divisible by 4 is also divisible by 4. Therefore this set is a ring.

(b) The set consisting of zero and the odd integers, $\{\ldots, -5, -3, -1, 0, 1, 3, 5, \ldots\}$, contains 0, and the negative of each element, and is closed under multiplication. However, the set is not closed under addition, so is not a ring. For example, $1 + 3 = 4$ and 4 is not in the set.

A set with two operations satisfying A1 through A4, M1 through M4, and A-M is called a *field*. If the set also has an order relation satisfying O1, O2, A-O, and M-O, it is an *ordered field*. If the completeness axiom also holds, the set is called a *complete ordered field*. The set of all rational numbers is contained in every field within the real number system. A system of real numbers (containing all rationals) will be a field provided it is closed under addition and multiplication, and contains the additive inverse of each of its elements and the multiplicative inverse of each of its nonzero elements. Any field within the real number system is automatically ordered by the order relation of the real number system.

Example 3 Show that the set S of all numbers of the form $r + s\sqrt{2}$, where r and s are rational, is a field.

Solution. The set S contains every rational r since $r + 0 \cdot s$ is in S. The sum and product of two elements $r + s\sqrt{2}$ and $r' + s'\sqrt{2}$ of S are given by

$$(r + s\sqrt{2}) + (r' + s'\sqrt{2}) = (r + r') + (s + s')\sqrt{2}$$

$$(r + s\sqrt{2})(r' + s'\sqrt{2}) = (rr' + 2ss') + (sr' + rs')\sqrt{2}$$

The numbers $r + r'$, $s + s'$, $rr' + 2ss'$, and $sr' + rs'$ are again rational numbers if r, s, r', and s' are, since the rationals are closed under addition and multiplication. Hence S is closed under addition and multiplication. The additive inverse of $r + s\sqrt{2}$ is $(-r) + (-s)\sqrt{2}$, which is in S. The multiplicative inverse of $r + s\sqrt{2}$ (if $r \neq 0$ or $s \neq 0$) is

$$\frac{r}{r^2 - 2s^2} - \frac{s}{r^2 - 2s^2}\sqrt{2}$$

since

$$(r + s\sqrt{2})\left(\frac{r}{r^2 - 2s^2} - \frac{s}{r^2 - 2s^2} \cdot \sqrt{2}\right) = (r + s\sqrt{2})\left(\frac{r - s\sqrt{2}}{r^2 - 2s^2}\right)$$

$$= \frac{r^2 - 2s^2}{r^2 - 2s^2} = 1$$

Notice that $r^2 - 2s^2 \neq 0$ since $\sqrt{2}$ is not rational. The numbers $r/(r^2 - 2s^2)$ and $-s/(r^2 - 2s^2)$ are rational and therefore S is a field.

So far we have considered groups, rings, and fields which are subsets of the real number system, because for such subsets we have a ready-made addition and multiplication. However, we can exhibit other kinds of addition and multiplication by giving a table.

The set $\{z, u\}$ with two elements is a field if we define "+" and "·" by the following tables:

+	z	u
z	z	u
u	u	z

·	z	u
z	z	z
u	z	u

The commutativity and associativity of these operations can be checked by writing out all possible sums or products. The distributive law can similarly be checked by looking at a finite number of cases. In fact six cases will suffice to check that since there are two possibilities for a, and two possibilities for $b + c$: b and c the same (two cases), or b and c different (one case). The element z is an *additive identity*, and u is a *multiplicative identity*. The existence of inverses can also be checked in the tables.

Example 4 Show that z is the additive identity and u is the multiplicative identity in the above field. What are the inverses $-u$, $-z$, and u^{-1}?

Solution. From the tables we see that

$$z + z = z \quad \text{and} \quad z + u = u$$

Hence z is an additive identity. Similarly,

$$u \cdot z = z \quad \text{and} \quad u \cdot u = u,$$

so u is a multiplicative identity. The inverse $-u$ will be the element which satisfies $u + (\ ) = z$. Since $u + u = z$, $u = -u$. The additive inverse of z is z since $z + z = z$; that is, $-z = z$. The multiplicative inverse of u will be the element which satisfies $u \cdot (\ ) = u$. Since $u \cdot u = u$, we conclude that $u^{-1} = u$.

Example 5 Show that the field of Example 4 cannot be ordered.

Solution. If there is an order, $<$, then either $z < u$ or $u < z$. If $z < u$, then we can add u to both sides, and we get $z + u < u + u$, or $u < z$. In other words, if $z < u$, then $u < z$, which is impossible.

A similar argument shows that $u < z$ is also impossible.

SECTION 1-7, REVIEW QUESTIONS

1. The completeness axiom is necessary to show that the real number system has numbers other than _____ numbers.   rational
2. Rational numbers are numbers of the form m/n, where m and n are _____.   integers
3. Any set with an operation which satisfies A1, A2, A3, and A4 is a commutative _____.   group
4. A set of real numbers which is closed under addition, contains the negatives of its elements, and zero, is a group under _____.   addition
5. A set of real numbers is a group under multiplication if it contains the identity 1, reciprocals of its elements, and is closed under _____.   multiplication
6. A commutative ring has two operations both of which are commutative and _____.   associative
7. A ring need not have a multiplicative identity, but must have an _____ identity.   additive
8. A field must have both identities, and inverses, and reciprocals for all elements except _____.   zero
9. Zero never has a _____ inverse.   multiplicative
10. If a field also has an order relation it is called an _____.   ordered field

SECTION 1-7, EXERCISES

1. Which of the following sets of numbers is a group under addition?
    (a) $\{0\}$
    (b) the positive even integers
    (c) the positive and negative integers divisible by 5
    (d) $\{-1, 0, 1\}$
    (e) the numbers of the form $m + n\pi$, where m, n are integers.
2. Which of the following sets of numbers is a group under multiplication?

(a) $\{0, 1\}$
(b) all the numbers of the form $m + n\sqrt{2}$, where m, n are integers
(c) $\{-1, 1\}$
(d) the even integers
(e) $\{1\}$
(f) the integers

3. Which of the groups in Exercise 1 is also a ring?
4. Which of the sets in Exercise 2 are rings?
5. Show that the set of numbers of the form $r + s\sqrt{3}$, where r and s are rational, is a field. Hint: Since $(r + s\sqrt{3})(r - s\sqrt{3}) = r^2 - 3s^2$, what do you multiply $r + s\sqrt{3}$ by to get 1?
6. The operations of "addition" and "multiplication" on the set $S = \{0, 1, 2, 3\}$ are defined by the following tables:

+	0	1	2	3
0	0	1	2	3
1	1	2	3	0
2	2	3	0	1
3	3	0	1	2

·	0	1	2	3
0	0	0	0	0
1	0	1	2	3
2	0	2	0	2
3	0	3	2	1

(a) Verify that 0 and 1 are the respective identities for + and ·.
(b) Is addition commutative?
(c) Is multiplication commutative?
(d) What is the additive inverse of 3? Does every element of S have an addition inverse?
(e) What is the multiplicative inverse of 3? Show that S is not a field by finding a non-zero element with no multiplicative inverse.

7. The operations of "addition" and "multiplication" on the set $\{z, u, t\}$ are defined in the following tables. Both operations are associative, and the distributive law holds.

+	z	u	t
z	z	u	t
u	u	t	z
t	t	z	u

·	z	u	t
z	z	z	z
u	z	u	t
t	z	t	u

(a) Is addition commutative?
(b) Is there an identity for addition? If so, what is it?
(c) List the additive inverse for each element that has one.
(d) Is $\{z, u, t\}$ a group under addition?
(e) Is multiplication commutative?
(f) Is there a multiplicative identity? If so, what is it?

(g) List the multiplicative inverse for each element that has one.
(h) Is $\{z, u, t\}$ a ring?
(i) Is $\{z, u, t\}$ a field?

2 ALGEBRAIC EXPRESSIONS

2-1 INTEGER EXPONENTS

In this chapter we develop the basic techniques for combining algebraic expressions. While some of these manipulations are not too exciting by themselves, they are fundamental to all subsequent work in mathematics. In particular, we will depend heavily on the facts developed here when we turn to solving equations in Chapter 3. We start with a discussion of how exponents are used to simplify algebraic expressions.

We have already used the following forms of exponent notation:

$$x^2 = x \cdot x,$$
$$x^3 = x \cdot x \cdot x,$$
$$x^4 = x \cdot x \cdot x \cdot x,$$

In general, we agree that for any positive integer n,

$$x^n = \underbrace{x \cdot x \cdot \ldots \cdot x}_{n \text{ factors}}.$$

For example, $x^1 = x$, and $x^7 = x \cdot x \cdot x \cdot x \cdot x \cdot x \cdot x$. We recall from Chapter 1 that $x^{-1} = 1/x$. Hence

$$(x^{-1})(x^{-1}) = \frac{1}{x} \cdot \frac{1}{x} = \frac{1}{x^2},$$

$$(x^{-1})(x^{-1})(x^{-1}) = \frac{1}{x} \cdot \frac{1}{x} \cdot \frac{1}{x} = \frac{1}{x^3},$$

and so on.

We now make the following definition to simplify the writing of expressions such as those above. For any positive integer n,

$$x^{-n} = \frac{1}{x^n}.$$

This identity also holds automatically if n is negative. For example, with $n = -3$ we have $x^{-(-3)} = 1/x^{-3}$, since

$$\frac{1}{x^{-3}} = \frac{1}{\frac{1}{x^3}} = x^3.$$

Example 1 Write each of the following with positive exponents: (a) x^{-5}; (b) $1/x^{-4}$; (c) x^{-2}/y^3; (d) xy^{-2}.

Solution.

(a) $$x^{-5} = \frac{1}{x^5}$$

(b) $$\frac{1}{x^{-4}} = \frac{1}{\frac{1}{x^4}} = x^4$$

(c) $$\frac{x^{-2}}{y^3} = \frac{\frac{1}{x^2}}{y^3} = \frac{1}{x^2 y^3}$$

(d) $$xy^{-2} = x \frac{1}{y^2} = \frac{x}{y^2}$$

As the examples above illustrate, any term x^k in a product or quotient can be moved from numerator to denominator or vice versa by changing the sign of the exponent.

To complete the definition so that x^n is defined for any integer n,

Integer Exponents 35

we agree that for any non zero number x,
$$x^0 = 1.$$

Simply by counting the number of factors when m and n are positive we see that
$$x^m \cdot x^n = x^{m+n}. \tag{1}$$

For example,
$$x^3 \cdot x^5 = (x \cdot x \cdot x) \cdot (x \cdot x \cdot x \cdot x \cdot x) = x^8.$$

Even if m or n is negative, the rule (1) still holds; that is, you simply add the signed exponents. For example, with $m = 5$ and $n = -2$, we have
$$x^5 \cdot x^{-2} = (x \cdot x \cdot x \cdot x \cdot x) \cdot \frac{1}{x \cdot x} = x \cdot x \cdot x = x^3,$$
or
$$x^5 \cdot x^{-2} = x^{5-2}.$$

Similarly, with $m = -3$ and $n = -1$, we get
$$x^{-3} \cdot x^{-1} = \frac{1}{x \cdot x \cdot x} \cdot \frac{1}{x} = \frac{1}{x^4} = x^{-4}.$$

The reason for defining x^0 to be 1 is seen by considering the law $x^m x^n = x^{m+n}$ in the case $m = -n$. For example, if $m = 3$ and $n = -3$, then (1) would give
$$x^3 \cdot x^{-3} = x^{3-3} = x^0.$$

But we know that
$$x^3 \cdot x^{-3} = x^3 \cdot \frac{1}{x^3} = 1,$$

so we agree that $x^0 = 1$.

From rule (1), and the fact that $x^{-n} = 1/x^n$, we can obtain a second rule:
$$\frac{x^m}{x^n} = x^m \cdot x^{-n} = x^{m-n}.$$

Formulas for $(x \cdot y)^n$ and $(x/y)^n$ can be obtained by simply counting the factors; thus
$$(x \cdot y)^n = \underbrace{(xy)(xy) \cdots (xy)}_{n \text{ factors}} \tag{3}$$

$$= \underbrace{(x \cdot x \cdot \ldots \cdot x)}_{n \text{ factors}} \cdot \underbrace{(y \cdot y \ldots \cdot y)}_{n \text{ factors}}$$

$$= x^n \cdot y^n$$

In exactly the same way we get

$$\left(\frac{x}{y}\right)^n = \frac{x^n}{y^n} \qquad (4)$$

The last rule for exponents is

$$(x^m)^n = x^{mn} \qquad (5)$$

If x^m is used as a factor n times (that is, $(x^m)^n$), then you have x used as a factor mn times altogether (that is, x^{mn}).

The definitions of x^{-n}, x^0, and the five basic laws of exponents are listed below. So far these laws have been verified only for integer values of m and n.

$$x^{-n} = \frac{1}{x^n}$$

$$x^0 = 1$$

$$x^m x^n = x^{m+n}$$

$$\frac{x^m}{x^n} = x^{m-n}$$

$$(x^m)^n = x^{mn}$$

$$(xy)^n = x^n y^n$$

$$\left(\frac{x}{y}\right)^n = \frac{x^n}{y^n}$$

Example 2 Simplify and write without negative exponents:

(a) $\dfrac{(xy)^4}{x^3}$; (b) $\dfrac{(a^{-2}b^{-1})^{-3}}{(ab)^{-1}}$; (c) $(xz)^5(2x^2)^{-2}$.

Solution.

(a) $$\frac{(xy)^4}{x^3} = \frac{x^4 y^4}{x^3} = x^{4-3} y^4 = xy^4$$

(b) $$\frac{(a^{-2}b^{-1})^{-3}}{(ab)^{-1}} = \frac{a^6 b^3}{a^{-1} b^{-1}} = a^7 b^4$$

Integer Exponents 37

(c) $(xz)^5(2x^2)^{-2} = x^5 z^5 2^{-2} x^{-4} = \dfrac{xz^5}{4}$

The laws of signs for multiplication can also be combined with exponent notation. We know that $(-a)^2 = a^2$, $(-a)^3 = -a^3$, $(-a)^4 = a^4$, $(-a)^5 = -a^5$, etc. In general,

$$(-a)^n = \begin{cases} a^n & \text{if } n \text{ is even} \\ -a^n & \text{if } n \text{ is odd} \end{cases}$$

Example 3 Simplify and write without negative exponents:

(a) $\dfrac{(-ab^2)^3}{(ab)^2}$;   (b) $\left(\dfrac{-x}{a^2}\right)^3 \left(\dfrac{ax}{-b}\right)^2$;

Solution.

(a) $\dfrac{(-ab^2)^3}{(ab)^2} = \dfrac{-(ab^2)^3}{a^2 b^2} = \dfrac{-a^3 b^6}{a^2 b^2} = -ab^4$

(b) $\left(\dfrac{-x}{a^2}\right)^3 \left(\dfrac{ax}{-b}\right)^2 = \dfrac{-x^3}{a^6} \cdot \dfrac{a^2 x^2}{b^2} = -\dfrac{x^5}{a^4 b^2}$

Example 4 Simplify and write as a single fraction with no negative exponents:

(a) $x + x^{-1}$ (b) $ab^{-1} + ba^{-1}$ (c) $(x^{-1} + y)^{-1}$

Solution.

(a) $x + x^{-1} = x + \dfrac{1}{x} = \dfrac{x^2 + 1}{x}$

(b) $ab^{-1} + ba^{-1} = \dfrac{a}{b} + \dfrac{b}{a} = \dfrac{a^2 + b^2}{ab}$

(c) $(x^{-1} + y)^{-1} = \left(\dfrac{1}{x} + y\right)^{-1} = \left(\dfrac{1 + xy}{x}\right)^{-1} = \dfrac{x}{1 + xy}$

SECTION 2-1, REVIEW QUESTIONS

1. The expression x^n means x used as a factor n times, and x^{-n} means _____ used as a factor n times.

2. Since x^{-n} stands for $1/x$ used as a factor n times, x^{-n} is the same as 1/_____ . x^n

3. For any integers m and n, $x^m \cdot x^n =$ _____ . x^{m+n}

4. We define x^0 to be _____ for $x \neq 0$.   1

5. For positive or negative integers m and n, $(xy)^n =$ _____, $x^n y^n$
   $(x/y)^n =$ _____, and $(x^m)^n =$ _____. x^n/y^n x^{mn}

6. For an even integer n, $(-x)^n = x^n$, and for an odd integer n, $(-x)^n =$ _____ . $-x^n$

SECTION 2-1, EXERCISES

Simplify each of the following and write without negative exponents.

1. $\dfrac{x^{-3}}{y^{-4}}$ 2. $\dfrac{(-x)^3}{y^{-2}}$ Example 1,2,3

3. $\dfrac{x^{-5}}{x^{-6}}$ 4. $\dfrac{(-xy)^6}{x^{-2}}$

5. $\dfrac{(xy)^5}{x^3}$ 6. $\dfrac{(x^{-2}y^{-3})^2}{xy}$

7. $\dfrac{(3x^{-2}y^{-3})^{-2}}{9xy}$ 8. $\dfrac{(2ab)^{-3}}{8a^{-1}b^{-1}}$

9. $(18x^{-1}y^{-2})(2xy)^{-3}$ 10. $(-2x^{-3}y)(-8xy^2)^{-1}$
11. $[(3x)^{-3}y](3xy^{-1})$ 12. $(3x^{-1}y^{-1})^0(5x^{-3}y^{-10})^0$
13. $(7x^0y^0)^{-1}(49x^{-1}y^{-1})^0$ 14. $(8x^{-3}y^0)[(2x)^{-1}y]^{-3}$

15. $\left(\dfrac{3y}{2x}\right)^{-1}\left(\dfrac{3y^{-1}}{2x^{-1}}\right)$ 16. $\dfrac{(27a)^{-1}}{(8b)^{-1}}\left(\dfrac{2b^3}{3a^3}\right)^{-1}$

17. $\dfrac{-64a^{-3}}{(4b)^2}\left(\dfrac{4ab}{16}\right)^{-2}$ 18. $2^3 \cdot 4^{-1}$

19. $3^{-4} \cdot 3^3$ 20. $(-5)^{-3}(-5)^2$
21. $(-7)^{-1}(-7)^3$ 22. $2^3 \cdot 3^2$
23. $3^4 \cdot 2^4$ 24. $x^{-n} \cdot x^{2n}$
25. $x^{-n+1} \cdot x^{n+1}$ 26. $y^{2n-1} \cdot y^{-2n}$

27. $y^{2n-1} \cdot y^{-2n+1}$ 28. $\dfrac{(xy)^{3n+1}}{(xy)^{2n-1}}$

29. $\left[\dfrac{(xy)^{-3n-1}}{(xy)^{4n-1}}\right]^{-1}$ 30. $\left[\dfrac{x^{n+1}y^n}{x^n y^{n+1}}\right]^{-n}$

Simplify the following and write as a single fraction with no negative exponents.

Example 4

31. $x^{-1} + y^{-1}$
32. $a + b^{-1}$
33. $(x + y)^{-1}$
34. $(x + x^{-1})^{-1}$
35. $(a^{-1} + b^{-1})^{-1}$
36. $(ab^{-1} + 1)^{-1}$
37. $a^{-2} + b^{-2}$
38. $(xy^{-1} + yx^{-1})^{-1}$
39. $(x^2 + x^{-2})^{-3}$
40. $(x^{-1} + y^{-1})(x + y)$

Give numerical examples to show that none of the following are true for all real numbers. [For example, in exercise 41, if $a = 1$, $b = 1$, then $(a + b)^{-1} = (1 + 1)^{-1} = \frac{1}{2}$, but $a^{-1} + b^{-1} = 1 + 1 = 2$. There are many possible "answers" for each question.]

41. $(a + b)^{-1} = a^{-1} + b^{-1}$
42. $(a + b)^2 = a^2 + b^2$
43. $(x^{-2} + y^{-2})^{-1} = x^2 + y^2$
44. $(x^{-1} + 1)^{-1} = x + 1$
45. $(x - y)^{-1} = x^{-1} - y^{-1}$
46. $\dfrac{x + y}{x^{-1} + y^{-1}} = x^2 + y^2$
47. $a^{-2} + b^{-2} = \dfrac{1}{a^2 + b^2}$
48. $(a^{-3} + b^{-3})(a + b) = a^{-2} + b^{-2}$

2-2 FRACTIONAL EXPONENTS

We are all familiar with what is meant by a square root, a cube root, and so on. As we shall see in this section, it turns out that these ideas are most conveniently handled by using fractional exponents. The rules for treating integer exponents can be extended to fractional exponents, giving a systematic way of dealing with the various roots of numbers.

We know that for any positive number a and any positive integer n, there is exactly one positive number r such that $r^n = a$. This fact is a consequence of the completeness axiom for the order relation, and is proved in more advanced courses. The number r is called the *positive nth root* of a, and is denoted

$$r = a^{1/n}.$$

This notation is suggested by the rule $(a^m)^n = a^{mn}$, for if we formally take $m = 1/n$ in this rule, we get

$$(a^{1/n})^n = a^{n/n} = a.$$

That is, $a^{1/n}$ is a number whose nth power equals a. As usual, we will call a the *base* and $1/n$ the *exponent* in the expression $a^{1/n}$.

Negative numbers have no (real) square roots, fourth roots, sixth roots, etc., since even powers of any number are nonnegative. On the other hand, negative numbers do have cube roots, fifth roots, etc. We will also use the fractional exponent notation to express these. The following are some examples of this notation: $9^{1/2} = 3$; $16^{1/4} = 2$; $64^{1/3} = 4$; $(-27)^{1/3} = -3$; $(-32)^{1/5} = -2$; and $(-1/8)^{1/3} = -1/2$.

Notice that there are two square roots of 9, and two fourth roots of 16, but the exponent notation indicates the positive one. Thus $(3)^2 = 9$ and $(-3)^2 = 9$, but $9^{1/2} = 3$; similarly, $(2)^4 = 16$ and $(-2)^4 = 16$, but $16^{1/4} = 2$.

If a and b are *positive* bases, then we have the following properties:

$$(ab)^{1/n} = a^{1/n} b^{1/n}, \tag{1}$$

and

$$\left(\frac{a}{b}\right)^{1/n} = \frac{a^{1/n}}{b^{1/n}}. \tag{2}$$

To verify (1), we raise the right side to the nth power, using the rule of Sec. 2-1 for a product, and recall that $(a^{1/n})^n = a$ by definition:

$$(a^{1/n} \cdot b^{1/n})^n = (a^{1/n})^n (b^{1/n})^n = ab.$$

Hence the nth power of the right side of (1) is ab, which means that the right side is $(ab)^{1/n}$. The proof of (2) is similar. It is important to remember that although we use notations like $(-8)^{1/3} = -2$ involving negative bases, *the rules for combining exponent expressions only hold for positive bases.*

For positive integers m and n, and a positive base a, we can check that

$$(a^m)^{1/n} = (a^{1/n})^m. \tag{3}$$

We will show that the nth power of the right side is a^m, and hence that the right side equals $(a^m)^{1/n}$:

$$[(a^{1/n})^m]^n = (a^{1/n})^{mn} = [(a^{1/n})^n]^m = a^m.$$

The numbers $(a^{1/n})^m$ and $(a^m)^{1/n}$ are equal, and it is natural to define $a^{m/n}$ to mean either one of these expressions; thus

$$a^{m/n} = (a^m)^{1/n} = (a^{1/n})^m. \tag{4}$$

Agreement (4) extends our definition of exponential expressions to cover the case of all positive rational exponents m/n. For negative rational exponents (that is, numbers $-m/n$, where m and n are positive integers), we define

$$a^{-m/n} = \frac{1}{a^{m/n}}. \tag{5}$$

We have now defined a^r for every *positive* number a and every positive or negative *rational* number r.

Example 1 Simplify: (a) $(36)^{3/2}$; (b) $(8)^{-2/3}$; (c) $(9/4)^{5/2}$; (d) $(81)^{-3/4}$; (e) $(1/25)^{-1/2}$.

Solution.

(a) $$36^{3/2} = 6^3 = 216$$

(b) $$8^{-2/3} = \frac{1}{8^{2/3}} = \frac{1}{2^2} = \frac{1}{4}$$

(c) $$\left(\frac{9}{4}\right)^{5/2} = \left(\frac{3}{2}\right)^5 = \frac{3^5}{2^5} = \frac{243}{32}$$

(d) $$81^{-3/4} = \frac{1}{81^{3/4}} = \frac{1}{3^3} = \frac{1}{27}$$

(e) $$\left(\frac{1}{25}\right)^{-1/2} = \frac{1}{(1/25)^{1/2}} = \frac{1}{1/5} = 5$$

By using the definitions above, we can check that all our earlier laws for integer exponents still hold for rational exponents, providing all bases are positive. We list these rules below, where r and s now stand for any positive or negative rational numbers, and a and b are arbitrary positive numbers.

$$a^0 = 1$$

$$a^{-r} = \frac{1}{a^r}$$

$$a^r a^s = a^{r+s}$$

$$\frac{a^r}{a^s} = a^{r-s}$$

$$(ab)^r = a^r b^r$$

$$\left(\frac{a}{b}\right)^r = a^r/b^r$$

$$(a^r)^s = a^{rs}$$

The reason for insisting on positive bases can be seen by considering the following example. We know that $(-8)^{1/3} = -2$, but if the last law above held for negative bases, we would also have

$$(-8)^{1/3} = (-8)^{2/6} = [(-8)^2]^{1/6} = [64]^{1/6} = 2.$$

We will henceforth assume that all literal numbers are positive where fractional exponents are involved.

Example 2 Simplify and write without negative exponents $(9a^4)^{-3/2}$.

Solution.

$$(9a^4)^{-3/2} = \frac{1}{(9a^4)^{3/2}}$$

$$= \frac{1}{9^{3/2}(a^4)^{3/2}}$$

$$= \frac{1}{3^3 a^6}$$

$$= \frac{1}{27 a^6}$$

Example 3 Simplify and write without negative exponents $-(81^{-1}x^0 y^2)^{-3/4}$.

Solution.

$$-(81^{-1}x^0 y^2)^{-3/4} = -[(81)^{3/4}(1)^{-3/4}(y^2)^{-3/4}]$$

$$= -(3)^3 \cdot 1 \cdot y^{-3/2}$$

$$= -\frac{27}{y^{3/2}}$$

Example 4 Simplify and write without negative exponents:

$$\frac{x^3 y^{1/4}}{xy^{1/2}}$$

Solution.

$$\frac{x^3 y^{1/4}}{xy^{1/2}} = x^{(3-1)} y^{(1/4 - 1/2)}$$

$$= x^2 y^{-1/4}$$

$$= \frac{x^2}{y^{1/4}}$$

Example 5 Simplify and write as a single fraction with no negative exponents.

(a) $1 + x^{-1/2}$ (b) $a^{1/3} + a^{-1/3}$

Solution.

(a) $1 + x^{-1/2} = 1 + \dfrac{1}{x^{1/2}}$

$= \dfrac{x^{1/2} + 1}{x^{1/2}}$

(b) $a^{1/3} + a^{-1/3} = a^{1/3} + \dfrac{1}{a^{1/3}}$

$= \dfrac{a^{2/3} + 1}{a^{1/3}}$

SECTION 2-2, REVIEW QUESTIONS

a	1. An nth root of a number a is a number r such that $r^n =$ _____.
one	2. Every positive number a has _____ positive nth root.
$a^{1/n}$	3. The positive nth root of a is denoted _____.
exponent	4. In the expression $a^{1/n}$, a is called the base and $1/n$ is the _____.
negative	5. Positive numbers have two square roots, one positive, and one _____.
positive	6. The expression $a^{1/2}$ stands for the _____ square root if a is positive.
$(a^{1/n})^m$	7. For $a > 0$, we can define $a^{m/n}$ to be either $(a^m)^{1/n}$ or _____ since these expressions are equal.
1 1/a^r a^{r+s}	8. For positive or negative rational numbers r and s, and positive bases a and b,
a^{r-s} $a^r b^r$ a^r/b^r	$a^0 =$ _____ ; $a^{-r} =$ _____ ; $a^r a^s =$ _____ ;
a^{rs}	$a^r/a^s =$ _____ ; $(ab)^r =$ _____ ; $(a/b)^r =$ _____ ; and $(a^r)^s =$ _____ .

SECTION 2-2, EXERCISES

Simplify and write without negative exponents.

1. $(2^{-4})(2^6)$
2. $(3^5)(3^{-2})$
3. $(5^2)(5^{-3})$ **Example 1**
4. $(4^{-7})(4^9)$
5. $(2^{-2})(8^{1/3})$
6. $(4^{1/2})(9^{1/2})$
7. $(9 \cdot 25)^{1/2}$
8. $\left(\dfrac{27}{8}\right)^{2/3}$
9. $\left(\dfrac{16}{9}\right)^{-3/2}$
10. $(32)^{-1/5}$
11. $(4x^2)^{3/2}$
12. $\left(\dfrac{a^2}{25}\right)^{1/2}$ **Example 2,3**
13. $(a^6)^{-2/3}$
14. $(25x^4)^{-3/2}$
15. $(16a^2 b^{-4})^{-1/2}$
16. $(81x^4 y^{-8})^{3/4}$
17. $\left(\dfrac{1}{36} x^{2/3}\right)^{1/2}$
18. $\left(\dfrac{1}{16} x^{-1} y^2\right)^{-3/4}$
19. $(-27 y^0 x^{-3})^{-2/3}$
20. $(64^{-5/6} xy^{15})^{2/5}$

Simplify and write without negative exponents.

21. $\dfrac{ab^2}{a^{1/2} b}$
22. $\dfrac{a^{3/4} b^2}{a^{1/2} b^{1/2}}$
23. $\dfrac{x^{1/3} y^{1/4}}{x^{-1/3} y^{1/2}}$ **Example 4**
24. $\dfrac{x^{2/5} y^{1/5}}{x^{-3/5} y^{2/5}}$
25. $\dfrac{(x^{1/2} y^{1/3})^2}{x^{1/3} y^{1/3}}$

Simplify and write as a single fraction with no negative exponents.

26. $2 - x^{-1/2}$
27. $x^{1/2} + x^{-1/2}$
28. $a^{1/3} + a^{-1}$ **Example 5**
29. $a^{1/3} - a^{-1/3}$
30. $x^{1/2} y^{-1/2} + x^{-1/2} y^{1/2}$

Give numerical examples to show that none of the following are true for all real numbers.

31. $(a + b)^{1/2} = a^{1/2} + b^{1/2}$
32. $(a + b)^{1/3} = a^{1/3} + b^{1/3}$
33. $(a + b)^{1/2} = a^{-1/2} + b^{-1/2}$
34. $(a + b)^{-1/2} = \dfrac{1}{a^{1/2} + b^{1/2}}$

2-3 RADICAL EXPRESSIONS

The following equations use another familiar notation for expressing roots: $\sqrt{4} = 2$, $\sqrt[3]{27} = 3$. These are called *radical expressions*, and they are also a standard part of mathematical notation. In this section we will develop some techniques for combining and simplifying such radical expressions.

We know that positive numbers have two roots of any even order (for example, 2 and -2 are fourth roots of 16); and negative numbers have no even order roots (for example, -4 has no real square root).

However, every number has exactly one root of any odd order; for example, -3 is the only cube root of -27, and 2 is the only fifth root of 32. We will use the term *principal nth root* to denote the positive nth root when there are two, and of course also to denote the unique nth root when there is just one.

The principal nth root is indicated either with a fractional exponent, $a^{1/n}$, or with the radical expression $\sqrt[n]{a}$. The sign $\sqrt{}$ is called the *radical sign*, the positive integer n is called the *index*, and the number a is called the *radicand*. When the index is 2, we write $\sqrt{a}$ instead of $\sqrt[2]{a}$. The following are examples of the radical sign used to designate principal roots: $\sqrt[3]{-8} = -2$; $\sqrt[4]{1/16} = 1/2$; $\sqrt{9} = 3$; $\sqrt[3]{-64} = -4$; $\sqrt[3]{1/64} = 1/4$; and $\sqrt[5]{32} = 2$.

The rules for combining radicals with positive radicands are immediate from the rules we already have for fractional exponents. For example, the following express the rules for the nth root of a product or quotient in both notations.

$$\sqrt[n]{ab} = \sqrt[n]{a}\sqrt[n]{b} \qquad (ab)^{1/n} = a^{1/n}b^{1/n} \qquad (1)$$

$$\sqrt[n]{\frac{a}{b}} = \frac{\sqrt[n]{a}}{\sqrt[n]{b}} \qquad \left(\frac{a}{b}\right)^{1/n} = \frac{a^{1/n}}{b^{1/n}} \qquad (2)$$

We can simplify radicals with a numerical radicand by using (1) and factoring the radicand into prime factors. Remember that a prime number is a positive integer with no factors other than 1 and the number itself. Any integer can be written as a product of primes; for example, $126 = 2 \cdot 3^2 \cdot 7$. We get this factorization as follows:

$$126 = 2 \cdot 63 = 2 \cdot 3 \cdot 21 = 2 \cdot 3 \cdot 3 \cdot 7 = 2 \cdot 3^2 \cdot 7.$$

Similarly, $300 = 2^2 \cdot 3 \cdot 5^2$, since we can write

$$300 = 2 \cdot 150 = 2 \cdot 2 \cdot 75 = 2 \cdot 2 \cdot 3 \cdot 25 = 2 \cdot 2 \cdot 3 \cdot 5 \cdot 5.$$

Example 1 Simplify (a) $\sqrt{125}$ and (b) $\sqrt[3]{3240}$.

Solution.

(a) $$\sqrt{125} = \sqrt{5^3} = \sqrt{5^2 \cdot 5} = \sqrt{5^2}\sqrt{5} = 5\sqrt{5}$$

(b) $$\sqrt[3]{3240} = \sqrt[3]{2^3 \cdot 3^4 \cdot 5} = \sqrt[3]{2^3 \cdot 3^3 \cdot 3 \cdot 5}$$
$$= \sqrt[3]{2^3 \cdot 3^3}\sqrt[3]{3 \cdot 5}$$
$$= 2 \cdot 3 \cdot \sqrt[3]{15}$$

The same technique works in simplifying radicals with non-numerical radicands. All literal numbers in radical expressions are always assumed to be positive.

Example 2 Simplify (a) $\sqrt{8a^3x^5}$ and (b) $\sqrt[3]{64x^6y^4}$.

Solution.

(a) $\sqrt{8a^3x^5} = \sqrt{4a^2x^4 \cdot 2ax} = \sqrt{4a^2x^4}\sqrt{2ax} = 2ax^2\sqrt{2ax}$
(b) $\sqrt[3]{64x^6y^4} = \sqrt[3]{64x^6y^3 \cdot y} = \sqrt[3]{64x^6y^3}\sqrt[3]{y} = 4x^2y\sqrt[3]{y}$

The next two laws for radicals are again just restatements of rules we have already developed for fractional exponents.

$$\sqrt[n]{a^m} = (\sqrt[n]{a})^m \qquad (a^m)^{1/n} = (a^{1/n})^m \qquad (3)$$

$$\sqrt[n]{\sqrt[k]{a}} = \sqrt[nk]{a} \qquad (a^{1/k})^{1/n} = a^{1/nk} \qquad (4)$$

These rules can sometimes be used to lower the index in a radical expression.

Example 3 Simplify (a) $\sqrt[6]{x^2}$; (b) $\sqrt[12]{8x^3y^6}$; and (c) $\sqrt[4]{a^2b^4/81}$.

Solution.

$$\sqrt[6]{x^2} = \sqrt[3]{\sqrt{x^2}} = \sqrt[3]{x}$$

Radical Expressions 47

(c) $\sqrt[12]{8x^3y^6} = \sqrt[4]{\sqrt[3]{8x^3y^6}} = \sqrt[4]{2xy^2}$

$\sqrt[4]{\dfrac{a^2b^4}{81}} = \sqrt[4]{\dfrac{b^4}{81}} \sqrt[4]{a^2} = \dfrac{b}{3}\sqrt{a}$

The examples above show how radical expressions can sometimes be simplified by removing factors from the radical sign and lowering the index. The one other simplification we make consistently is to write radical expressions so no fraction appears under the radical sign, and no radical occurs in the denominator. This last simplification is called *rationalizing the denominator*.

The next examples show the method.

Example 4 Simplify (a) $\sqrt[5]{5/ab^2}$; (b) $\sqrt{2x/3y^2}$; and (c) $1/\sqrt[3]{4xy^2}$.

Solution.

(a) $\sqrt[5]{\dfrac{5}{ab^2}} = \sqrt[5]{\dfrac{5a^4b^3}{(ab^2)(a^4b^3)}} = \dfrac{\sqrt[5]{5a^4b^3}}{\sqrt[5]{a^5b^5}} = \dfrac{\sqrt[5]{5a^4b^3}}{ab}$

(b) $\sqrt{\dfrac{2x}{3y^3}} = \sqrt{\dfrac{(2x)(3y)}{(3y^3)(3y)}} = \dfrac{\sqrt{6xy}}{\sqrt{9y^4}} = \dfrac{\sqrt{6xy}}{3y^2}$

(c) $\dfrac{1}{\sqrt[3]{4xy^2}} = \dfrac{\sqrt[3]{2x^2y}}{\sqrt[3]{4xy^2}\sqrt[3]{2x^2y}} = \dfrac{\sqrt[3]{2x^2y}}{\sqrt[3]{8x^3y^3}} = \dfrac{\sqrt[3]{2x^2y}}{2xy}$

To summarize, the three steps in simplifying radicals are:

1. Remove factors from under the radical: $\sqrt{x^3} = x\sqrt{x}$.
2. Make the index as small as possible: $\sqrt[6]{x^3} = \sqrt{x}$.
3. Rationalize the denominator: $1/\sqrt[3]{x} = \sqrt[3]{x^2}/x$.

SECTION 2-3, REVIEW QUESTIONS

1. When a number has both a positive and a negative nth root, the principal nth root is the _____ one.
2. The radical expression for the principal nth root of a is _____.
3. In the expression $\sqrt[n]{a}$, n is the _____ and a is the _____.

positive

$\sqrt[n]{a}$

index   radicand

4. If the index is 2, we write _____ instead of $\sqrt[2]{a}$.

5. The rules for combining radicals follow from the rules for fractional _____.

exponents

6. The rules for combining radicals (like those for fractional exponents) depend on the radicand being _____.

positive

7. For positive radicands we have: $\sqrt{ab} =$ _____ ; $\sqrt{a/b} =$ _____ ; $\sqrt[n]{a^m} =$ _____ ; and $\sqrt[n]{\sqrt[m]{a}} =$ _____ .

$\sqrt{a}\sqrt{b}$ $\sqrt{a}/\sqrt{b}$

$(\sqrt[n]{a})^m$ $\sqrt[nm]{a}$

SECTION 2-3, EXERCISES

Simplify each of the following

1. $\sqrt{8}$
2. $\sqrt{\dfrac{4}{25}}$
3. $\sqrt{\dfrac{27}{4}}$ Example 1

4. $\sqrt{48}$
5. $\sqrt{3}\cdot\sqrt{12}$
6. $\sqrt{2}\cdot\sqrt{50}$

7. $\sqrt[3]{16}$
8. $\sqrt[3]{81}$
9. $\sqrt{a^3}$ Example 2

10. $\sqrt{a^4 x^5}$
11. $\sqrt{a^5 x^8}$
12. $\sqrt[3]{a^4}$

13. $\sqrt[3]{x^5}$
14. $\sqrt[3]{a^4 x^7}$
15. $\sqrt[4]{x^2}$ Example 3

16. $\sqrt[4]{a^2 x^4}$
17. $\sqrt[4]{a^2 x^8}$
18. $\sqrt[4]{16 a^5 b^8}$

19. $\sqrt[3]{9 a^9 b^{11}}$
20. $\sqrt[3]{\sqrt{64}}$
21. $\sqrt{\sqrt{81}}$

22. $\sqrt{\sqrt[3]{64 a^9 b^{15}}}$
23. $\sqrt{\sqrt{81 a^6 b^{12}}}$
24. $\sqrt{\dfrac{x}{2}}$ Example 4

25. $\sqrt{\dfrac{3}{a}}$
26. $\sqrt{\dfrac{x}{8}}$
27. $\sqrt{\dfrac{8a}{3b}}$

28. $\sqrt{\dfrac{x^3}{27}}$
29. $\sqrt{\dfrac{8}{x^5}}$
30. $\sqrt{\dfrac{1}{y^3}}$

31. $\sqrt{\dfrac{x^5}{y^3}}$
32. $\sqrt{\dfrac{a^7 b^3}{c^4}}$
33. $\sqrt{\dfrac{27x}{9y}}$

34. $\sqrt{\dfrac{24x}{a^2 y}}$
35. $\sqrt{\dfrac{x^2 y}{18 a^4}}$
36. $\sqrt[3]{\dfrac{x^5 y^7}{4}}$

37. $\sqrt[3]{\dfrac{1}{12 a b^5}}$
38. $\sqrt{\dfrac{5a}{8 b^3}}\cdot\sqrt{\dfrac{5 a^3}{2b}}$
39. $\sqrt{\dfrac{15 x^3 y}{a^5 b^3}}\cdot\sqrt{\dfrac{3y}{a b^2}}$

40. $\sqrt{\dfrac{12a^3b^5}{5xy^3}} \cdot \sqrt{\dfrac{x^3y}{15ab}}$ 41. $\sqrt{\dfrac{15x^5}{7y^7}} \cdot \sqrt{\dfrac{21y^3}{x^3}}$ 42. $\sqrt{\dfrac{x^{-3}y^5}{ab}} \cdot \sqrt{\dfrac{x^3y^3}{a^{-1}b^{-1}}}$

43. $\sqrt{\dfrac{x^{-1/2}y^{-1/3}}{a^{-1}b^{-1}}} \cdot \sqrt{\dfrac{x^{3/2}y^{7/3}}{a^3b^3}}$

2-4 SPECIAL PRODUCTS

The distributive laws tell us how to multiply a sum of two terms by a third term. The left and right distributive laws are, respectively:

$$a(b + c) = ab + ac, \quad (b + c)a = ba + ca.$$

Of course we can replace $a, b,$ and c by any expressions; for example,

$$2x(3x + 5y) = 6x^2 + 10xy; \quad (2a^2b - 5b)(3ab) = 6a^3b^2 - 15ab^2$$

In a product of the form

$$(A + B)(C + D)$$

we first regard $(A + B)$ as the single term a of the left distributive law, and distribute it through the sum $C + D$. We get

$$(A + B)(C + D) = (A + B)C + (A + B)D.$$

If we now use the right-side distributive law on each of the terms on the right, we get

$$(A + B)(C + D) = (A + B)C + (A + B)D$$
$$= AC + BC + AD + BD$$
$$= AC + AD + BC + BD.$$

That is, to multiply one sum by another, you multiply each term of the first sum by each term of the second sum. For example,

$$(2x - y)(3x - 1) = (2x)(3x) + (2x)(-1) + (-y)(3x) + (-y)(-1)$$
$$= 6x^2 - 2x - 3xy + y.$$

The same principal holds for sums of more than two terms. For example,

$$(A + B)(C + D + E) = AC + AD + AE + BC + BD + BE.$$

Here we multiplied A by each of the terms $C, D,$ and E, and then multiplied B by each of these terms. The product

$(A + B + C)(D + E + F)$ would be expanded similarly, and there would be nine terms altogether.

Example 1 Multiply and simplify $(2x + y)(x^2 + y^2)$.

Solution.

$$(2x + y)(x^2 + y^2) = 2x \cdot x^2 + 2x \cdot y^2 + y \cdot x^2 + y \cdot y^2$$
$$= 2x^3 + 2xy^2 + yx^2 + y^3$$

Example 2 Multiply and simplify $(3x + 2)(x^2 + x + 1)$.

Solution.

$(3x + 2)(x^2 + x + 1)$
$= 3x \cdot x^2 + 3x \cdot x + 3x \cdot 1 + 2 \cdot x^2 + 2 \cdot x + 2 \cdot 1$
$= 3x^3 + 3x^2 + 3x + 2x^2 + 2x + 2$
$= 3x^3 + 5x^2 + 5x + 2$

To multiply two polynomials, we frequently display the work in an array which makes it easy to add like powers.

Example 3 Multiply and simplify $(3x^2 + 2x + 1)(5x^3 + x + 4)$.

Solution.

$$\begin{array}{r}
5x^3 + 0 \cdot x^2 + x + 4 \\
3x^2 + 2x + 1 \\
\hline
5x^3 + + x + 4 \\
10x^4 + + 2x^2 + 8x \\
15x^5 + + 3x^3 + 12x^2 \\
\hline
15x^5 + 10x^4 + 8x^3 + 14x^2 + 9x + 4
\end{array}$$

We multiplied each of the terms $5x^3$, x, and 4 first by 1, then by $2x$, and then by $3x^2$. Writing each of these products in a separate row makes it easier to collect the like powers of x.

Certain products occur so frequently that it is a good idea to memorize the form of the answer. The following identities are easily verified:

$$(x + y)(x - y) = x^2 - y^2, \tag{1}$$

$$(x + y)^2 = x^2 + 2xy + y^2, \tag{2}$$

$$(x - y)^2 = x^2 - 2xy + y^2. \tag{3}$$

Example 4 Multiply (a) $(2x - 3y)(2x + 3y)$; (b) $(4a + x)^2$; and (c) $(y - 2x)^2$.

Solution. We write the answers directly from the pattern above:

(a) $\quad (2x - 3y)(2x + 3y) = (2x)^2 - (3y)^2 = 4x^2 - 9y^2$

(b) $\quad (4a + x)^2 = (4a)^2 + 2(4a)(x) + x^2 = 16a^2 + 8ax + x^2$

(c) $\quad (y - 2x)^2 = y^2 - 2(y)(2x) + (2x)^2 = y^2 - 4xy + 4x^2$

Another type of product which occurs repeatedly is

$$(ax + b)(cx + d) = acx^2 + (ad + bc)x + bd.$$

The two terms with x to the first power always combine in this sort of product, so we can omit the step of writing out all four products.

Example 5 Multiply (a) $(2x - 3)(4x + 5)$; and (b) $(x + 3)(5x + 2)$

Solution.

(a) $\quad (2x - 3)(4x + 5) = 8x^2 + (-12 + 10)x - 15$
$\quad\quad\quad\quad\quad\quad\quad\quad = 8x^2 - 2x - 15$

(b) $\quad (x + 3)(5x + 2) = 5x^2 + (15 + 2)x + 6$
$\quad\quad\quad\quad\quad\quad\quad\quad = 5x^2 + 17x + 6$

Formula (1) above provides a method for rationalizing denominators which contain a sum or difference of two square roots.

Example 6 Simplify (a) $1/(\sqrt{x} - \sqrt{y})$ and (b) $2\sqrt{ab}/(\sqrt{ab} + 1)$.

Solution.

(a)
$$\frac{1}{\sqrt{x} - \sqrt{y}} = \frac{1}{\sqrt{x} - \sqrt{y}} \cdot \frac{\sqrt{x} + \sqrt{y}}{\sqrt{x} + \sqrt{y}}$$
$$= \frac{\sqrt{x} + \sqrt{y}}{(\sqrt{x})^2 - (\sqrt{y})^2}$$
$$= \frac{\sqrt{x} + \sqrt{y}}{x - y}$$

(b)
$$\frac{2\sqrt{ab}}{\sqrt{ab} + 1} = \frac{2\sqrt{ab}}{\sqrt{ab} + 1} \cdot \frac{\sqrt{ab} - 1}{\sqrt{ab} - 1}$$
$$= \frac{2\sqrt{ab}(\sqrt{ab} - 1)}{(\sqrt{ab})^2 - 1^2}$$

$$= \frac{2ab - 2\sqrt{ab}}{ab - 1}$$

SECTION 2-4, REVIEW QUESTIONS

1. The two forms of the distributive law are $a(b + c) = ab + ac$, and $(a + b)c =$ _____ .   ac + bc

2. To multiply two sums, you multiply each term of the first sum by each _____ of the second sum.   term

3. The product $(A + B)(C + D + E)$ is found by multiplying A by each of the terms C, D, and E, and then multiplying _____ by each of the terms C, D, and E.   B

4. The following special products are expanded as follows:
   $(x + y)(x - y) =$ _____ $x^2 - y^2$
   $(x + y)^2 =$ _____ $x^2 + 2xy + y^2$
   $(x - y)^2 =$ _____ $x^2 - 2xy + y^2$

5. To rationalize a denominator of the form $\sqrt{x} + \sqrt{y}$, multiply numerator and denominator by _____ . $\sqrt{x} - \sqrt{y}$

6. $(\sqrt{x} + \sqrt{y})(\sqrt{x} - \sqrt{y}) =$ _____ .   x−y

SECTION 2-4, EXERCISES

Multiply and simplify each of the following.

1. $(x - y)(x^2 + y^2)$
2. $(x + 1)(y - 1)$ Example 1
3. $(2x + y)(x^3 + y)$
4. $(3x - y)(y - 2)$
5. $(a - 2b)(x + 3y)$
6. $(3x + y)(2a - 3b)$
7. $(x + y)(x^2 - y^2)$
8. $(x^3 - y^3)(x - y)$
9. $(x + 1)(2x^2 + x + 3)$
10. $(2x - 1)(x^2 - 3x + 1)$ Example 2,3
11. $(x + 3)(-x^2 - x + 4)$
12. $(4x - 2)(3x^2 + x - 5)$
13. $(5x - 3)(x^4 + x^2 - 2)$
14. $(x - y)(x^2 - xy + y)$
15. $(x^2 + 2x + 1)(x^2 - x + 3)$
16. $(x^2 - 3x - 1)(2x^2 + x + 1)$
17. $(2x^2 + x + 2)(x^3 - x^2 + 1)$
18. $(x^3 + x + 1)(x^4 - x^2 + 1)$
19. $(2x^2 - 3x + 2)(4x^2 - x + 2)$
20. $(3x^3 - x - 1)(x + x^2 - 2)$
21. $(a + b)(a - b)$
22. $(2x - y)(2x + y)$ Example 4

23. $(x^2 + 1)(x^2 - 1)$
24. $(a^3 + b)(a^3 - b)$
25. $(4x - 3y)(4x + 3y)$
26. $(x^n + 1)(x^n - 1)$
27. $(x + 2y)^2$
28. $(3x - 2)^2$
29. $(3a + b)^2$
30. $(x^2 + 2)^2$
31. $(x^2 - y^2)^2$
32. $(5x + 7y)^2$
33. $(3a^2 - 3b^2)^2$
34. $(x^n + 1)^2$

Example 5
35. $(x + 1)(2x + 3)$
36. $(2x - 3)(x + 2)$
37. $(3x - 2)(x - 5)$
38. $(5x + 2)(2x - 7)$
39. $(x - 2b)(2x + b)$
40. $(a - b)(2a + b)$

Example 4
41. $(x + 1)(1 - x)$
42. $(x^{1/2} + y^{1/2})^2$
43. $(x^{1/3} + 1)(x^{1/3} - 1)$
44. $(x^{1/4} - a^{1/3})^2$
45. $(a^{2/3} + b^{2/3})(a^{2/3} - b^{2/3})$
46. $(x^{n/2} + 1)(x^{n/2} - 1)$
47. $(\sqrt{x} + 1)(\sqrt{x} - 1)$
48. $(\sqrt{a} + \sqrt{b})(\sqrt{a} - \sqrt{b})$
49. $(\sqrt{x} - \sqrt{y})^2$
50. $(\sqrt{x} + 1)^2$

Simplify and write without radicals in the denominator.

Example 6
51. $\dfrac{1}{\sqrt{x} - 1}$
52. $\dfrac{1}{\sqrt{a} + \sqrt{b}}$
53. $\dfrac{1}{\sqrt{xy} + 2}$

54. $\dfrac{\sqrt{xy}}{\sqrt{x} + \sqrt{y}}$
55. $\dfrac{\sqrt{x}}{\sqrt{3x} + 5}$
56. $\dfrac{\sqrt{x} - \sqrt{y}}{\sqrt{x} + \sqrt{y}}$

2-5 FACTORING

We have seen how the distributive law allows us to express a product of polynomials as a single polynomial. Frequently we want to reverse the process, and write polynomials as products. This process is called *factoring*. The simplest kind of factoring is removing a monomial (one term) factor from a polynomial. This simply amounts to reading the distributive law backward: $ab + ac = a(b + c)$.

Example 1 Factor $2x^3y^4z + 6x^2y^5 + 14x^4y^3z^3$.

Solution. We check each variable which occurs in each term, and find the highest power of that number which occurs in all terms. The letter x occurs in each term, and x^2 is the highest power which divides each term. Similarly, y occurs in each term, and y^3 is the highest power which divides each term. Each term is also divisible by 2, so $2x^2y^3$ is the largest common monomial factor:

$$2x^3y^4z + 6x^2y^5 + 14x^4y^3z^3 = 2x^2y^3(xyz + 3y^2 + 7x^2z^3).$$

Example 2 Factor $6a^2b^3 + 15a^2b^2 + 9a$.

Solution. The largest numerical factor of each term is 3, and the only other common factor is a; therefore $3a$ is the largest monomial factor and
$$6a^2b^3 + 15a^2b^2 + 9a = 3a(2ab^3 + 5ab^2 + 3).$$

The next most common type of factoring problem is to factor a polynomial of the form $x^2 + px + q = 0$ (called a *quadratic polynomial in one variable*). Suppose such a polynomial can be written as a product of linear factors $x + a$ and $x + b$, where a and b are integers. Then we would have
$$(x + a)(x + b) = x^2 + (a + b)x + ab = x^2 + px + q.$$
Therefore given a polynomial $x^2 + px + q$, we search for integers a and b such that $a + b$ will be the coefficient of x (that is, p), and ab will be equal to the constant term q.

Example 3 Factor $x^2 + 5x + 6$.

Solution. We search for a and b such that $ab = 6$, and $a + b = 5$. The possible factors of 6 are 1 and 6, -1 and -6, 2 and 3, and -2 and -3. The factors whose sum is 5 are 2 and 3. Hence
$$x^2 + 5x + 6 = (x + 2)(x + 3).$$

Example 4 Factor $x^2 + 2x + 3 = 0$.

Solution. Here we are looking for integers a and b such that $ab = 3$, and $a + b = 2$. The only integer factorizations of 3 are $(3)(1)$ and $(-3)(-1)$. Since $3 + 1 = 4 \neq 2$, and $-3 - 1 = -4 \neq 2$, the given polynomial has no linear factors with integer coefficients.

Example 5 Factor $x^2 - 3x - 18$.

Solution. We write all factorizations of -18: 1, -18; -1, 18; 2, -9; -2, 9; 3, -6; and -3, 6. The pair whose sum is -3 is 3 and -6; hence
$$x^2 - 3x - 18 = (x + 3)(x - 6).$$

The common products
$$(x - a)(x + a) = x^2 - a^2,$$
$$(x + a)^2 = x^2 + 2ax + a^2,$$
$$(x - a)^2 = x^2 - 2ax + a^2,$$

all have this form, and one should learn to look for one of these special cases when the constant term is a square.

Example 6 Factor (a) $x^2 - 36$ and (b) $x^2 - 6x + 9$.

Solution.

(a) $\qquad\qquad x^2 - 36 = (x + 6)(x - 6)$

(b) $\qquad\qquad x^2 - 6x + 9 = (x - 3)^2$

For quadratic polynomials where the coefficient of x^2 is not 1, we have the pattern

$$(ax + b)(cx + d) = acx^2 + (bc + ad)x + bc.$$

To factor such a polynomial, we first write all factors of the coefficient of x^2, and then all factors of the constant term. Then we try all possible combinations to see if one of them gives the correct coefficient of x.

Example 7 Factor $2x^2 - 11x - 6$.

Solution. The possible factorizations of 2 are $(2)(1)$; and $(-2)(-1)$. The possible factorizations of -6 are $(-6)(1)$, $(6)(-1)$, $(-3)(2)$, $(3)(-2)$. We will see that it is sufficient to consider the factorization $(2)(1)$ of 2, along with each of the possible factorizations of 6.

We will try each factorization of 6 in both ways in the form

$$(2x + \underline{\qquad})(x + \underline{\qquad}).$$

That is, for the factors -6 and 1, we try both

$$(2x - 6)(x + 1) \quad \text{and} \quad (2x + 1)(x - 6).$$

The eight possibilities we get are:

(i) $(2x - 6)(x + 1)$ $\qquad$ (i)′ $(2x + 1)(x - 6)$

(ii) $(2x + 6)(x - 1)$ $\qquad$ (ii)′ $(2x - 1)(x + 6)$

(iii) $(2x - 3)(x + 2)$ $\qquad$ (iii)′ $(2x + 2)(x - 3)$

(iv) $(2x + 3)(x - 2)$ $\qquad$ (iv)′ $(2x - 2)(x + 3)$

By expanding each of the above, we see that (i)′ is the correct factorization:

$$2x^2 - 11x - 6 = (2x + 1)(x - 6).$$

We need only consider the factors 2 and 1 of 2 in Example 7, and not -2 and -1 because of the identity

$$(-2x - 1)(-x + 6) = (2x + 1)(x - 6).$$

The factors on the right are the negatives of those on the left. Similarly, we find the factorization

$$-3x^2 + 7x - 2 = (-3x + 1)(x - 2) \tag{1}$$

by considering the factors -3 and 1 of -3. The possibility corresponding to 3 and -1 is

$$-3x^2 + 7x - 2 = (3x - 1)(-x + 2),$$

which is gotten by just changing the signs of both factors in (1).

In any problem of factoring polynomials, it is important to *look for monomial factors first.*

Example 8 Factor (a) $12x^3 - 27x$, and (b) $5x^4 + 15x^3 + 10x^2$.

Solution. (a) We first remove the monomial factor $3x$, and then factor the resulting difference of squares.

$$12x^3 - 27x = 3x(4x^2 - 9) = 3x(2x + 3)(2x - 3)$$

(b) First remove the monomial factor $5x^2$, and then factor the remaining quadratic expression.

$$5x^4 + 15x^3 + 10x^2 = 5x^2(x^2 + 3x + 2) = 5x^2(x + 2)(x + 1)$$

The difference of two squares, $a^2 - b^2$, is one of the forms for which we have a standard factorization: $a^2 - b^2 = (a - b)(a + b)$. The difference of two cubes, $a^3 - b^3$, and the sum of two cubes, $a^3 + b^3$, are also frequently occurring forms for which the factors should be memorized. The following identities are easily checked by expanding the products on the right:

$$a^3 - b^3 = (a - b)(a^2 + ab + b^2),$$
$$a^3 + b^3 = (a + b)(a^2 - ab + b^2).$$

Example 9 Factor (a) $8 - x^3$ and (b) $27x^6 + y^9$.

Solution.

(a) $\quad 8 - x^3 = 2^3 - x^3$
$$= (2 - x)(2^2 + 2x + x^2)$$
$$= (2 - x)(4 + 2x + x^2)$$

(b) $\quad 27x^6 + y^9 = (3x^2)^3 + (y^3)^3$
$$= (3x^2 + y^3)[(3x^2)^2 - (3x^2)(y^3) + (y^3)^2]$$

$$= (3x^2 + y^3)(9x^4 - 3x^2y^3 + y^6)$$

SECTION 2-5, REVIEW QUESTIONS

distributive   1. Factoring a monomial factor from a sum depends on the _____ law.

a   2. The distributive law $ab + ac = a(b + c)$ can be regarded as factoring the monomial _____ from the sum $ab + ac$.

all   3. The largest monomial factor in a sum contains each letter which occurs in every term, to the highest power which occurs in _____ terms.

P   S   4. The quadratic $x^2 + Sx + P$ will have the factors $(x + a)(x + b)$ provided $ab =$ _____ and $a + b =$ _____.

5. The following special quadratics have the factorizations

(x+a)(x−a) $x^2 - a^2 =$ _____

(x+a)² $x^2 + 2ax + a^2 =$ _____

(x−a)² $x^2 - 2ax + a^2 =$ _____

square   6. You look for one of the forms above if the constant term in a quadratic polynomial is a perfect _____.

C   7. To factor $Ax^2 + Bx + C$, you first find all factorizations of A and of _____.

c′   c   8. For the factorizations $aa' = A$ and $cc' = C$, you would try both possibilities $(ax + c)(a'x + c')$ and $(ax +$ ___$)(a'x +$ ___$)$.

(a+b)(a²− ab+b²)   9. The sum of two cubes is factored by the formula $a^3 + b^3 =$ _____.

10. The difference of two cubes is factored by the formula

(a−b)(a²+ab+b²) $a^3 - b^3 =$ _____.

SECTION 2-5, EXERCISES

Factor each of the following.

Example 1,2

1. $3ax^2 + 15a^2x$
2. $2x^2y + 16xz - 10x^2z^2$
3. $24a^4b^2 - 18a^3b^3c^3 + 12a^2b^4$
4. $34x^2y + 51xy^2$
5. $7ac + 35bc - 21c^2$
6. $x^4y^3 + x^2y^3$

Example 3,4,5

7. $x^2 - 5x + 6$
8. $x^2 - 3x + 2$

58 *Algebraic Expressions*

9. $x^2 + 3x + 2$
10. $x^2 - 7x + 12$
11. $x^2 + 6x - 16$
12. $x^2 - 6x - 7$
13. $x^2 + 9x + 18$
14. $x^2 - x - 30$
15. $x^2 - 13x - 30$
16. $x^2 - 10x - 24$
17. $4x^2 - 1$
18. $x^2 - 9a^2$ Example 6
19. $25x^2 - 16y^2$
20. $36x^4 - 9$
21. $x^2 - 4x + 4$
22. $x^2 + 6x + 9$
23. $x^2 + 8x + 16$
24. $x^2 - 10x + 25$
25. $x^2 + 18x + 81$
26. $x^2 - 14x + 49$
27. $4x^2 + 4x + 1$
28. $9x^2 + 12x + 4$ Example 7
29. $2x^2 - 5x - 3$
30. $6x^2 - 2x - 4$
31. $10x^2 - 11x - 6$
32. $12x^3 - 27x$
33. $3x^2 + 15x + 18$
34. $x^5 - x^3$
35. $5x^2 - 20x + 20$
36. $6ax^2 - 36ax + 30a$
37. $a^2bx^2 + 2abx + b$
38. $x^3 + 1$
39. $27 - a^3$
40. $x^3 - y^3$
41. $8x^3 - 27$
42. $a^2x^3 - a^5$
43. $3xy^4 + 81x^4y$

2-6 RATIONAL EXPRESSIONS

The only difference between algebra and the arithmetic we all learn in grammar school is the use of letters to stand for numbers in algebra. Expressions involving only numbers always simplify to give just a single number; for example,

$$(4 \times 8 + 31) \div 21 = 3.$$

However, expressions containing literal numbers must be treated as formulas, using algebraic laws, until values are specified for the letters. In this section we study one of the standard types of formula which arises wherever literal numbers are used.

A formula which is the result of performing the four basic operations on one or more literal numbers is called a *rational algebraic expression*. The following are examples of such expressions:

$$\frac{2xz}{a+b}, \quad \frac{x + 1/x}{x^{10} + 1}, \quad \frac{x^5 + 3x^4 + 2x + 1}{4x^4 + x^2 - 3}.$$

In their simplest form, rational expressions are just quotients of polynomials. In this section we will review the techniques for handling formulas of this kind.

The basic property of the real numbers which is used to simplify fractions is the identity

$$\frac{a}{b} = \frac{ac}{bc}, \quad \text{where } b \neq 0, c \neq 0. \tag{1}$$

In words, (1) says that a fraction is unchanged in value if you multiply or divide both numerator and denominator by the same nonzero number. When we simplify a fraction by dividing numerator and denominator by the same factor, we say we have *canceled* the factor.

Example 1 Simplify by canceling all factors common to numerator and denominator.

(a) $\dfrac{x^3yz}{xz^2 + x^2z}$ (b) $\dfrac{x-1}{x^2-1}$ (c) $\dfrac{x^2 + 5x + 6}{x^2 + 4x + 4}$

Solution. The numerators and denominators are first written in factored form.

(a) $\dfrac{x^3yz}{xz^2 + x^2z} = \dfrac{x^3yz}{xz(z+x)} = \dfrac{x^2y}{z+x} \quad (x \neq 0, z \neq 0)$

(b) $\dfrac{x-1}{x^2-1} = \dfrac{x-1}{(x+1)(x-1)} = \dfrac{1}{x+1} \quad (x \neq 1, -1)$

(c) $\dfrac{x^2 + 5x + 6}{x^2 + 4x + 4} = \dfrac{(x+2)(x+3)}{(x+2)^2} = \dfrac{x+3}{x+2} \quad (x \neq -2)$

Rational expressions are multiplied and divided like any other fractions. The basic rules are:

$$\frac{a}{b} \cdot \frac{c}{d} = \frac{ac}{bd},$$

and $\dfrac{a}{b} \div \dfrac{c}{d} = \dfrac{a}{b} \cdot \dfrac{d}{c} = \dfrac{ad}{bc}.$

When multiplying or dividing rational expressions, we first write them in factored form. This ensures that the expressions are in simplest form before we start, and facilitates simplification of the answer.

Example 2 Simplify

(a) $\dfrac{x^2 - 7x + 6}{x^2 - 5x - 6} \cdot \dfrac{x^2 - 1}{x^2 + 4x - 5}$ (b) $\dfrac{x^2 - 7x + 6}{x^2 - 5x - 6} \div \dfrac{x^2 - 1}{x^2 + 4x - 5}$

Solution. First factor each quadratic expression, and simplify the fraction. Then multiply or divide and further simplify.

(a) $\dfrac{x^2 - 7x + 6}{x^2 - 5x - 6} \cdot \dfrac{x^2 - 1}{x^2 + 4x - 5} = \dfrac{(x - 6)(x - 1)}{(x - 6)(x + 1)} \cdot \dfrac{(x + 1)(x - 1)}{(x + 5)(x - 1)}$

$= \dfrac{x - 1}{x + 1} \cdot \dfrac{x + 1}{x + 5}$

$= \dfrac{x - 1}{x + 5}$

(b) $\dfrac{x^2 - 7x + 6}{x^2 - 5x - 6} \div \dfrac{x^2 - 1}{x^2 + 4x - 5} = \dfrac{(x - 6)(x - 1)}{(x - 6)(x + 1)} \div \dfrac{(x + 1)(x - 1)}{(x + 5)(x - 1)}$

$= \dfrac{x - 1}{x + 1} \div \dfrac{x + 1}{x + 5} = \dfrac{x - 1}{x + 1} \cdot \dfrac{x + 5}{x + 1}$

$= \dfrac{(x - 1)(x + 5)}{(x + 1)^2}$

To add two fractions with the same denominator, we simply add their numerators. This is a consequence of the distributive law, as we saw in Chapter 1.

$$\dfrac{a}{c} + \dfrac{b}{c} = ac^{-1} + bc^{-1}$$

$$= (a + b)c^{-1}$$

$$= \dfrac{a + b}{c}$$

To add two fractions with different denominators, we multiply numerator and denominator of each by the appropriate factor so the resulting denominators will be the same.

$$\dfrac{a}{b} + \dfrac{c}{d} = \dfrac{ad}{bd} + \dfrac{cb}{db}$$

$$= \dfrac{ad + cb}{bd}$$

Before adding two rational expressions, we write them in factored form so we can obtain the simplest common denominator. The least common denominator will have each factor which occurs in any of the denominators, and will have each such factor to the highest power to which it occurs in any denominator.

Example 3 Add and simplify

$$\frac{1}{x^2 + 3x + 2} + \frac{x^2 - 5x + 6}{x^2 - 4}$$

Solution.

$$\frac{1}{x^2 + 3x + 2} + \frac{x^2 - 5x + 6}{x^2 - 4} = \frac{1}{(x + 2)(x + 1)} + \frac{(x - 3)(x - 2)}{(x + 2)(x - 2)}$$

$$= \frac{1}{(x + 2)(x + 1)} + \frac{x - 3}{x + 2}$$

$$= \frac{1}{(x + 2)(x + 1)} + \frac{(x - 3)(x + 1)}{(x + 2)(x + 1)}$$

$$= \frac{1 + (x - 3)(x + 1)}{(x + 2)(x + 1)}$$

$$= \frac{1 + x^2 - 2x - 3}{(x + 2)(x + 1)}$$

$$= \frac{x^2 - 2x - 2}{(x + 2)(x + 1)}$$

Example 4 Perform the indicated operations and simplify

$$\frac{1}{x - 2} - \frac{x^2 - 1}{x^2 + 2x + 1} + \frac{x - 1}{x^2 - 3x + 2}$$

Solution. We first factor numerators and denominators and simplify each fraction. We get

$$\frac{1}{x - 2} - \frac{(x + 1)(x - 1)}{(x + 1)^2} + \frac{x - 1}{(x - 2)(x - 1)} =$$

$$\frac{1}{x - 2} - \frac{x - 1}{x + 1} + \frac{1}{x - 2}.$$

The least common denominator is $(x + 1)(x - 2)$, and so the above expression is equal to

$$\frac{x + 1}{(x + 1)(x - 2)} - \frac{(x - 1)(x - 2)}{(x + 1)(x - 2)} + \frac{x + 1}{(x + 1)(x - 2)} =$$

$$= \frac{x+1-(x-1)(x-2)+x+1}{(x+1)(x-2)}$$

$$= \frac{x+1-(x^2-3x+2)+x+1}{(x+1)(x-2)}$$

$$= \frac{-x^2+5x}{(x+1)(x-2)}.$$

Compound fractions involving rational expressions are treated just like ordinary fractions. We can either simplify the numerator and denominator first, and then divide, or we can multiply both numerator and denominator by an expression which will clear them of fractions. We illustrate both techniques in the following example.

Example 5 Simplify
$$\frac{1+\dfrac{x}{x+1}}{\dfrac{1}{x}+\dfrac{1}{x+1}}$$

Solution. (a) We will first express the numerator and denominator as single fractions and then perform the division.

$$1+\frac{x}{x+1} = \frac{x+1}{x+1}+\frac{x}{x+1}$$

$$= \frac{2x+1}{x+1}$$

$$\frac{1}{x}+\frac{1}{x+1} = \frac{x+1}{x(x+1)}+\frac{x}{x(x+1)}$$

$$= \frac{2x+1}{x(x+1)}$$

Therefore
$$\frac{1+\dfrac{x}{x+1}}{\dfrac{1}{x}+\dfrac{1}{x+1}} = \frac{2x+1}{x+1} \div \frac{(2x+1)}{x(x+1)}$$

$$= \frac{2x+1}{x+1} \cdot \frac{x(x+1)}{(2x+1)}$$

$$= x.$$

Solution. (b) As an alternate technique, we multiply numerator and denominator of the original expression by $x(x + 1)$. This will clear both numerator and denominator of fractions.

$$\frac{1 + \dfrac{x}{x+1}}{\dfrac{1}{x} + \dfrac{1}{x+1}} = \frac{x(x+1)\left[1 + \dfrac{x}{(x+1)}\right]}{x(x+1)\left[\dfrac{1}{x} + \dfrac{1}{(x+1)}\right]}$$

$$= \frac{x(x+1) + x^2}{x + 1 + x}$$

$$= \frac{2x^2 + x}{2x + 1}$$

$$= \frac{x(2x + 1)}{2x + 1}$$

$$= x$$

SECTION 2-6, REVIEW QUESTIONS

rational

1. A quotient of two polynomials is called a _____ algebraic expression.

2. Rational expressions are fractions, and the basic property used to simplify

a/b

 fractions is $ac/bc = $ _____ .

divide

3. You can multiply or _____ the numerator and denominator of any fraction by any nonzero number without changing its value.

4. Dividing numerator and denominator by the same factor is called

cancelling

 _____ that factor.

factor

5. Before combining rational expressions, you first _____ numerators and denominators.

6. If the numerators and denominators are in factored form, it is easy to find

denominator

 the least common _____ .

7. The easiest way to simplify a compound fraction is to multiply both numerator and denominator by an expression that will clear both the

fractions

 numerator and denominator of _____ .

SECTION 2-6, EXERCISES

Complete any indicated operation and simplify.

1. $\dfrac{9ab}{12a^2b^2}$

2. $\dfrac{27x^3y}{36xy^2}$

3. $\dfrac{3x - 9}{9x - 9}$ Example 1

4. $\dfrac{10 - 5x}{10 + 5x}$

5. $\dfrac{x^2 - y^2}{x + y}$

6. $\dfrac{c + d}{c^2 - d^2}$

7. $\dfrac{x^2 - 25}{x^2 + 10x + 25}$

8. $\dfrac{x^3 - 4x}{2x^3 + 8x^2 + 8x}$

9. $\dfrac{a^2 - a - 6}{2a^2 - a - 10}$

10. $\dfrac{b^2 - 3b - 10}{b^2 - 2b - 15}$

11. $\dfrac{x^3 - 1}{x^3 + 4x^2 - 5x}$

12. $\dfrac{x^2 + x}{x^3 + 1}$

13. $\dfrac{4a^2}{6b^2} \cdot 24ab$

14. $36xy \cdot \dfrac{9x}{4y^2}$

15. $\dfrac{x + y}{3a - 9b} \cdot \dfrac{12}{x + y}$ Example 2

16. $\dfrac{3x}{x^2 - xy} \cdot \dfrac{4y^2 - 4x^2}{12y}$

17. $\dfrac{2x - 3}{x^2 - 1} \cdot \dfrac{2x^2 + x - 3}{4x^2 - 9}$

18. $\dfrac{3x^2 - 2xy - y^2}{x^2 - y^2} \cdot \dfrac{1}{3x^2 + 4xy + y^2}$

19. $\dfrac{y^2 - 5y + 6}{y^2 - 4} \cdot \dfrac{y^2 + 11y + 18}{y^2 - 2y - 3}$

20. $\dfrac{x^3 - 27y^3}{x + 3y} \cdot \dfrac{x^2 - 9y^2}{x^2 + 3xy + 9y^2}$

21. $\dfrac{2x^2 - 8}{x^2 - 3x - 10} \div \dfrac{x^2 - 3x + 2}{x^2 - 6x + 5}$

22. $\dfrac{2xy + y^2}{y^2 - x^2} \div \dfrac{xy + 2x^2}{x + y}$

23. $\dfrac{1 - a}{1 + a^2} \div \dfrac{(a - 1)^2}{1 - a^2}$

24. $\dfrac{x}{x^2 - 1} - \dfrac{1}{x - 1}$

25. $\dfrac{x}{3x - 2y} - \dfrac{y}{2x + 3y}$ Example 3,4

26. $\dfrac{x - 3}{x + 3} + \dfrac{x + 3}{3 - x}$

27. $\dfrac{2}{xy + y^2} + \dfrac{3}{x^2 + xy}$

28. $\dfrac{3}{a + b} + \dfrac{2a}{(a - b)^2}$

29. $\dfrac{2x}{x^2 + 3x + 2} - \dfrac{1}{x^2 - 1}$

30. $\dfrac{x}{x^2 - 5x + 6} + \dfrac{x}{x^2 - x - 6}$

31. $\dfrac{x + 3}{3 - x} + \dfrac{x^2}{x^2 - 9}$

32. $\dfrac{\dfrac{x}{2y}}{\dfrac{4}{yz}}$

33. $\dfrac{1 + \dfrac{1}{x}}{\dfrac{1}{x}}$

34. $\dfrac{\dfrac{a}{a + b}}{\dfrac{b}{a + b}}$ Example 5

35. $\dfrac{\dfrac{2x}{x+1}}{\dfrac{x-1}{x+1}}$
36. $\dfrac{3a^2 - 2a - 1}{\dfrac{3a+1}{a-1}}$
37. $\dfrac{\dfrac{1}{a} + \dfrac{1}{b}}{\dfrac{1}{a} - \dfrac{1}{b}}$

38. $\dfrac{1 + \dfrac{1}{a}}{1 - \dfrac{1}{a}}$
39. $x - \dfrac{x}{x + \dfrac{1}{4}}$
40. $1 - \dfrac{1}{1 - \dfrac{1}{x-2}}$

2-7 REVIEW FOR CHAPTER

Simplify each of the following and write without negative exponents.

1. $\dfrac{a^{-3}}{(-x)^2}$
2. $\dfrac{-5x^{-1}}{15y^{-2}}$

3. $\dfrac{(-7a^0 b^{-2})^{-1}}{(7a^{-1})^2 b^{-1}}$
4. $x^{-2} y^{-2}$

5. $x^{-2} + y^{-2}$
6. $(x^{-2} + y^{-2})(x + y)$

7. $x^{-1} + y$
8. $x^{-1/2} + y^{-1/2}$

9. $(x^{-1} + y^{-1})(x + y)$
10. $(x^{-1/2} y^{-1/2})^2$

11. $(-2yx^{-3}y^6)^{-1/3}$
12. $-(16x^{-4}y^{-8})^{-3/4}$

Simplify each of the following.

13. $\dfrac{\sqrt{4y^2}}{\sqrt{x^4}}$
14. $\dfrac{\sqrt{14}}{\sqrt{2x^5}}$

15. $\sqrt{\dfrac{y^3}{3x^3}}$
16. $\sqrt{\dfrac{27a^5}{3b^3}}$

17. $\sqrt{\dfrac{x+y}{x-y}}$
18. $\sqrt{\dfrac{a-b}{a+b}}$

19. $\sqrt{\dfrac{5a}{3a^{-1}}}$
20. $\sqrt{\sqrt{81a^5 b^4}}$

21. $\sqrt[3]{\sqrt{64a^7 b^8}}$
22. $\dfrac{1}{\sqrt{3} - 2}$

66 *Algebraic Expressions*

23. $\dfrac{\sqrt{x} + \sqrt{y}}{\sqrt{x} - \sqrt{y}}$
24. $\dfrac{a^{1/2} - b^{1/2}}{a^{1/2} + b^{1/2}}$

Multiply and simplify each of the following.

25. $(8x - 5)(x - 3)$
26. $(3x^2 - 2x + 4)(2x - 5)$
27. $(x + 2)(x^2 - 2x + 4)$
28. $(3a - b)(a^2 - 3a + 7)$
29. $(x^2 - 1)(x^4 - 5x^2 - 8)$
30. $(x^2 + x + 1)(x - 1)$
31. $(a + 5)(a^2 - 5a + 25)$
32. $(2x - 3)(4x^2 + 6x + 9)$
33. $(3x + 5)(2x - 7) - (7x - 4)(x - 1)$
34. $(3a - 2)(3a + 2) - (3a - 2)^2$
35. $(x - 1)(x^3 + x^2 + x + 1)$
36. $(x + 1)(x^3 - x^2 + x - 1)$
37. $(x^{2n} - y^{2n})^2$
38. $(x^{2n} + y^{2n})(x^{2n} - y^{2n})$
39. $(x^{1/3} - y^{1/3})^2$
40. $(a^{1/3} + b^{1/3})(a^{1/3} - b^{1/3})$

Factor each of the following.

41. $121x^2 - 49$
42. $3x^7 + x$
43. $15x^5 - 20x^3 + 45x$
44. $9x^2 + 30x + 25$
45. $-3x^3 - 27x$
46. $x^4 - 81$
47. $27x^3 - 64y^6$
48. $8a^6 + 27y^3$
49. $2(2 + x) - 3(2 + x)$
50. $(x + 17)^2 - 7(x + 17) + 6$

Complete any indicated operation and simplify.

51. $\dfrac{2x + x^2}{4x - 5} \cdot \dfrac{16x - 20}{4x^2 + 2x^3}$
52. $(a + b)^3 \div (a + b)^2$

53. $\dfrac{a^2 - 5a}{2a^3 - a^2} \cdot \dfrac{2a^2 - 5a + 2}{a^2 - 7a + 10}$
54. $\dfrac{3x^2 + 6x - 24}{x^2 - 7x + 10} \div \dfrac{3x^2 + 4x}{x^3 - 5x^2}$

55. $\dfrac{x^2 + 13xy + 12y^2}{3xy^2} \div (x + y)$
56. $\dfrac{7x - 2}{x(b + c)} - \dfrac{2}{b + c}$

57. $\dfrac{1}{x - 3} + \dfrac{2}{9 - x^2}$
58. $\dfrac{5}{x + 7} + \dfrac{25}{x^2 + 9x + 14}$

59. $\dfrac{7}{x^2 - y^2} - \dfrac{5}{3x + 3y}$
60. $\dfrac{5}{xy + y^2} + \dfrac{2}{x^2 + xy}$

61. $\dfrac{2 - \dfrac{x}{y}}{2 + \dfrac{x}{y}}$
62. $\dfrac{1 - \dfrac{3}{x - 2}}{1 + \dfrac{3}{2 - x}}$

63. $\dfrac{\dfrac{y}{x} - \dfrac{4x}{y}}{\dfrac{y^2}{4x^2} - \dfrac{4x^2}{y^2}}$

64. $\dfrac{\dfrac{1}{x} - \dfrac{3}{y} + \dfrac{1}{z}}{\dfrac{5}{xy} - \dfrac{2}{yz} + \dfrac{1}{xz}}$

3 EQUATIONS AND INEQUALITIES

3-1 LINEAR EQUATIONS

In the applications of mathematics to other fields, the usual first step is to express certain physical facts as equations or inequalities. In this chapter we will study techniques for treating the kinds of equations and inequalities which arise in this way.

There are two basically different kinds of equations involving literal numbers, or variables. The following are examples of these two types:

$$\frac{x^2 - 1}{x - 1} = x + 1 \qquad (1)$$

$$2x = x + 4 \qquad (2)$$

Equation (1) is called an *identity*, and equation (2) is called a *conditional equation*.

An identity is an equation which is a true sentence for all values of the letters for which both sides make sense. In (1), the left side is not defined for $x = 1$, but the two sides are equal for any other value of x. The following are some examples of identities, with the admissible values of the letters indicated in parentheses.

Identities

$$(x + 1)^2 = x^2 + 2x + 1 \qquad \text{(all } x\text{)}$$

$$\sqrt{a^2} = |a| \qquad \text{(all } a\text{)}$$

$$\frac{1}{x-1} - \frac{1}{x+1} = \frac{2}{x^2 - 1} \qquad (x \neq 1, -1)$$

$$(\sqrt{x-1})^2 = x - 1 \qquad (x \geq 1)$$

A conditional equation is one which is not true for all values of the variables. Thus the truth of the equation constitutes a condition on the variables. For example, equation (2) is true for $x = 4$, but for no other value of x. The values of the variable for which a conditional equation becomes a true statement are *roots* or *solutions* of the equation. To solve a conditional equation means to find *all* the roots. For an equation to be conditional, there must be some number which is not a root, but a conditional equation need not have any roots. That is, the solution set of a conditional equation can be empty, but must not contain all real numbers.

Example 1 Tell which equation is an identity and which is a conditional equation. For the conditional equation, give one value of x which is *not* a solution.

(a) $\dfrac{1-x}{x-1} = \dfrac{x-1}{1-x}$ (b) $x^2 + 1 = x$

Solution. Equation (a) is an identity, since both sides equal -1 for all admissible values of x. Equation (b) is conditional; for example, 0 is not a root: $0^2 + 1 \neq 0$.

To solve a conditional equation, we try to simplify it in some way that does not change the solution set. Two conditional equations are called *equivalent* if they have exactly the same solutions. Hence one way to solve an equation is to find, by successive simplifications of one kind or another, an equivalent equation whose roots are obvious.

If we add the same number or expression to both sides of an equation, we obtain an equivalent equation. We can also multiply both sides by the same expression, provided that expression is never zero. We show below how these two operations can be used to solve a simple but important class of equations called *linear equations*. A linear equation is an equation which is equivalent to one of the form

$$ax + b = 0$$

for some numbers a and b, with $a \neq 0$. The examples below treat linear equations and some simple variants.

Example 2 Solve
$$5x - 4 = 3x + 6$$

Solution. The following equations are equivalent.

$5x - 4 = 3x + 6$

$5x = 3x + 10$ (Add 4 to both sides)

$2x = 10$ (Subtract $3x$ from both sides)

$x = 5$ (Divide both sides by 2)

It is clear that the final equation has just one solution, 5, so 5 is also the only solution of the given equation.

The same method used in Example 2 shows that any linear equation, $ax + b = 0$, has exactly the one root $x = -b/a$.

Example 3 Solve
$$\frac{2x}{x-1} = 1 + \frac{2}{x-1}$$

Solution. First we multiply both sides of the equation by $x - 1$. This does not necessarily give an equivalent equation since $x - 1$ can equal zero. The resulting equation, however, will have all the roots of the given equation as solutions, plus possibly one more.

$$\frac{2x}{x-1} = 1 + \frac{2}{x-1}$$

$$2x = x - 1 + 2$$

$$2x = x + 1$$

$$x = 1$$

We know that $x = 1$ is the only possible solution of the given equation. However, since neither side of the given equation makes sense when $x = 1$, we conclude that there are no solutions. Although we arrived at the linear equation $2x = x - 1 + 2$ in the solution process, the given equation is not linear.

Example 4 Solve
$$\frac{1}{x} + \frac{2}{x} + \frac{3}{x} = 4$$

Solution.

$$\frac{1}{x} + \frac{2}{x} + \frac{3}{x} = 4$$

$1 + 2 + 3 = 4x$ (Multiply by x)

$6 = 4x$ (Simplify)

$x = \frac{3}{2}$ (Divide by 4)

Since the second equation is not necessarily equivalent to the given equation, we must check the only possible root, $\frac{3}{2}$, in the original equation.

$$\frac{2}{3} + \frac{4}{3} + \frac{6}{3} = \frac{12}{3} = 4$$

Hence $\frac{3}{2}$ is a solution, and the only one.

Example 5 Solve

$$\frac{2x + 2}{3x - 5} = \frac{4x + 1}{6x - 7}$$

Solution. We can clear the equation of fractions by multiplying both sides by $(3x - 5)(6x - 7)$.

$$\frac{2x + 2}{3x - 5} = \frac{4x + 1}{6x - 7}$$

$$(3x - 5)(6x - 7)\left(\frac{2x + 2}{3x - 5}\right) = (3x - 5)(6x - 7)\left(\frac{4x + 1}{6x - 7}\right)$$

$$(6x - 7)(2x + 2) = (3x - 5)(4x + 1)$$

$$12x^2 - 2x - 14 = 12x^2 - 17x - 5$$

$$-2x - 14 = -17x - 5$$

$$15x - 14 = -5$$

$$15x = 9$$

$$x = \frac{3}{5}$$

The number 3/5 is the only possible solution, and we check it in the given equation to see if it actually is a solution.

$$\frac{2\left(\frac{3}{5}\right)+2}{3\left(\frac{3}{5}\right)-5} \stackrel{?}{=} \frac{4\left(\frac{3}{5}\right)+1}{6\left(\frac{3}{5}\right)-7}$$

$$\frac{2\cdot 3 + 10}{3\cdot 3 - 25} \stackrel{?}{=} \frac{4\cdot 3 + 5}{6\cdot 3 - 35}$$

$$\frac{16}{-16} \stackrel{?}{=} \frac{17}{-17}$$

$$-1 \stackrel{\checkmark}{=} -1$$

Hence 3/5 is the only solution.

The following example shows how linear equations can arise in word problems.

Example 6. If a number is increased by 10, the result is three times the original number. What is the number?

Solution. We let x stand for any number with the given property. Then the condition leads to the following linear equation.

$$x + 10 = 3x$$
$$2x = 10$$
$$x = 5$$

Hence 5 is the unique number with the stated property.

Linear Equations 73

SECTION 3-1, REVIEW QUESTIONS

conditional   1. The two kinds of equations involving letters are identities and _____ equations.

all   2. A conditional equation is one which is not true for _____ values of the variables.

3. A value of the variable for which an equation is a true sentence is called a

solution   root or _____ of the equation.

equivalent   4. Two equations with the same solutions are called _____.

add   5. You get an equivalent equation if you _____ the same number to both sides of an equation.

ax+b   6. A linear equation has the form _____ = 0.

one   7. The equation $ax + b = 0$ has exactly _____ root.

more   8. If you multiply both sides of an equation by an expression containing the unknown, you may get an equation with _____ roots than the original.

SECTION 3-1, EXERCISES

Tell which of the following equations are identities, and which are conditional equations. For each conditional equation give one value of the variable which is not a solution. For each identity give the admissible values of the variables.

Example 1

1. $x^2 - 4 = (x + 2)(x - 2)$ 2. $x^2 - 4 = -4$

3. $2x = 0$ 4. $1 + \dfrac{1}{x} = \dfrac{x + 1}{x}$

5. $\dfrac{a^2 - b^2}{a - b} = a + b$ 6. $a^2 - 1 = a - 1$

7. $(\sqrt{x})^2 = x$ 8. $\sqrt{a^2} = a$

9. $\sqrt{x^2} = |x|$ 10. $\dfrac{x - 1}{1 - x} = -1$

Solve the following equations.

Example 2

11. $3x - 4 = 2x + 7$ 12. $4 - 2x = 2x + 1$

13. $7x + 8 = 3x$ 14. $\dfrac{1}{2}x - 3 = \dfrac{1}{2}$

74 Equations and Inequalities

15. $\dfrac{2}{3} - \dfrac{3}{2}x = \dfrac{1}{6}x - \dfrac{7}{3}$

16. $\dfrac{3}{4} - \dfrac{5}{12}x - \dfrac{7}{12} = \dfrac{5}{12}x$

17. $\dfrac{1}{3}x - \dfrac{5}{6} = \dfrac{3}{2} + \dfrac{4}{3}x$

18. $\dfrac{x}{3} - \dfrac{4x+1}{2} = x - \dfrac{5}{6}$

19. $\dfrac{1}{3}(x+8) - \dfrac{1}{4}(3-2x) = \dfrac{1}{6}$

20. $0.6(1 - 0.8x) = 0.5 - 0.4(2 - 1.8x)$

21. $5 - \dfrac{3}{2(x+1)} = \dfrac{4x}{x+1}$

22. $\dfrac{2x}{x-1} = 3 - \dfrac{4}{2(1-x)}$

Example 3,4,5

23. $\dfrac{5x^2}{x^2-7} - 3 = 2 - \dfrac{x}{x^2-7}$

24. $\dfrac{3x}{x^2-4} + 5 = \dfrac{4x^2}{x^2-4} + 1$

25. $\dfrac{1}{x} + \dfrac{1}{3} = 4$

26. $\dfrac{1}{x} + \dfrac{3}{2x} - 4 = 0$

21. $\dfrac{2}{x} + \dfrac{3}{x} - \dfrac{1}{4} = 2$

28. $\dfrac{5}{3x} + \dfrac{5}{x} = 7$

29. $\dfrac{1}{2} + \dfrac{3}{5} + \dfrac{1}{x} = \dfrac{4}{3}$

30. $\dfrac{x-1}{x} + \dfrac{3}{2} = \dfrac{x+1}{2x}$

Solve the following equations for x in terms of the other variables.

31. $x + xa = 3$

32. $ax - 3bx = (a - 3b)^2$

33. $ax + c = bx - d$

34. $5ax + 5x - b = 2x - 3ax + c$

35. $\dfrac{x}{a} - b = \dfrac{b}{a} - x$

36. $\dfrac{1}{x} + \dfrac{1}{a} = 1$

37. $\dfrac{1}{x} + \dfrac{1}{R} = \dfrac{1}{r}$

38. $ax + 1 = \dfrac{1}{a} - x$

39. If a number is decreased by 8, the result is $\dfrac{3}{5}$ of the given number. What is the number?   Example 6

40. If a number is decreased by 9, the result is four times the number. What is the number?

41. If a number is increased by 14, the number is tripled. What is the number?

42. If a number is increased by 5, the result is two-thirds of the original number. What is the number?

43. If a rectangle has a perimeter of 70 feet, and it is 6 times as long as it is wide, how wide is the rectangle?

44. If a rectangle has a perimeter of 100 feet, and the length is 4 feet greater than the width, how wide is the rectangle?

45. The sum of half a number and three-fourths of the same number is 25. What is the number?

46. Find the number such that if you multiply it by 3, subtract 7, and then divide by 4, you get 2.

47. The sum of two consecutive odd numbers is 48. Find the numbers.
48. The sum of two consecutive even numbers is 194. Find the numbers.
49. A man has four sons, each of whom is 2 years older than his next younger brother. If the sum of boys' ages is 52 years, how old is each of the boys?
50. What fraction equals its own reciprocal, and has denominator two more than its numerator?
51. If Jim can run a mile in 6 minutes and Jeff can run a mile in 10 minutes, when will Jim lap Jeff if they start from the same place at the same time on a circular mile track?
52. In Exercise 51, assume that the boys run in opposite directions, and tell how long they must run before they meet.

3-2 HIGHER DEGREE POLYNOMIAL EQUATIONS

In the last section we saw how to solve linear equations and equations which can be reduced to linear equations. Now we will consider slightly more complicated equations. The general form of a polynomial equation is

$$a_n x^n + a_{n-1} x^{n-1} + \cdots + a_1 x + a_0 = 0$$

where $a_0, a_1, \ldots, a_n$ are given numbers, $a_n \neq 0$, and $n \geqslant 1$. The number n is called the *degree* of the equation.

The following are some examples of polynomial equations.

$$3x^3 - x^2 + 1 = 0$$
$$x^4 - x^3 + x^2 - x + 1 = 0$$
$$x^{10} - 1 = 0$$
$$3x - 4 = 0$$

The basic fact used to solve polynomial equations of degree greater than one is the following: *For any numbers* A *and* B,

$$AB = 0 \text{ if and only if } A = 0 \text{ or } B = 0.$$

It follows from this that if the polynomial $P(x)$ can be factored as $P(x) = Q(x) \cdot R(x)$, then the roots of $P(x) = 0$ consist of the roots of $Q(x) = 0$ together with those of $R(x) = 0$. A similar statement holds for products of three or more factors. Thus the product $Q(x) \cdot R(x) \cdot S(x) = 0$ if and only if $Q(x) = 0$ or $R(x) = 0$ or $S(x) = 0$.

76 *Equations and Inequalities*

Example 1 Solve
$$(x - 1)(x + 3) = 0$$
Solution. For the product to be zero, x must satisfy
$$x - 1 = 0 \quad \text{or} \quad x + 3 = 0$$
Conversely, if x satisfies either of these equations, it satisfies the given equation. The roots are therefore 1 and -3.

Example 2 Solve
$$x^2(x - 1) - 4x(x - 1) = 0$$
Solution. We write the polynomial in factored form by first factoring out the term $x - 1$.
$$x^2(x - 1) - 4x(x - 1) = 0$$
$$(x - 1)(x^2 - 4x) = 0$$
$$(x - 1)x(x - 4) = 0$$
The last equation is equivalent to the given equation, and its solutions are the solutions of any of the three equations
$$x - 1 = 0, \quad x = 0, \quad x - 4 = 0.$$
Hence the solutions are 1, 0, and 4.

A linear equation is a polynomial equation of degree one. A polynomial equation of degree two is called a *quadratic equation*. Hence a quadratic equation has the form
$$ax^2 + bx + c = 0, \quad \text{where } a \neq 0.$$
If the second degree polynomial $ax^2 + bx + c$ can be factored, then the solutions can be found as in the examples above. For example, the equation
$$x^2 - 5x + 6 = 0$$
can be written in factored form as
$$(x - 3)(x - 2) = 0.$$
Hence the roots are the roots of $x - 3 = 0$ and $x - 2 = 0$, so $x = 3$ or $x = 2$.

If a quadratic polynomial cannot be factored by inspection, the roots can nevertheless always be found by *completing the square*. We illustrate this technique with the equation
$$2x^2 - 9x - 5 = 0$$

We collect the x terms together, and complete the square as follows.

$$2x^2 - 9x - 5 = 0$$
$$2x^2 - 9x = 5$$
$$2\left(x^2 - \frac{9}{2}x + \frac{81}{16}\right) = 5 + \frac{81}{8} \quad \text{($\frac{81}{16}$ is the square of half the coefficient of x)}$$
$$2\left(x - \frac{9}{4}\right)^2 = \frac{121}{8}$$
$$\left(x - \frac{9}{4}\right)^2 = \frac{121}{16}$$
$$x - \frac{9}{4} = \pm\frac{11}{4} \quad \text{(take the square root at both sides)}$$
$$x = \frac{9}{4} \pm \frac{11}{4}$$
$$x = 5 \quad \text{or} \quad x = -1/2$$

The two roots are 5 and $-1/2$.

The process of completing the square will work for any quadratic equation. If we go through the above process for the general equation $ax^2 + bx + c = 0$, we obtain a formula for the roots of any quadratic equation in terms of the coefficients a, b, and c.

$$ax^2 + bx + c = 0 \tag{1}$$
$$x^2 + \frac{b}{a}x = -\frac{c}{a}$$
$$x^2 + \frac{b}{a}x + \frac{b^2}{4a^2} = \frac{b^2}{4a^2} - \frac{c}{a}$$
$$\left(x + \frac{b}{2a}\right)^2 = \frac{b^2 - 4ac}{4a^2} \tag{2}$$
$$x + \frac{b}{2a} = \pm\frac{1}{2a}\sqrt{b^2 - 4ac}$$
$$x = \frac{-b \pm \sqrt{b^2 - 4ac}}{2a}$$

Equations and Inequalities

All of the equations above are equivalent. The right-hand side of the final equation, which gives an explicit formula for the roots of (1), is called the *quadratic formula*.

From (2) we see that if $b^2 - 4ac < 0$, the equation has no real roots since the left side is non-negative, and the right side is negative. If $b^2 - 4ac = 0$, then the equation has exactly one root, $x = -b/2a$. If $b^2 - 4ac > 0$, the equation has two distinct real roots.

Example 3 Use the quadratic formula to solve
$$3x^2 - 7x + 4 = 0$$

Solution. Here $a = 3$, $b = -7$, and $c = 4$, so we get
$$x = \frac{-(-7) \pm \sqrt{(-7)^2 - 4(3)(4)}}{2(3)}$$
$$= \frac{7 \pm \sqrt{49 - 48}}{6}$$
$$= \frac{7 \pm 1}{6}$$

The solutions are 4/3 and 1.

Equations involving rational functions can also lead to quadratic or higher degree equations. In solving this type of equation, it is sometimes necessary to multiply both sides of the equation by an expression which is zero for some values of the variable. This can lead to an equation with more roots than the given equation. Hence we must either check the roots in the original equation, or verify that none of the expressions we multiplied by are zero for the values of the roots we find.

Example 4 Solve
$$\frac{8}{x+1} + x = 5$$

Solution. We multiply both sides by $x + 1$ to clear the fractions.
$$\frac{8}{x+1} + x = 5 \qquad (3)$$
$$8 + x^2 + x = 5x + 5 \qquad (4)$$

Every root of (3) is a root of (4), and every root of (4) is a root of (3) unless -1 happens to be a root of (4). We solve (4) as follows:

$$x^2 - 4x + 3 = 0$$
$$(x - 3)(x - 1) = 0$$
$$x = 3 \quad \text{or} \quad x = 1$$

The two roots of (4) are 1 and 3, and these are also the roots of (3).

SECTION 3-2, REVIEW QUESTIONS

1. If A and B are any numbers, then $A \cdot B = 0$ if and only if $A = 0$ _____ $B = 0$.

2. If $P(x)$, $Q(x)$, and $R(x)$ are polynomials, and $P(x) = Q(x) \cdot R(x)$ is a factorization of $P(x)$, then the roots of $P(x) = 0$ are the roots of $Q(x) = 0$ together with the roots of _____ $= 0$.

3. Any equation of the form $ax^2 + bx + c = 0$, where $a \neq 0$, is called a _____ equation.

4. Any quadratic equation can be solved by _____ the _____.

5. If you complete the square in the general quadratic equation $ax^2 + bx + c = 0$, you get the quadratic formula for the roots: $x =$ _____.

6. If $b^2 - 4ac < 0$, the quadratic equation has _____ real roots; if $b^2 - 4ac = 0$, the equation has _____ root; if $b^2 - 4ac > 0$, the equation has _____ real roots.

Margin notes: or; $R(x)$; quadratic; completing square; $\frac{-b \pm \sqrt{b^2 - 4ac}}{2a}$; no; one; two

SECTION 3-2, EXERCISES

Solve the following polymonial equations.

Example 1
1. $(x + 1)(x - 2) = 0$
2. $(x + 5)(x + 2) = 0$
3. $(2x + 1)(x - 4) = 0$
4. $(3x - 2)(5x + 7) = 0$

Example 2
5. $x^2 - 3x = 0$
6. $2x^2 + 8x = 0$
7. $x^2 - 9 = 0$
8. $x^2 - 16 = 0$
9. $x^2 - 5 = -1$
10. $x^2 - 28 = -3$
11. $x^3 - 9x = 0$
12. $2x^3 - 32x = 0$
13. $x(x + 1) - 2(x + 1) = 0$
14. $2x(x - 3) - 8(x - 3) = 0$
15. $x^2(x - 2) - 4(x - 2) = 0$
16. $x^2(x + 1) - 9(x + 1) = 0$
17. $x^2 - 2x + 1 = 0$
18. $x^2 + 6x + 9 = 0$
19. $x^2 - 5x + 6 = 0$
20. $x^2 + 5x + 4 = 0$

21. $x^2 - 6x + 8 = 0$
22. $x^2 + 7x + 12 = 0$
23. $x^3 + 2x^2 + x = 0$
24. $x^3 - 10x^2 + 25x = 0$
25. $x^4 + x^2 - 2 = 0$
26. $x^4 - 3x^2 - 4 = 0$
27. $x^4 - 2x^2 + 1 = 0$
28. $x^4 - 6x^2 + 9 = 0$
29. $x^4 + 4x^2 + 4 = 0$
30. $x^4 + 10x^2 + 25 = 0$
31. $x^2 - 2ax - 24a^2 = 0$
32. $x^2 - 3bx - 4b^2 = 0$
33. $x^2 + (a + b)x + ab = 0$
34. $x^2 - (a + b)x - ab = 0$
35. $x - \dfrac{9}{x} = 0$
36. $x + \dfrac{4}{x} = 0$ Example 4
37. $x - \dfrac{35}{x + 2} = 0$
38. $(x + 5) - \dfrac{9}{x + 5} = 0$

Use the quadratic formula to solve the following.

39. $2x^2 + 5x - 3 = 0$
40. $3x^2 + 2x - 1 = 0$ Example 3
41. $2x^2 + x - 3 = 0$
42. $4x^2 + 3x - 1 = 0$
43. $2x^2 + x - 5 = 0$
44. $x^2 + 3x - 1 = 0$
45. $2x^2 + 6x + 3 = 0$
46. $3x^2 + 6x + 2 = 0$
47. $cx^2 + bx + a = 0$
48. $bx^2 + cx + a = 0$

If r and s are the roots of the quadratic equation $ax^2 + bx + c = 0$, use the quadratic formula to show that:

49. $r + s = -\dfrac{b}{a}$
50. $rs = \dfrac{c}{a}$

51. A 16 by 20 foot rug is placed on a floor that contains 480 square feet so that there is a uniform border around the rug. How far is the rug from the wall?
52. Find the dimensions of a rectangle if its area is 96 square feet and the length is 4 feet more than the width.
53. Find all the numbers that are 20 less than their squares.
54. Find all the numbers that are 12 less than their squares.
55. Find the largest value of c so that $x^2 + 8x + c = 0$ will have at least one solution.
56. Find the smallest value of a so that $ax^2 + 4x - 1 = 0$ will have at least one solution.
57. For what values of a will $ax^2 + 4x + a = 0$ have exactly one solution?
58. For what values of b will $x^2 + bx + b = 0$ have exactly one solution?

3-3 SPECIAL TECHNIQUES

In this section we will look at some techniques for solving several different kinds of conditional equations.

Equations that contain radicals can frequently be simplified by squaring both sides. This process may introduce new roots, but will not lead to an equation with fewer roots than the original. For example, $\sqrt{x} = -1$ has no roots since the radical is always nonnegative, but squaring gives $x = 1$, which does have a root. Since squaring both sides can introduce new roots, we must check the roots found this way in the *original* equation.

Example 1 Solve

$$\sqrt{x+1} + 5 = x$$

Solution. We put the radical by itself on one side of the equation, and square both sides.

$\sqrt{x+1} + 5 = x$	
$\sqrt{x+1} = x - 5$	Add -5 to both sides
$x + 1 = x^2 - 10x + 25$	Square both sides
$x^2 - 11x + 24 = 0$	Add $-x - 1$ to both sides
$(x - 8)(x - 3) = 0$	Factor.
$x = 8 \quad \text{or} \quad x = 3$	

Now check both possibilities in the given equation:

$$\sqrt{8+1} + 5 \stackrel{?}{=} 8 \qquad \text{(checks)}$$
$$\sqrt{3+1} + 5 \stackrel{?}{=} 3 \qquad \text{(does not check)}$$

Hence the only root is 8.

If more than one radical occurs in an equation, it may be necessary to square both sides of the equation more than once.

Example 2 Solve

$$\sqrt{x+6} - \sqrt{x+1} = 1$$

Solution. We first put the radicals on opposite sides of the equation so that squaring does not introduce a product of the two radicals.

$\sqrt{x+6} - \sqrt{x+1} = 1$	
$\sqrt{x+6} = 1 + \sqrt{x+1}$	
$x + 6 = 1 + 2\sqrt{x+1} + x + 1$	Square both sides
$4 = 2\sqrt{x+1}$	

$$2 = \sqrt{x+1}$$
$$4 = x+1 \quad \text{Square again}$$
$$x = 3$$

Check: $\sqrt{3+6} - \sqrt{3+1} \stackrel{?}{=} 1$

$3 - 2 \stackrel{\checkmark}{=} 1$

If radicals other than square roots occur, it may be necessary to raise both sides of the equation to a higher power.

Example 3 Solve

$$\sqrt{x+2} = \sqrt[4]{5x+6}$$

Solution. Here we raise both sides of the equation to the fourth power to clear the radicals.

$$\sqrt{x+2} = \sqrt[4]{5x+6}$$
$$(\sqrt{x+2})^4 = 5x+6$$
$$(x+2)^2 = 5x+6$$
$$x^2 + 4x + 4 = 5x + 6$$
$$x^2 - x - 2 = 0$$
$$(x-2)(x+1) = 0$$
$$x = 2 \quad \text{or} \quad x = -1$$

Check: $\sqrt{2+2} \stackrel{?}{=} \sqrt[4]{5 \cdot 2 + 6}$

$\sqrt{4} \stackrel{\checkmark}{=} \sqrt[4]{16}$

$\sqrt{-1+2} \stackrel{?}{=} \sqrt[4]{5(-1)+6}$

$\sqrt{1} \stackrel{\checkmark}{=} \sqrt[4]{1}$

The two roots are -1 and 2.

Some equations can be put in the form of a linear or quadratic equation in a new variable by making a substitution.

Example 4 Solve

(a) $$x^4 + x^2 - 6 = 0$$

Special techniques

(b) $$\frac{1}{(x^2-1)^2} + \frac{1}{x^2-1} - 6 = 0$$

Solution. Equation (a) is a quadratic equation in x^2, and (b) is the same quadratic equation in $1/(x^2 - 1)$. To solve (a) we let $u = x^2$.

$$u^2 + u - 6 = 0$$

$$(u + 3)(u - 2) = 0$$

$$u = -3 \quad \text{or} \quad u = 2$$

Therefore,

$$x^2 = -3 \quad \text{or} \quad x^2 = 2.$$

Since $x^2 = -3$ is impossible, $\pm\sqrt{2}$ are the roots of (a).

To solve (b) we let $u = 1/(x^2 - 1)$. This gives the same quadratic equation as (a), and hence

$$u = -3 \quad \text{or} \quad u = 2,$$

$$\frac{1}{x^2 - 1} = -3 \quad \text{or} \quad \frac{1}{x^2 - 1} = 2.$$

We solve these two equations separately.

$$1 = -3x^2 + 3 \qquad\qquad 1 = 2x^2 - 2$$
$$3x^2 = 2 \qquad\qquad\qquad 2x^2 = 3$$
$$x^2 = \frac{2}{3} \qquad\qquad\qquad x^2 = \frac{3}{2}$$
$$x = \pm\sqrt{\frac{2}{3}} \qquad\qquad x = \pm\sqrt{\frac{3}{2}}$$
$$= \pm\frac{\sqrt{6}}{3} \qquad\qquad\quad = \pm\frac{\sqrt{6}}{2}$$

The four roots of (b) are thus $\sqrt{6}/3$, $-\sqrt{6}/3$, $\sqrt{6}/2$, and $-\sqrt{6}/2$.

Equations involving the absolute value sign are solved by remembering that $|x| = a$ means that $x = a$ or $-x = a$. Here, of course, we assume that $a \geq 0$; if $a < 0$, then $|x| = a$ has no solution.

Example 5 Solve

$$|3x + 2| = |x - 8|.$$

Solution. Two numbers whose absolute values are the same are either equal or are negatives of each other. Therefore the equation is the same as

$$3x + 2 = x - 8 \quad \text{or} \quad 3x + 2 = -(x - 8),$$
$$2x = -10 \quad \text{or} \quad 4x = 6,$$
$$x = -5 \quad \text{or} \quad x = \frac{3}{2}.$$

Therefore the roots are -5 and $3/2$.

SECTION 3-3, REVIEW QUESTIONS

1. To solve an equation with a radical, put the radical by itself on one side and _____ both sides. **square**
2. Squaring both sides of an equation may introduce new _____. **roots**
3. Higher order radicals can sometimes be simplified by raising both sides of the equation to a higher _____. **power**
4. Complicated equations can sometimes be reduced to linear or quadratic form by making a _____. **substitution**
5. To solve $(\sqrt[3]{x})^2 - 3\sqrt[3]{x} + 2 = 0$, you would make the substitution $u = $ _____. $\sqrt[3]{x}$
6. Since $|P(x)| = |-P(x)|$; the equation $|P(x)| = a$ is the same as $P(x) = a$ or _____ $= a$. $-P(x)$
7. The roots of $P(x) = a$ together with the roots of $-P(x) = a$ give all the _____ of $|P(x)| = a$. **roots**

SECTION 3-3, EXERCISES

1. $\sqrt{5x} = x$ 2. $\sqrt{7x} = x$ 3. $\sqrt{6x} = 2x$ **Example 1**
4. $\sqrt{5x} = 3x$ 5. $\sqrt{6x + 7} = x$ 6. $\sqrt{3x + 4} = x$
7. $\sqrt{x - 12} = x$ 8. $\sqrt{3x - 10} = x$ 9. $\sqrt{2x - 3} = x - 1$

	10. $\sqrt{3x+7} = x+3$
	12. $\sqrt{x-1} = x-3$
Example 2	14. $\sqrt{3x-2} - 2 = x$
	16. $\sqrt{3x+3} = \sqrt{2x+1}$
Example 3	18. $\sqrt{x+5} + \sqrt{x} = 5$
	20. $\sqrt[4]{7x+18} = \sqrt{x}$
	22. $\sqrt[4]{12x+21} = \sqrt{x+4}$
Example 4	24. $\sqrt{3x} = \sqrt[4]{4-9x}$

10. $\sqrt{3x+7} = x+3$
11. $\sqrt{x-3} = x-5$
12. $\sqrt{x-1} = x-3$
13. $\sqrt{2x+6} - 1 = -x$

Example 2
14. $\sqrt{3x-2} - 2 = x$
15. $\sqrt{5x-1} = \sqrt{x}+1$
16. $\sqrt{3x+3} = \sqrt{2x}+1$
17. $\sqrt{x+1} - \sqrt{x-2} = 1$

Example 3
18. $\sqrt{x+5} + \sqrt{x} = 5$
19. $\sqrt[4]{3x+4} = \sqrt{x}$
20. $\sqrt[4]{7x+18} = \sqrt{x}$
21. $\sqrt[4]{5x+1} = \sqrt{x+1}$
22. $\sqrt[4]{12x+21} = \sqrt{x+4}$
23. $\sqrt{5x} = \sqrt[4]{3-10x}$

Example 4
24. $\sqrt{3x} = \sqrt[4]{4-9x}$
25. $x^4 - 4x^2 + 3 = 0$
26. $x^4 - 6x^2 + 8 = 0$
27. $x^4 - 13x^2 + 36 = 0$
28. $x^4 + x^2 - 20 = 0$
29. $x^4 + 7x^2 + 12 = 0$
30. $x^4 + 6x^2 + 5 = 0$
31. $(x^2 - 3x)^2 - 2(x^2 - 3x) - 8 = 0$
32. $(x^2 + 5x)^2 - 2(x^2 + 5x) - 24 = 0$
33. $(x^2 - 2x)^2 - 2(x^2 - 2x) - 3 = 0$
34. $(x^2 - 4x)^2 + 8(x^2 - 4x) + 15 = 0$
35. $\dfrac{1}{(x-1)^2} - \dfrac{4}{x-1} + 3 = 0$
36. $\dfrac{1}{(x-2)^2} + \dfrac{5}{x-2} - 36 = 0$
37. $\left(\dfrac{x}{x-1}\right)^2 - 3\left(\dfrac{x}{x-1}\right) + 2 = 0$
38. $\left(\dfrac{x+1}{x}\right)^2 + 4\left(\dfrac{x+1}{x}\right) + 3 = 0$

Example 5
39. $|x-2| = 3$
40. $|x+5| = 4$
41. $|x| = |x+2|$
42. $|x+4| = |x|$
43. $|1-x| = |x+7|$
44. $|x-6| = |3-x|$
45. $|2x+3| = |x-3|$
46. $|3x-5| = |2x+7|$

47. Find a number such that 3 times its square root equals 45.
48. Find all the numbers such that the square root of twice the number is equal to the number.
49. Find all the numbers such that the square root of 3 less than the number is equal to 5 less than the number.

50. Find all the numbers such that the square root of one less than 5 times the number equals one more than the square root of the number.

3-4 LINEAR INEQUALITIES

So far we have treated equations, and learned techniques for solving several common types of conditional equations. In many applications of mathematics, particularly in the social sciences, the empirical facts lead to statements which are *inequalities* rather than equations. In this section and the next we will study several kinds of inequalities that occur frequently.

We have seen that equations containing variables are either identities or conditional equations. We make the same sort of distinction for inequalities, and speak of absolute inequalities and conditional inequalities. An *absolute inequality* is one which holds for all values of the variable for which all the expressions make sense. The following are absolute inequalities:

$$\frac{1}{x^2} > 0 \qquad x + 2 > x + 1 \qquad \sqrt{3x - 5} \geq 0$$

A *conditional inequality* is one which is not true for all possible values of the variable. For example,

$$2x > 3, \quad (x - 1)(x - 2) \leq 0, \quad x^3 < x,$$

are conditional inequalities. Specific values of x for which these inequalities do *not* hold are easy to find. For example, the first inequality fails for $x = 1$, the second fails for $x = 0$, and the third fails for $x = 1$.

To solve an inequality means to find all the numbers for which the inequality does hold. These numbers are called roots or solutions as in the case of equations. For inequalities, however, the solution set does not typically consist of one or two numbers, but is more likely to be a set consisting of one or more intervals. The respective solution sets of the examples above are listed below:

$2x > 3$, solution set consists of all x such that $x > \frac{3}{2}$;

$(x - 1)(x - 2) \leq 0$, solution set consists of all x such that $1 \leq x \leq 2$;

$x^3 < x$, solution set consists of all x such that $x < -1$ or $0 < x < 1$.

The basic properties of the order relation which we use to solve inequalities were demonstrated in Section 1-6; they are the following facts:

1. If $A < B$, then $A + C < B + C$.
2. (a) If $A < B$ and $C > 0$, then $AC < BC$.
   (b) If $A < B$ and $C < 0$, then $AC > BC$.

The first property implies that we obtain an equivalent inequality if we add (or subtract) the same expression from both sides of an inequality. Property 2 says that an inequality is preserved if we multiply (or divide) both sides by a *positive* number, but reversed if we multiply (or divide) both sides by a *negative* number.

Example 1 Solve
$$2x - 3 > 4x + 7$$

Solution. The following inequalities are equivalent:

$$2x - 3 > 4x + 7$$
$$-2x - 3 > 7 \quad \text{Add } -4x \text{ to both sides}$$
$$-2x > 10 \quad \text{Add 3 to both sides}$$
$$x < -5 \quad \text{Divide by } -2$$

Notice that in the final step, the inequality sign must be reversed since we divide both sides by the negative number -2. Hence the solution set of the inequality consists of all x such that $x < -5$.

We recall that $|x|$ can be interpreted as the distance from x to 0. It is clear then that $|x| < a \ (a > 0)$ means the same as the two simultaneous inequalities $-a < x$ and $x < a$. Similarly, $|x| \leq a$ means $-a \leq x$ and $x \leq a$.

Example 2 Solve
$$|3x - 5| < x + 1$$

Solution. The given inequality is equivalent to the simultaneous inequalities below, which we solve separately.

$$|3x - 5| < x + 1$$

$-(x + 1) < 3x - 5$	and	$3x - 5 < x + 1$
$-x + 4 < 3x$	and	$3x < x + 6$
$4 < 4x$	and	$2x < 6$
$1 < x$	and	$x < 3$

The solution set consists of all x such that $1 < x < 3$.

The inequality $|x| > a$ means the same as $x > a$ or $x < -a$. Inequalitites involving the absolute value sign are therefore handled differently when the sign is $>$ instead of $<$. Thus

$|P(x)| < a$ means $-a < P(x)$ and $P(x) < a$,
$|P(x)| > a$ means $P(x) < -a$ or $P(x) > a$.

The solution set of $|P(x)| > a$ is the union of the solution set of $P(x) > a$ and the solution set of $P(x) < -a$. Similar statements hold for inequalities of the form $|P(x)| \geq a$.

Example 3 Solve
$$|2x + 3| \geq 5$$

Solution.
$$|2x + 3| \geq 5$$

$2x + 3 \leq -5$	or	$2x + 3 \geq 5$
$2x \leq -8$	or	$2x \geq 2$
$x \leq -4$	or	$x \geq 1$

The solution set consists of all x such that $x \leq -4$ or $x \geq 1$.

We cannot multiply both sides of an inequality by an expression containing the unknown without taking into account the sign of the expression. We therefore treat inequalities involving fractions somewhat differently from equalities.

The solutions of an inequality of the form $P(x)/Q(x) > 0$ will consist of the values of x for which $P(x)$ and $Q(x)$ have the same sign. Similarly, $P(x)/Q(x) < 0$ has as solutions the values of x such that $P(x)$ and $Q(x)$ have opposite signs. Suppose that $P(x)$ and $Q(x)$ are linear expressions. Then there is some number x_1 such that $P(x)$ is positive on one side of x_1 and negative on the other side. There is also a number x_2 such that $Q(x)$ changes sign at x_2. If, for example, $x_1 < x_2$, then there will be a different combination of signs for $P(x)$ and $Q(x)$, to the left of x_1, between x_1 and x_2 and to the right of x_2. If we indicate the signs of $P(x)$ and $Q(x)$ on a graph (see Fig. 3-1), then we can read off the sign of $P(x)/Q(x)$ directly.

Example 4 Solve
$$\frac{x-1}{x+1} > 2$$

Solution. The following inequalities are equivalent.

$$\frac{x-1}{x+1} > 2$$

$$\frac{x-1}{x+1} - 2 > 0$$

$$\frac{x-1-2x-2}{x+1} > 0$$

$$\frac{-x-3}{x+1} > 0$$

$$\frac{x+3}{x+1} < 0 \qquad \text{The sign } > \text{ changes to } < \text{ when you multiply by } -1$$

The inequality will be satisfied whenever $x+3$ and $x+1$ have opposite signs. The numerator $x+3$ is positive to the right of -3 and negative to the left of -3; and the denominator $x+1$ is positive to the right of -1 and negative to the left of -1 as shown in Fig. 3-1.

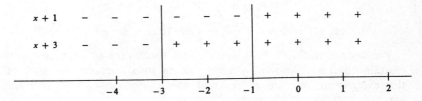

Figure 3-1

The numerator and denominator have opposite signs ($x+1 < 0$, $x+3 > 0$) between -3 and -1, so that the solution set consists of the numbers x such that $-3 < x < -1$.

SECTION 3-4, REVIEW QUESTIONS

1. Inequalities containing variables are either absolute inequalities or _____ inequalities.

conditional

2. A conditional inequality is one which is not true for all values of the _____.

variables

3. If an inequality is true for all values of the variables, it is an _____ inequality.

absolute

90 *Equations and Inequalities*

4. The solution sets of inequalities are usually unions of _____.   **intervals**

5. The solution of the inequality $|x| < 2$ is the open interval from _____ to _____.   **−2, 2**

6. If you add or subtract the same expression to both sides of an inequality, you get an _____ inequality.   **equivalent**

7. An equivalent inequality also results if you multiply both sides by a _____ number.   **positive**

8. If you multiply both sides of an inequality by a negative number, you must reverse the _____ sign.   **inequality**

9. The inequality $|P(x)| < a$ means $-a < P(x)$ and _____.   **P(x)<a**

10. The inequality $|P(x)| > a$ means $P(x) < -a$ _____ $P(x) > a$.   **or**

SECTION 3-4, EXERCISES

Solve the following inequalities.

1. $2x + 7 > 5$
2. $3x + 10 > 1$
3. $3x - 5 < 1$ Example 1
4. $5x - 8 < 7$
5. $2 - x > 3$
6. $5 - 2x < 9$
7. $3x + 4 \leq 8 - 5x$
8. $6 - 2x \geq 5x - 8$
9. $2 - x \geq 10 - 7x$
10. $5 + x \leq 5x - 9$
11. $\frac{1}{2}x + \frac{3}{4} < 2$
12. $\frac{2}{3}x + \frac{5}{2} > -1$
13. $\frac{2}{5} - \frac{1}{5}x < 3$
14. $\frac{3}{4} - 2x < -1$
15. $|x - 3| < 4$ Example 2
16. $|x + 2| < 1$
17. $|2x - 3| < 5$
18. $|3x + 5| < 4$
19. $|3x + 4| < 4 + x$
20. $|2x + 5| < 1 - x$
21. $|2x + 5| \leq x + 1$
22. $|3x + 11| \leq 2x - 1$
23. $|5x| \leq 6 - x$
24. $|2x| \leq 2 + x$
25. $|3x| \leq 1 + 4x$
26. $|x| \leq x + 2$
27. $|x - 2| > 1$ Example 3
28. $|x - 5| > 3$
29. $|2x - 1| > 3$
30. $|3x - 4| > 5$
31. $|8 - 5x| \geq 2$
32. $|2 - 3x| \geq 6$
33. $|7x + 4| \geq 3$
34. $|5x + 2| \geq 3$
35. $\left|\frac{1}{2}x + 2\right| > 5$
36. $\left|\frac{1}{3}x - 2\right| > 5$
37. $\frac{x - 1}{x - 2} < 0$
38. $\frac{x - 3}{x - 2} < 0$
39. $\frac{x + 1}{x - 2} > 0$ Example 4
40. $\frac{x - 5}{x + 3} > 0$
41. $\frac{x - 2}{x + 2} > 3$
42. $\frac{x + 5}{x - 3} > 5$
43. $\frac{2x - 1}{x + 1} > 1$
44. $\frac{3x + 1}{x - 2} > 2$

Further Inequalities 91

3-5 FURTHER INEQUALITIES

Inequalities involving higher degree polynomials can be solved by considering the factors separately. Any polynomial with real coefficients can be factored into real linear and irreducible (non-factorable) quadratic factors. If we find the intervals on which the individual factors are positive or negative, then it is easy to tell where the polynomial itself is positive or negative. We know that linear factors change sign at exactly one point on the real line. It can also be shown that irreducible quadratic factors have the same sign for all values of x.

Example 1 Solve

$$(2x + 3)(x - 1) \leq 0$$

Solution. The product is negative where the linear factors have opposite signs. The factor $2x + 3$ is positive for $x > -3/2$, and negative for $x < -3/2$. Similarly, $x - 1$ is positive for $x > 1$ and negative for $x < 1$. We indicate the signs of two linear factors schematically in Fig. 3-2. The factors have opposite signs between $-\frac{3}{2}$ and 1. The numbers $-3/2$ and 1, where the polynomial is zero, are also solutions, so the solution set consists of all x such that $-\frac{3}{2} \leq x \leq 1$.

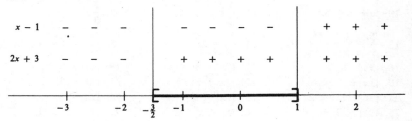

Figure 3-2

Polynomials with more than two factors can be treated in much the same way.

Example 2 Solve

$$x^2(x + 1)(x - 1)(x - 3) > 0$$

Solution. We again indicate where the individual factors are positive and negative. Since $x^2 \geq 0$ for all x, the only effect of the factor x^2 is to exclude 0 as a solution. The intervals on which the other factors are positive and negative are shown in Fig. 3-3. The product is

92 *Equations and Inequalities*

positive if all the factors are positive or if two are positive and two negative. Hence the solution set consists of all x such that $-1 < x < 0$ or $0 < x < 1$ or $x > 3$.

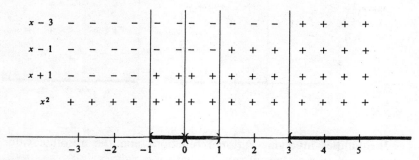

Figure 3-3

Inequalities equivalent to those of the form $R(x) > 0$, for a rational function $R(x)$, can also be solved by considering the signs of the factors in the numerator and denominator. To solve *equalities* of this form $[R(x) = 0]$, we would clear the fractions by multiplying both sides by the appropriate factor. We cannot do this with inequalities because the factors are sometimes negative and sometimes positive, and thus we would not know when to reverse the inequality.

Example 3 Solve

$$\frac{1}{x} < \frac{3}{x + 2}$$

Solution. We collect all the terms on one side of the inequality, and write the resulting rational function in factored form.

$$\frac{1}{x} < \frac{3}{x + 2}$$

$$\frac{1}{x} - \frac{3}{x + 2} < 0$$

$$\frac{x + 2 - 3x}{x(x + 2)} < 0$$

$$\frac{2(1 - x)}{x(x + 2)} < 0$$

The signs of the factors $1 - x$, x, and $x + 2$ are shown in Fig. 3-4. The solutions will be those values of x for which exactly one factor is negative, or all three factors are negative. The latter case does not

Further Inequalities 93

happen here, and the solution set consists of all x such that $-2 < x < 0$ or $x > 1$.

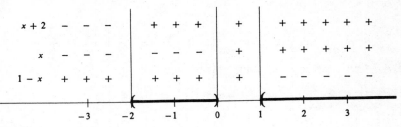

Figure 3-4

Inequalities involving rational functions and the absolute value sign reduce to a system of two ordinary inequalities.

$|R(x)| > S(x)$ means $R(x) > S(x)$ or $R(x) < -S(x)$,
$|R(x)| < S(x)$ means $-S(x) < R(x)$ and $R(x) < S(x)$.

Example 4 Solve
$$|x^2 - 3| < 1$$

Solution. The inequality is equivalent to the system
$$-1 < x^2 - 3 \quad \text{and} \quad x^2 - 3 < 1$$
Hence $\quad 2 < x^2 \quad \text{and} \quad x^2 < 4$

The solution set of the first inequality consists of all x such that $x < -\sqrt{2}$ or $x > \sqrt{2}$, and the solution set of the second consists of all x such that $-2 < x < 2$. The two sets are shown in Fig. 3-5,

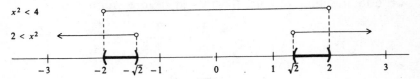

Figure 3-5

and the solution set of the given inequality will be the points common to both, that is, the intersection of the two solution sets. Hence x is a solution if and only if $-2 < x < -\sqrt{2}$, or $\sqrt{2} < x < 2$.

SECTION 3-5, REVIEW QUESTIONS

1. A polynomial with real coefficients can be factored into real linear factors and irreducible _____ factors.   **quadratic**

2. Irreducible quadratic factors do not change sign, and linear factors change sign _____.   **once**

3. If we know the intervals on which the factors of a polynomial are positive or negative, then we can find the _____ where the polynomial itself is positive or negative.   **intervals**

4. We cannot clear the fractions from an inequality by multiplying by some factor because the factor may not always have the same _____.   **sign**

5. An inequality involving the absolute value sign is equivalent to a _____ of ordinary inequalities.   **system**

6. The inequality $|P(x)| < a$ is equivalent to the system $-a < P(x)$ and _____.   **P(x)<a**

7. The inequality $|P(x)| > a$ is equivalent to the system $P(x) < -a$ _____ $P(x) > a$.   **or**

SECTION 3-5, EXERCISES

Solve the following inequalities.

1. $(x - 1)^2 \leq 0$
2. $(x + 2)^2 \geq 0$ Example 1
3. $(x - 1)(x - 2) \leq 0$
4. $(x + 1)(x - 3) < 0$
5. $(x - 2)(x - 5) > 0$
6. $(x + 3)(x - 1) > 0$
7. $(2x - 1)(x + 4) < 0$
8. $(3x + 2)(x - 3) < 0$
9. $(5x + 1)(2x - 3) \geq 0$
10. $(3x + 1)(2x + 7) \geq 0$
11. $(2x + 1)(x - 2) \leq 0$
12. $(3x + 5)(2x + 1) \leq 0$
13. $x^2(x - 1) > 0$
14. $x^2(x + 2) > 0$ Example 2
15. $(x - 1)(x - 2)(x - 3) > 0$
16. $(x + 1)(x - 3)(x - 4) < 0$
17. $(x + 3)(x - 1)(x + 2) \leq 0$
18. $(x - 2)(x + 5)(x + 1) \geq 0$
19. $(2x - 3)(x + 1)(3x - 7) > 0$
20. $(3x + 2)(2x - 5)(x - 2) > 0$
21. $(5x - 1)(2x + 3)(x + 1) < 0$
22. $(3x + 10)(x - 7)(4x + 15) < 0$
23. $x(x - 1)^2(x - 2) > 0$
24. $(x + 1)^2(x - 1)(x - 3) > 0$
25. $x^3(x + 1)^2(x - 2) \leq 0$
26. $x^3(x + 3)(x - 1)^2 \geq 0$
27. $\dfrac{1}{x} > \dfrac{3}{x}$
28. $\dfrac{2}{x} > \dfrac{1}{x}$
29. $\dfrac{6}{x} < 3$ Example 3

Further Inequalities 95

30. $\dfrac{4}{x} > 1$ 31. $\dfrac{5}{x-1} < \dfrac{4}{x}$ 32. $\dfrac{6}{x-2} < \dfrac{3}{x}$

33. $\dfrac{3}{1-x} < \dfrac{2}{x+1}$ 34. $\dfrac{5}{2-x} < \dfrac{3}{x+2}$ 35. $|x^2 - 4| < 2$

Example 4
36. $|x^2 - 3| < 2$ 37. $|x^2 - 1| < 3$ 38. $|x^2 + 1| < 10$
39. $|x^2 + 5| < 3$ 40. $|x^2 + 4| < 4$ 41. $|x^2 + 2| > 1$
42. $|x^2 + 9| > 3$ 43. $|x^2 - 1| > 3$ 44. $|x^2 - 2| > 7$
45. $|x^2 - 8| > 7$ 46. $|x^2 - 9| > 5$

3-6 REVIEW FOR CHAPTER

Solve the following equations.

1. $x = 20 - 3(x + 4)$ 2. $3(x - 2) = 7(2x + 1)$

3. $\dfrac{1}{2}x - \dfrac{6}{5} = 12 - \dfrac{3}{2}x$ 4. $5x + \dfrac{1}{3} = 2x - \dfrac{3}{2}$

5. $\dfrac{1}{5}(x + 2) = \dfrac{2}{3} + \dfrac{1}{9}x$

6. $\dfrac{3}{5}\left(1 - \dfrac{4}{5}x\right) = \dfrac{1}{2} - \dfrac{2}{5}\left(2 - \dfrac{9}{5}x\right)$

7. $0.13x + 1.17 = 1.23x - 0.04$

8. $6.1 - 5x = 4.6x + 3.22$ 9. $\dfrac{2x - 9}{3} = \dfrac{3x + 4}{2}$

10. $\dfrac{2x + 3}{2x - 4} = \dfrac{x - 1}{x + 1}$ 11. $\dfrac{3}{x} - \dfrac{4}{5x} = \dfrac{1}{10}$

12. $\dfrac{2x + 1}{x} + \dfrac{x - 4}{x + 1} = 3$

13. $\dfrac{5}{x - 1} - \dfrac{5}{x + 1} = \dfrac{2}{x - 2} - \dfrac{2}{x + 3}$

14. $\dfrac{2}{x - 3} - \dfrac{4}{x + 3} = \dfrac{16}{x^2 - 9}$

Solve each of the following equations for x in terms of the other variables.

15. $2(x - P) = 3(6P - x)$ 16. $2bx - 2a = ax - 4b$

96 Equations and Inequalities

17. $\dfrac{2x - a}{b} = \dfrac{2x - b}{a}$ 18. $\dfrac{x - a}{x - b} = \dfrac{x - c}{x - d}$

19. $\dfrac{1}{ax} + \dfrac{1}{bx} = \dfrac{1}{c}$ 20. $x + rtx = A$

Solve the following equations.

21. $\dfrac{3(x + 5)}{4} = \dfrac{2(3x - 1)}{5}$ 22. $\dfrac{2x + 3}{5} - \dfrac{x + 1}{2} = 6x$

23. $1 - \dfrac{2(1 - x)}{x - 1} = \dfrac{x + 2}{3}$ 24. $3 - (x + 2) = 2 - \dfrac{3x - 3}{x - 1}$

25. $3x^2 + 5x = 0$ 26. $5x^2 - 7x = 0$

27. $11x = 2 + 15x^2$ 28. $9x^2 - 16 = 0$

29. $4x^2 - 25 = 0$ 30. $7x^2 - 5 = 2x$

31. $x^2 + 2x = 9$ 32. $x^2 - 5x - 3 = 0$

33. $7x^2 - 10x + 5 = 0$ 34. $2x^2 + 4x - 7 = 0$

35. $6x^2 - x - 3 = 0$ 36. $5 + 6x - x^2 = 0$

37. $2x^2 = 8x - 7$ 38. $x - 5 = 3x^2$

39. $x - \dfrac{4}{x} + 3 = 0$ 40. $x - \dfrac{9}{x} = 0$

41. $x + 5 + \dfrac{18}{x + 5} - 9 = 0$ 42. $x - \dfrac{35}{x + 2} = 0$

43. $x - 9 = \dfrac{72}{x - 8}$ 44. $\dfrac{x^2 + 10}{x - 5} = \dfrac{7x}{x - 5}$

45. $\sqrt{2x - 3} = 5 + x$ 46. $\sqrt{x - 2} = x - 4$

47. $\sqrt{8x + 5} = \sqrt{2x + 2}$ 48. $\sqrt{5x + 6} + \sqrt{3x - 2} = 6$

49. $\sqrt{6x + 7} - \sqrt{3x + 3} = 1$ 50. $\sqrt{4x - 3} = \sqrt{8x + 1} - 2$

51. $x^4 - 8x^2 - 9 = 0$ 52. $x^4 - 26x^2 + 25 = 0$

53. $(x^2 - x)^2 - 14(x^2 - x) + 24 = 0$

54. $(x^2 + 1)^2 + 6(x^2 + 1) + 8 = 0$

55. $|x + 3| = 5$ 56. $|x + 1| = 7$

57. $|3x - 2| = 2$ 58. $|2x - 5| = 5$

59. $|2x - 3| = |x - 1|$
60. $|2 - 5x| = |x + 2|$

Solve the following inequalities.

61. $2x + 3 < 7$
62. $3x - 2 < 10$

63. $\dfrac{x}{2} - 1 < 3 - x$
64. $x + 2 < 5 - \dfrac{x}{3}$

65. $\dfrac{2x}{3} + 1 < \dfrac{5(x - 1)}{3}$
66. $\dfrac{7x - 5}{2} < \dfrac{5x + 4}{5}$

67. $1 \leq \dfrac{2 - 3x}{3}$
68. $\dfrac{x - 3}{x} < 1$

69. $|3x - 1| + 3 < 2$
70. $|x + 2| + x < 6$

71. $(1 - x)(1 - x) < 0$
72. $(2x + 1)(2x + 1) < 0$

73. $(3x + 2)(x - 3) > 0$
74. $(x - 2)(4x + 5) > 0$

75. $(x + 3)(x - 2)(x + 4) > 0$
76. $(x - 5)(x - 3)(x + 1) > 0$

77. $x^2(x - 6)(x - 3) < 0$
78. $x^2(x - 5)(x - 4) < 0$

79. If a rectangle has a perimeter of 104 centimeters and the length is three times the width, how long is the rectangle?

80. If a rectangle has an area of 120 square inches and the length is seven inches longer than the width, how long is the rectangle?

81. The sum of a number, 1/3 the number, and 3/5 the number is 29. What is the number?

82. The sum of the squares of two consecutive odd numbers is 130. What are the numbers?

83. If Jim can run a mile in 5 minutes and Jeff can run a mile in 4 minutes, in how many minutes will they meet if they start from the same place on a 1-mile oval track, but run in opposite directions?

84. A piece of wire 40 inches long was cut into two pieces. Each piece was then bent to form a square. If the sum of the areas of the two squares was 58 square inches, how long was each piece of wire?

85. When a border of uniform width was added to a garden having dimensions 30 meters by 20 meters, the total area of the garden was doubled. Find the width of the border.

86. Find a number such that the square root of four less than four times the number equals the number.

4 ANALYTIC GEOMETRY

4-1 INTERVALS

In many parts of mathematics the interplay between algebraic and geometric ideas is particularly important. Algebraic methods can frequently be used to provide simple solutions for geometric problems. At the same time, a geometric interpretation often helps to make abstract ideas easier to visualize.

In this chapter we will see how coordinate systems on a line or plane can be used to relate algebraic equations with geometric figures. In this section we study some simple graphs on the number line.

Recall that $a < b$ has the geometric interpretation that a is to the left of b on the number line. Similarly, $a > b$ means that a is to the right of b (see Fig. 4-1).

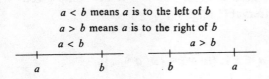

Figure 4-1

The notation $a \leq b$ (a is less than or equal to b) means that either $a < b$ or $a = b$. Similarly, $a \geq b$ means that $a > b$ or $a = b$.

We frequently want to talk about the set of points between two given points on the number line. Such a set is called an *interval*. An interval which contains its two end points is called a *closed interval*, and is denoted with brackets:

$$[a, b] = \{x: a \leq x \leq b\}, \quad \text{(the closed interval from } a \text{ to } b\text{)}.$$

The interval of points strictly between two points is called an *open interval*, and is denoted with parentheses:

$$(a, b) = \{x: a < x < b\}, \quad \text{(the open interval from } a \text{ to } b\text{)}.$$

We indicate open and closed intervals graphically as shown in Figure 4-2.

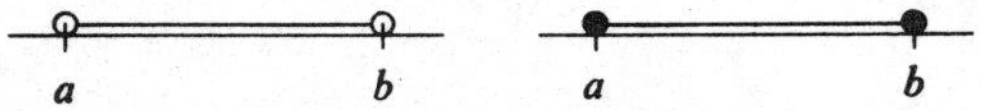

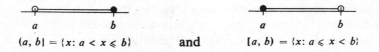

Open interval	Closed interval
$(a, b) = \{x: a < x < b\}$	$[a, b] = \{x: a \leq x \leq b\}$

Figure 4-2

Combinations of one parenthesis and one bracket can be used to indicate intervals that contain one endpoint but not the other. Such intervals are called *half-open* or *half-closed* (Fig. 4-3).

$(a, b] = \{x: a < x \leq b\}$ **and** $[a, b) = \{x: a \leq x < b\}$

Figure 4-3

The set of all numbers greater than (or less than) some given number is also called an interval (Fig. 4-4):

$$(a, \infty) = \{x: a < x\}, \quad [a, \infty) = \{x: a \leq x\},$$
$$(-\infty, b) = \{x: x < b\}, \quad (-\infty, b] = \{x: x \leq b\}.$$

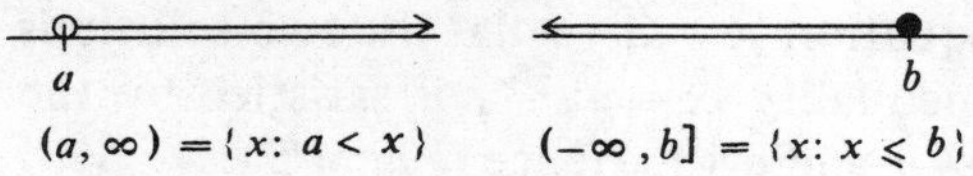

a b
$(a, \infty) = \{x: a < x\}$ $(-\infty, b] = \{x: x \leq b\}$

Figure 4-4

The symbols ∞ and $-\infty$ (infinity and minus infinity) do not stand for numbers, but simply indicate that the interval only has one endpoint.

The distance from the point x on the number line to the origin, 0, is called the *absolute value* of x, and is denoted $|x|$. Thus $|x| = x$ if $x \geq 0$, and $|x| = -x$, if $x < 0$. For any number x, $|x|$ is nonnegative, and if $x \neq 0$, both $|x|$ and $|-x|$ stand for the positive one of the numbers $x, -x$. For example, $|0| = 0$, $|-2| = 2$, $|5| = 5$, $|-3/2| = 3/2$.

If r is a positive number, then $|x| < r$ means the same as $-r < x < r$. Hence we can write the open interval from $-r$ to r in any of the following three ways:

$$(-r, r) = \{x: |x| < r\} = \{x: -r < x < r\}.$$

Similarly, the closed interval from $-r$ to r can be written

$$[-r, r] = \{x: |x| \leq r\} = \{x: -r \leq x \leq r\}.$$

Example 1 Write each of the following intervals in two other ways:

(a) $(-5, 5)$ (b) $\{x: |x| \leq 7\}$ (c) $\{x: -1 \leq x \leq 1\}$.

Solution.

(a) $(-5, 5) = \{x: |x| < 5\} = \{x: -5 < x < 5\}.$

(b) $\{x: |x| \leq 7\} = [-7, 7] = \{x: -7 \leq x \leq 7\}.$

(c) $\{x: -1 \leq x \leq 1\} = [-1, 1] = \{x: |x| \leq 1\}.$

In the same way that $|x|$ gives the distance from x to 0, the quantity $|x - a|$ gives the distance from x to a. Thus $|x - a| = 1$ means that x is one unit to the left or right of a; hence $x = a - 1$ or

$x = a + 1$. In general, the equation $|x - a| = r$ has the two solutions $a - r$ and $a + r$, and the inequality $|x - a| < r$ is satisfied by the numbers strictly between $a - r$ and $a + r$. Hence (Fig. 4-5)

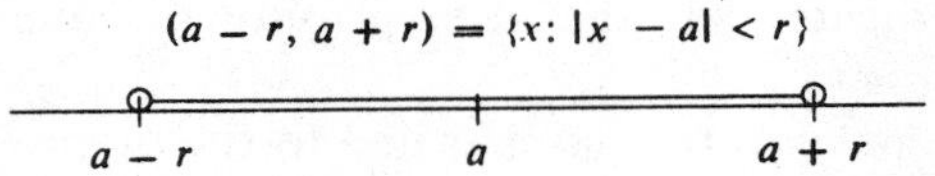

$(a - r, a + r) = \{x : |x - a| < r\}$

a - r   a   a + r

Figure 4-5

$$|x - a| < r \quad \text{if and only if} \quad a - r < x < a + r,$$

and

$$|x - a| \leq r \quad \text{if and only if} \quad a - r \leq x \leq a + r.$$

Example 2   Graph each of the following inequalities:

      (a) $|x + 1| < 2$ (b) $|x - 3| \leq 1$.

Solution.

(a) The inequality can be written $|x - (-1)| < 2$, and hence is satisfied by all numbers x strictly between $-1 - 2$ and $-1 + 2$. Hence the graph is the open interval $(-3, 1)$ (Fig. 4-6).

(b) The inequality $|x - 3| \leq 1$ is satisfied by $3 - 1$ and $3 + 1$ and the numbers in between. Hence the graph is the closed interval $[2, 4]$ (Fig. 4-7).

Any open or closed interval is the graph of some inequality of the form $|x - a| < r$ or $|x - a| \leq r$. To find the inequality, given the interval, we notice that the midpoint, a, is the average of the endpoints $a - r$ and $a + r$, and the number r represents half the length of the interval.

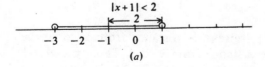

(a)

Figure 4-6

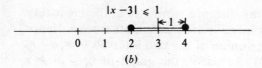
(b)

Figure 4-7

Example 3 Write a single inequality using the absolute value notation whose graph is the given interval:

(a) (1, 5)   (b) [−2, 1].

Solution.

(a) The midpoint, a, of (1, 5) is $\frac{(1+5)}{2} = 3$, and the half-length, r, is $\frac{(5-1)}{2} = 2$. Therefore (1, 5) is the graph of $|x - 3| < 2$.

(b) The midpoint, a, of [−2, 1] is $\frac{(-2+1)}{2} = -\frac{1}{2}$, and the half-length, r, is $\frac{(1-(-2))}{2} = \frac{3}{2}$. Therefore [−2, 1] is the graph of $\left|x - \left(-\frac{1}{2}\right)\right| \leq \frac{3}{2}$, or $\left|x + \frac{1}{2}\right| \leq \frac{3}{2}$.

An inequality of the form $|x| > r$ expresses the fact that x is more than r units from 0; that is, $x < -r$, or $x > r$. The graph of $|x| > r$ is therefore the union of the two open intervals $(-\infty, -r)$ and (r, ∞), as shown in Fig. 4-8. The inequality $|x| \geq r$ would include the points $-r$ and r in its graph, as shown in Fig. 4-9.

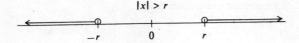

Figure 4-8

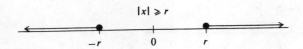

Figure 4-9

Since $|x - a|$ represents the distance from x to a, the inequality $|x - a| > r$ is satisfied by points x more than r units from a. Hence the graph of $|x - a| > r$ is the union of the two intervals $(-\infty, a - r)$ and $(a + r, \infty)$ (Fig. 4-10). Similarly, the graph of $|x - a| \geq r$ consists of two closed intervals $(-\infty, a - r]$ and $[a + r, \infty)$ (Fig. 4-11).

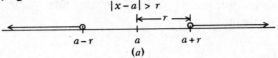

Figure 4-10

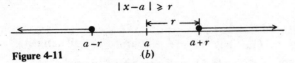

Figure 4-11

Example 4 Graph the inequalities

$$(a) \ |x + 2| \geq 1 \qquad (b) \ |x - 2| > \tfrac{3}{2}.$$

Solution.

(a) The graph contains the points $-2 - 1$ and $-2 + 1$, and the points more than one unit from -2, hence consists of the intervals $(-\infty, -3]$ and $[-1, \infty)$ (Fig. 4-12).

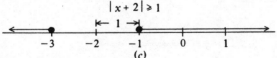

Figure 4-12

(b) The graph consists of the points to the left of $2 - \tfrac{3}{2} = \tfrac{1}{2}$, and the points to the right of $2 + \tfrac{3}{2} = \tfrac{7}{2}$; i.e., the intervals $\left(-\infty, \tfrac{1}{2}\right)$ and $\left(\tfrac{7}{2}, \infty\right)$ (Fig. 4-13).

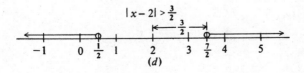

Figure 4-13

106 *Analytic Geometry*

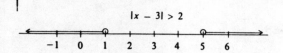

Figure 4-14

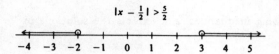

Figure 4-15

SECTION 4-1, REVIEW QUESTIONS

1. We picture the real numbers system by identifying each number with a point on a _____ .   line
2. The positive number x is associated with the point x units to the _____ of 0.   right
3. Positive numbers are to the right of 0, and negative numbers are to the _____ of 0.   left
4. The expression $a < b$ is read a is _____ _____ b.   less than
5. Geometrically, a is less than b if a is to the _____ of b on the number line.   left
6. The open interval of points strictly between a and b is denoted _____ .   (a,b)
7. Open intervals are denoted with parentheses, and closed intervals with _____ .   brackets
8. The closed interval $[a, b]$ consists of (a, b) together with the endpoints _____ and _____ .   a   b
9. The absolute value of x is denoted by _____ . |x|
10. $|x|$ is defined to be x if $x \geq 0$, and _____ if $x < 0$. $-x$
11. The absolute value of x is always _____ .   non-negative
12. Geometrically, $|x|$ gives the _____ from x to 0.   distance
13. $|x|$ is the distance from x to 0, and $|x - a|$ is the distance from x to _____ .   a

Intervals

SECTION 4-1, EXERCISES

Write each of the following intervals in two other ways, and graph them.

Example 1
1. $(-3, 3)$
2. $[-10, 10]$
3. $\{x: |x| < 6\}$
4. $\{x: |x| < \frac{1}{2}\}$
5. $\{x: -\frac{3}{2} < x < \frac{3}{2}\}$
6. $\{x: -4 < x < 4\}$

Graph each of the following inequalities, and express the solutions in the form (a, b) or $[a, b]$.

Example 2
7. $|x - 2| < 3$
8. $|x - 3| < 1$
9. $|x + 2| < 3$
10. $|x + 5| < 2$
11. $|x - 7| \leq 1$
12. $|x - 1| \leq 3$
13. $|x + 1| \leq 2$
14. $|x + 2| \leq 1$

Write a single inequality involving the absolute value which has each of the following intervals as solutions.

Example 3
15. $(1, 7)$
16. $(3, 5)$
17. $(-2, 4)$
18. $(-6, -2)$
19. $[5, 11]$
20. $[1, 9]$
21. $[-3, 5]$
22. $[-9, 3]$
23. $(-10, -5)$
24. $(-5, -2)$
25. $[-5, -1]$
26. $[-3, -\frac{1}{2}]$

Graph the following inequalities, and express the solutions as a union of two intervals.

Example 4
27. $|x - 2| > 2$
28. $|x - 5| \geq 3$
29. $|x - 1| > 2$
30. $|x - 4| > 6$
31. $|x + 3| \geq 1$
32. $|x + 1| \geq 3$
33. $|x + 5| > 2$
34. $|x + 2| > 4$

Write a single inequality using the absolute value sign whose solutions are the points *outside* the given interval.

35. $[-5, 5]$
36. $(-1, 1)$
37. $(1, 3)$
38. $[-1, 5]$
39. $[-6, -2]$
40. $(-5, 1)$

4-2 COORDINATES IN THE PLANE

The system of cartesian coordinates is a device that allows us to associate the points of a plane with pairs of real numbers. In this way we can bring algebraic methods to bear on geometric problems. At the same time, we have a way of interpreting many algebraic ideas geometrically. This procedure frequently permits us to use our geometric intuition to help solve non-geometric problems.

To establish a system of cartesian coordinates, start with two perpendicular number lines, called *coordinate axes* as shown in Fig.

4-16. The intersection of the coordinate lines is called the *origin*. Each point in the plane is associated with a pair of numbers (x, y) called the *coordinates* of the point. The first coordinate, x, determines the distance of a point to the right or left of the vertical axis. The second coordinate, y, determines the distance of the point

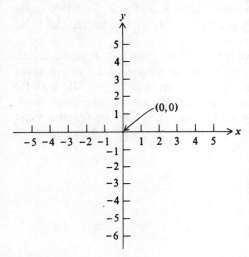

Figure 4-16

above or below the horizontal axis. The horizontal axis is called the x-axis and the vertical axis is called the y-axis.

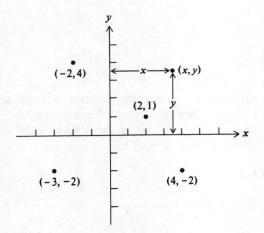

Figure 4-17

The first coordinate of a point is called the *x-coordinate* or *abscissa* of the point. The second coordinate of a point is called the *y-coordinate* or *ordinate* of the point. A positive abscissa denotes a point to the right of the *y*-axis and a positive ordinate denotes a point above the *x*-axis. For example (see Fig. 4-17), the point (2, 1) is two units to the right of the vertical axis, and (−2, 4) is two units to the left. The point (2, 1) is one unit above the horizontal axis, and (4, −2) is two units below it.

The two coordinate axes divide the plane into four regions, called *quadrants*. The first quadrant is the set of points whose coordinates are both positive. The other quadrants are labeled II, III, and IV in the counterclockwise direction around the origin. The four quadrants correspond respectively to the four possibilities for the signs of the coordinates (+, +), (−, +), (−, −), (+, −) as shown in Fig. 4-18.

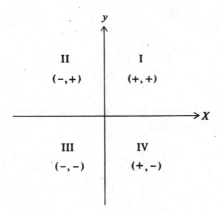

Figure 4-18

Now let us turn to the problem of finding the distance between two points (x_1, y_1) and (x_2, y_2) in the coordinate plane. First construct a right triangle as shown in Fig. 4-19. By using the Pythagorean theorem we can find the distance d from (x_1, y_1) to (x_2, y_2):

$$d^2 = (x_2 - x_1)^2 + (y_2 - y_1)^2$$

$$d = \sqrt{(x_2 - x_1)^2 + (y_2 - y_1)^2}$$

Notice that it does not matter which point is designated as the

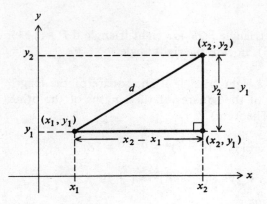

Figure 4-19

first point (x_1, y_1) and which is designated as the second point (x_2, y_2), since $(x_2 - x_1)^2 = (x_1 - x_2)^2$ and $(y_2 - y_1)^2 = (y_1 - y_2)^2$.

Example 1 Find the distance between $(1, 4)$ and $(-2, 2)$.

Solution. Applying the formula $d = \sqrt{(x_2 - x_1)^2 + (y_2 - y_1)^2}$, with $(x_2, y_2) = (-2, 2)$ and $(x_1, y_1) = (1, 4)$, we have

$$d = \sqrt{(-2 - 1)^2 + (2 - 4)^2}$$
$$= \sqrt{(-3)^2 + (-2)^2}$$
$$= \sqrt{9 + 4} = \sqrt{13}.$$

Example 2 Show that the points $P(0, 1)$, $Q(2, 2)$, and $R(4, 3)$ lie on a line.

Solution. We know from geometry that the points P, Q, and R are colinear if $PQ + QR = PR$.

$$PQ = \sqrt{(2 - 0)^2 + (2 - 1)^2} \quad = \quad \sqrt{5}$$
$$QR = \sqrt{(4 - 2)^2 + (3 - 2)^2} \quad = \quad \sqrt{5}$$
$$PR = \sqrt{(4 - 0)^2 + (3 - 1)^2}$$
$$= \sqrt{20} = 2\sqrt{5}$$

Since $PQ + QR = PR$, the points are colinear.

Example 3 Show that triangle PQR is a right triangle if $P = (1, 1)$, $Q = (-2, 4)$, $R = (-3, -3)$. Locate the right angle.

Solution. A triangle is a right triangle if the square of the length of one side is the sum of the squares of the lengths of the other two (the Pythagorean Theorem).

$$PQ^2 = (-2-1)^2 + (4-1)^2 = 9 + 9 = 18$$

$$QR^2 = (-3+2)^2 + (-3-4)^2 = 1 + 49 = 50$$

$$RP^2 = (1+3)^2 + (1+3)^2 = 16 + 16 = 32$$

Since $PQ^2 + RP^2 = QR^2$, PQR is a right triangle; QR is the hypotenuse, and the right angle is at P.

Example 4 Write an equation in x and y which holds for all points (x, y) which are equidistant from $(0, 1)$ and $(2, -3)$.

Solution. The distances from (x, y) to $(0, 1)$ and $(2, -3)$, respectively, are $\sqrt{(x-0)^2 + (y-1)^2}$ and $\sqrt{(x-2)^2 + (y+3)^2}$. Therefore the equation is

$$\sqrt{(x-0)^2 + (y-1)^2} = \sqrt{(x-2)^2 + (y+3)^2}.$$

Squaring both sides and simplifying, we get:

$$(x-0)^2 + (y-1)^2 = (x-2)^2 + (y+3)^2,$$

$$x^2 + y^2 - 2y + 1 = x^2 - 4x + 4 + y^2 + 6y + 9,$$

$$-4x + 8y + 12 = 0,$$

$$-x + 2y + 3 = 0.$$

SECTION 4-2, REVIEW QUESTIONS

1. We can bring algebraic methods to bear on geometric problems by labeling the points of a plane with ordered _____ of numbers.

pairs

2. The numbers x and y (of the pair (x, y)) are called the _____ of the point.   **coordinates**

3. The first coordinate, x, determines distance _____ or _____ from the y-axis.   **right** / **left**

4. The second coordinate, y, determines distance _____ or _____ from the x-axis.   **up** / **down**

5. The vertical axis is the y-axis, and the horizontal axis is the _____-axis.   **x**

6. The point of intersection of the axes is called the _____.   **origin**

7. The coordinates of the origin are (_____, _____).   **(0,0)**

8. The x-coordinate of a point is also called its _____.   **abscissa**

9. The abscissa of (x, y) is x and y is the _____.   **ordinate**

10. The ordinate of a point is its distance above or below the _____-axis.   **x**

11. The axes divide the plane into four regions called _____.   **quadrants**

12. The x-coordinate of a point is positive in quadrants _____ and _____.   **I   IV**

13. The y-coordinate of a point is negative in quadrants _____ and _____.   **III   IV**

14. The distance between the points (x_1, y_1) and (x_2, y_2) is given by:
    $d =$ _____. $\sqrt{(x_2-x_1)^2 + (y_2-y_1)^2}$

SECTION 4-2, EXERCISES

1. What is the x-coordinate of any point on the y-axis?
2. What is the y-coordinate of any point on the x-axis?
3. In which quadrants is the x-coordinate of every point
   (a) positive?   (b) negative?
4. In which quadrants is the y-coordinate of every point
   (a) positive?   (b) negative?

Without graphing, state which quadrant each of the following points is in.

5. $(-3, -4)$ 6. $(-3, 4)$ 7. $(4, -3)$
8. $(3, -4)$ 9. $(3, 4)$ 10. $(4, 3)$
11. $(-4, 3)$ 12. $(-4, -3)$

Coordinates in the Plane 113

Find the distance between each pair of points.

Example 1
13. (5, 1), (1, 4)
14. (3, 2), (6, 6)
15. (−2, 1), (−6, −2)
16. (−1, 0), (2, −4)
17. (0, 5), (1, 4)
18. (2, −3), (3, −4)
19. (−1, 3), (3, −1)
20. (5, −1), (2, −4)
21. (−1, 2), (4, 3)
22. (2, −5), (8, −3)

Show that the following points are collinear.

Example 2
23. (0, 3), (6, 6), (2, 4)
25. (−5, 0), (7, −2), (1, −1)
24. (−1, −3), (1, 1), (2, 3)
26. (−3, 2), (3, −4), (0, −1)

Show that the following triangles are right triangles, and locate the right angle.

Example 3
27. (2, 4), (6, 8), (−1, 7)
28. (7, 5), (3, 1), (6, −2)
29. (6, 7), (−1, 6), (2, 3)
30. (5, 2), (1, 6), (8, 5)
31. (3, 4), (−2, 4), (−1, 2)
32. (−1, −1), (2, −5), (−2, −3)
33. (3, −4), (2, −1), (9, −2)
34. (−2, 5), (−1, 8), (4, 3)

Write an equation in x and y which expresses the fact that (x, y) is equidistant from each pair of points.

Example 4
35. (0, 1), (0, −5)
36. (3, 0), (−1, 0)
37. (2, −1), (−2, 1)
38. (1, 2), (3, −2)
39. Write an equation in x and y which expresses the fact that (x, y) is the same distance from (0, 4) as from the x-axis.
40. Write an equation in x and y which expresses the fact that (x, y) is the same distance from (2, 0) as from the y-axis.
41. Write an equation in x and y which is satisfied if (x, y) is twice as far from (1, 0) as from (4, 0).
42. Write an equation in x and y which is satisfied if (x, y) is half as far from (1, 2) as from (4, 8).

43. Show that (3, 5) is the midpoint of the segment from (2, 3) to (4, 7).
44. Show that $\left(\dfrac{1-3}{2}, \dfrac{-2+2}{2}\right) = (-1, 0)$ is the midpoint of the segment from (1, −2) to (−3, 2).
45. Show that $(\tfrac{1}{2}x_1 + \tfrac{1}{2}x_2, \tfrac{1}{2}y_1 + \tfrac{1}{2}y_2)$ is the midpoint of the segment from (x_1, y_1) to (x_2, y_2).
46. Show that $(\tfrac{1}{3}x_1 + \tfrac{2}{3}x_2, \tfrac{1}{3}y_1 + \tfrac{2}{3}y_2)$ is $\tfrac{2}{3}$ of the way from (x_1, y_1) to (x_2, y_2).

4-3 CIRCLES

In the last section we developed a formula for finding the distance between two points in the coordinate plane. From this *distance formula* we see that if (x, y) is any point in the plane, then its distance from the origin is given by

$$d = \sqrt{(x - 0)^2 + (y - 0)^2}$$
$$= \sqrt{x^2 + y^2}.$$

Now, a circle with its center at the origin and radius 2 is defined to be the set of all points 2 units from the origin. Therefore (x, y) is a point on this circle if and only if

$$\sqrt{x^2 + y^2} = 2.$$

Equivalently,

$$x^2 + y^2 = 4. \tag{1}$$

For example, $(2, 0)$ is a point on the circle, since $x = 2$ and $y = 0$ satisfies (1) above:

$$2^2 + 0^2 = 4.$$

The point $(\sqrt{2}, \sqrt{2})$ has coordinates that also satisfy the equation:

$$(\sqrt{2})^2 + (\sqrt{2})^2 = 2 + 2 = 4.$$

Thus $(\sqrt{2}, \sqrt{2})$ is a point on the circle (see Fig. 4-20).

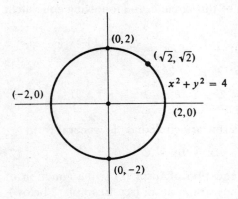

Figure 4-20

Circles 115

In this example, we described a geometric property (a point is on a circle) by means of an algebraic equation involving the coordinates of the point. This is the basic idea behind the use of coordinates in treating geometric concepts. The *graph* of an equation in x and y is the set of points (x, y) whose coordinates satisfy the equation. An equation like $x^2 + y^2 = 4$ is called *the equation of a geometric figure* if its graph is the given figure.

Now let us look at more general circles centered at points other than the origin. A point (x, y) is on the circle with center (h, k) and radius r (Fig. 4-21) provided that

$$\sqrt{(x - h)^2 + (y - k)^2} = r.$$

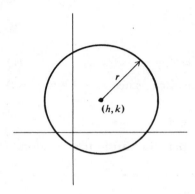

Figure 4-21

We can rewrite the equation of this circle in the following equivalent forms:

$$(x - h)^2 + (y - k)^2 = r^2,$$
$$x^2 - 2hx + h^2 + y^2 - 2ky + k^2 = r^2,$$
$$x^2 + y^2 - 2hx - 2ky + (h^2 + k^2 - r^2) = 0.$$

The above equation of an arbitrary circle has the general form

$$x^2 + y^2 + Ax + By + C = 0 \qquad (2)$$

It is easy to show that any equation of form (2) is the equation of some circle, provided it has any graph at all (see Example 2 below). We will now illustrate the method of finding the center and radius of any such circle.

Example 1 Find the center and radius of the circle $x^2 + y^2 - 2x + 3y + 1 = 0$, and graph the circle.

Solution. We complete the squares in the x and y terms so that the left side appears as the distance from (x, y) to some point.

$$x^2 + y^2 - 2x + 3y + 1 = 0$$
$$x^2 - 2x \quad + y^2 + 3y \quad = -1$$
$$x^2 - 2x + 1 + y^2 + 3y + \frac{9}{4} = -1 + \frac{9}{4} + 1$$
$$(x - 1)^2 + \left(y + \frac{3}{2}\right)^2 = \frac{9}{4}$$
$$\sqrt{(x - 1)^2 + \left(y + \frac{3}{2}\right)^2} = \frac{3}{2}$$

From this last equation we see that the graph is the circle of radius $3/2$ and center $(1, -3/2)$. See Fig. 4-22.

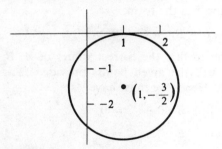

Figure 4-22

Example 2 Identify the graphs of

(a) $x^2 + y^2 - 2x - 4y + 9 = 0$

(b) $x^2 + y^2 - 2x - 4y + 5 = 0$

Solution.

(a) We complete the squares as follows:

$$x^2 + y^2 - 2x - 4y + 9 = 0$$

$$x^2 - 2x \quad + y^2 - 4y \quad = -9$$

$$x^2 - 2x + 1 + y^2 - 4y + 4 = -9 + 1 + 4$$

$$(x - 1)^2 + (y - 2)^2 = -4.$$

In this form it is clear that no pairs (x, y) satisfy the equation, since the sum of two squares can not be negative.

(b) This equation differs from (a) only in the constant term. After completing the squares we get

$$(x - 1)^2 + (y - 2)^2 = 0.$$

There is exactly one pair (x, y) which satisfies this equation; $x = 1$ and $y = 2$. The graph is the single point $(1, 2)$. We sometimes refer to a single point as a degenerate circle, or circle of radius zero.

Example 3 Find the equation of the circle passing through the three points $(0, 0)$, $(0, 2)$, and $(1, 1)$.

Solution. We know the equation has the form

$$x^2 + y^2 + Ax + By + C = 0$$

so we use the given information to find the correct values of A, B, and C. The circle will go through the given points provided their coordinates satisfy the equation. Hence we must have

$$0^2 + 0^2 + A \cdot 0 + B \cdot 0 + C = 0$$
$$0^2 + 2^2 + A \cdot 0 + B \cdot 2 + C = 0$$
$$1^2 + 1^2 + A \cdot 1 + B \cdot 1 + C = 0$$

These simplify to

$$C = 0$$
$$2B + C = -4$$
$$A + B + C = -2$$

Solving this system for A, B, C, we get $A = 0$, $B = -2$, $C = 0$, and the equation is

$$x^2 + y^2 - 2y = 0$$

or
$$x^2 + (y - 1)^2 = 1$$
which is the equation of a circle with center at (0, 1) and radius 1.

Example 4 Find the equation of the circle that is tangent to the *y*-axis at (0, 0), and passes through the point (3, 1).

Solution. Since the circle is tangent to the *y*-axis, the center must be at some point $(h, 0)$ on the *x*-axis (see Fig. 4-23). Since (0, 0) is on the circle and $(h, 0)$ is the center, the radius must also be h. Hence the equation is
$$(x - h)^2 + y^2 = h^2$$

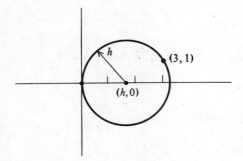

Figure 4-23

To find the value of h such that (3, 1) is on the circle, we substitute 3 for x and 1 for y:
$$(3 - h)^2 + 1^2 = h^2$$
$$9 - 6h + h^2 + 1 = h^2$$
$$-6h = -10$$
$$h = 5/3$$

Thus, we have
$$(x - 5/3)^2 + y^2 = (5/3)^2$$

Example 5 Identify the set of points (x, y) whose distance from (4, 4) is twice the distance from (1, 1).

Solution. The distances of (x, y) from $(4, 4)$ and $(1, 1)$ are

$$\sqrt{(x-4)^2 + (y-4)^2} \text{ and } \sqrt{(x-1)^2 + (y-1)^2}.$$

Hence the points (x, y) must satisfy the following equations.

$$\sqrt{(x-4)^2 + (y-4)^2} = 2\sqrt{(x-1)^2 + (y-1)^2}$$

$$(x-4)^2 + (y-4)^2 = 4[(x-1)^2 + (y-1)^2]$$

$$x^2 - 8x + 16 + y^2 - 8y + 16 = 4[x^2 - 2x + 1 + y^2 - 2y + 1]$$

$$3x^2 + 3y^2 = 24$$

$$x^2 + y^2 = 8.$$

The graph is the circle of radius $2\sqrt{2}$ centered at the origin.

SECTION 4-3, REVIEW QUESTIONS

$\sqrt{x^2+y^2}$

1. The distance from (x, y) to $(0, 0)$ is _____.

2. The equation of the circle of radius r and center $(0, 0)$ is $\sqrt{x^2 + y^2} = $ _____.

r

graph

3. The set of points whose coordinates satisfy an equation is called the _____ of the equation.

4. The graph of $x^2 + y^2 + Ax + By + C = 0$ is either the empty set, a single point, or a _____.

circle

5. The center and radius of the circle $x^2 + y^2 + Ax + By + C = 0$ are found by completing the _____ in the x and y terms.

squares

6. The graph of $(x - h)^2 + (y - k)^2 = r^2$ is a circle of radius r, with center at (_____, _____).

(h,k)

SECTION 4-3, EXERCISES

Write the equations of the following circles, and put the equation in the form $x^2 + y^2 + Ax + By + C = 0$

1. center $(0, 0)$, radius 1
2. center $(0, 0)$, radius 3
3. center $(0, 0)$, radius 5
4. center $(0, 0)$, radius 4
5. center $(1, 2)$, radius 1
6. center $(2, -1)$, radius 3
7. center $(-3, 2)$, radius 5
8. center $(-1, -3)$, radius 2

Find the center and radius of each of the following, and graph.

9. $x^2 + y^2 = 36$
10. $x^2 + y^2 - 49 = 0$ Example 1, 2
11. $x^2 + y^2 + 4 = 0$
12. $x^2 + y^2 = 0$
13. $x^2 + y^2 - 2x - 6y + 9 = 0$
14. $x^2 + y^2 - 4x - 4y + 4 = 0$
15. $x^2 + y^2 + 4x + 2y - 20 = 0$
16. $x^2 + y^2 - 6x + 4y + 4 = 0$
17. $x^2 + y^2 + 6x - 2y + 14 = 0$
18. $x^2 + y^2 - 4x + 10y + 30 = 0$
19. $x^2 + y^2 - 6x + 8y + 25 = 0$
20. $x^2 + y^2 + 2x + 2y + 2 = 0$

Find the equation of the following circles.

21. Circle through $(0, 0)$, $(2, 2)$ and $(-4, 2)$.   Example 3
22. Circle through $(0, 0)$, $(2, -1)$ and $(-1, 1)$.
23. Circle through $(0, 1)$, $(2, 0)$ and $(1, 1)$.
24. Circle through $(-1, 0)$, $(0, -1)$ and $(1, -1)$.
25. Circle through $(0, 0)$, center $(1, 3)$.   Example 4
26. Circle through $(0, 0)$, center $(2, -1)$.
27. Circle through $(1, 2)$, center $(3, 4)$.
28. Circle through $(3, -1)$, center $(1, 1)$.
29. Circle tangent to the y-axis at $(0, 0)$, and passing through $(2, 1)$.
30. Circle tangent to the x-axis at $(0, 0)$, and passing through $(1, -3)$.
31. Circle tangent to the y-axis at $(0, 1)$, and passing through $(3, 0)$.
32. Circle tangent to the x-axis at $(2, 0)$, and passing through $(0, 1)$.

Show that the following graphs are circles; find the centers and radii.

33. All points whose distance from $(4, -4)$ is twice the distance from $(1, -1)$.   Example 5

34. All points whose distance from (−2, 2) is one-half the distance from (−8, 8).
35. All points whose distance from (8, 4) is twice the distance from (2, 1).
36. All points whose distance from (1, −2) is half the distance from (4, −8).
37. All points whose distance from (4, 0) is three times the distance from (0, 0).
38. All points whose distance from (−1, 3) is three times the distance from (−1, −1).

39. (a) Complete the squares and show that $x^2 + y^2 + Ax + By + C = 0$ can be written:

$$\sqrt{\left(x + \frac{A}{2}\right)^2 + \left(y + \frac{B}{2}\right)^2} = \frac{1}{2}\sqrt{A^2 + B^2 - 4C}$$

(b) Write a condition on the quantity $A^2 + B^2 - 4C$ which insures that (i) the graph is a proper circle, (ii) the graph is a single point, (iii) the graph is the empty set (no graph).

(c) Use the condition of (b) to determine what the graph is in exercises 15 - 20.

4-4 LINEAR EQUATIONS

In this section we continue our study of the relationship between geometric figures and their algebraic equations. We will show that every equation of the form

$$Ax + By + C = 0 \qquad (1)$$

has a straight line as its graph and, conversely, that every line has an equation of this form. Of course, we assume that A and B are not both zero, so that the equation actually involves x or y (or both).

Consider any line that is not perpendicular to the x-axis, and let $P_1 = (x_1, y_1)$ and $P_2 = (x_2, y_2)$ be any two points on the line (Fig. 4-24). The ratio $(y_2 - y_1)/(x_2 - x_1)$ provides a measure of the steepness of the line. If $P'_1 = (x'_1, y'_1)$ and $P'_2 = (x'_2, y'_2)$ are any other two points on the line, then the ratio determined by P'_1 and P'_2 is the same as the ratio determined by P_1 and P_2 since the two triangles shown in Fig. 4-24 are similar, That is,

$$\frac{y'_2 - y'_1}{x'_2 - x'_1} = \frac{y_2 - y_1}{x_2 - x_1}. \qquad (2)$$

The ratio (2) determined by *any* pair of points on the line is called its *slope*, and is denoted m:

$$m = \frac{y_2 - y_1}{x_2 - x_1}. \tag{3}$$

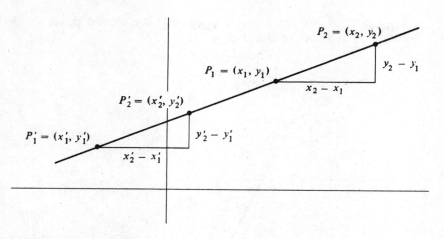

Figure 4-24

If the line slopes upward to the right, as in Fig. 4-24, the slope is positive. If the line slopes downward to the right, the slope is negative. The smaller the absolute value of the slope, the more nearly horizontal the line is, and a line with zero slope is parallel to the x-axis. The larger the absolute value of the slope, the more nearly vertical the line is. The slope of a vertical line is not defined.

Now we will derive an equation that characterizes the points on a line. Let (x_0, y_0) be any fixed point on a given line with slope m, and let (x, y) be any other point on the line. The slope as determined by (x, y) and (x_0, y_0) must be m, which gives

$$\frac{y - y_0}{x - x_0} = m,$$

or

$$y - y_0 = m(x - x_0). \tag{4}$$

Equation (4) is necessary and sufficient in order for (x, y) to lie on the line through (x_0, y_0) with slope m. Equation (4) is called the *point-slope* form of the equation of the line.

Example 1 Write the equation of the line through (1, 3) with slope -2, and sketch the graph.

Solution. If (x, y) is any point on the line, then calculating the slope, -2, from (x, y) and (1, 3) we get

Linear Equations

$$\frac{y-3}{x-1} = -2 \quad \text{or} \quad y - 3 = -2(x-1)$$
$$y = -2x + 5$$

Letting $x = 4$, we get $y = -3$, so $(4, -3)$ is a second point which allows us to draw the line (Fig. 4-25).

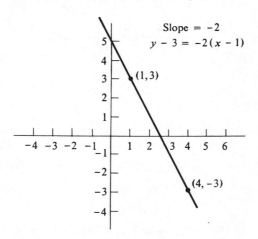

Figure 4-25

If we write the equation of the line with slope m, which goes through the point $(0, b)$ on the y-axis, we get

$$\frac{y-b}{x-0} = m$$
$$y - b = mx$$
$$y = mx + b \tag{5}$$

Equation (5) is called the *slope-intercept* form of the equation of a line. We see from (5) that whenever we have an equation in which y is a linear function of x (i.e., a first-degree polynomial, $mx + b$), the graph is a line, the coefficient of x is the slope, and the constant is the y-intercept.

Example 2 Find the slope, and graph the line $y = 2x - 1$.

Solution. By inspection, the slope is 2, and the y-intercept is -1. If $x = 3$, $y = 5$; therefore $(3, 5)$ is a second point on the line (see Fig. 4-26).

If we are given any two points, we can determine the slope of the

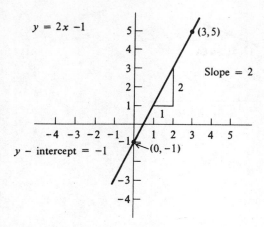

Figure 4-26

line through these points by (3). Hence we can use either point and the slope to obtain the equation from the point-slope form (4).

Example 3 Write the equation and graph the line through $(-2, 1)$ and $(3, 3)$. Is $(1, 2)$ on this line?

Solution. The slope is given by

$$m = \frac{3-1}{3-(-2)} = \frac{2}{5}.$$

Using the point $(-2, 1)$ and the slope, we get

$$y - 1 = \tfrac{2}{5}(x - (-2))$$

$$= \tfrac{2}{5}x + \tfrac{4}{5};$$

$$y = \tfrac{2}{5}x + \tfrac{9}{5}.$$

If we substitute $x = 1$ and $y = 2$ in the equation, we get the false statement

$$2 \stackrel{?}{=} \frac{2}{5}(1) + \frac{9}{5} = \frac{11}{5}.$$

Hence $(1, 2)$ does not lie on the line (see Fig. 4-27).

Now consider any equation of the form

$$Ax + By + C = 0. \tag{6}$$

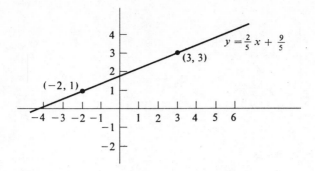

Figure 4-27

If $B = 0$, the equation is simply

$$x = -\frac{C}{A}$$

which we recognize as the equation of a vertical line through $(-C/A, 0)$. If $B \neq 0$, Equation (6) can be written as

$$y = -\frac{A}{B}x - \frac{C}{B}. \tag{7}$$

We recognize (7) as the equation of the line with slope $-(A/B)$ and y-intercept $-(C/B)$. Hence any equation having the form of Equation (6) represents a line. Conversely, any line has an equation that can be put in this form.

Example 4 Find the equation of the line which is the perpendicular bisector of the segment joining (2, 0) and (0, 4). What point on the y-axis is equidistant from the two given points?

Solution. A point (x, y) is on the perpendicular bisector if and only if it is equidistant from (2, 0) and (0, 4). Hence we have the equation

$$\sqrt{(x-0)^2 + (y-4)^2} = \sqrt{(x-2)^2 + (y-0)^2}$$

Squaring both sides and simplifying, we get

$$x^2 + y^2 - 8y + 16 = x^2 - 4x + 4 + y^2,$$
$$4x - 8y + 12 = 0,$$
$$x - 2y + 3 = 0,$$
$$y = \tfrac{1}{2}x + \tfrac{3}{2}.$$

126 *Analytic Geometry*

The y-intercept is $\frac{3}{2}$, so $(0, \frac{3}{2})$ is the point on the y-axis that is equidistant from the two points.

The slope of a line determines its angle, α, with any horizontal line (Fig. 4-28), and hence its angle with the x-axis. It follows that two lines are parallel if and only if they have the same slope. Now let us see how the slopes of perpendicular lines are related.

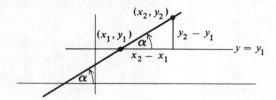

Figure 4-28

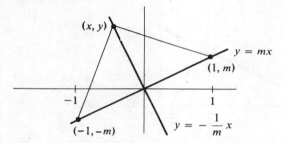

Figure 4-29

Consider a line with slope m and, for convenience, let the line go through the origin. Its equation is then

$$y = mx.$$

The line through the origin which is perpendicular to this line will be the perpendicular bisector of the segment joining $(1, m)$ and $(-1, -m)$. See Fig. 4-29. The equation of the perpendicular bisector is determined as in Example 4.

$$\sqrt{(x-1)^2 + (y-m)^2} = \sqrt{(x+1)^2 + (y+m)^2}$$
$$x^2 - 2x + 1 + y^2 - 2my + m^2 = x^2 + 2x + 1 + y^2 + 2my + m^2$$
$$-4my = 4x$$
$$y = -\frac{1}{m}x$$

Hence *the slope of the perpendicular line is the negative reciprocal of the slope of the given line*. Since parallel lines have the same slope, any two lines are perpendicular if and only if their slopes are negative reciprocals of each other.

Example 5 Find the equation of the line through (1, 2) and perpendicular to $y = \frac{1}{2}x + 3$.

Solution. The line must have slope $-1/\frac{1}{2} = -2$. We get the equation immediately from the point-slope form:

$$y - 2 = (-2)(x - 1)$$
$$y = -2x + 4$$

The two lines are shown in Fig. 4-30.

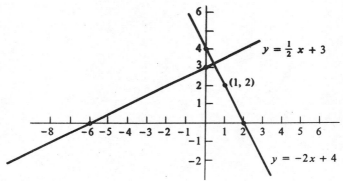

Figure 4-30

SECTION 4-4, REVIEW QUESTIONS

linear   1. Equations of the form $Ax + By + C = 0$ are called _____ equations.

line   2. The graph of every linear equation is a straight _____.

    3. If $B = 0$, the equation has the form $x = -(C/A)$ and the line is *parallel*

y   to the _____ - axis.

    4. Any nonvertical line has an equation that can be written in the slope–

mx+b   intercept form $y =$ _____.

    5. In the equation $y = mx + b$, the constant m is the slope of the line and the

intercept   constant b is the y-_____.

6. If $m > 0$, the line slopes up to the right and if $m < 0$, the line slopes _____ to the right. down

7. The slope of the line between (x_0, y_0) and (x_1, y_1) is given by

   $m =$ _____. $\dfrac{y_1 - y_0}{x_1 - x_0}$

8. The point-slope form of the equation of the line through (x_0, y_0) with slope m is $y - y_0 =$ _____. $m(x - x_0)$

9. Two lines are parallel if and only if their slopes are equal, and two lines are perpendicular if and only if their slopes are _____ reciprocals of each other. negative

SECTION 4-4, EXERCISES

Write the equations of the lines through the given point, with the given slope.

1. Through $(2, 3)$; slope 4.   Example 1
2. Through $(1, -2)$; slope 3.
3. Through $(2, -5)$; slope -2.
4. Through $(-1, -4)$; slope -3.
5. Through $(-2, -5)$; slope $\frac{3}{4}$.
6. Through $(-1, -2)$; slope $-\frac{2}{3}$.

Find the slope, the y-intercept, and graph the line.

7. $y = 3x + 1$ 8. $y = 4x - 2$ Example 2
9. $y - 2x = 5$ 10. $y + 7x = -1$
11. $y + 3x - 4 = 0$ 12. $y - 5x + 8 = 0$
13. $2y = 4x - 5$ 14. $3y = -4x + 1$

Write the equation and graph the line through the given two points.

15. $(0, 0), (3, 4)$; is $(6, 8)$ on the line?   Example 3
16. $(0, 0), (2, -1)$; is $(-6, 3)$ on the line?
17. $(2, 3), (4, 7)$; is $(3, 6)$ on the line?
18. $(3, 1), (2, 8)$; is $(3, 0)$ on the line?
19. $(-2, 3), (1, -1)$; is $(0, 0)$ on the line?
20. $(-3, -1), (2, 3)$; is $(-2, 0)$ on the line?

Find the perpendicular bisector of the segments between the following pairs of points. Find the points on the y-axis and x-axis which lie on the perpendicular bisector.

Example 4 21. (0, 2), (3, 0)   22. (−2, 0), (0, 4)
23. (3, 1), (−1, 2)   24. (1, −3), (4, 1)

Find the equations of the following lines.

Example 5 25. Through (2, 1), perpendicular to $y = x + 3$.
26. Through (2, 3), perpendicular to $y = 2x - 5$.
27. Through (−3, 4), perpendicular to $y - 3x + 4 = 0$.
28. Through (−1, −3), perpendicular to $y + 4x - 7 = 0$.
29. Through (5, 2), perpendicular to $2y - x + 1 = 0$.
30. Through (−3, 4), perpendicular to $4y + 2x - 3 = 0$.

31. Through (4, 2), parallel to the x-axis.
32. Through (3, −5), parallel to the y-axis.
33. Through (−1, −2), parallel to $y = 5x - 1$.
34. Through (3, −4), parallel to $y = -2x + 4$.
35. Through (5, −1), parallel to $2x - y + 3 = 0$.
36. Through (−6, −2), parallel to $3x + 6y - 5 = 0$.
37. Through (0, 4), and parallel to the line through (2, 4) and (7, 8).
38. Through (1, 3) and parallel to the line through (1, −3) and (4, −5).
39. Through (−2, 1) and perpendicular to the line through (4, 1) and (0, 2).
40. Through (−1, −1) and perpendicular to the line through (−5, 1) and (−2, −4).

41. Show that the equation of the line with y-intercept b and x-intercept a is $\frac{x}{a} + \frac{y}{b} = 1$. (This is called the 2-intercept form.)
42. Show that the equation with x-intercept a and slope m can be written $x = -\frac{1}{m}y + a$.
43. Find the equation of the line tangent to the circle $x^2 + y^2 = 25$ at the point (4, 3). (Hint: the tangent is perpendicular to the radius.)
44. Find the equation of the line tangent to $x^2 + y^2 = 1$ at $\left(\frac{1}{2}, \frac{\sqrt{3}}{2}\right)$.

4-5 INEQUALITIES

In the last two sections we investigated two types of equations:

$$x^2 + y^2 + Ax + By + C = 0,$$
$$Ax + By + C = 0.$$

The first equation has a circle (or point or empty set) as its graph. The second equation always has a straight line as its graph. In general, the graph of any equation in x and y is usually (but not always) some sort of curve in the plane.

Now let's turn to the graphs of *inequalities* in x and y. The graph of an inequality in x and y is the set of points (x, y) whose coordinates satisfy the inequality. Consider first the linear inequality

$$y > x \quad \text{or} \quad y - x > 0.$$

The corresponding linear equation, $y = x$, is the straight line with $m = 1$ as shown in Fig. 4-31. The graph of $y > x$ consists of all

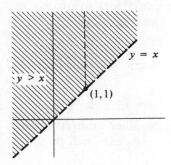

Figure 4-31

points above this line. For example $(1, 1)$ is on the graph of $y = x$, and all points $(1, y)$ for $y > 1$ are on the graph of the inequality. The graph of $y > x$ is therefore the half-plane of points above the line $y = x$.

Similarly, the graph of the equation $x = 2$ is a vertical straight line, and the graph of the inequality $x \leq 2$ is this line together with all points to its left (i.e., points with smaller abscissa). See Fig. 4-32.

Any point (x, y) that is not on the graph of $Ax + By + C = 0$ will clearly satisfy

$$Ax + By + C > 0 \quad \text{or} \quad Ax + By + C < 0$$

but not both. All the points on one side of the line will satisfy one of these inequalities and all the points on the other side will satisfy the other. There are four possible types of linear inequality depending on the choice of $<, >, \leq,$ or $\geq$. The graphs of

$$Ax + By + C \geq 0 \quad \text{and} \quad Ax + By + C \leq 0$$

will include the corresponding line. Such a graph is called a *closed half-plane*. The graphs of

$$Ax + By + C > 0 \quad \text{and} \quad Ax + By + C < 0$$

will exclude the line and are called *open half-planes*. Open half-planes are indicated with a *broken* boundary line, and closed half-planes are indicated with a *solid* boundary line.

Example 1 Graph the set of points whose coordinates satisfy $x + 2y - 4 > 0$.

Solution. First graph the line $x + 2y - 4 = 0$, indicating it with a dotted line, since it will not be part of the graph. The x-intercept is 4 and the y-intercept is 2.

To see which half-plane is the graph of the inequality, check a convenient point, say $(0, 0)$, to see if it will satisfy $x + 2y - 4 > 0$ or not. Since

$$0 + 2 \cdot 0 - 4 \not> 0$$

the origin is not in the desired half-plane. Therefore the graph of $x + 2y - 4 > 0$ is the half-plane not containing $(0, 0)$; i.e., the half-plane above the line $x + 2y - 4 = 0$. See Fig. 4-33.

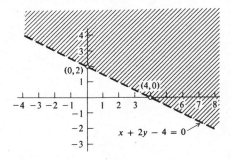

Figure 4-33

We can also consider inequalities corresponding to the equation

$$x^2 + y^2 + Ax + By + C = 0 \qquad (1)$$

Again we see that every point not on the graph of (1) will satisfy either

$$x^2 + y^2 + Ax + By + C < 0 \qquad (2)$$

or

$$x^2 + y^2 + Ax + By + C > 0 \qquad (3)$$

We will refer to inequalities of the forms (2) or (3), or the corresponding form using $\leq$ or $\geq$, as *circular inequalities*.

Example 2 Graph the set of points whose coordinates satisfy the inequality $x^2 + y^2 - 2x < 3$.

Solution. Complete the square as you would for an equation, to identify the boundary circle.

$$x^2 + y^2 - 2x < 3$$
$$x^2 - 2x + 1 + y^2 < 4$$
$$(x - 1)^2 + y^2 < 4$$
$$\sqrt{(x - 1)^2 + y^2} < 2$$

The geometric interpretation of the last inequality is that the distance from (x, y) to $(1, 0)$ must be less than 2. That is, the graph consists of all points strictly inside the circle $(x - 1)^2 + y^2 = 4$, which has radius 2 and center at $(1, 0)$. We plot the boundary circle with a broken line to indicate that the circle is not part of the graph. See Fig. 4-34.

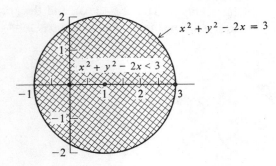

Figure 4-34

The same type of argument that was used in Example 2 can be applied to any circular inequality. The graph is either all points inside the boundary circle or all points outside the boundary circle. Graphs of the circular inequalities

$$x^2 + y^2 + Ax + By + C \leq 0 \quad \text{and} \quad x^2 + y^2 + Ax + By + C \geq 0$$

will include the boundary circle, and we indicate this by drawing a solid circle.

Recall that some equations of the form $x^2 + y^2 + Ax + By + C = 0$ have graphs that are empty, or reduce to a single point. The possibilities for circular inequalities corresponding to these situations are indicated below.

(a) $x^2 + y^2 < 0$ no graph

(b) $x^2 + y^2 + 1 > 0$ whole plane

(c) $(x - h)^2 + (y - k)^2 > 0$ whole plane except center (h, k)

Example 3   Graph the set of points whose coordinates satisfy both inequalities $x^2 + y^2 - 2x - 2y + 1 \leq 0$ and $x - y + 1 \leq 0$.

Solution.   The graph of the simultaneous inequalities will be the intersection of their graphs. We identify the circle by completing the square:

$$x^2 - 2x + y^2 - 2y \leq -1$$
$$x^2 - 2x + 1 + y^2 - 2y + 1 \leq -1 + 2$$
$$(x - 1)^2 + (y - 1)^2 \leq 1$$

The graph of the circular inequality is the inside of the circle centered at (1, 1) with radius 1 together with the circle (Fig. 4-35). The graph of $x - y + 1 = 0$ has intercepts (0, 1) and (−1, 0), and the graph of $x - y + 1 \leq 0$ or $y \geq x + 1$ will be the set of points above $y = x + 1$.

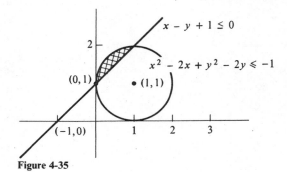

Figure 4-35

134   Analytic Geometry

Example 4 Graph the points whose coordinates satisfy all three inequalities: $y \leq \frac{3}{2}x, y \leq -\frac{1}{2}x + 4, y \geq \frac{1}{2}x$.

Solution. The three lines $y = \frac{3}{2}x$, $y = -\frac{1}{2} + 4$, and $y = \frac{1}{2}x$ are shown in Fig. 4-36. The graph is the triangular area below the first two lines, and above the third.

Many plane areas bounded by lines or line segments can be characterized by a finite number of linear inequalities. For example, the three lines of Fig. 4-36 divide the plane into six regions other than the shaded triangular region. Each of these six regions, labeled I, II, ..., VI in Fig. 4-36, can be characterized by changing one or more of the given inequalities from "$\geq$" to "$\leq$" or vice versa. For example, region I is the solution of the three inequalities $y \geq \frac{3}{2}x$, $y \geq -\frac{1}{2}x + 4$, $y \geq \frac{1}{2}x$. Since the first two of these inequalities imply the third (this is clear from the figure) this region is in fact the solution of the system of the first two inequalities.

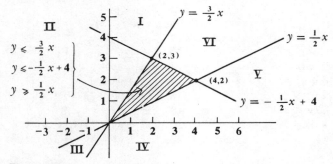

Figure 4-36

SECTION 4-5, REVIEW QUESTIONS

1. The graph of $x^2 + y^2 + Ax + By + C = 0$ is a circle, and the graph of $Ax + By + C = 0$ is a _____ . **line**

2. $Ax + By + C < 0$ is a linear inequality, and $x^2 + y^2 + Bx + Cy + D < 0$ is a _____ inequality. **circular**

3. The graph of an inequality is the set of points whose coordinates _____ the inequality. **satisfy**

4. The graph of a linear inequality is a _____-_____. **half-plane**

Inequalities 135

closed	5. A half-plane which includes the boundary line is a _____ half-plane.
open	6. A closed half-plane includes the boundary line, and an _____ half-plane does not.
circle	7. The graph of a circular inequality is generally the region inside or outside a _____.
squares	8. To find the boundary circle for the graph of a circular inequality we complete the _____ in both the x-terms and the y-terms.
intersection	9. The graph of two or more simultaneous inequalities is the _____ of their individual graphs.
all	10. The intersection of two sets or more is the set of points that are in _____ the sets.

SECTION 4-5, EXERCISES

Graph the following inequalities.

Example 1
1. $x \geq 0$
2. $y \leq 0$
3. $y < 3$
4. $x \geq -1$
5. $y \geq x + 1$
6. $y \geq 2x - 1$
7. $y < -3x + 2$
8. $y < -2x + 4$
9. $2x - y + 3 > 0$
10. $x - 3y + 4 > 0$
11. $3x + y - 2 \leq 0$
12. $4x - 3y + 2 \leq 0$

Example 2
13. $x^2 + y^2 < 4$
14. $x^2 + y^2 \leq 1$
15. $(x - 1)^2 + y^2 \geq 1$
16. $(x + 1)^2 + (y - 2)^2 \leq 4$
17. $x^2 + y^2 - 2y < 3$
18. $x^2 - 4x + y^2 < 8$
19. $x^2 + y^2 - 4x - 6y < -12$
20. $x^2 + y^2 - 2x - 6y \geq -6$
21. $x^2 + 2x + y^2 \geq 1$
22. $x^2 + y^2 - 4y < -3$

Graph the points which satisfy all the inequalities in the following systems.

Example 3, 4
23. $x \geq 0, y \leq 0$
24. $x > 1, y < -2$
25. $x^2 + y^2 \leq 4, y \geq 1$
26. $x^2 + y^2 < 9, x < 1$
27. $(x - 1)^2 + (y - 1)^2 \leq 1, y \geq x$
28. $(x - 2)^2 + (y + 1)^2 \leq 1, y \leq -\frac{1}{2}x$
29. $x^2 + y^2 - 2y \leq 0, x - y + 1 \geq 0$
30. $x^2 + 4y + y^2 < 0, x + y + 2 \leq 0$

Give the smallest system of inequalities which characterize the given region of Fig. 4-36.

31. Region I
32. Region II
33. Region III
34. Region IV
35. Region V
36. Region VI

136 Analytic Geometry

Graph the points which satisfy the given inequalities simultaneously.

37. $y \leq x, y \geq 0, x \leq 1$ Example 4
38. $y \geq x, x \geq 0, y \leq 1$
39. $y \leq x + 1, y \geq 2x, y \geq 0$
40. $y \geq -x, y \geq x - 1, y \leq -\frac{1}{2}x + \frac{1}{2}$
41. $y \geq \frac{1}{2}x, y \leq 2x, y \geq -x + 2$
42. $y \geq \frac{1}{2}x + 1, y \geq -x + 1, y \geq -2x$
43. $y \leq x - 1, y \geq \frac{1}{2}x$
44. $y \geq -x + 2, y \geq x$

4-6 PARABOLAS

The parabola, ellipse, and hyperbola are curves that occur frequently in calculus and its applications. These curves are called *conic sections*, because it can be shown that they are the curves formed by the intersection of a plane with a cone (see Fig. 4-37.) We will study the parabola first, and then look at the ellipse and hyperbola in the next section.

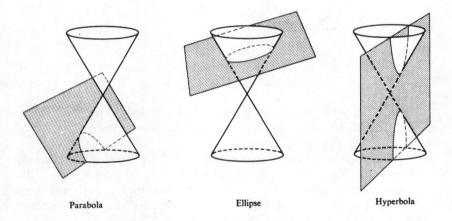

Parabola   Ellipse   Hyperbola

Figure 4-37

The graph of any equation of the form

$$y = ax^2 + bx + c, \quad (a \neq 0) \quad (1)$$

is called a *parabola*. The simplest form of (1) is $y = x^2$. We find some points on the curve by assigning values to x and computing the corresponding values of y:

x	0	$\pm\frac{1}{2}$	± 1	$\pm\frac{3}{2}$	± 2
y	0	$\frac{1}{4}$	1	$\frac{9}{4}$	4

We plot the points found above, and connect them with a smooth curve to obtain the graph shown in Fig. 4-38. Because the same y-value corresponds to both x and $-x$, the curve is symmetric about

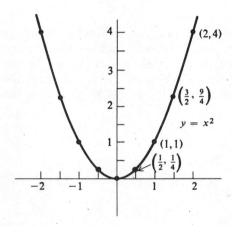

Figure 4-38

the y-axis, which is called the *axis* of the parabola. The point where a parabola intersects its axis is called the *vertex*, so the vertex of $y = x^2$ is at the origin.

The graph of $y = ax^2$, for $a > 0$, has the general appearance of Fig. 4-37: Any such curve opens upward, is symmetric about the y-axis, and has its vertex at the origin. Parabolas with equations $y = ax^2$, where a is a negative number, open downward as illustrated by the graph of $y = -\frac{1}{2}x^2$ shown in Fig. 4-39.

The general parabola $y = ax^2 + bx + c$ has the same *shape* as $y = ax^2$, but it has a different position with respect to the axes. The graph of $y = ax^2 + bx + c$ has the vertical line $x = -(b/2a)$ as its axis, and its vertex is at the point $(-b/2a, (4ac - b^2)/4a)$. To show this we complete the square as follows:

$$y = ax^2 + bx + c$$

138 Analytic Geometry

$$= a\left(x^2 + \frac{b}{a}x\right) + c$$

$$= a\left(x^2 + \frac{b}{a}x + \frac{b^2}{4a^2}\right) + c - \frac{b^2}{4a}$$

$$= a\left(x + \frac{b}{2a}\right)^2 + \frac{4ac - b^2}{4a}$$

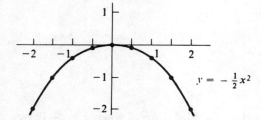

$y = -\frac{1}{2}x^2$

Figure 4-39

The expression $(x + b/2a)$ indicates the distance from (x, y) to the line $x = -b/2a$ in the same way that x indicates the distance from (x, y) to the y-axis. Hence the curve is symmetric about the line $x = -b/2a$, and therefore has its vertex on this line. When $x = -b/2a$, $y = (4ac - b^2)/4a$, so that $(-b/2a, (4ac - b^2)/4a)$ is the vertex.

Example 1 Find the axis and vertex of $y = 2x^2 - 4x - 3$ and sketch the curve.

Solution.
$$y = 2x^2 - 4x - 3$$
$$= 2(x^2 - 2x) - 3$$
$$= 2(x^2 - 2x + 1) - 3 - 2$$
$$= 2(x - 1)^2 - 5$$

The axis is the line $x = 1$, and the vertex is the point $(1, -5)$. The curve opens upward since the coefficient of the squared term is positive. See Fig. 4-40.

Example 2 Find the equation of the parabola that has $x = -1$ as its axis, $(-1, 2)$ as its vertex, and passes through the origin. Sketch the graph.

Solution. We know the equation will have the form
$$y = a(x + 1)^2 + d$$

Parabolas 139

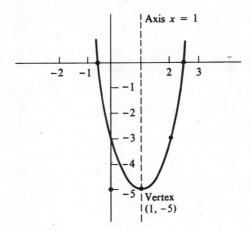

Figure 4-40

because the line $x = -1$ is the axis. Since $y = 2$ when $x = -1$, we must have $d = 2$, and the equation has the form

$$y = a(x + 1)^2 + 2$$

We also know that $x = 0$, $y = 0$ must satisfy the equation because the curve passes through the origin, and we use this fact to find the coefficient a:

$$0 = a(0 + 1)^2 + 2$$
$$0 = a + 2$$
$$a = -2$$

The equation is therefore

$$y = -2(x + 1)^2 + 2 \quad \text{or} \quad y = -2x^2 - 4x$$

See Fig. 4-41.

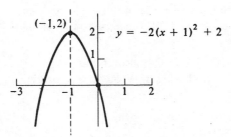

Figure 4-41

140 *Analytic Geometry*

The inequalities
$$y > ax^2 + bx + c \quad \text{and} \quad y < ax^2 + bx + c$$
will have as their respective graphs the set of all points above or below the corresponding parabola $y = ax^2 + bx + c$. Inequalities involving $\geq$ or $\leq$ would also include the parabola as part of the graph.

Example 3 Graph the inequality $y > -x^2 + 4x - 3$.

Solution. Graph the parabola $y = -x^2 + 4x - 3$, and indicate it with a broken line, since the curve itself is not part of the graph of the inequality:
$$\begin{aligned} y &= -x^2 + 4x - 3 \\ &= -(x^2 - 4x) - 3 \\ &= -(x^2 - 4x + 4) - 3 + 4 \\ &= -(x - 2)^2 + 1 \end{aligned}$$

The parabola is symmetric about the line $x = 2$, has its vertex at $(2, 1)$, and opens downward, since the coefficient of the squared term is -1. The graph of the inequality is the region above the curve. This is shown in Fig. 4-42.

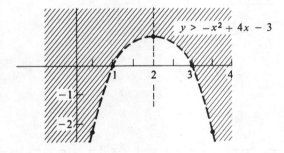

Figure 4-42

If we interchange x and y in the general equation (1), we obtain the equation of a parabola with its axis horizontal. Thus any equation of the form
$$x = ay^2 + by + c \quad (a \neq 0)$$
is such a parabola. The axis and vertex of these parabolas can be found by completing the square in the y-terms. If $a > 0$ the curve opens to the right, and if $a < 0$ the curve opens to the left.

Example 4 Graph the equation $x + 2y - y^2 = 0$.

Solution. Solve for x in terms of y, and then complete the square:

$$x + 2y - y^2 = 0$$
$$x = y^2 - 2y$$
$$= y^2 - 2y + 1 - 1$$
$$= (y - 1)^2 - 1$$

Hence the curve is a parabola whose axis is the horizontal line $y = 1$. The vertex is the point of the parabola on the axis; namely $(-1, 1)$. The parabola opens to the right since the coefficient of the squared term is positive. See Fig. 4-43.

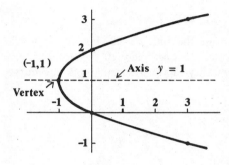

Figure 4-43

Example 5 Graph the curves $y = x + 1$ and $x = 3 - (y - 2)^2$; give two inequalities which characterize the region bounded by the two curves.

Solution. The line and the parabola are shown in Fig. 4-44. The line $y = 2$ is the axis of the parabola, and the parabola opens to the left because the coefficient of $(y - 2)^2$ is negative. The region inside the two curves (excluding the boundary curves) is characterized by the two inequalities

$$y < x + 1 \text{ and } x < 3 - (y - 2)^2.$$

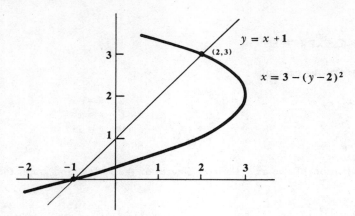

Figure 4-44

SECTION 4-6, REVIEW QUESTIONS

1. The parabola, ellipse, and hyperbola are called _____ sections.   **conic**

2. A conic section is a curve which is the intersection of a cone and a _____.   **plane**

3. The graph of any equation of the form $y = ax^2 + bx + c$, $(a \neq 0)$, is a _____.   **parabola**

4. Parabolas are symmetric about a line called the _____ of the parabola.   **axis**

5. The axis of $y = ax^2 + bx + c$ is parallel to the _____-axis.   **y**

6. The point where a parabola intersects its axis is called its _____.   **vertex**

7. The vertex of $y = x^2$ is at the _____.   **origin**

8. The curve $y = ax^2 + bx + c$ opens upward if $a > 0$, and opens downward if _____.   **a < 0**

9. The graph of the inequality $y < ax^2 + bx + c$ consists of all points _____ the graph of $y = ax^2 + bx + c$.   **below**

10. The graph of $x = ay^2 + by + c$ is a parabola whose axis is parallel to the _____-axis.   **x**

SECTION 4-6, EXERCISES

Graph the following parabolas.

1. $y = \frac{3}{2}x^2$
2. $y = 2x^2$
3. $y = -x^2$
4. $y = -\frac{1}{2}x^2$
5. $y = x^2 + 1$
6. $y = x^2 - 2$
7. $y = -x^2 - 4$
8. $y = -x^2 + 1$
9. $y = 2x^2 - 3$
10. $y = -2x^2 + 5$

Find the equation of the axis, the coordinates of the vertex, and graph the parabola.

Example 1

11. $y = (x - 1)^2$
12. $y = (x + 2)^2$
13. $y = 2(x + 2)^2 - 3$
14. $y = -(x + 1)^2 + 4$
15. $y = x^2 - 4x + 1$
16. $y = x^2 - 2x - 1$
17. $y = -x^2 - 6x - 8$
18. $y = -x^2 + 4x - 5$
19. $y = 2x^2 - 4x + 5$
20. $y = -\frac{1}{2}x^2 - 2x - 1$

Find the equations of the following parabolas.

Example 2

21. axis: $x = 0$; vertex: $(0, 0)$; through $(1, -1)$.
22. axis: $x = 0$; vertex: $(0, 0)$; through $(-2, 8)$.
23. axis: $x = 3$; vertex: $(3, 1)$; through $(4, 5)$.
24. axis: $x = -2$; vertex: $(-2, -1)$; through $(3, 49)$.
25. axis: $x = \frac{3}{2}$; vertex: $(\frac{3}{2}, 3)$; through $(\frac{5}{2}, -2)$.
26. axis: $x = -1$; vertex: $(-1, -5)$; through $(2, 22)$.

27. axis parallel to the y-axis; through $(0, 0), (1, 1), (-1, 0)$.
28. axis parallel to the y-axis; through $(0, 0), (2, 2), (1, 2)$.

Graph the following inequalities.

Example 3

29. $y \leq -2(x - 1)^2 + 4$
30. $y \geq \frac{1}{2}(x + 1)^2 - 2$
31. $y < x^2 - 4x + 8$
32. $y < x^2 - 6x$
33. $y \geq \frac{1}{2}x^2 + 2x + 1$
34. $y \geq -\frac{3}{2}x^2 - 12x - 4$
35. $x > (y - 1)^2 - 2$
36. $x > (y + 2)^2 - 1$
37. $x > y^2 - 4y + 5$
38. $x > y^2 - 6y + 8$

Graph the two curves, and give two inequalities which characterize the region bounded by the curves (excluding the boundary curves themselves).

Example 4, 5

39. $x = y^2; x = 4$
40. $x = 1 - y^2; x = 0$
41. $x = y^2 - 1; y = x - 1$
42. $x = (y - 1)^2 + 1; y = x - 2$
43. $y = \frac{1}{2}x^2 - 2; y = \frac{1}{2}x$
44. $y = 1 - x^2; y = -x - 1$
45. $x = y^2; x = 2 - y^2$
46. $y = x^2 - 1; y = 1 - x^2$
47. $y = \frac{1}{2}x^2 + 1; y = 2x^2$
48. $y = 2 - 2x^2; y = 1 - x^2$

49. Show that the points (x, y) which are equidistant from $(0, 1)$ and the line $y = -1$ lie on the parabola $y = \frac{1}{4}x^2$.

144   Analytic Geometry

50. Show that the points (x, y) which are equidistant from $(0, p)$ and the line $y = -p$ $(p > 0)$ lie on the parabola $x^2 = 4py$. (The point $(0, p)$ is called the *focus* of the parabola, and the line $y = -p$ is called its *directrix*.)

4-7 ELLIPSE AND HYPERBOLA

We continue the study of conic sections by looking at the graph of the equations of the form

$$\frac{x^2}{a^2} + \frac{y^2}{b^2} = 1 \quad (a > 0, b > 0). \tag{1}$$

The graph of (1) above is called an *ellipse*. Notice that a circle is a special kind of ellipse since, if $a = b$, the equation becomes

$$x^2 + y^2 = a^2.$$

The curve of Equation (1) intersects the x-axis at the points $(-a, 0)$, $(a, 0)$ and intersects the y-axis at the points $(0, -b)$ and $(0, b)$. Because both terms on the left of (1) are non-negative and their sum is 1, we must have

$$\frac{x^2}{a^2} \leq 1 \quad \text{and} \quad \frac{y^2}{b^2} \leq 1.$$

Hence $|x| \leq a$ and $|y| \leq b$ for any point (x, y) on the curve, and the curve lies inside the box formed by the lines $x = a$, $x = -a$, $y = b$, $y = -b$.

Since x is squared in (1), the coordinates of $P' = (-x_0, y_0)$ satisfy the equation whenever the coordinates of $P = (x_0, y_0)$ do. The points P' and P are symmetric about the y-axis, and consequently the curve is symmetric about the y-axis. A similar argument shows that $P'' = (x_0, -y_0)$ is on the curve whenever $P = (x_0, y_0)$ is. P'' and P are symmetric about the x-axis, and hence the curve is symmetric about the x-axis.

This principle holds quite generally: *Any equation in which x occurs only to even powers has a graph that is symmetric about the y-axis; if y occurs only to even powers the graph is symmetric about the x-axis.*

The graph of the ellipse (1) is shown in Fig. 4-45.

Example 1 Graph the ellipse $x^2/16 + y^2/9 = 1$.

Solution. First note that the intercepts on the axis are the points

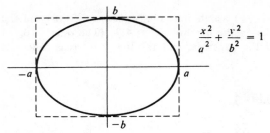

Figure 4-45

$(-4, 0)$, $(4, 0)$ and $(0, -3)$, $(0, 3)$. Solve for y in terms of x and compute the values of y corresponding to $x = \pm 1, \pm 2, \pm 3$.

$$\frac{x^2}{16} + \frac{y^2}{9} = 1$$

$$\frac{y^2}{9} = 1 - \frac{x^2}{16} = \frac{16 - x^2}{16}$$

$$y^2 = \frac{9}{16}(16 - x^2)$$

$$y = \pm \frac{3}{4}\sqrt{16 - x^2}$$

x	± 1	± 2	± 3
y	$\pm\frac{3}{4}\sqrt{15} \doteq \pm 2.9$	$\pm\frac{3}{4}\sqrt{12} \doteq \pm 2.6$	$\pm\frac{3}{4}\sqrt{7} \doteq \pm 2.0$

We use the positive coordinates to plot the part of the curve in the first quadrant, and then complete the graph using its symmetry about the axes, as shown in Fig. 4-46.

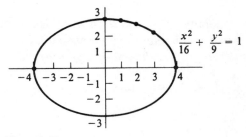

Figure 4-46

146 *Analytic Geometry*

Now we look at the *hyperbola*, which is the graph of equations of the following forms:

$$\frac{x^2}{a^2} - \frac{y^2}{b^2} = 1 \qquad (2)$$

$$-\frac{x^2}{a^2} + \frac{y^2}{b^2} = 1 \qquad (3)$$

We will look first at the graph of (2). We can then determine the graph of (3) by reversing the roles of x and y.

The hyperbola (2) intersects the x-axis at $(-a, 0)$ and $(a, 0)$. If we set $x = 0$, we get the impossible equation $-(y^2/b^2) = 1$; thus the curve does not intersect the y-axis. Since both x and y occur as squares in (2) and (3), the curves are symmetric about both axes. Solving (2) for y we get

$$\frac{x^2}{a^2} - \frac{y^2}{b^2} = 1$$

$$\frac{y^2}{b^2} = \frac{x^2}{a^2} - 1$$

$$= \frac{x^2 - a^2}{a^2}$$

$$y = \pm \frac{b}{a} \sqrt{x^2 - a^2}$$

For large values of x, $\frac{b}{a}\sqrt{x^2 - a^2}$ becomes close to $\frac{b}{a}x$. Hence the curve approaches the two lines $y = \pm \frac{b}{a} x$ as x gets large. These two lines are called the *asymptotes* of the hyperbola. The graph of (2) is shown in Fig. 4-47. Notice that the asymptotes are the diagonals of the box through the points $(-a, 0)$, $(a, 0)$, $(0, -b)$, $(0, b)$.

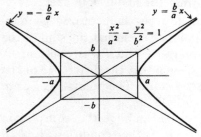

Figure 4-47

Hyperbolas of the form (3) intersect the y-axis instead of the x-axis, and open upward and downward rather than to the right and left. The lines $y = \pm \dfrac{b}{a} x$ are again the asymptotes of these hyperbolas.

Example 2 Graph the hyperbola $-(x^2/4) + y^2/1 = 1$.

Solution. The curve will intersect the y-axis ($x = 0$) at $(0, 1)$ and $(0, -1)$. The lines $y = \pm\tfrac{1}{2}x$ are the asymptotes [these are the diagonals of the box through $(0, 1)$, $(0, -1)$, $(2, 0)$, $(-2, 0)$]. We solve for y in terms of x, and compute the values of y corresponding to $x = 0, 1, 2, 3$:

$$\frac{-x^2}{4} + \frac{y^2}{1} = 1$$

$$y^2 = 1 + \frac{x^2}{4}$$

$$= \frac{4 + x^2}{4}$$

$$y = \pm\tfrac{1}{2}\sqrt{4 + x^2}$$

x	0	± 1	± 2	± 3	± 4
y	± 1	$\pm\tfrac{1}{2}\sqrt{5} \doteq \pm 1.1$	$\pm\tfrac{1}{2}\sqrt{8} \doteq 1.4$	$\pm\tfrac{1}{2}\sqrt{13} \doteq 1.8$	$\pm\tfrac{1}{2}\sqrt{20} \doteq 2.2$

The part of the curve in the first quadrant is graphed from the points of the table with positive coordinates (Fig. 4-48), and the curve is completed by making it symmetric about both axes.

The ellipse and hyperbola can also be characterized geometrically. An ellipse is the set of all points P such that the sum of the distances from P to two given points F_1 and F_2 is constant. A hyperbola is the set of all points P such that the difference of the distances from P to F_1 and F_2 is constant. The points F_1, F_2 are called the *foci* of the ellipse or hyperbola.

Let $F_1 = (c, 0)$ and $F_2 = (-c, 0)$, and let the constant in question for the ellipse be $2a$. Then the condition is $F_1 P$

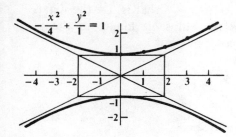

Figure 4-48

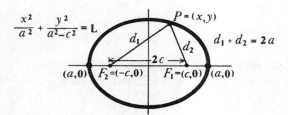

Figure 4-49

$+ F_2P = 2a$, where we necessarily have $2a > 2c$ (see Fig. 4-49). If $P = (x, y)$, then this gives the equation

$$\sqrt{(x-c)^2 + y^2} + \sqrt{(x+c)^2 + y^2} = 2a.$$

We put one radical on the right, and square; this gives

$$x^2 - 2cx + c^2 + y^2 = x^2 + 2cx + c^2 + y^2 + 4a^2 - 4a\sqrt{(x+c)^2 + y^2},$$

$$-4cx - 4a^2 = -4a\sqrt{(x+c)^2 + y^2},$$

$$cx + a^2 = a\sqrt{(x+c)^2 + y^2}.$$

Squaring again gives

$$c^2x^2 + 2ca^2x + a^4 = a^2[x^2 + 2cx + c^2 + y^2],$$

$$a^4 - a^2c^2 = (a^2 - c^2)x^2 + a^2y^2,$$

$$1 = \frac{x^2}{a^2} + \frac{y^2}{a^2 - c^2}. \tag{4}$$

If the foci are at $(0, c)$ and $(0, -c)$ --- that is, on the y-axis instead of the x-axis --- then the ellipse is longer in the y-direction, and the equation becomes

$$\frac{x^2}{a^2 - c^2} + \frac{y^2}{a^2} = 1. \tag{5}$$

Example 3 Find the foci of the ellipses

$$(a) \frac{x^2}{9} + \frac{y^2}{4} = 1; \qquad (b) \frac{x^2}{6} + \frac{y^2}{10} = 1.$$

Solution.

(a) Since the ellipse is longer in the x-direction, the foci are on the x-axis. If $(c, 0)$ and $(-c, 0)$ are the foci, then we have

$$a^2 = 9, a^2 - c^2 = 4.$$

Hence $c^2 = 9 - 4 = 5$, $c = \sqrt{5}$; the foci are at $(\sqrt{5}, 0)$ and $(-\sqrt{5}, 0)$.

(b) In (b), the ellipse is longer in the y-direction and the foci must be on the y-axis. If we let $(0, c)$, $(0, -c)$ be the foci, we have $a^2 = 10$, $a^2 - c^2 = 6$; so $c^2 = 4$, $c = 2$. The foci are $(0, 2)$ and $(0, -2)$.

The geometric condition defining a hyperbola is

$$|F_1 P - F_2 P| = 2a.$$

If the foci are at $(c, 0)$ and $(-c, 0)$, which means that $2a < 2c$, we get the equation

$$\sqrt{(x - c)^2 + y^2} - \sqrt{(x + c)^2 + y^2} = \pm 2a. \tag{6}$$

Squaring twice and simplifying gives

$$\frac{x^2}{a^2} - \frac{y^2}{c^2 - a^2} = 1. \tag{7}$$

If the foci are $(0, c)$ and $(0, -c)$, we get

$$-\frac{x^2}{c^2 - a^2} + \frac{y^2}{a^2} = 1. \tag{8}$$

Thus for a hyperbola the foci are on the x-axis if the x^2-term has the plus sign (so the curve intersects the x-axis), and on the y-axis if the y^2-term has the plus sign (so the curve intersects the y-axis).

Example 4 Find the foci and intercepts of

$$\text{(a) } \frac{x^2}{16} - \frac{y^2}{25} = 1; \qquad \text{(b) } -\frac{x^2}{2} + \frac{y^2}{7} = 1.$$

Solution.

(a) The intercepts are $(\pm 4, 0)$, so the foci are on the x-axis. We have $a^2 = 16$, $c^2 - a^2 = 25$, so $c^2 = 41$, and the foci are $(\sqrt{41}, 0), (-\sqrt{41}, 0)$.

(b) The intercepts are $(0, \pm\sqrt{7})$, so the foci are on the y-axis. Here $a^2 = 7$ and $c^2 - a^2 = 2$, so $c^2 = 9$. The foci are $(0, 3)$ and $(0, -3)$.

SECTION 4-7, REVIEW QUESTIONS

1. The graph of $\frac{x^2}{a^2} + \frac{y^2}{b^2} = 1$ is called an _____.   ellipse

2. An ellipse in which $a = b$ is a _____.   circle

3. The point symmetric to (x_0, y_0) about the y-axis is (_____, _____). $(-x_0, y_0)$

4. The points $(-x_0, y_0)$, (x_0, y_0) are symmetric about the y-axis and the points (x_0, y_0), (_____, _____) are symmetric about the x-axis. $(x_0, -y_0)$

5. If the x-terms are all squared in an equation its graph is symmetric about the _____-axis.   y

6. The ellipse $\frac{x^2}{a^2} + \frac{y^2}{b^2} = 1$ intersects the x-axis at _____ and _____, and the y-axis at _____ and _____. $(a, 0)$ $(-a, 0)$ $(0, b)$ $(0, -b)$

3	7. The ellipse, $\dfrac{x^2}{2} + \dfrac{y^2}{9} = 1$, lies inside the box formed by the four lines $x = \pm\sqrt{2}$, $y = \pm$ _____ .
hyperbola	8. The graph of $\dfrac{x^2}{a^2} - \dfrac{y^2}{b^2} = 1$ or $-\dfrac{x^2}{a^2} + \dfrac{y^2}{b^2} = 1$ is called a _____ .
x	9. The hyperbola $\dfrac{x^2}{a^2} - \dfrac{y^2}{b^2} = 1$ intersects the _____-axis but
y	not the _____-axis.
	10. Both hyperbolas $\dfrac{x^2}{a^2} - \dfrac{y^2}{b^2} = 1$ and $-\dfrac{x^2}{a^2} + \dfrac{y^2}{b^2} = 1$ have the lines $y = \pm\dfrac{b}{a}x$
asymptotes	as _____ .
	11. The asymptotes are the diagonals of the box through the points $(a, 0)$,
(−a,0)(0,−b)	$(0, b)$, _____ , _____ .
	12. A point which moves so the sum of its distances to two given points is
ellipse	constant traces out an _____ .
foci	13. The two given points are called the _____ of the ellipse.
	14. An ellipse with foci at $(\pm c, 0)$ has an equation of the form $\dfrac{x^2}{a^2} +$
$a^2 - c^2$	$\dfrac{y^2}{(\text{_____})} = 1$.
	15. The foci of $\dfrac{x^2}{a^2} + \dfrac{y^2}{a^2 - c^2} = 1$ are at $(\pm c, 0)$, and the foci of $\dfrac{x^2}{a^2 - c^2} +$
$(0, \pm c)$	$\dfrac{y^2}{a^2} = 1$ are at (_____, _____).
	16. A point which moves so the difference of its distances from two given
hyperbola	points is a constant is a _____ .
	17. A hyperbola with foci at $(\pm c, 0)$ has an equation of the form
$c^2 - a^2$	$\dfrac{x^2}{a^2} - \dfrac{y^2}{(\text{_____})} = 1$.
$(0, \pm c)$	18. A hyperbola $-\dfrac{x^2}{(c^2 - a^2)} + \dfrac{y^2}{a^2} = 1$ has its foci at (_____, _____).
	19. In the ellipse, $\dfrac{x^2}{a^2} + \dfrac{y^2}{b^2} = 1$, c^2 is the difference of a^2 and b^2, and in the
$a^2 + b^2$	hyperbola, $\pm\dfrac{x^2}{a^2} \mp \dfrac{y^2}{b^2} = 1$, $c^2 =$ _____ .

SECTION 4-7, EXERCISES

Graph the following ellipses.

1. $\dfrac{x^2}{9}+\dfrac{y^2}{4}=1$
2. $\dfrac{x^2}{16}+\dfrac{y^2}{4}=1$
3. $\dfrac{x^2}{9}+\dfrac{y^2}{25}=1$ Example 1
4. $\dfrac{x^2}{1}+\dfrac{y^2}{4}=1$
5. $\dfrac{x^2}{3}+\dfrac{y^2}{9}=1$
6. $\dfrac{x^2}{4}+\dfrac{y^2}{8}=1$
7. $\dfrac{x^2}{4}+\dfrac{y^2}{4}=1$
8. $\dfrac{x^2}{7}+\dfrac{y^2}{7}=1$

Write the equation of the ellipses which are centered at the origin and pass through the following points.

9. $(\pm 2, 0), (0, \pm 1)$
10. $(\pm 3, 0), (0, \pm 2)$
11. $(\pm\sqrt{5}, 0), (0, \pm\sqrt{7})$
12. $(\pm\sqrt{10}, 0), (0, \pm\sqrt{2})$
13. $(1, 1), (2, 0)$
14. $(2, 1), (3, 0)$
15. $(2, \sqrt{3/5}), (\sqrt{10/3}, 1)$
15. $(1, 2/\sqrt{3}), (\sqrt{3/2}, 1)$

Graph the following hyperbolas along with their asymptotes.

17. $x^2 - y^2 = 1$
18. $x^2 - y^2 = 4$ Example 2
19. $\dfrac{x^2}{4}-\dfrac{y^2}{9}=1$
20. $\dfrac{x^2}{16}-\dfrac{y^2}{4}=1$
21. $\dfrac{x^2}{5}-\dfrac{y^2}{4}=1$
22. $\dfrac{x^2}{8}-\dfrac{y^2}{3}=1$
23. $-\dfrac{x^2}{4}+y^2=1$
24. $-\dfrac{x^2}{4}+\dfrac{y^2}{9}=1$
25. $-\dfrac{x^2}{3}+\dfrac{y^2}{5}=1$
26. $-\dfrac{x^2}{4}+\dfrac{y^2}{7}=1$
27. $-x^2 + y^2 = 1$
28. $-x^2 + y^2 = 9$

Write the equation of the hyperbolas which are centered at the origin and satisfy the following conditions.

29. intercepts $(\pm 2, 0)$; asymptotes $y = \pm \tfrac{3}{2}x$
30. intercepts $(\pm 1, 0)$; asymptotes $y = \pm 2x$
31. intercepts $(\pm 3, 0)$; asymptotes $y = \pm \tfrac{1}{2}x$
32. intercepts $(\pm 2, 0)$; asymptotes $y = 5x$
33. intercepts $(0, \pm 1)$; asymptotes $y = \pm x$
34. intercepts $(0, \pm 2)$; asymptotes $y = \pm x$
35. intercepts $(0, \pm 5)$; asymptotes $y = \pm 2x$
36. intercepts $(0, \pm 3)$; asymptotes $y = \pm 3x$

Ellipse and Hyperbola   153

Example 3, 4   Find the foci of the following ellipses and hyperbolas.

37. $\dfrac{x^2}{16} + \dfrac{y^2}{9} = 1$ 38. $\dfrac{x^2}{25} + \dfrac{y^2}{9} = 1$ 39. $\dfrac{x^2}{7} + \dfrac{y^2}{3} = 1$

40. $\dfrac{x^2}{10} + \dfrac{y^2}{5} = 1$ 41. $\dfrac{x^2}{4} + \dfrac{y^2}{9} = 1$ 42. $\dfrac{x^2}{5} + \dfrac{y^2}{6} = 1$

43. $\dfrac{x^2}{4} + \dfrac{y^2}{29} = 1$ 44. $\dfrac{x^2}{2} + \dfrac{y^2}{11} = 1$ 45. $\dfrac{x^2}{4} + \dfrac{y^2}{9} = 1$

46. $\dfrac{x^2}{16} - \dfrac{y^2}{9} = 1$ 47. $\dfrac{x^2}{3} - \dfrac{y^2}{6} = 1$ 48. $\dfrac{x^2}{2} - \dfrac{y^2}{4} = 1$

49. $-\dfrac{x^2}{4} + \dfrac{y^2}{9} = 1$ 50. $-\dfrac{x^2}{25} + \dfrac{y^2}{4} = 1$ 51. $-\dfrac{x^2}{3} + \dfrac{y^2}{5} = 1$

52. $-\dfrac{x^2}{7} + \dfrac{y^2}{4} = 1$

53. Do the algebraic simplification to show that the equation of the ellipse with foci at $(0, \pm c)$ is given by (5).
54. Simplify (6) and get (7).
55. Show that the hyperbola with foci at $(0, \pm c)$ is given by (8).

4-8 REVIEW FOR CHAPTER

Graph the following intervals. Express the fact that x is in the interval with a single inequality using absolute value notation.

1. $(-5, 5)$ 2. $[-2, 2]$ 3. $(1, 5)$ 4. $[-3, 1]$

Express the solutions of the following inequalities in the form $[a, b]$ or (a, b).

5. $|x - 5| < 1$ 6. $|x - 2| \leqslant 3$
7. $|x + 1| \leqslant 3$ 8. $|x + 6| < 3$

Express the solutions of the following inequalities as a union of two intervals.

9. $|x| \geqslant 3$ 10. $|x - 1| > 1$
11. $|x + 2| > 5$ 12. $|x - 5| \geqslant 2$

Find the distance between the following pairs of points.

13. $(6, 2), (2, 5)$ 14. $(1, 0), (4, 4)$

154 Analytic Geometry

15. (3, 8), (4, 7)   16. (1, −3), (−2, 3)

Use the distance formula to tell whether the three points are collinear.

17. (1, 4), (7, 7), (3, 5)   18. (−1, 2), (5, 6), (1, 3)

Write the equation of the perpendicular bisector of the segments between the points.

19. (3, 1), (1, 3)   20. (2, 4), (6, 0)

Find the center and radius of the following circles, and graph.

21. $x^2 + y^2 - 2x - 6y + 6 = 0$
22. $x^2 + y^2 + 4x - 2y - 4 = 0$
23. $x^2 + y^2 - 4x + 8y + 11 = 0$
24. $x^2 + y^2 + 10x - 6y + 34 = 0$

Write the equation of the following circles.

25. center (−1, −4); radius 5
26. center (−1, 0); tangent to y-axis
27. through (0, 2), (2, 0), (0, −2)
28. center (1, 2); through (2, 3)

Graph the following lines.

29. $y = 2x - 3$ 30. $y = -\frac{1}{2}x + 2$
31. $2x - y + 3 = 0$ 32. $x + y = 0$

Write the equations of the following lines.

33. through (1, 3); slope 2
34. through (3, 1) and (4, 5)
35. slope −3; y-intercept 5
36. slope −2; x-intercept 2
37. through (2, 5); parallel to $2y - x + 7 = 0$
38. through (−1, 4); perpendicular to $x - 3y + 1 = 0$

Graph the following inequalities and systems of inequalities.

39. $y < 2x - 3$
40. $y \geqslant -x + 1$
41. $y < x + 5, y > 2x - 3$
42. $y \leqslant x, y \geqslant 0, y \leqslant -2x + 4$
43. $x^2 + (y - 3)^2 < 9$
44. $(x + 1)^2 + (y + 2)^2 \geqslant 4$
45. $x^2 + (y - 2)^2 \leqslant 9, y \leqslant -x - 1$
46. $x^2 + y^2 \geqslant 1, (x - 1)^2 + y^2 \leqslant 1$

Graph the conic sections. For parabolas, give the axis and vertex; for ellipses and hyperbolas, find the foci.

47. $y = (x + 2)^2$
48. $y = (x - 3)^2 - 2$
49. $y = x^2 - 4x + 3$
50. $y = -x^2 + 6x + 7$
51. $x = (y - 2)^2 + 1$
52. $x = -y^2 - 6y + 1$
53. $\dfrac{x^2}{3} + \dfrac{y^2}{2} = 1$
54. $\dfrac{x^2}{4} + \dfrac{y^2}{25} = 1$
55. $\dfrac{x^2}{3} - \dfrac{y^2}{6} = 1$
56. $-\dfrac{x^2}{4} + \dfrac{y^2}{2} = 1$

156 Analytic Geometry

5 FUNCTIONS

5-1 RELATIONS

We use the word "relation" in ordinary language to indicate some property connecting two people. Consider the question: "What relation is person x to person y?" We might expect answers such as "x is the father of y," "x works for y," or "x and y live in the same town."

In mathematics we also use this word to indicate some property connecting two mathematical objects x and y. Ordinarily the objects x and y will be *numbers*, and when we wish to stress this we will speak of a *numerical relation*. The phrases below define three different numerical relations.

(1) x is greater than y
(2) x and y are coordinates of a point in the first quadrant
(3)   The point (x, y) is one unit from the origin

It is obviously necessary to consider the order of the objects x and y in describing a relation. The statement "$x > y$" is not the same as

"$y > x$," and the statement "x is the father of y" is not the same as "y is the father of x."

We formalize our definition by agreeing that a *relation* is any *set of ordered pairs* (x, y). In a numerical relation, x and y are numbers. The *graph* of a numerical relation R is the set of points whose coordinates (x, y) form one of the pairs in the set R. The graphs of the relations defined by (1), (2), and (3) above are shown in Fig. 5-1.

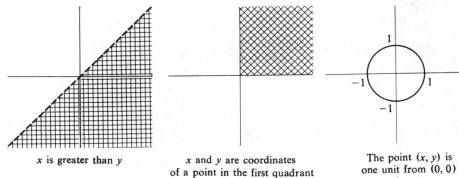

x is greater than y x and y are coordinates of a point in the first quadrant   The point (x, y) is one unit from $(0, 0)$

Figure 5-1

Any set of pairs (x, y) is a relation, and therefore any set of points in the plane is the graph of some relation. However, we are usually interested in relations defined by equations or inequalities in x and y. The relation defined by such an equation or inequality is just the set of all pairs (x, y) so that x and y satisfy the equation or inequality.

The set of all first elements of pairs in R is called the *domain* of R, and the *range* of R is the set of all second elements. For a numerical relation, the domain is the set of numbers gotten by projecting the graph onto the x-axis, and the range is the set gotten by projecting the graph onto the y-axis.

Example 1   Graph the relations:

$$(a)\ xy \leq 0 \qquad (b)\ y^2 = x$$

Solution.

(a) The graph of $xy = 0$ consists of the two coordinate axes: $x = 0$ and $y = 0$. For the product xy to be negative, x and y

must have opposite signs. Hence, the graph of $xy \leq 0$ consists of the points in the second and fourth quadrants (Fig. 5-2).

(b) The graph of the relation defined by $y^2 = x$ is the same as the graph of the equation; namely, the parabola shown in Fig. 5-2.

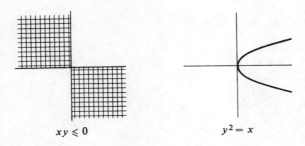

Figure 5-2

Example 2 Graph the relation $y \geq |x|$ and give the domain and range.

Solution. Since $|x| = x$ if $x \geq 0$, and $|x| = -x$ if $x < 0$, the graph of $y = |x|$ is the same as the line $y = x$ for $x \geq 0$ and the line $y = -x$ for $x < 0$. The graph of $y \geq |x|$ consists of all points on and above the graph of $y = |x|$ (Fig. 5-3). The domain of $y \geq |x|$ is all real numbers x, and the range is the set of all non-negative numbers y.

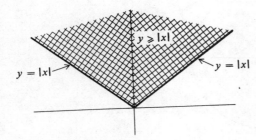

Figure 5-3

Example 3 Graph the relation $(x - 1)^2 + y^2 \leq 4$, and give the domain and range.

Relations 159

Solution. We recognize that $(x - 1)^2 + y^2 = 4$ is the equation of the circle with center at $(1, 0)$ and radius 2. The graph of the inequality is therefore the set of points on this circle, and those inside it (see Fig. 5-4). The domain is the interval $[-1, 3]$ and the range is the interval $[-2, 2]$.

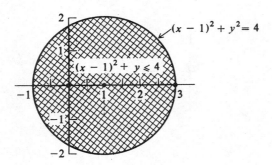

Figure 5-4

Example 4 Graph $(y - x)(x^2 + y^2 - 1) = 0$.

Solution. A product is zero provided one of the factors is zero. Therefore, the graph consists of the graph of $y - x = 0$ together with the graph of $x^2 + y^2 - 1 = 0$ (Fig. 5-5).

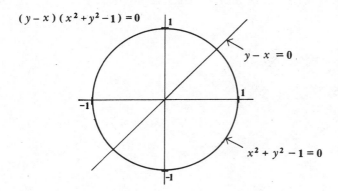

Figure 5-5

Example 5 Graph $y \leq \sqrt{1 - x^2}$ and give the domain and range of the relation.

Solution. The equation $y = \sqrt{1-x^2}$ represents the top half of the circle $x^2 + y^2 = 1$. The graph of $y \leq \sqrt{1-x^2}$ consists of all points (x, y), with $-1 \leq x \leq 1$, which lie under the circle. Since the expression $\sqrt{1-x^2}$ makes no sense if $|x| > 1$, the domain is $[-1, 1]$. The range is $[-\infty, 1]$ (Fig. 5-6).

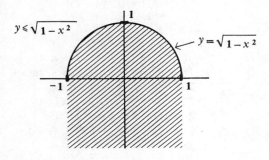

Figure 5-6

SECTION 5-1, REVIEW QUESTIONS

1. Any set of ordered pairs is called a _____.   **relation**

2. The graph of a relation is the set of _____ whose coordinates (x, y) are in the set.   **points**

3. Any set of points is the _____ of some relation.   **graph**

4. The graph of any equation in x and y is the same as the graph of the _____ defined by the equation.   **relation**

5. The domain of a relation is the set of _____ elements of pairs in the relation.   **first**

6. The domain is the set of first elements, and the _____ is the set of second elements.   **range**

7. The range of a numerical relation is the set gotten by projecting its graph on the _____-axis.   **y**

8. The range is the projection on the y-axis, and the projection on the x-axis is the _____.   **domain**

SECTION 5-1, EXERCISES

Graph the following relations; give the domain and range.

Example 1,2
1. $y = |x|$
2. $y = -|x - 1|$
3. $y = x^2 + 1$
4. $y = 1 - 2x^2$
5. $x = -y^2$
6. $x = 1 + y^2$
7. $x(y - 1) > 0$
8. $(x + 1)(y + 1) < 0$
9. $y \leqslant |x|$
10. $y > |x - 1|$
11. $y \leqslant 1 - x^2$
12. $x \leqslant -1 - y^2$

Example 3
13. $x^2 + y^2 < 1$
14. $x^2 + y^2 \geqslant 9$
15. $(x - 2)^2 + (y - 3)^2 \leqslant 1$
16. $(x + 1)^2 + (y - 1)^2 \leqslant 4$
17. $x^2 + y^2 + 6x - 2y - 6 < 0$
18. $x^2 + y^2 - 10x + 4y + 28 > 0$

Example 4
19. $(y - x)(x - 1) = 0$
20. $(y + x)(y + 1) = 0$
21. $(y - 2x)(y + x) = 0$
22. $(x + 2y + 1)(2y - x) = 0$
23. $(x + y + 1)(y - x^2) = 0$
24. $(y - 2x)(x - y^2) = 0$
25. $(y + x^2)(x^2 + y^2 - 1) = 0$
26. $(x - y^2)(x^2 + y^2 - 4) = 0$
27. $x^2 - y^2 = 0$
28. $y^2 - 4x^2 = 0$
29. $y^2 - x^4 = 0$
30. $x^2 - y^4 = 0$

Example 5
31. $y \leqslant \sqrt{4 - x^2}$
32. $y \geqslant \sqrt{1 - x^2}$
33. $x \geqslant \sqrt{9 - y^2}$
34. $x \leqslant \sqrt{16 - y^2}$
35. $y \leqslant 3\sqrt{1 - \dfrac{x^2}{4}}$
36. $y \geqslant -4\sqrt{1 - \dfrac{x^2}{25}}$

5-2 FUNCTIONS

In the last section we studied relations. Now we will look at the most important kind of relation, the *function*. The idea of functional relationship is fundamental in all of mathematics and its applications. Scientific language is full of such phrases as "the force is a function of the displacement," or "the pressure is a function of the temperature." The idea expressed in these phrases is clear—one quantity depends on, or is determined by the values of another quantity.

We define a *function* to be a relation in which there is exactly one range element associated with each element of the domain. That is, a function f is a set of ordered pairs (x, y) in which y is uniquely determined by x. If f is a function, then we write $y = f(x)$ (read, y equals f of x) to indicate that (x, y) is the unique pair in f with first element x. The graph of f is therefore the same thing as the graph of the equation $y = f(x)$. A graph is the graph of a function provided each vertical line intersects the graph one time at most. See Fig. 5-7.

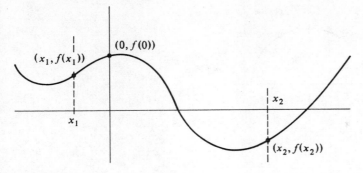

Figure 5-7

The most common way of defining a function is to give a formula for $f(x)$. For example, functions f and g are defined by

$$f(x) = \sqrt{4 - x^2}; \qquad g(x) = \frac{1}{1 - x}.$$

If no domain is specified, we understand that the domain is the set of values of x for which the formula makes sense. The domain of f is the interval $[-2, 2]$ where the radicand is non-negative. The domain of g is the set of all numbers except 1, where the denominator is zero. The graphs of f and g, which are the same as the graphs of $y = \sqrt{4 - x^2}$ and $y = 1/(1 - x)$, respectively, are shown in Fig. 5-8.

A function with a finite domain (and hence a finite range also) can be defined simply by exhibiting the set of ordered pairs in the function. For example, the set

$$F = \{(1, 2), (2, 3), (3, 71)\}$$

is a function whose graph consists of three points. This function could equally well be defined by giving the value of the function for each number in the domain:

$$F(1) = 2, \qquad F(2) = 3, \qquad F(3) = 71$$

Example 1 Which of the following relations is a function and which is not? For the function, specify the value of the function for each number in the domain.

$$R = \{(-1, 1), (0, 1), (1, 0), (0, -1)\}$$
$$S = \{(5, 1), (6, 1), (7, 1), (8, 2), (9, 3)\}$$

Solution. R is not a function, since both $(0, 1)$ and $(0, -1)$ are in

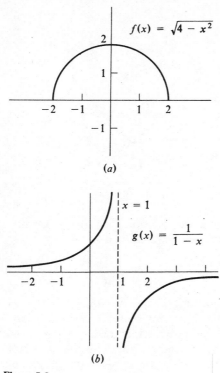

Figure 5-8

R. S is a function since no two pairs in S have the same first element. The value of S for each number in its domain is

$$S(5) = 1, \quad S(6) = 1, \quad S(7) = 1, \quad S(8) = 2, \quad S(9) = 3$$

Example 2 Graph the relations (a) $4x^2 + 9y^2 = 36$; (b) $y - 3 + x^2 = 0$; (c) $xy = 1$. Tell which of these relations are functions, and for the functions give a formula for y in terms of x.

Solution. The graphs are shown in Fig. 5-9.

The relation (a) is not a function, because some vertical lines intersect the graph twice, but (b) and (c) are functions. In equations (b) and (c) we can solve for y in terms of x; we get the formulas:

$$y = 3 - x^2$$

$$y = \frac{1}{x} \quad (x \neq 0)$$

Note that if we attempt to solve (a) for y in terms of x we get

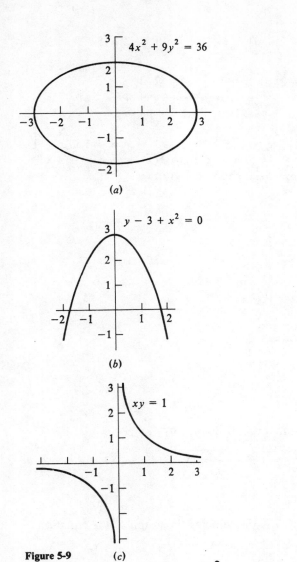

Figure 5-9

$$y = \pm \frac{2}{3}\sqrt{9 - x^2}$$

which shows that two values of y correspond to each x in the interval $(-3, 3)$.

Example 3 Graph the function F defined by $F(x) = |x - 1| - x$.

Solution. First calculate the values of $F(x)$ for $x = -2, -1, 0, 1, 2, 3$, and plot the points $(x, F(x))$: for example,

Functions 165

$$F(-2) = |-2 - 1| - (-2) = |-3| + 2 = 5$$

x	-2	-1	0	1	2	3
$F(x)$	5	3	1	-1	-1	-1

Since $|x - 1| = x - 1$ if $x \geq 1$, $F(x) = x - 1 - x = -1$ if $x \geq 1$. Similarly, $|x - 1| = -(x - 1)$ if $x < 1$, so $F(x) = -x + 1 - x = -2x + 1$ if $x < 1$. Figure 5-10 shows the graph.

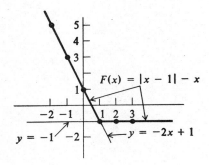

Figure 5-10

Example 4 Graph the function G defined by

$$G(x) = 1 - \sqrt{4 - (x - 1)^2}.$$

Solution. The graph of G is the same as the graph of the equation

$$y = 1 - \sqrt{4 - (x - 1)^2}\,.$$

This can be written

$$y - 1 = -\sqrt{4 - (x - 1)^2}.$$

Since the radical is non-negative, $y - 1 \leq 0$, and the equation above is equivalent to

$$(y - 1)^2 = 4 - (x - 1)^2, \quad y - 1 \leq 0;$$
$$(x - 1)^2 + (y - 1)^2 = 4, \quad y \leq 1.$$

Hence the graph is the part of the circle $(x - 1)^2 + (y - 1)^2 = 4$, which also satisfies $y \leq 1$ (i.e., the bottom half of the circle). See Fig. 5-11.

We use the word *continuous* to describe a function whose graph has no jumps or wild oscillations. The concept of continuity is subtle, and is studied extensively in higher courses in analysis. In this text we will use the term in an intuitive way to express the idea of a function whose graph has no jumps. All of the functions discussed previously in this section other than those with finite domains are continuous at all points of their domains. The functions of exercises 37-40 are not everywhere continuous.

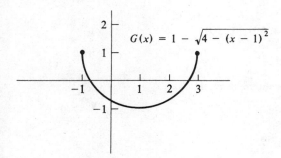

Figure 5-11

SECTION 5-2, REVIEW QUESTIONS

1. A relation in which there is exactly one range element associated with each domain element is called a _____. **function**

2. If f is a function and (x, y) is an element of f, we write $y = $ _____. **f(x)**

3. The expression "$f(x)$" is read "_____ _____ _____." **f of x**

4. The graph of a function f is the same as the graph of the equation $y = $ _____. **f(x)**

5. The most common way of defining a function is to specify a _____ for $f(x)$. **formula**

6. The set of values of x for which the formula for $f(x)$ makes sense is the _____ of the function. **domain**

Functions   167

7. Any graph is the graph of a function provided each vertical line intersects the graph at most _____ time.

one

SECTION 5-2, EXERCISES

Tell whether the following relations are functions. For the functions, give the function value for each number in the domain.

Example 1
1. $F = (1, 2), (-2, 3), (-3, 4), (4, -5)$
2. $G = (3, 1), (1, 3), (2, 4), (4, 2)$
3. $H = (-1, 0), (0, 1), (1, 0), (-1, 1), (0, 0)$
4. $I = (2, 1), (10, 1), (2, 10), (1, 10)$
5. $J = (1, 0), (2, 0), (3, 0)$
6. $K = (0, 1), (0, 2), (0, 3)$

For F, G, J in the exercises above, evaluate

7. $G(F(-2))$
8. $F(G(2))$
9. $J(G(3))$
10. $J(F(1))$

Graph the following relations and tell which are functions. For each function, find a formula for y in terms of x.

Example 2
11. $y - 2x = 3$
12. $x + 2y - 4 = 0$
13. $x^2 - y = 1$
14. $\frac{1}{2}y + 2x^2 = x$
15. $y^2 - x = 0$
16. $x + y^2 + 3 = 0$
17. $y + \sqrt{1 - x^2} = 0$
18. $\sqrt{1 - \frac{x^2}{4}} = \frac{y}{3}$
19. $\frac{x^2}{4} + \frac{y^2}{9} = 1$
20. $\frac{y^2}{2} - \frac{x^2}{3} = 1$
21. $|x|y = 1$
22. $|x - 1||y - 1| = 0$
23. $x|y| = 1$
24. $|xy| = 1$
25. $|x + y| = 0$
26. $|x - y| = 0$

Graph the following functions. Find the indicated function values.

Example 3
27. $f(x) = |x| - x; f(-2)$
28. $f(x) = x + |x|; f(-3)$
29. $f(x) = |x + 1| + x; f(-2)$
30. $f(x) = x + 1 - |x - 1|; f(-1)$

Example 4
31. $F(x) = \sqrt{4 - x^2}; F(0); F(F(0))$
32. $G(x) = -\sqrt{1 - x^2}; G(0); G(G(0))$

168 *Functions*

33. $H(x) = -2 + \sqrt{1 - (x - 1)^2}$; $H(2)$
34. $I(x) = 2 - \sqrt{4 - (x + 3)^2}$; $I(-1)$
35. $J(x) = 1 + \sqrt{x - 1}$; $J(3)$
36. $K(x) = 1 - \sqrt{-1 - x}$; $K(-1)$

The function $y = [x]$ is called the greatest integer function and is defined as the largest integer less than or equal to x (see graph).

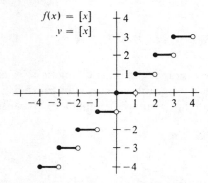

Graph the following functions and give the domain and range.

37. $y = [2x]$ 38. $y = [-2x + 1]$ 39. $y = |[x]|$ 40. $y = [|x|]$

5-3 SUM, PRODUCT, AND COMPOSITION OF FUNCTIONS

In this section we see how new functions can be formed from others by certain algebraic rules. By natural extensions of the algebraic operations we arrive at definitions for the sum, difference, product, and quotient of two functions. We also introduce here the important concept of the *composition* of two functions.

Let f and g be any two functions. We define new functions $f + g, f - g, fg,$ and f/g by the following formulas:

$$(f + g)(x) = f(x) + g(x),$$

$$(f - g)(x) = f(x) - g(x),$$

$$(fg)(x) = f(x)g(x),$$

$$(f/g)(x) = f(x)/g(x).$$

The domains of the new functions will consist of all the numbers x for which the formulas on the right make sense. The domains of $f + g$, $f - g$, and fg are all the numbers x for which both $f(x)$ and $g(x)$ are defined. The domain of f/g is the set of numbers x for which both $f(x)$ and $g(x)$ are defined, and $g(x) \neq 0$.

The sum and difference functions are easily graphed from the graphs of f and g by simply adding (or subtracting) the ordinates at each point.

Example 1 If $f(x) = x + 1$, $g(x) = \sqrt{4 - x^2}$, graph f, g, $f + g$, and $f - g$.

Solution. We first graph the linear function f and the circular function g on the same graph as shown in Fig. 5-12. The domain of f is the whole real line, but the domain of g consists only of the

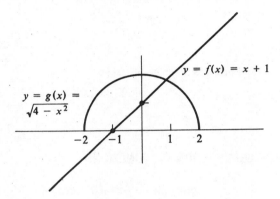

Figure 5-12

interval $[-2, 2]$. Therefore the functions $f + g$ and $f - g$ are only defined on $[-2, 2]$. We can obtain the graphs by adding or subtracting ordinates at selected points as shown in Figs. 5-13 and 5-14.

The product and quotient of two functions can also be sketched by using the individual graphs. However, the approximate numerical values of the ordinates at various points, as read from the graph, must be multiplied or divided to obtain the ordinates for the new functions fg and f/g.

Perhaps the most important idea involved in combining two functions to form a new function is *composition*. If f and g are two functions, then the *composition* of f and g, denoted $f \circ g$, is defined by the formula:

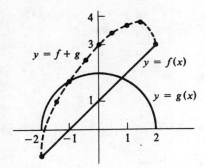

Figure 5-13

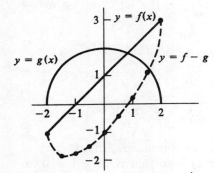

Figure 5-14

$$f \circ g(x) = f(g(x)).$$

The notation $f \circ g$ is read "f of g."

Consider the functions f and g, defined by $g(x) = x - 1$, $f(x) = x^2$. By definition, f of any number is that number squared; therefore, $f(g(x))$ is $g(x)$ squared:

$$f(g(x)) = [g(x)]^2.$$

Now replace $g(x)$ by its formula $x - 1$, and get

$$f \circ g(x) = f(g(x)) = [g(x)]^2 = (x - 1)^2.$$

The domain of $f \circ g$ is all real numbers x, since $g(x)$ is defined for all x, and f is defined for every number and hence every value of $g(x)$.

Example 2 Write formulas for $f \circ g(x)$ and $g \circ f(x)$ if $f(x) = 2x + 1$ and $g(x) = \sqrt{x}$. Describe the domains of $f \circ g$ and $g \circ f$.

Solution.

$$f \circ g(x) = f(g(x))$$
$$= 2g(x) + 1$$
$$= 2\sqrt{x} + 1.$$

Since $f(x)$ is defined for all x, $f(g(x))$ is defined whenever $g(x)$ is defined; therefore, the domain of $f \circ g$ is the set of non-negative numbers.

$$g \circ f(x) = g(f(x))$$
$$= \sqrt{f(x)}$$
$$= \sqrt{2x + 1}$$

Since g is only defined for non-negative numbers, the domain of $g \circ f$ is all numbers x such that $f(x) \geqslant 0$; that is $2x + 1 \geqslant 0$, or $x \geqslant -\frac{1}{2}$.

Example 3 Write a formula for $f \circ g(x)$ and $g \circ f(x)$ if $f(x) = \sqrt{4 - x^2}$ and $g(x) = x^2 - 3$. Describe the domains of $f \circ g$ and $g \circ f$.

Solution.

$$f \circ g(x) = f(g(x))$$
$$= \sqrt{4 - (g(x))^2}$$
$$= \sqrt{4 - (x^2 - 3)^2}$$

The domain of f is $[-2, 2]$; therefore the domain of $f \circ g$ consists of the numbers x such that $-2 \leqslant g(x) \leqslant 2$; i.e.,

$$-2 \leqslant x^2 - 3 \leqslant 2,$$

$$1 \leq x^2 \leq 5.$$

Hence the domain of $f \circ g$ is $[1, \sqrt{5}] \cup [-\sqrt{5}, -1]$.

$$g \circ f(x) = g(f(x))$$
$$= (f(x))^2 - 3$$
$$= (4 - x^2) - 3$$
$$= 1 - x^2$$

Since $g(x)$ is defined for all x, $g(f(x))$ is defined whenever $f(x)$ is defined. The domain of $g \circ f$ is therefore the same as the domain of f; i.e., the interval $[-2, 2]$. We have seen that the domain of $g \circ f$ is $[-2, 2]$, and that for $x \in [-2, 2]$, $g \circ f(x) = 1 - x^2$. Notice that $g \circ f(x)$ is not defined for all x even though the formula $1 - x^2$ is defined for all x.

Example 4 Let $s(x) = x^2$, $r(x) = \sqrt{x}$, $f(x) = x + 5$, and $t(x) = 2x$. Express each of the following formulas as compositions of functions chosen from $s, r, f,$ and t:

(a) $2x^2$; (b) $\sqrt{x + 5}$; (c) $4x^2$; (d) $2\sqrt{x} + 5$; (e) $\sqrt{2x + 5}$

Solution

(a) $2x^2 = 2s(x) = t(s(x))$

(b) $\sqrt{x + 5} = \sqrt{f(x)} = r(f(x))$

(c) $4x^2 = (2x)^2 = (t(x))^2 = s(t(x))$

(d) $2\sqrt{x} + 5 = 2r(x) + 5 = t(r(x)) + 5 = f(t(r(x)))$

(e) $\sqrt{2x + 5} = \sqrt{t(x) + 5}$ $\sqrt{f(t(x))} = r(f(t(x)))$

SECTION 5-3, REVIEW QUESTIONS

1. The function $f + g$ is defined by _____ the values of the two functions at each point: $(f + g)(x) = f(x) + g(x)$.

adding

Sum, Product, and Composition of Functions

0   2. The value of $(f/g)(x)$ is $f(x)/g(x)$ at all points where both $f(x)$ and $g(x)$ are defined and $g(x) \neq$ _____.

domains   3. The domain of fg consists of all points that are in the _____ of both f and g.

composition   4. The function $f \circ g$ defined by $f \circ g\,(x) = f(g(x))$ is called the _____ of f and g.

5. The composition $f \circ g$ is defined for all x in the domain of g such that f is

g(x)   also defined at _____.

6. If $g(x)$ is defined and f is defined at $g(x)$, then $f(g(x))$ is defined, so x is in

f ∘ g   the domain of _____.

SECTION 5-3, EXERCISES

Graph f, g, $f + g$, and $f - g$ on the same coordinate system for each of the following pairs of functions.

Example 1
1. $f(x) = x; g(x) = x^2$
2. $f(x) = x - 1; g(x) = \frac{1}{2}x^2$
3. $f(x) = \sqrt{1 - x^2}; g(x) = x$
4. $f(x) = -x; g(x) = \sqrt{4 - x^2}$
5. $f(x) = \sqrt{1 - x^2}; g(x) = x^2$
6. $f(x) = 1 - x^2; g(x) = -\sqrt{1 - x^2}$

Write a single formula for $f + g, f - g, fg$, and f/g.

7. $f(x) = x^2; g(x) = x$
8. $f(x) = 1 + x; g(x) = \sqrt{x}$
9. $f(x) = 1 + x; g(x) = 1 - x$
10. $f(x) = x^2; g(x) = \sqrt{x}$
11. $f(x) = 1 - x^2; g(x) = 1 + x$
12. $f(x) = 1 + 2x; g(x) = x^3$

Example 2, 3   Write a single formula for $f \circ g\,(x)$ and give the domain of $f \circ g$.

13. $f(x) = \sqrt{x}; g(x) = x - 1$
14. $f(x) = x^2; g(x) = x + 2$
15. $f(x) = x^2; g(x) = \sqrt{x}$
16. $f(x) = \sqrt{x}; g(x) = x^2$
17. $f(x) = 2x + 1; g(x) = 2x - 1$
18. $f(x) = 2x + 3; g(x) = 3x + 5$
19. $f(x) = \sqrt{x - 1}; g(x) = x + 1$
20. $f(x) = \sqrt{2x}; g(x) = \frac{1}{2}x^2$
21. $f(x) = 1 - x^2; g(x) = \sqrt{1 - x^2}$

22. $f(x) = 2 + x^2 ; g(x) = \sqrt{4 - x^2}$
23. $f(x) = \sqrt{9 - x^2} ; g(x) = \sqrt{9 - x^2}$
24. $f(x) = 1 - x^2 ; g(x) = \sqrt{1 - x}$

Let $t(x) = 3x$, $m(x) = x - 2$, $r(x) = \sqrt{x}$, and $s(x) = x^2$. Express each of the following formulas as compositions of functions chosen from t, m, r, and s.

Example 4

25. $3\sqrt{x}$ 26. $\sqrt{x - 2}$
27. x^4 28. $x - 4$
29. $3x - 6$ 30. $(x - 2)^2$
31. $(3x)^2$ 32. $\sqrt{3x}$
33. $x^2 - 2$ 34. $3x - 2$

Let $A(x) = x + 1$, $D(x) = 2x$, and $K(x) = x^3$, $S(x) = x - 3$. Express each of the following formulas as compositions of functions chosen from A, D, K, and S.

Example 4

35. $2x + 2$ 36. $2x^3$
37. $2x + 1$ 38. $x^3 + 1$
39. $(x + 1)^3$ 40. $8x^3$
41. $x - 2$ 42. $2x - 3$
43. $x + 2$ 44. $x - 6$
45. $4x$ 46. x^9

5-4 POLYNOMIAL FUNCTIONS

A *polynomial* function is one of the form

$$f(x) = a_n x^n + a_{n-1} x^{n-1} + \cdots + a_1 x + a_0$$

If $a_n \neq 0$, then $f(x)$ is an *nth-degree* polynomial. We have already studied first-degree polynomials (linear functions), and second-degree polynomials (quadratic functions). Recall that the graph of the linear function $f(x) = mx + b$ is the line through $(0, b)$ with slope m. The graph of the quadratic function $f(x) = ax^2 + bx + c$ is a parabola symmetric about the line $x = -(b/2a)$.

One of the basic facts we need is the following *division theorem* for polynomials: If $A(x)$ and $B(x)$ are any two polynomials, then there are unique polynomials $Q(x)$ and $R(x)$, with the degree of $R(x)$ less than the degree of $B(x)$, such that

$$A(x) = B(x)Q(x) + R(x) \qquad (1)$$

For values of x such that $B(x) \neq 0$, Equation (1) is equivalent to

$$\frac{A(x)}{B(x)} = Q(x) + \frac{R(x)}{B(x)} \qquad (2)$$

The polynomial $Q(x)$ is the *quotient* of $A(x)$ and $B(x)$, and $R(x)$ is the *remainder*. The quotient and remainder can be found by the familiar division algorithm as illustrated in Example 1.

Example 1 Divide $A(x) = 2x^4 - 8x^3 + 5x^2 - 10x + 2$ by $B(x) = 2x^2 + 3$; that is, find the quotient $Q(x)$ and remainder $R(x)$ satisfying Equation (1).

Solution

$$\begin{array}{r}
x^2 - 4x + 1 \\
2x^2 + 3 \overline{\smash{)}2x^4 - 8x^3 + 5x^2 - 10x + 2} \\
\underline{2x^4 + 3x^2 } \\
-8x^3 + 2x^2 - 10x + 2 \\
\underline{-8x^3 - 12x } \\
2x^2 + 2x + 2 \\
\underline{2x^2 + 3} \\
2x - 1
\end{array}$$

Hence $Q(x) = x^2 - 4x + 1$, and $R(x) = 2x - 1$. The identity (1) in this case would read

$$2x^4 - 8x^3 + 5x^2 - 10x + 2 = (2x^2 + 3)(x^2 - 4x + 1) + (2x - 1)$$

The most important case of the division theorem for us is when $B(x)$ is a first degree polynomial of the form $x - c$. Then the degree of the remainder must be zero, which means that the remainder is a constant r:

$$A(x) = (x - c)Q(x) + r. \qquad (3)$$

From Equation (3) we immediately see that if $A(c) = 0$, then $r = 0$ and $(x - c)$ divides $A(x)$. The converse is also true: if $x - c$ divides $A(x)$ (i.e., $r = 0$), then $A(c) = 0$. Hence we have the following important criterion for divisibility:

$x - c$ divides $A(x)$ if and only if $A(c) = 0$.

Example 2 Factor the polynomial $A(x) = x^3 + 3x^2 - 13x - 15$.

Solution. We compute some values of $A(x)$ to attempt to find by trial a value of c for which $A(c) = 0$:

$$A(0) = 0 + 0 + 0 - 15 = -15,$$
$$A(1) = 1 + 3 - 13 - 15 = -24,$$
$$A(-1) = -1 + 3 + 13 - 15 = 0.$$

Since $A(-1) = 0$, we know that $x - (-1)$ divides $A(x)$.

$$\begin{array}{r}
x^2 + 2x - 15 \\
x+1 \overline{\smash{\big)} x^3 + 3x^2 - 13x - 15} \\
\underline{x^3 + x^2 } \\
2x^2 - 13x - 15 \\
\underline{2x^2 + 2x } \\
-15x - 15 \\
\underline{-15x - 15}
\end{array}$$

Hence we have the factorization

$$x^3 + 3x^2 - 13x - 15 = (x + 1)(x^2 + 2x - 15)$$

We can find the roots -5 and 3 of the quadratic equation

$$x^2 + 2x - 15 = 0$$

by the quadratic formula. This gives us the linear factors of $x^2 + 2x - 15$:

$$x^2 + 2x - 15 = (x + 5)(x - 3).$$

We thus have the final factorization.

$$x^3 + 3x^2 - 13x - 5 = (x + 1)(x + 5)(x - 3).$$

Some quadratic functions, like $x^2 + 1$, cannot be factored into real linear factors. However, every polynomial can be written as a product of real linear factors and real quadratic factors. We will now illustrate the way in which we can use the factored form to graph a polynomial function.

Example 3 Graph $f(x) = x^3 - 2x^2 - 5x + 6$.

Solution. By trial we find that $f(1) = 0$, so $(x - 1)$ is a factor.

Division gives us the other (quadratic) factor, which can then be factored into linear factors:

$$f(x) = (x - 1)(x^2 - x - 6)$$
$$= (x - 1)(x - 3)(x + 2)$$

The graph of f intersects the x-axis at the three points $(-2, 0)$, $(1, 0)$, $(3, 0)$. One factor in the product changes sign as x passes through each of the numbers $-2, 1, 3$, and consequently $f(x)$ changes sign as x passes through these three values. For $x > 3$, all three factors are positive, and $f(x)$ gets large as x gets large. We will express this fact with the notation

$$\lim_{x \to +\infty} f(x) = \lim_{x \to +\infty} (x^3 - 2x^2 - 5x + 6) = +\infty$$

(read, the limit as x approaches plus infinity is plus infinity). For $x < -2$, all three factors are negative, and hence $f(x) < 0$. For numerically large negative values of x, $f(x)$ is a numerically large negative number, which we express as follows:

$$\lim_{x \to -\infty} f(x) = -\infty$$

The graph of f is shown in Fig. 5-15.

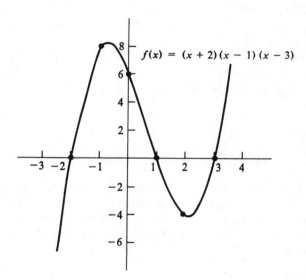

Figure 5-15

Example 4 Graph $f(x) = -x^3 + 2x^2 - x$.

Solution. The polynomial has the factorization

$$f(x) = -x(x^2 - 2x + 1) = -x(x - 1)^2$$

Hence $f(x) = 0$ for $x = 0$ and $x = 1$. The function changes sign as x passes through 0, but nowhere else. For $x > 0$, $f(x) \leq 0$, and $\lim_{x \to +\infty} f(x) = -\infty$; for $x < 0$, $f(x) > 0$ and $\lim_{x \to -\infty} f(x) = +\infty$. See Fig. 5-16.

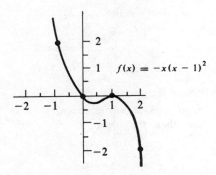

Figure 5-16

As illustrated in the last two examples, the behavior of any polynomial for numerically large values of x is the same as that of the highest degree term. We can thus find $\lim_{x \to +\infty} f(x)$ and $\lim_{x \to -\infty} f(x)$ even though we do not have the factors of $f(x)$.

Example 5 Find $\lim_{x \to +\infty} (-3x^5 + x^3 - 2x + 1)$ and $\lim_{x \to -\infty} (-3x^5 + x^3 - 2x + 1)$.

Solution. Write the formula as follows:

$$-3x^5 + x^3 - 2x + 1 = x^5\left(-3 + \frac{1}{x^2} - \frac{2}{x^4} + \frac{1}{x^5}\right)$$

For numerically large values of x, $1/x^2$, $-2/x^4$, and $1/x^5$ are small, and the quantity in the parentheses is approximately -3. Hence for numerically large values of x, the polynomial behaves like $-3x^5$, and we have

$$\lim_{x \to +\infty} (-3x^5 + x^3 - 2x + 1) = -\infty;$$
$$\lim_{x \to -\infty} (-3x^5 + x^3 - 2x + 1) = +\infty$$

SECTION 5-4, REVIEW QUESTIONS

1. A function of the form $f(x) = a_n x^n + a_{n-1} x^{n-1} + \cdots + a_1 x + a_0$ is a _____.

   polynomial

2. If $a_n \neq 0$ in the formula $a_n x^n + \cdots + a_0$, then n is the _____ of the polynomial.

   degree

3. First-degree polynomials are called _____ functions, and second-degree polynomials are called _____ functions.

   linear
   quadratic

4. The graph of a quadratic function is a _____.

   parabola

5. For any polynomials $A(x)$, $B(x)$, there are unique polynomials $Q(x)$, $R(x)$ such that the degree of $R(x)$ is less than the degree of $B(x)$, and
   $A(x) = B(x)Q(x) +$ _____.

   R(x)

6. $Q(x)$ is called the _____, and $R(x)$ is the *remainder*.

   quotient
   remainder

7. The degree of the _____ is less than the degree of divisor.

8. If $A(x)$ is a polynomial and $A(c) = 0$, then $A(x)$ has as a factor the linear function _____.

   x − c

9. If we find the linear factor $x - c$, we get the other factor of $A(x)$ by _____ $A(x)$ by $x - c$.

   dividing

10. For numerically large values of x, the polynomial $a_n x^n + \cdots + a_0$ behaves like the term _____.

    $a_n x^n$

11. If $a_n > 0$, then $\lim\limits_{x \to +\infty} a_n x^n =$ _____.

    + ∞

SECTION 5-4, EXERCISES

Perform the indicated division and give the quotient and remainder.

Example 1

1. $(3x^5 + 5x^3 + x^2 + 5x + 1) \div (x^2 + 1)$
2. $(x^4 - 5x^3 + x - 1) \div (x^2 + 2)$

3. $(2x^4 - 2x^3 + 3x^2 + 3x - 5) \div (2x^2 - 1)$
4. $(x^5 + 2x^3 + 3) \div (x^3 + 2x)$
5. $(3x^3 + 4x^2 - 10x - 15) \div (x - 2)$
6. $(2x^4 + 5x^3 - 2x^2) \div (x + 3)$

Factor the following polynomials.

7. $x^3 + 6x^2 + 8x$
9. $x^3 - 6x^2 + 11x - 6$
11. $x^3 - x^2 + x - 1$
13. $x^2 + x + 1$
15. $x^3 + 2x^2 + 5x$

8. $x^3 - 4x^2 - 21x$ Example 2
10. $x^3 - 4x^2 + x + 6$
12. $2x^3 + 4x^2 + x + 2$
14. $x^2 - 3x + 3$
16. $x^3 - 3x^2 + 7x$

Graph the following polynomial functions.

17. $f(x) = x^3 - 2x^2 - x + 2$
19. $f(x) = x^3 - x$
21. $f(x) = x^3 + 2x^2$
23. $f(x) = x^3 + 2x^2 + x$

18. $f(x) = x^3 + 3x^2 - x - 3$ Example 3, 4
20. $f(x) = x^3 - x^2 - 2x$
22. $f(x) = x^3 - x^2$
24. $f(x) = -x^3 + 6x^2 - 9x$

Find $\lim_{x \to +\infty} f(x)$ and $\lim_{x \to -\infty} f(x)$ for each of the following functions f.

25. $f(x) = x^3 - 10x^2 - 100x - 1000$ Example 5
26. $f(x) = 3x^4 - 4x^3 + 5x^2 - 6x + 7$
27. $f(x) = -3x^3 - x^2 + 19x - 3$
28. $f(x) = -x^4 - 2x^3 + x^2 + 5x - 3$
29. $f(x) = x^2 - x^4 + x - x^3$
30. $f(x) = 2 - x^5 + 6x^3 - 3x$

5-5 RATIONAL FUNCTIONS

Apart from polynomials the simplest algebraic functions are the *rational functions*. A rational function is simply a quotient of two polynomials, $A(x)/B(x)$. The polynomials themselves are a special type of rational function, since we can take $B(x) = 1$ for the denominator. We will usually assume that $A(x)$ and $B(x)$ have no common factors—otherwise we could divide them out.

The domain of the rational function $R(x) = A(x)/B(x)$ is the set of numbers x such that $B(x) \neq 0$. Hence the domain of any rational function consists of the whole real line, except possibly for a finite number of points.

We consider the graph of $R(x) = (x + 1)/(x - 1)$ to get some idea of the possible behavior of a rational function. The *numerator* is zero at -1, and changes sign as x passes through -1. Hence $R(-1) = 0$, and $R(x)$ change sign as x passes through -1. The *denominator* is zero at 1 and changes sign there. For values of x close to 1, the denominator is numerically small, and the numerator is near 2. Hence $|R(x)|$ is large if x is near 1. As x approaches 1 from the right, both numerator and denominator are positive so $R(x)$ approaches $+\infty$; we write this as follows:

$$\lim_{x \to 1+} \frac{x+1}{x-1} = +\infty$$

(read, the limit as x approaches 1 from the right is $+\infty$). As x approaches 1 from the left, the numerator and denominator have different signs, and

$$\lim_{x \to 1-} \frac{x+1}{x-1} = -\infty.$$

The vertical line $x = 1$ through the one point not in the domain of R is called a *vertical asymptote*. The behavior of $R(x)$ for numerically large values of x can be seen by writing

$$R(x) = \frac{x+1}{x-1} = \frac{1 + 1/x}{1 - 1/x}.$$

If $|x|$ is large, $R(x)$ is nearly 1, and hence

$$\lim_{x \to +\infty} \frac{x+1}{x-1} = 1; \quad \lim_{x \to -\infty} \frac{x+1}{x-1} = 1.$$

The horizontal line $y = 1$ is called a *horizontal asymptote*. See Fig. 5-17. Here the existence of a *horizontal asymptote* depends on the fact that *the degree of the denominator is greater than or equal to the degree of the numerator. If the degree of the denominator is strictly larger, the x-axis will be the horizontal asymptote.*

Example 1 Graph $R(x) = \dfrac{4x}{1 + x^2}$.

Solution. The denominator is never zero, so the domain of R is the whole line, and there is no vertical asymptote. The function is zero at $x = 0$ and changes sign as x passes through 0. For $x \neq 0$, we write

$$R(x) = \frac{4x}{1 + x^2} = \frac{4/x}{1/x^2 + 1}.$$

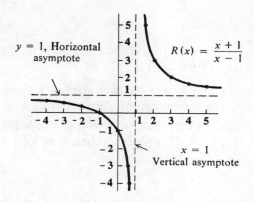

Figure 5-17

If $|x|$ is large, the denominator is close to one, and the numerator is close to zero; hence

$$\lim_{x \to +\infty} \frac{4x}{1+x^2} = 0; \quad \lim_{x \to -\infty} \frac{x^4}{1+x^2} = 0.$$

See Fig. 5-18.

If the numerator has larger degree than the denominator in a rational function $R(x)$, then $|R(x)|$ will become large as $|x|$ gets large.

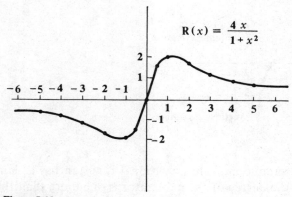

Figure 5-18

Example 2 Graph $R(x) = \dfrac{x^2 - 1}{2x - 4}$.

Solution. First write numerator and denominator in factored form, and cancel common factors if there are any:

Rational Functions 183

$$R(x) = \frac{x^2 - 1}{2x - 4}$$

$$= \frac{(x + 1)(x - 1)}{2(x - 2)}$$

The numerator is zero at -1 and 1, and changes sign as x passes through each of these points. The denominator is zero at 2, so the line $x = 2$ is a vertical asymptote. To see what happens for large values of $|x|$, we divide $x^2 - 1$ by $2x - 4$. This gives a quotient of $\frac{1}{2}x + 1$, and a remainder of 3; hence

$$R(x) = \frac{x^2 - 1}{2x - 4} = \frac{1}{2}x + 1 + \frac{3}{2x - 4}.$$

In this form it is clear that as $|x|$ gets large, $R(x)$ becomes close to the linear function $f(x) = \frac{1}{2}x + 1$. Therefore $y = \frac{1}{2}x + 1$ is a slant asymptote. Figure 5-19 shows the graph of $R(x)$, along with the two asymptotic lines $x = 2$ and $y = \frac{1}{2}x + 1$.

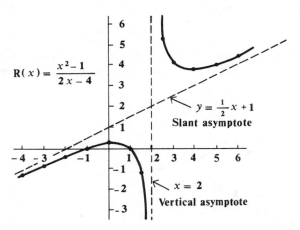

Figure 5-19

As the above example indicates, a rational function has a slant asymptote when the degree of the numerator is one more than the degree of the denominator. In this case the quotient is a linear polynomial whose graph is the asymptotic line.

Example 3 Graph $R(x) = \dfrac{2x^2 + 1}{x^2 - 1}$

Solution. In factored form we have $R(x) = \dfrac{2x^2 + 1}{(x+1)(x-1)}$. Hence there are vertical asymptotes at $x = 1$ and $x = -1$. For $-1 < x < 1$, the denominator is negative, the numerator positive, and

$$\lim_{x \to 1-} R(x) = \lim_{x \to -1+} R(x) = -\infty.$$

For $x > 1$ and $x < -1$, $R(x) > 0$ and we have

$$\lim_{x \to 1+} R(x) = \lim_{x \to -1-} R(x) = +\infty.$$

To see what happens for large values of $|x|$ we write

$$R(x) = \frac{2x^2 + 1}{x^2 - 1} = \frac{2 + 1/x^2}{1 - 1/x^2}.$$

Hence $\lim\limits_{x \to +\infty} R(x) = \lim\limits_{x \to -\infty} R(x) = 2$, and $y = 2$ is a horizontal asymptote. Notice finally that in $R(x)$, x occurs only to even powers, so $R(x)$ is a symmetric about the y-axis. (See Fig. 5-20).

We can summarize the behavior of $R(x)$ for large values of $|x|$ by looking at the general form

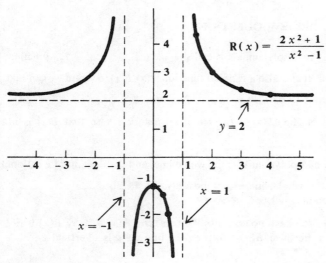

Figure 5-20

Rational Functions

$$R(x) = \frac{A(x)}{B(x)} = \frac{a_n x^n + a_{m-1} x^{n-1} + \ldots + a_0}{b_m x^m + b_{m-1} x^{m-1} + \ldots + b_0}.$$

In general,

$$\lim_{x \to +\infty} R(x) = \lim_{x \to +\infty} \frac{a_n x^n}{b_m x^m}, \text{ and } \lim_{x \to -\infty} R(x) = \lim_{x \to -\infty} \frac{a_n x^n}{b_m x^m}.$$

If $m = n$, then,

$$\lim_{x \to +\infty} R(x) = \lim_{x \to -\infty} R(x) = \frac{a_n}{b_n}.$$

If $m > n$, then

$$\lim_{x \to +\infty} R(x) = \lim_{x \to -\infty} R(x) = 0.$$

If $n = m + 1$, there is a slant asymptote whose slope is a_n / b_m.

SECTION 5-5, REVIEW QUESTIONS

rational

1. A quotient of two polynomials is called a _____ function.

2. A polynomial $A(x)$ is also a rational function $A(x)/B(x)$, because we can take

1

the denominator to be $B(x) = $ _____ .

3. The domain of $A(x)/B(x)$ is the set of all numbers x so that $B(x)$ is not

zero

_____ .

4. The domain of a polynomial is the whole line, and the domain of a rational

finite

function is the whole line except possibly for a _____ number of points.

5. If $A(x)$ and $B(x)$ have no common factors, and $R(x) = A(x)/B(x)$ is not defined at a because $B(a) = 0$, then the line $x = a$ is a vertical

asymptote

_____ .

186 *Functions*

6. A rational function has a vertical asymptote through each point not in its _____.   **domain**

7. A rational function will have a horizontal asymptote if the degree of the _____ is larger than or equal to the degree of the _____.   **denominator**   **numerator**

8. If the degrees of numerator and denominator are the same, the horizontal asymptote will be different from the _____-axis.   **x**

9. If the degree of the denominator of $R(x)$ is strictly larger than that of the numerator, the x-axis is the horizontal asymptote, and $\lim\limits_{x \to +\infty} R(x) = $ _____.   **0**

10. If the degree of the numerator is one more than the degree of the denominator, the quotient is a _____ polynomial.   **linear**

11. If the quotient is a linear polynomial, then its graph is a _____ asymptote of the rational function.   **slant**

12. If $R(x) = (ax^{n+1} + \ldots + a_0)/(bx^n + \ldots + b_0)$ then $R(x)$ has a slant asymptote whose slope is _____.   **a/b**

SECTION 5-5, EXERCISES

Find all asymptotes and graph the following rational functions.

1. $R(x) = \dfrac{1}{x}$ 2. $R(x) = \dfrac{1}{x-1}$ **Example 1, 3**

3. $R(x) = \dfrac{2}{x+2}$ 4. $R(x) = \dfrac{-1}{x+3}$

5. $R(x) = \dfrac{x-1}{x+1}$ 6. $R(x) = \dfrac{x+3}{x+2}$

7. $R(x) = \dfrac{3x+1}{2x-1}$ 8. $R(x) = \dfrac{-x+2}{2x+3}$

9. $R(x) = \dfrac{-2x+1}{2x-5}$ 10. $R(x) = \dfrac{2x+4}{3x-1}$

11. $R(x) = \dfrac{2x}{1+x^2}$ 12. $R(x) = \dfrac{-3x}{x^2+2}$

13. $R(x) = \dfrac{2-x}{x^2-1}$ 14. $R(x) = \dfrac{3+x}{x^2-4}$

15. $R(x) = \dfrac{3x^2+x}{2x}$ 16. $R(x) = \dfrac{-x^2+2x}{x}$ **Example 2**

17. $R(x) = \dfrac{2x^2 + x}{x - 1}$ 18. $R(x) = \dfrac{-4x^2 + 4x}{2x - 3}$

19. $R(x) = \dfrac{x^2 + 1}{x - 1}$ 20. $R(x) = \dfrac{2x^2 - 1}{x + 1}$

Example 3 21. $R(x) = \dfrac{x^2 + 1}{x^2 - 1}$ 22. $R(x) = \dfrac{2x^2 + 1}{x^2 - 4}$

23. $R(x) = \dfrac{2x^2 + 3}{3x^2 + 2}$ 24. $R(x) = \dfrac{4x^2 + 1}{2x^2 + 3}$

25. $R(x) = \dfrac{x^2 + 1}{2x^2 + 1}$ 26. $R(x) = \dfrac{2x^2 + x}{x^2 + 3}$

27. $R(x) = \dfrac{3x^2 + x}{2x^2 + x + 1}$ 28. $R(x) = \dfrac{x^2 - 2x}{x^2 - x + 2}$

5-6 IMPLICIT FUNCTIONS

We have seen that the graph of any function is also the graph of an equation in x and y. For example, the graph of the function f defined by $f(x) = x^2 + 1/x$ is the same as the graph of the equation $y = x^2 + 1/x$. On the other hand, we know that there are equations whose graphs are not the graphs of functions. For example the graph of

$$x^2 + y^2 = 1$$

is a circle of radius 1, centered at the origin. This is the graph of a relation that is not a function because there are two values of y associated with each value of x in the interval $(-1, 1)$.

Although the equation $x^2 + y^2 = 1$ does not define a function, we can divide up the graph of this equation into two pieces, each of which is the graph of a function. The top half of the circle, which is the graph of $y = \sqrt{1 - x^2}$, represents a function. Similarly, the bottom half of the circle, which is the graph of $y = -\sqrt{1 - x^2}$, also represents a function (see Fig. 5-21).

If we define two functions y_1, y_2 on $[-1, 1]$ by

$$y_1(x) = \sqrt{1 - x^2}, \qquad y_2(x) = -\sqrt{1 - x^2},$$

then we have

$$x^2 + (y_1(x))^2 \equiv 1, \qquad x^2 + (y_2(x))^2 \equiv 1.$$

In other words, both y_1 and y_2 are functions that yield an identity when substituted for y in the equation

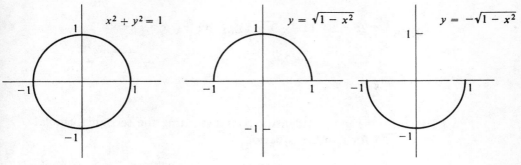

Figure 5-21

$$x^2 + y^2 = 1.$$

Now consider the general question: When does a polynomial equation in x and y have one or more functions $y(x)$ that will satisfy the equation identically on some interval. A general polynomial equation in x and y can be written

$$P_n(x)y^n + P_{n-1}(x)y^{n-1} + \cdots + P_1(x)y + P_0(x) = 0 \quad (1)$$

where $P_0, P_1, \ldots, P_n$ are polynomials in x. For any fixed value of x, Equation (1) is an nth-degree polynomial equation in y. Such an equation has at most n roots: $y_1, y_2, \ldots$. The values of these roots, however, depend on the coefficients $P_0(x), \ldots, P_n(x)$, and so they depend on x. Thus we would expect that there would be one or more functions $y = y_1(x)$, $y = y_2(x), \ldots$, that satisfy Equation (1) identically on appropriate intervals.

Let's say that any function $y = y(x)$ that satisfies (1) identically on some interval is *defined implicitly* by the equation. This nomenclature is meant to suggest that Equation (1) does determine y as a function of x even though we may not be able to solve the equation for y. That is, y is a function of x even if we cannot write y *explicitly* in terms of x.

Example 1 Show that the functions $y = 1 + \sqrt{x+1}$ and $y = 1 - \sqrt{x+1}$ are defined implicitly by the equation $y^2 - 2y - x = 0$.

Solution. We substitute $1 + \sqrt{x+1}$ for y in the equation and check that an identity results.

$$y^2 - 2y - x = (1 + \sqrt{x+1})^2 - 2(1 + \sqrt{x+1}) - x$$
$$= 1 + 2\sqrt{x+1} + x + 1 - 2 - 2\sqrt{x+1} - x$$
$$\equiv 0.$$

Similarly, we get an identity by substituting the second formula $1 - \sqrt{x+1}$ for y in the equation.

$$y^2 - 2y - x = (1 - \sqrt{x+1})^2 - 2(1 - \sqrt{x+1}) - x$$
$$= 1 - 2\sqrt{x+1} + x + 1 - 2 + 2\sqrt{x+1} - x$$
$$\equiv 0.$$

The two functions $y = 1 + \sqrt{x+1}$ and $y = 1 - \sqrt{x+1}$ are respectively the top and bottom halves of the parabola which is the graph of $y^2 - 2y - x = 0$ (Fig. 5-22).

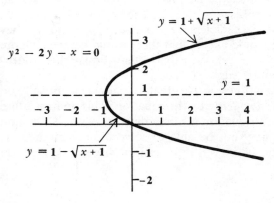

Figure 5-22

Example 2 Find explicit formulas for two functions defined implicitly by the equation $y^2 - 2y + x^2 - 3 = 0$.

Solution. The equation is quadratic in y, so we can use the quadratic formula (with the coefficients $a = 1$, $b = -2$, $c = x^2 - 3$) to solve for y in terms of x.

$$y = \frac{-(-2) \pm \sqrt{4 - 4(1)(x^2 - 3)}}{2(1)}$$

$$= 1 \pm \sqrt{1 - x^2 + 3}$$

$$= 1 \pm \sqrt{4 - x^2}$$

Hence $y = 1 + \sqrt{4 - x^2}$ and $y = 1 - \sqrt{4 - x^2}$ are defined implicitly by the equation. The graphs of these functions are respectively the top and bottom halves of the circle $y^2 - 2y + x^2 - 3 = 0$ (Fig. 5-23).

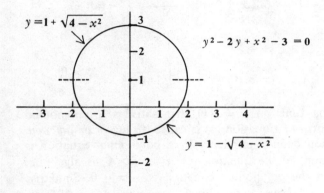

Figure 5-23

Example 3 Graph the equation $(1 - y - yx^2)(y - x^2 - 1) = 0$ and write explicit formulas for some of the functions defined implicitly by the equations.

Solution. A product is zero if and only if one of the factors is zero. Therefore the graph is the union of the graphs of

$$1 - y - yx^2 = 0 \quad \text{and} \quad y - x^2 - 1 = 0.$$

The equations can be solved for y and are equivalent to

$$y = \frac{1}{1 + x^2} \quad \text{and} \quad y = x^2 + 1.$$

The graph is shown in Fig. 5-24. Notice that even though the equation has degree 2 as a polynomial in y, there are four continuous functions defined implicitly by the equation.

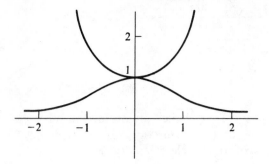

Figure 5-24

$$y = x^2 + 1 \quad \text{(all } x\text{)} \qquad y = \frac{1}{1 + x^2} \quad \text{(all } x\text{)}$$

$$y = \begin{cases} \dfrac{1}{1 + x^2} & (x \leq 0) \\ x^2 + 1 & (x > 0) \end{cases} \qquad y = \begin{cases} x^2 + 1 & x \leq 0 \\ \dfrac{1}{1 + x^2} & x > 0 \end{cases}$$

Any continuous function $y = y(x)$ that satisfies a polynomial equation of the form of Equation (1) is called an *algebraic function*. Hence any function defined implicitly by a polynomial equation is algebraic. For example, the function $y = \sqrt{x}$, $x \geq 0$, is algebraic, because it satisfies the polynomial equation $y^2 - x = 0$. Similarly, $y = \sqrt[3]{x}$ is algebraic because it satisfies the equation $y^3 - x = 0$.

Example 4 Show that $y = \sqrt{1 - 1/x} + 2$ is an algebraic function.

Solution. To show that y satisfies a polynomial equation, we must first eliminate the radical by squaring:

$$y = \sqrt{1 - \frac{1}{x}} + 2,$$

$$y - 2 = \sqrt{1 - \frac{1}{x}},$$

$$(y - 2)^2 = 1 - \frac{1}{x},$$

$$x(y - 2)^2 = x - 1,$$

$$x(y - 2)^2 - x + 1 = 0.$$

SECTION 5-6, REVIEW QUESTIONS

1. The graph of an equation in x and y always represents a relation, but may not represent a _____.   function
2. The graph of a function f is the same as the graph of the equation $y = $ _____.   f(x)
3. Any function $y = y(x)$ which identically satisfies an equation in x and y is said to be defined _____ by the equation.   implicitly
4. An equation in x and y may define one or more functions implicitly even if you cannot solve for y in terms of _____.   x
5. If $y = y(x)$ is a function which satisfies a polynomial equation, then y is an _____ function.   algebraic
6. To show that a function y is algebraic you show that it satisfies a polynomial equation _____.   identically
7. If you have a formula for y in terms of x, then y is defined _____.   explicitly

SECTION 5-6, EXERCISES

Show that the function y is defined implicitly by the given equation.

1. $y^2 + x - 2 = 0; y = -\sqrt{2 - x}$ Example 1
2. $x - y^2 + 5 = 0; y = \sqrt{x + 5}$
3. $y^2 - 2y - x + 1 = 0; y = 1 + \sqrt{x}$
4. $y^2 - 4y - x + 3 = 0; y = \sqrt{1 + x} + 2$
5. $y^2 + 2xy - x^4 + x^2 = 0; y = x^2 - x$
6. $y^2 + x^3 y - 1 + x^3 = 0; y = 1 - x^3$
7. $x^2(y^2 - 1) - 2xy + 1 = 0; y = 1 + \dfrac{1}{x}$
8. $xy - \dfrac{1}{(y - x)^2} - 1 = 0; y = x + \dfrac{1}{x}$

Find explicit formulas for one or more functions which are defined implicitly by the following equations.

9. $xy - 1 = 0$ 10. $xy + y = 1$ Example 2
11. $x^2 y + 2y - 4 = 0$ 12. $y + 2x - xy = 0$
13. $\dfrac{x}{y} + \dfrac{2}{x} - 1 = 0$ 14. $x^2 + \dfrac{x}{y} - \dfrac{3}{y} = 0$
15. $xy^2 + x^2 - 1 = 0$ 16. $3y^2 + x^3 - 2 = 0$
17. $y^2 + 2y + x = 0$ 18. $y^2 - x^2 - 2y = 0$

19. $\dfrac{x^2}{4} + \dfrac{y^2}{9} - 1 = 0$ 20. $\dfrac{x^2}{16} + \dfrac{y^2}{25} - 1 = 0$

21. $\dfrac{x^2}{4} - \dfrac{y^2}{9} - 1 = 0$ 22. $\dfrac{x^2}{9} - \dfrac{y^2}{25} - 1 = 0$

Graph each of the following, and find explicit formulas for some of the functions defined implicitly by the equations.

Example 3

23. $(y - 3x)(y + 2x - 1) = 0$ 24. $(y + x)(2y + x + 3) = 0$

25. $y - \dfrac{x^2}{y} = 0$ 26. $y - \dfrac{x^4}{y} = 0$

27. $y^2 + y - xy = 0$ 28. $2xy + y^2 - 3y = 0$
29. $y^2 - yx^2 = 0$ 30. $x^2 y + y(y - 1)^2 - y = 0$
31. $(y + 1)(y + x^2 + 1) = 0$ 32. $(y - x^2 - 1)(x^2 + y^2 - 1) = 0$
33. $y^2 - 3xy + 2x^2 = 0$ 34. $y^2 + xy - 6x^2 = 0$

Show that each of the following is an algebraic function by giving a polynomial equation which it satisfies.

Example 4

35. $y = \sqrt{x + 1}$ 36. $y = 1 - \sqrt{x}$

37. $y = x + \dfrac{1}{x}$ 38. $y = 1 + \dfrac{1}{x - 1}$

39. $y = \sqrt{x + \dfrac{1}{x}}$ 40. $y = 1 - \dfrac{1}{\sqrt{x}}$

41. $y = \sqrt[3]{x^2 + x}$ 42. $y = \sqrt[5]{1 - x^2}$

5-7 INVERSE FUNCTIONS

Recall that a function is a relation that associates exactly one element of the range with each element of the domain. A function may, of course, associate the same range element with more than one domain element. For example, if $f(x) = x^2$, then $f(-2) = 4$ and $f(2) = 4$. Both pairs $(-2, 4), (2, 4)$ are in the function. The same range element, 4, is associated with 2 and -2.

Those functions that pair only one domain element with each range element are called *one-to-one functions*. That is, f is a one-to-one function if for each y in the range, there is exactly one x in the domain such that (x, y) is in f. Another way of saying the same thing is that f is one-to-one provided $f(x_1) \neq f(x_2)$ when $x_1 \neq x_2$.

A function is called *strictly increasing* provided $f(x_2) > f(x_1)$ whenever $x_2 > x_1$. A *strictly decreasing* function is one in which

$f(x_2) < f(x_1)$ whenever $x_2 > x_1$. If these conditions hold only for values of x in some interval (a, b), we say that f is strictly increasing or decreasing on (a, b). For example, if $f(x) = x^2$, then f is strictly decreasing on $(-\infty, 0)$ and strictly increasing on $(0, \infty)$. A function that is either strictly increasing or strictly decreasing is clearly one-to-one.

If f is a one-to-one function, then the set of pairs (y, x), for (x, y) in f, is also a function. This function is called the inverse of f and is denoted by f^{-1}. The domain of f^{-1} is the range of f, and the range of f^{-1} is the domain of f. If $f(x) = y$, then $f^{-1}(y) = x$. Hence $f^{-1}(f(x)) = x$ for all x in the domain of f, and $f(f^{-1}(x)) = x$ for all x in the range of f.

Example 1 Tell whether the following functions are one-to-one, and give the inverse function when there is one:

(a) $f = \{(1, 2), (2, -1), (-1, 0), (0, 6), (3, 5)\}$

(b) $g = \{(0, 1), (2, 4), (-1, 3), (3, 1)\}$

Solution.

(a) f is one-to-one, since the function values $2, -1, 0, 6, 5$ are different for different values of x. The inverse function is $f^{-1} = \{(2, 1), (-1, 2), (0, -1), (6, 0), (5, 3)\}$.
(b) g is not a one-to-one function since $g(0) = 1$ and $g(3) = 1$. There is therefore no inverse function.

Note that the same notation is used to indicate the two different ideas of reciprocal number ($a^{-1} = 1/a$) and inverse function. Although these ideas are distinct, there is an obvious parallel suggested by the fact that for all x,

$$f^{-1}(f(x)) = x = f(f^{-1}(x)).$$

Since f^{-1} is the set of all pairs (y, x) for (x, y) in f, it is easy to see the connection between the graphs of f and f^{-1}. The points (x, y) and (y, x) are symmetric about the line $y = x$ as seen in Fig. 5-25. Hence the graphs of f and f^{-1} are symmetric about the line $y = x$. Conversely if f and g are any two functions such that the graph of f is symmetric to the graph of g about the line $y = x$, then $f^{-1} = g$ and $g^{-1} = f$.

Recall that for an equation to be a function, every *vertical line*

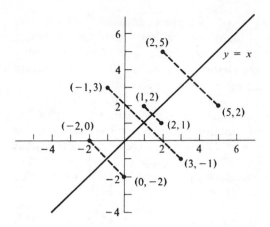

Figure 5-25

must intersect its graph once at most. If a function is one-to-one, each *horizontal line* must also intersect its graph at most once.

Because the function $f(x) = x^2$ is not one-to-one [see Fig. 5-26(a)], we cannot talk about its inverse. However, if we let $g(x) = x^2$ for $x \geq 0$, the g is *strictly* increasing and, therefore, is one-to-one. [See Fig. 5-26(b).] The inverse function is

$$g^{-1}(x) = \sqrt{x}.$$

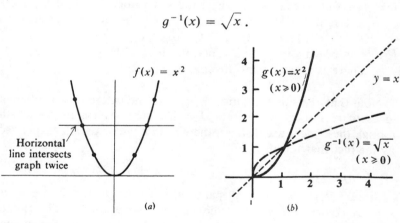

Figure 5-26

Note that

$$g^{-1}(g(x)) = \sqrt{x^2} = x \quad (x \geq 0)$$

and

$$g(g^{-1}(x)) = (\sqrt{x})^2 = x \quad (x \geq 0).$$

The function $h(x) = x^3$ is one-to-one, and its inverse is $h^{-1}(x) = \sqrt[3]{x}$. These graphs are shown in Fig. 5-27.

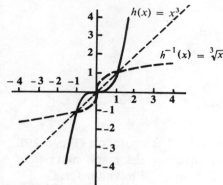

Figure 5-27

Example 2 Graph the functions and tell whether they have inverses. If the inverse exists, write a formula for it, and graph f^{-1} on the same axes as f.

(a) $f(x) = 2x + 3$ (b) $f(x) = x^2 - 2x$

Solution.

(a) The graph of $f(x) = 2x + 3$ is a straight line (Fig. 5-28). Since every horizontal line intersects the graph at most once, the function is one-to-one, and hence has an inverse. For simplicity of notation, let $g(x) = f^{-1}(x)$. Then $f(g(x)) = x$, and we can solve this to get a formula for $g(x)$.

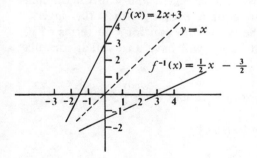

Figure 5-28

Inverse Functions   197

$$f(g(x)) = x$$

$$2g(x) + 3 = x$$

$$2g(x) = x - 3$$

$$g(x) = \frac{1}{2}x - \frac{3}{2}$$

Hence $f^{-1}(x) = \frac{1}{2}x - \frac{3}{2}$ (Fig. 5-28).

(b) The graph of $x^2 - 2x = (x - 1)^2 - 1$ is a parabola (Fig. 5-29). Since some horizontal lines intersect the graph more than once, the function is not one-to-one, and has no inverse.

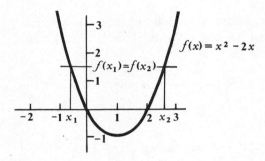

Figure 5-29

Suppose a one-to-one function is defined by an equation in x and y. The inverse function consists of the pairs (y, x) for (x, y) in the function. Therefore, if we interchange x and y in the original equation we obtain a new equation that defines the inverse function. If we first solve the given equation for x in terms of y, and then interchange x and y, we will have an explicit formula for the inverse function.

Example 3 Find the inverse to the function $y = \frac{1}{x} + 1$.

Solution. First solve for x in terms of y.

$$y = \frac{1}{x} + 1$$

$$xy = 1 + x$$

$$x(y - 1) = 1$$

$$x = \frac{1}{y - 1}$$

This last equation is equivalent to the original, and hence defines (implicitly) the same function. If we now interchange x and y we get the equation which defines the inverse function:

$$y = \frac{1}{x - 1}.$$

Notice that if $f(x) = x^{-1}$, then f is its own inverse. Algebraically, this is the statement that

$$f(f(x)) = \frac{1}{\frac{1}{x}} = x.$$

Geometrically, the fact that the function $y = 1/x$ is its own inverse reflects the fact that the graph of $xy = 1$ is symmetric about the line $y = x$.

In Example 3, notice how the inverse function undoes the operations the function does, to get its value. To get the function value, $\frac{1}{x} + 1$, you first take the inverse, and then add one. To undo this, you first subtract one, then take the reciprocal: $(x - 1)^{-1}$, or $\frac{1}{x - 1}$. As another example, consider $f(x) = (x - 3)^2$, for $x \geq 3$. To get $f(x)$ you first subtract 3, then square. To get $f^{-1}(x)$ you go backwards and first take the square root; and then add 3; $f^{-1}(x) = \sqrt{x} + 3$ $(x \geq 0)$. Below we give some other simple functions f where f^{-1} can be written by inspection.

$\begin{cases} f(x) = 2x - 7 \\ f^{-1}(x) = \frac{x + 7}{2} \end{cases}$ (multiply by 2, subtract 7)

(add 7, divide by 2)

$\begin{cases} f(x) = 5(x + 3) \\ f^{-1}(x) = \frac{1}{5}x - 3 \end{cases}$ (add 3, multiply by 5)

(divide by 5, subtract 3)

$$\begin{cases} f(x) = 2x^3 \\ f^{-1}(x) = \sqrt[3]{x/2} \end{cases}$$ (cube, multiply by 2)

(divide by 2, take cube root)

$$\begin{cases} f(x) = \frac{1}{x} + 6 \\ f^{-1}(x) = (x-6)^{-1} = \frac{1}{x-6} \end{cases}$$ (take inverse, add 6)

(subtract 6, take inverse)

SECTION 5-7, REVIEW QUESTIONS

one-to-one

1. A function which has only one pair (x, y) for any given value of y is called _____-_____-_____.

\>

2. A function f is strictly increasing if $f(x_2) > f(x_1)$ when x_2 _____ x_1.

decreasing

3. If $f(x_2) < f(x_1)$ whenever $x_2 > x_1$ then f is a strictly _____ function.

4. Every strictly increasing or decreasing function is _____.

one-to-one

_____-_____-_____.

inverse

5. If f is a one-to-one function, then the set of all pairs (y, x) for (x, y) in f is called the _____ of f.

f⁻¹

6. The inverse of f is written _____.

range

7. The domain of f^{-1} is the _____ of f.

y = x

8. The graph of f^{-1} is symmetric to the graph of f about the line _____.

f⁻¹

9. If f and g are any functions whose graphs are symmetric about the line $y = x$, then $f = g^{-1}$ and $g = $ _____.

interchanging

10. If a function is defined by an equation in x and y, then the inverse function is defined by the equation you get by _____ x and y.

x

11. Functions f and g are inverses of each other provided $f(g(x)) = g(f(x)) = $ _____.

SECTION 5-7, EXERCISES

Tell whether or not the following functions are one-to-one. If the function is one-to-one give its inverse.

1. $f = \{(0, 2), (1, 1), (2, 4), (3, 3)\}$ Example 1
2. $f = \{(0, 0), (1, 2), (2, 1), (3, 3)\}$
3. $f = \{(1, 0), (0, 1), (2, 0), (3, 2)\}$
4. $f = \{(7, 1), (2, 7), (3, 5), (8, 9), (10, 7)\}$
5. $f = \{(-1, 2), (-2, 3), (3, -2), (0, -1), (2, 1)\}$
6. $f = \{(10, 9), (9, 8), (7, 10), (8, 11)\}$

Graph each of the following functions and tell whether it has an inverse. If the function has an inverse, write a formula for it and graph it on the same axes with f.

7. $f(x) = 2x$
8. $f(x) = \frac{1}{3}x$ Example 2
9. $f(x) = 3x - 2$
10. $f(x) = 2x - 1$
11. $f(x) = x^4$
12. $f(x) = x(x - 1)(x - 2)$
13. $f(x) = \frac{1}{x - 1}$
14. $f(x) = \frac{1}{x} - 2$
15. $f(x) = x^2 - 3$
16. $f(x) = 4 - x^2$
17. $f(x) = \sqrt{x + 2}$ $(x \geq -2)$
18. $f(x) = \sqrt{1 - x}$ $(x \leq 1)$
19. $f(x) = \begin{cases} x^2 & \text{if } x < 0 \\ -\sqrt{x} & \text{if } x \geq 0 \end{cases}$
20. $f(x) = \begin{cases} \sqrt{-x} & \text{if } x < 0 \\ -x^2 & \text{if } x \geq 0 \end{cases}$

The following functions are one-to-one on the indicated sets; write a formula for the inverse function.

21. $y = \frac{2}{x} - 1$ $(x \neq 0)$
22. $y = 3 - \frac{1}{x}$ $(x \neq 0)$ Example 3
23. $y = (x - 3)^2$ $(x \geq 3)$
24. $y = -(x + 2)^2$ $(x \leq -2)$
25. $y = x^2 + x$ $(x \geq 0)$
26. $y = x^2 + x + 1$ $(x \geq 0)$
27. $y = \sqrt{9 - x^2}$ $(0 \leq x \leq 3)$
28. $y = \frac{1}{2}\sqrt{7 - 4x^2}$ $\left(0 \leq x \leq \frac{\sqrt{7}}{2}\right)$

Write a formula for $f^{-1}(x)$ by inspection.

29. $f(x) = 4x$
30. $f(x) = 7x$
31. $f(x) = x + 5$
32. $f(x) = x - 1$
33. $f(x) = 2x + 1$
34. $f(x) = 3x - 2$
35. $f(x) = \frac{1}{x} - 9$
36. $f(x) = \frac{1}{x} + 3$
37. $f(x) = \frac{1}{x - 1}$
38. $f(x) = \frac{1}{x + 4}$

39. $f(x) = (x + 2)^3$ 40. $f(x) = (x - 3)^3$
41. $f(x) = \sqrt[3]{x - 2}$ 42. $f(x) = \sqrt[3]{x + 5}$
43. (a) Write by inspection the inverses of these linear functions
    (i) $f(x) = x$;   (ii) $f(x) = -x$;
    (iii) $f(x) = -x + 1$;   (iv) $f(x) = -x - 8$.
    (b) What straight lines other than $y = x$ itself are symmetric about the line $y = x$?
    (c) What linear functions $f(x) = mx + b$ (other than $f(x) = x$) are their own inverses?
    (d) If $f(x) = mx + b$, what is $f(f(x))$?
    (e) From (d) find conditions on m and b so that $f(f(x)) = x$. Does this agree with your answer to (c)?
44. Show algebraically that the function whose graph is the part of the circle $x^2 + y^2 = a^2$ in the first quadrant is its own inverse.

5-8 REVIEW FOR CHAPTER

Which of the following sets of ordered pairs describe a function?

1. $A = \{(1, 1), (2, 2), (3, 3)\}$
2. $B = \{(-2, 7), (-3, 5), (-4, 7), (-5, 5)\}$
3. $C = \{(3, 5), (3, 4), (3, 6), (3, 7)\}$
4. $D = \{(5, 3), (4, 3), (6, 3), (7, 3)\}$
5. $E = \{(0, -1), (0, -2), (0, -3)\}$
6. $F = \{(-1, -1), (-2, -2), (-3, -3)\}$
7. Give the domain and range of each relation or function described in Exercises 1 through 6.

In Exercises 8 through 14: give formulas which define the functions indicated, and specify their domains, if $f(x) = \sqrt{25 - x^2}$ and $g(x) = 5 - x$.

8. $f + g, g + f$ 9. $f - g, g - f$
10. fg, gf 11. $f/g, g/f$
12. $f \circ g, g \circ f$ 13. f^{-1} if $x \geq 0$
14. g^{-1}

15. Give a formula for $(f \circ g)(x)$ if $f(x) = x^2 + 3x - 1$ and $g(x) = \sqrt{x}$.

16. If $f(x) = \sqrt{9 - x^2}$, $x \geq 0$, graph f and f^{-1} on the same coordinate axes.

In Exercises 17 through 22, let $t(x) = x - 3$; $g(x) = x^3$; $r(x) = \sqrt[3]{x^2}$ and $f(x) = 4x$. Then express each of the following formulas as composition of functions chosen from f, g, t, and r.

17. $\sqrt[3]{x^2 - 6x + 9}$ 18. x^2 19. $4x - 3$
20. $x^3 - 9x^2 + 27x - 27$ 21. $2\sqrt[3]{2x^2}$ 22. $64x^3$

Perform the indicated division and give the quotient and remainder.

23. $(2x^4 - 17x^2 - 4) \div (x + 3)$
24. $(x^3 - 6x^2 - 2x + 40) \div (x - 4)$
25. $(x^3 + 2x^2 - 23x - 60) \div (x - 5)$

Graph each of the following polynomial functions.

26. $f(x) = x^3 - 4x^2 - x + 4$ 27. $f(x) = -3x^3 + 7x - 5$
28. $f(x) = x^4 + 3x^2 - 8x$ 29. $f(x) = 3x - 4x^2 - 2x^4$
30. $f(x) = -x^4 + 9x - 12$ 31. $f(x) = x^4 + 6x^2 - 13$
32. Find $\lim_{x \to +\infty} (x^3 - 4x^2 - x + 4)$ and $\lim_{x \to -\infty} (x^3 - 4x^2 - x + 4)$
33. Find $\lim_{x \to +\infty} (x^4 + 3x^2 - x)$ and $\lim_{x \to -\infty} (x^4 + 3x - 8x)$
34. Find $\lim_{x \to +\infty} (-x^4 + 9x - 12)$ and $\lim_{x \to -\infty} (-x^4 + 9x - 12)$

Find all the asymptotes and graph the following functions:

35. $f(x) = \dfrac{5}{x - 7}$ 36. $f(x) = \dfrac{-4}{3 - x}$

37. $f(x) = \dfrac{x}{x + 5}$ 38. $f(x) = \dfrac{3x}{2 - 3x}$

39. $f(x) = \dfrac{x + 1}{x - 4}$ 40. $f(x) = \dfrac{2 - x}{x + 8}$

41. $f(x) = \dfrac{3x}{x^2 + 5x + 6}$ 42. $f(x) = \dfrac{x + 2}{x^2 + 3x - 4}$

43. $f(x) = \dfrac{x + 4}{x^3 - 27}$ 44. $f(x) = \dfrac{x^2 - 4}{1 - x^3}$

Show that each of the following functions is an algebraic function by finding a polynomial equation that it satisfies:

45. $y = 2 - \sqrt{x}$ 46. $y = \sqrt{x - 4} + 7$
47. $y = \sqrt{25 - x^2} + 2$ 48. $y = \sqrt{9 + x^2}$

49. $y = \sqrt{1/x} - 3$ 50. $y = \sqrt{\dfrac{x + 1}{x - 2}}$

Find the inverse of each of the following functions, and graph the function and its inverse on the same coordinate system:

51. $f(x) = x + 3$
52. $f(x) = 7x - \frac{1}{4}$
53. $f(x) = \sqrt{4 - x^2}$ $(x \geq 0)$
54. $f(x) = (x - 2)^2 + 3$ $(x \geq 2)$

6 EXPONENTIAL AND LOGARITHMIC FUNCTIONS

6-1 THE EXPONENTIAL FUNCTION a^x

All the functions studied so far have been algebraic functions; that is, functions that satisfy an algebraic equation. The exponential and logarithmic functions introduced in this chapter are examples of a class of functions called *transcendental*. The trigonometric functions introduced in Chapter 7 are also examples of transcendental functions.

Recall that a^n, for a positive integer n, is defined to be the product of the base a by itself n times.

$$a^n = \overbrace{a \cdot a \cdot \ldots \cdot a}^{n \text{ factors}}$$

We then define a^x for x a negative integer or zero.

$$a^0 = 1; \qquad a^{-n} = \frac{1}{a^n}$$

Now extend the definition to the case of an exponent of the form $\pm 1/n$. *To do this we require that* a *be positive*, and define

$$a^{1/n} = \sqrt[n]{a} \qquad \text{the positive } n\text{th root of } a$$

$$a^{-1/n} = \frac{1}{a^{1/n}}, \quad \text{or} \quad \frac{1}{\sqrt[n]{a}}.$$

For exponents of the form $\pm m/n$ we then have the natural definitions

$$a^{m/n} = (a^{1/n})^m \quad (\text{or } a^{m/n} = (a^m)^{1/n}),$$

$$a^{-m/n} = \frac{1}{(a^{1/n})^m} \quad \left(\text{or } a^{-m/n} = \frac{1}{(a^m)^{1/n}}\right).$$

As examples of these definitions, consider the following:

$$8^{1/3} = \sqrt[3]{8} = 2,$$

$$8^{-1/3} = \frac{1}{\sqrt[3]{8}} = \frac{1}{2},$$

$$8^{2/3} = (\sqrt[3]{8})^2 = \sqrt[3]{(8)^2} = 4,$$

$$8^{-2/3} = \frac{1}{(\sqrt[3]{8})^2} = \frac{1}{\sqrt[3]{(8)^2}} = \frac{1}{4}.$$

We now have a^x defined for every positive or negative *rational* number x. The definition is finally extended to include irrational values of x. This is done by approximating an irrational value of x by a rational number r, and then using a^r to approximate the value of a^x. The details of this approximation process will not concern us here. However, it is important to know that it provides a definition of a^x for all real numbers x if $a > 0$, so that a^x is a continuous function, and that all the standard rules below hold for all real values of the exponent x.

$$a^x a^y = a^{x+y}$$

$$(a^x)^y = a^{xy}$$

$$\frac{a^x}{a^y} = a^{x-y} \qquad (a \text{ and } b \text{ must be positive numbers})$$

$$(ab)^x = a^x b^x$$

$$\left(\frac{a}{b}\right)^x = \frac{a^x}{b^x}$$

$$a^{-x} = \frac{1}{a^x}$$

$$a^0 = 1$$

206 *Exponential and Logarithmic Functions*

As we already know, the domain of a^x is the set of all real numbers x, and the range of a^x is some set of positive numbers. The behavior of a^x depends on whether $0 < a < 1$, $a = 1$, or $a > 1$. Of course $a^x = 1$ for all x if $a = 1$, so we really have only two cases of interest. First let's consider the case $a > 1$.

If $a > 1$, then $a^x > 1$ for all positive values of x. This fact implies that a^x is a strictly increasing function. Suppose $x_1 < x_2$, so that $x_2 = x_1 + p$ for some positive number p. Then

$$a^{x_2} = a^{x_1+p} = a^{x_1} \cdot a^p.$$

Since $a^p > 1$, it follows that $a^{x_2} > a^{x_1}$ if $x_2 > x_1$. For example, if $a = 2$, $x_1 = 3$, and $x_2 = 5$, then we have $2^5 > 2^3$.

We can also show that a^x takes on arbitrarily large values, for large values of x, if $a > 1$. To see this, we write $a = 1 + p$, for $p > 0$, and consider integer exponents n:

$$a^n = (1+p)^n = 1 + np + \frac{n(n-1)}{2}p^2 + \cdots + p^n.$$

Because all the terms on the right are positive,

$$a^n > 1 + np,$$

and hence a^n becomes arbitrarily large for large values of n. Therefore

$$\lim_{x \to +\infty} a^x = \infty \quad (a > 1).$$

Because we also have $a^{-x} = 1/a^x$, we know that for a large negative value of the exponent, the function value will be a small positive number. Hence

$$\lim_{x \to -\infty} a^x = 0 \quad (a > 1).$$

It follows that a^x is bounded below by 0 and is not bounded above. Because a^x is a continuous function, its graph has no breaks in it, and the range of a^x consists of *all* positive numbers.

If $0 < a < 1$, then $a^x < 1$ for all positive numbers x. Hence a^x is strictly decreasing, and $\lim_{x \to +\infty} a^x = 0$. The domain of a^x for $0 < a < 1$ is, of course, the whole line, and the range is still the set of all positive numbers. The graphs of $y = 2^x$ and $y = (\frac{1}{2})^x$ are shown in Fig. 6-1. The symmetry of the two graphs about the y-axis is a consequence of the fact that

$$2^{-x} = (2^{-1})^x = (\tfrac{1}{2})^x.$$

For applications, the most important base for an exponential

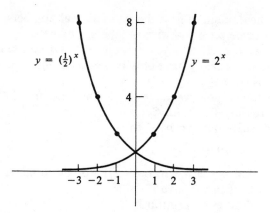

Figure 6-1

function is the irrational number $e = 2.7182818285\ldots$. (We will use the approximation $e = 2.72$ for the purposes of calculation.) The function e^x occurs over and over again in physical application. For example, a perfectly flexible cord suspended at two points will hang in the curve, called a catenary*, whose equation is

$$y = \tfrac{1}{2}(e^x + e^{-x}) .$$

The curve $y = e^x$ has the interesting property that the tangent line at any point on the curve has a slope equal to the y-value of the point. For example, the tangent at the point $(0, 1)$ has slope 1, and the tangent at the point $(1, e)$ has slope e (Fig. 6-2).

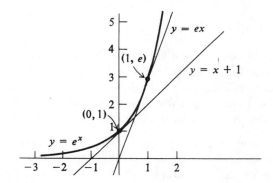

Figure 6-2

---

*From the Latin word *catena* for chain.

208 *Exponential and Logarithmic Functions*

SECTION 6-1, REVIEW QUESTIONS

1. The formula a^n is defined to be the product of a taken _____ times.   n

2. For n positive, we define a^{-n} to be _____. $1/a^n$

3. For n positive, $a^{1/n}$ is defined to be the positive nth _____ of a.   root

4. If x is an irrational number, then a^x is approximated by a^r, where r is a rational approximation to _____.   x

5. The positive number a is called the _____ of the exponential function a^x.   base

6. In the expression a^x, a is the base and x is the _____.   exponent

7. The domain of a^x, for $a > 0$, is the whole real line and the range, provided $a \neq 1$, is the set of _____ numbers.   positive

8. If $a > 1$, the function a^x is strictly _____.   increasing

9. If $0 < a < 1$, the function a^x is strictly _____.   decreasing

10. If $a > 0$, $a \neq 1$, the function a^x is bounded _____ by zero but not bounded _____.   below   above

11. If $a > 1$, $\lim_{x \to +\infty} a^x = $ _____ and $\lim_{x \to -\infty} a^x = $ _____. $+\infty$ 0

12. The number e to two decimal places is _____.   2.72

SECTION 6-1, EXERCISES

Write the following without exponents or radicals.

1. 2^4
2. 5^3
3. 3^{-2}
4. 2^{-5}
5. 4^0
6. 2^0
7. $27^{1/3}$
8. $4^{1/2}$
9. $125^{-1/3}$
10. $36^{-1/2}$
11. $8^{2/3}$
12. $4^{3/2}$
13. $16^{3/4}$
14. $27^{2/3}$
15. $64^{-2/3}$
16. $81^{-3/4}$
17. $125^{-4/3}$
18. $81^{-5/4}$
19. $32^{3/5}$
20. $128^{-2/7}$
21. $(1/27)^{2/3}$
22. $(1/16)^{3/4}$
23. $(9/16)^{-3/2}$
24. $(25/36)^{-3/2}$

Write the following with positive exponents and simplify where possible.

25. $x^2 y^{-1} \cdot 2x^3 y^{-1}$
26. $(x^{-1}y)^3 \cdot x^2 y$
27. $2x^2 \cdot (2x)^{-2}$
28. $x^2 y^3 \cdot (xy)^{-1}$
29. $e^2 \cdot e^3$
30. $e^3 \cdot e^{-1}$
31. $e^x \cdot e^{2x}$
32. $e^{-3x} \cdot e^{2x}$
33. $(e^x)^2$
34. $(e^3)^x$
35. e^{-5x}
36. e^{-3x}
37. e^{4x}/e^x
38. e^{4x}/e^{5x}

Graph the following functions and their inverses on the same axes.

39. $y = 2^x$
40. $y = 3^x$
41. $y = (1/2)^x$
42. $y = (1/3)^x$

Graph the following equations.

43. $x = 2^y$
44. $x = 3^y$
45. $x = (1/2)^y$
46. $x = (1/3)^y$

6-2 THE LOGARITHMIC FUNCTION $\text{LOG}_a x$

In the last section, we saw that for any positive base a other than one, a^x is either strictly increasing or strictly decreasing. Therefore, a^x is a one-to-one function if $a \neq 1$, and consequently it has an inverse function.

We introduce the notation $\log_a x$ for the function which is inverse to a^x. That is,

$$x = \log_a y \quad \text{means the same as} \quad a^x = y.$$

The function $\log_a x$ is called the logarithm of x to the base a. Because the domain of the exponential function is the set of real numbers, the range of the logarithm is also the set of real numbers. Similarly, the range of a^x is the set of positive numbers, so the domain of $\log_a x$ is the set of positive numbers. In other words, $\log_a x$ is *only defined for positive numbers*. The graphs of $y = 2^x$ and $y = \log_2 x$ are shown in Fig. 6-3. Notice that the graphs are symmetric about the line $y = x$ as is the case for any function and its inverse.

The notation $y = \log_a x$ is superfluous since we can express the same relationship with the equation $a^y = x$. However, it is convenient to have a separate notation for both the function and its inverse in the same way that we do for x^3 and $\sqrt[3]{x}$. The following pairs of equations express the same facts in exponential and logarithmic notations.

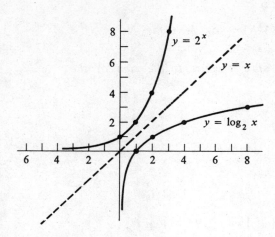

Figure 6-3

$$\log_2 8 = 3 \qquad 2^3 = 8$$
$$e^2 = 7.4 \qquad \log_e 7.4 = 2$$
$$\log_{1/3}\left(\frac{1}{9}\right) = 2 \qquad \left(\frac{1}{3}\right)^2 = \frac{1}{9}$$
$$10^{-3} = 0.001 \qquad \log_{10} 0.001 = -3$$
$$\log_4 \frac{1}{16} = -2 \qquad 4^{-2} = \frac{1}{16}$$

For any base a, $(a > 0, a \neq 1)$, we have the following identities, which are consequences of the definitions.

$$\log_a 1 = 0$$
$$\log_a a^r = r$$
$$a^{\log_a N} = N$$

The fundamental properties of logarithms are simply laws of exponents written in logarithmic notation. Consider first the exponential identity

$$a^r a^s = a^{r+s}.$$

Suppose $M = a^r$, $N = a^s$, so that $MN = a^{r+s}$. Then $r = \log_a M$, $s = \log_a N$ and $r + s = \log_a (MN)$. This gives us the first property of logarithms.

The Logarithmic Function $\log_a x$ 211

$$\log_a(MN) = \log_a M + \log_a N$$

The logarithm of a product is the sum of the logarithms.

In a similar way we have

$$\frac{M}{N} = \frac{a^r}{a^s} = a^{r-s}.$$

Consequently,

$$\log_a\left(\frac{M}{N}\right) = \log_a M - \log_a N.$$

The logarithm of a quotient is the difference of the logarithms.

Finally, if $a^r = M$, so that $r = \log_a M$, we have

$$M^k = (a^r)^k = a^{rk} = a^{kr}.$$

Hence $$kr = \log_a M^k,$$

or $$k \log_a M = \log_a M^k.$$

The logarithm of a number to a power is the power times the logarithm of the number.

The following examples illustrate the use of these principles.

Example 1 Express $\log_a 5xy$ as a sum of logarithms of single terms.

Solution. Use the law that the logarithm of a product is the sum of the logarithms.

$$\begin{aligned}\log_a 5xy &= \log_a 5 + \log_a xy \\ &= \log_a 5 + \log_a x + \log_a y\end{aligned}$$

Example 2 Express $\log_a \dfrac{3x^2}{a^3 y^3}$ as a sum and difference of logarithms.

Solution.

$$\log_a \dfrac{3x^2}{a^3 y^3} = \log_a 3x^2 - \log_a a^3 y^3$$
$$= \log_a 3 + \log_a x^2 - (\log_a a^3 + \log_a y^3)$$
$$= \log_a 3 + 2\log_a x - 3\log_a a - 3\log_a y$$
$$= \log_a 3 + 2\log_a x - 3 - 3\log_a y$$

Example 3 Simplify $\log_a \sqrt[3]{\dfrac{a^2 x}{y^5}}$

Solution. Since

$$\sqrt[3]{\dfrac{a^2 x}{y^5}} = \left(\dfrac{a^2 x}{y^5}\right)^{1/3},$$

we have

$$\log_a \sqrt[3]{\dfrac{a^2 x}{y^5}} = \log_a \left(\dfrac{a^2 x}{y^5}\right)^{1/3}$$
$$= \dfrac{1}{3} \log_a \dfrac{a^2 x}{y^5}$$
$$= \tfrac{1}{3}[\log_a a^2 + \log_a x - \log_a y^5]$$
$$= \tfrac{1}{3}[2 + \log_a x - 5\log_a y].$$

The final property of logarithms we need is a formula that allows us to change from one base to another. Suppose $r = \log_a N$ and $s = \log_b N$; that is,

$$a^r = N = b^s.$$

It follows that $s = \log_b a^r$ or $s = r \log_b a$. Substituting for r and s in $s = r \log_b a$, gives us the formula

$$\log_b N = (\log_a N)(\log_b a).$$

Notice also that if $\ell = \log_b a$, so that $b^\ell = a$, then $a^{1/\ell} = b$, and hence $1/\ell = \log_a b$. That is, $\log_b a = 1/\log_a b$, so that $\log_a b$ and $\log_b a$ are reciprocals of each other. Consequently, the formula above

also be written
$$\log_b N = \log_a N / \log_a b.$$

Example 4 Given that $\log_e 10 = 2.3026$, find $\log_{10} e$.

Solution. Since $\log_e 10 = 2.3026$,
$$\log_{10} e = \frac{1}{2.3026} = 0.4343.$$

Example 5 Convert the following base 10 logarithms to base e and the base e logarithms to base 10. (Use 2.303 for $\log_e 10$.)

(a) $\log_{10} 0.001 = -3$ (b) $\log_{10} 15 = 1.1761$
(c) $\log_e 5 = 1.6095$ (d) $\log_e (0.5) = -0.6933$

Solution. Since $\log_b N = \log_a N \log_b a$,
$$\log_e N = (\log_{10} N)(\log_e 10)$$
and

(a) $\log_e 0.001 = (-3)(2.303) = -6.909$,
(b) $\log_e 15 = (1.176)(2.303) = 2.707$.

Similarly,
$$\log_{10} N = \log_e N \log_{10} e$$
and

(c) $\log_{10} 5 = (1.6095)(0.4343) = 0.6990$,
(d) $\log_{10} (0.5) = (-0.6933)(0.4343) = -0.3011$.

SECTION 6-2, REVIEW QUESTIONS

increasing 1. If $a > 1$, then a^x is a strictly _____ function.

2. A strictly increasing function is one-to-one, and therefore has an

inverse _____ function.

$\log_a x$ 3. The inverse to a^x is denoted _____.

a^y 4. The equation $y = \log_a x$ means the same as $x =$ _____.

5. The functions y defined by the equations $y = a^x$ and $x = a^y$ are

inverses _____ of each other.

r 6. For any positive base a, ($a \neq 1$), $\log_a a^r =$ _____ and

0 $\log_a 1 =$ _____.

214 *Exponential and Logarithmic Functions*

7. The logarithm of a product is the _____ of the logarithms of the factors.   **sum**

8. The logarithm of a product is the sum of the logarithms, and the logarithm of a quotient is the _____ of the logarithms.   **difference**

9. For any exponent r, $\log_a N^r = ($_____$) \cdot \log_a N$.

10. The numbers $\log_a b$ and $\log_b a$ are _____ of each other.   **reciprocals**

11. To change $\log_a N$ to $\log_b N$ you multiply $\log_a N$ by _____.   **$\log_b a$**

SECTION 6-2, EXERCISES

Find the following logarithms.

1. $\log_6 36$
2. $\log_2 32$
3. $\log_2 \frac{1}{4}$
4. $\log_5 \frac{1}{25}$
5. $\log_{10} 10{,}000$
6. $\log_{10} .001$
7. $\log_9 \frac{1}{9}$
8. $\log_e 1$

In the following exercises, if the equation is written in exponential notation, rewrite it in logarithmic notation, and vice versa.

9. $\log_2 16 = 4$
10. $\log_3 27 = 3$
11. $\log_{10} 1{,}000 = 3$
12. $\log_{10} 1{,}000{,}000 = 6$
13. $6^2 = 36$
14. $7^2 = 49$
15. $8^{1/3} = 2$
16. $81^{1/4} = 3$
17. $10^{-2} = .01$
18. $2^{-3} = 1/8$
19. $9^{1/2} = 3$
20. $64^{1/2} = 8$

Express the following in terms of logarithms of single terms.

21. $\log_a 9xy$
22. $\log_a 8rs$ **Example 1,2,3**
23. $\log_a 3x^2 y$
24. $\log_a 4x^2 y^3$
25. $\log_a 2ax^2$
26. $\log_a a^2 x^3$
27. $\log_a (x^3/yz^2)$
28. $\log_a 5x^2 y/3z^3$
29. $\log_a \sqrt{xy}$
30. $\log_a \sqrt[3]{xy}/z^2$
31. $\log_a 1/\sqrt[5]{x^3}$
32. $\log_a \sqrt[3]{x}/\sqrt[4]{y^3}$

Use the information in Examples 4 and 5 to convert the following base 10 logarithms to base e, and vice versa.

33. $\log_{10} 100 = 2$
34. $\log_{10} (0.1) = -1$ **Example 4,5**

The Logarithmic Function $\log_a x$ 215

35. $\log_{10} 350 = 2.5441$
36. $\log_{10} 2.15 = 0.3324$
37. $\log_e 7.389 = 2$
38. $\log_e 1.4 = 0.3365$
39. $\log_e 207 = 5.3327$
40. $\log_e 10.2 = 2.3224$

6-3 TABLES OF COMMON LOGARITHMS

The exponential function e^x and its corresponding logarithm function $\log_e x$ occur frequently in applications. Base-e logarithms are accordingly called *natural logarithms* and are designated $\ln x$ instead of $\log_e x$. Logarithms to base 10 are easiest to use for computation because we write numbers in decimal notation. Base-10 logarithms are called *common logarithms*. When no base is specified, it is assumed that we are talking about common logarithms. Thus we will henceforth write $\log N$ instead of $\log_{10} N$ and $\ln N$ instead of $\log_e N$.

Powers of 10 have integer-valued common logarithms.

$$\log 0.001 = \log 10^{-3} = -3$$
$$\log 0.01 = \log 10^{-2} = -2$$
$$\log 0.1 = \log 10^{-1} = -1$$
$$\log 1 = \log 10^0 = 0$$
$$\log 10 = \log 10^1 = 1$$
$$\log 100 = \log 10^2 = 2$$

Since $\log x$ is a strictly increasing function, it follows that every number between 1 and 10 has a logarithm between 0 and 1. Any positive number can be written as some number between 1 and 10 times some power of 10. We use this fact in the following way.

From the Table of Common Logarithms in the Appendix we find that $\log 2.310$ equals 0.3636. This allows us to find the logarithm of any number whose digits are 2, 3, 1, 0, regardless of where the decimal point is:

$$\log 23.10 = \log [(2.310)(10)]$$
$$= \log (2.310) + \log 10$$
$$= 0.3636 + 1$$
$$\log 0.002310 = \log [(2.310)(10^{-3})]$$
$$= \log (2.310) + \log 10^{-3}$$
$$= 0.3636 - 3$$

216 Exponential and Logarithmic Functions

$$\log 23{,}100 = \log[(2.310)(10^4)]$$
$$= \log(2.310) + \log 10^4$$
$$= 0.3636 + 4$$

The decimal part of the logarithm is called the *mantissa*, and the integer part is called the *characteristic*. The mantissa is the same for all numbers with the same digits, and does not depend on the location of the decimal point. The characteristic depends only on the position of the decimal point. The mantissa is a *positive* decimal, but the characteristic can be either a positive or a negative integer.

The Table of Common Logarithms gives the four place logarithms of the numbers from 1.00 to 9.99, at intervals of 0.01. In other words, it gives the mantissas of the logarithms of all positive three-digit numbers. The first five rows of this table are reproduced in Table 6-1.

TABLE 6-1   TABLE OF COMMON LOGARITHMS

N	0	1	2	3	4	5	6	7	8	9
10	0000	0043	0086	0128	0170	0212	0253	0294	0334	0374
11	0414	0453	0492	0531	0569	0607	0645	0682	0719	0755
12	0792	0828	0864	0899	0934	0969	1004	1038	1072	1106
13	1139	1173	1206	1239	1271	1303	1335	1367	1399	1430
14	1461	1492	1523	1553	1584	1614	1644	1673	1703	1732

The numbers in the first row are the logarithms of 1.00, 1.01, 1.02, 1.03, ..., 1.09. The entries are all decimals, with the decimal point omitted for simplicitly. These are read from the first row:

$$\log 1.00 = 0.0000,$$
$$\log 1.01 = 0.0043,$$
$$\log 1.02 = 0.0086,$$
$$\cdot$$
$$\cdot$$
$$\cdot$$
$$\log 1.09 = 0.0374$$

Similarly, the second row gives the logarithms of the numbers 1.10, 1.11, ..., 1.19; thus

$$\log 1.10 = 0.0414$$
$$\log 1.11 = 0.0453$$
$$\log 1.19 = 0.0755$$

We find the logarithm of a four-place number by a process called linear interpolation. Suppose we want to find log 1.063. From the table we find

$$\log 1.07 = 0.0294$$
$$\log 1.06 = \underline{0.0253}$$
$$\text{difference} = 0.0041$$

Because 1.063 is 3/10 of the way from 1.06 to 1.07, the logarithm of 1.063 will be approximately 3/10 of the way from log 1.06 to log 1.07. The difference of these two logarithms is 0.0041, and 0.3 of this difference is 0.00123, which we round off to 0.0012. Hence we have

$$\log 1.06 = 0.0253$$
$$\tfrac{3}{10}(\log 1.07 - \log 1.06) = \underline{0.0012}$$
$$\log 1.063 = 0.0265$$

What we do in this kind of interpolation is to replace the curve $y = \log x$ by a straight line segment between the points where $x = 1.06$ and $x = 1.07$, and then find the y-value on the segment at the point 1.063. An exaggerated picture of the graphs is shown in Fig. 6-4. The

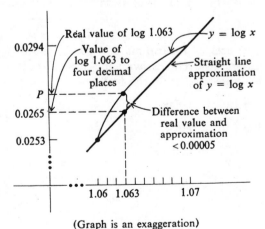

(Graph is an exaggeration)

Figure 6-4

figure obtained this way (log 1.063 = 0.0265) is accurate to four decimal places.

Example 1 Use linear interpolation to find log 143.6.

Solution. Since $14.36 = 1.436 \times 10^2$, the characteristic is 2. The mantissa is found from the table.

$$\log 1.44 = 0.1584$$
$$\log 1.43 = 0.1553$$
$$\text{difference} = 0.0031$$
$$\tfrac{6}{10} \text{ of difference} = 0.0019$$

$$\log 1.43 = 0.1553$$
$$\tfrac{6}{10} \text{ of difference} = \underline{0.0019}$$
$$\log 1.436 = 0.1572$$

Hence log 143.6 = 0.1572 + 2

The logarithm table can also be used to find x when we know log x. The number whose logarithm is ℓ is called the *antilog* of ℓ. Of course, the number whose log is ℓ is 10^ℓ, so antilog ℓ is just another way of writing 10^ℓ. It is usually necessary to use interpolation to find antilogs to four decimal places.

Example 2 Find antilog (0.1683−4).

Solution. Find the two mantissas in the table that are closest to 0.1683.

$$\log 1.48 = 0.1703$$
$$\log x = 0.1683$$
$$\log 1.47 = 0.1673$$
$$\log x - \log 1.47 = 0.0010$$
$$\log 1.48 - \log 1.47 = 0.0030$$

The given logarithm is therefore 0.0010/0.0030, or approximately 0.3 of the way from log 1.47 to log 1.48. Therefore the antilog is approximately 0.3 of the way from 1.47 to 1.48; that is,

$$\text{antilog } 0.1683 = 1.473$$
$$\text{antilog } (0.1683-4) = 0.0001473$$

REVIEW 6-3, REVIEW QUESTIONS

e
common

1. Natural logarithms have base _____, and logarithms to base 10 are called _____ logarithms.

ln N
log N

2. The natural logarithm of N is written _____, and the common logarithm of N is written _____.

mantissa
characteristic

3. The decimal part of log N is called the _____, and the integer part is called the _____.

decimal point

4. The mantissa depends only on the digits in the number, and not on the position of the _____ _____.

characteristic

5. The position of the decimal point determines the _____ of a logarithm.

integers

6. All mantissas are positive decimals between 0 and 1, but characteristics are positive or negative _____.

interpolation

7. We find the logarithm of a four-place number by linear _____.

1.24

8. The logarithm of 1.239 is the number 0.9 of the way from log 1.23 to log _____.

ℓ

9. The number whose logarithm is ℓ is 10^ℓ, which is also denoted antilog (_____).

exponents

10. Antilog ℓ is the same as 10^ℓ, since common logarithms are _____ of 10.

SECTION 6-3, EXERCISES

Use the table of common logarithms to find x in the following equations.

1. log 100 = x
2. log 0.001 = x
3. log 0.0001 = x
4. log 10,000 = x
5. log 300 = x
6. log 40 = x
7. log 15.6 = x
8. log 2.51 = x
9. log .502 = x
10. log .00406 = x

Example 1
11. log 3.025 = x
12. log 4.505 = x
13. log 2.331 = x
14. log 7.042 = x
15. log 23.18 = x
16. log 54.77 = x

Example 2
17. log x = .3010
18. log x = .8451
19. log x = 3.7016
20. log x = 4.9777
21. log x = .8759
22. log x = .9066

23. $\log x = 2.4803$
24. $\log x = 1.6062$
25. $\log x = .9661 - 3$
26. $\log x = .8910 - 5$
27. $\log x = .7813 - 2$
28. $\log x = .4411 - 1$

6-4 COMPUTATION WITH LOGARITHMS

We now know the three basic properties of logarithms:

$$\log MN = \log M + \log N,$$

$$\log \frac{M}{N} = \log M - \log N,$$

$$\log N^k = k \log N.$$

By means of the table, which allows us to determine $\log N$ from N and vice versa, we use these properties of logarithms to replace multiplication by addition, division by subtraction, and exponentiation by multiplication. Calculations that were formerly done with logarithms are now commonly done with a calculator. Nevertheless, it is important to understand the methods involved in calculating with logarithms.

To use a logarithm table, it is necessary to express logarithms in such a way that we have a positive decimal mantissa and an integer characteristic. For example, suppose $\log x = 0.5319 - 2$ and we want to find $\sqrt[3]{x}$. Since

$$\log \sqrt[3]{x} = \log x^{1/3} = \tfrac{1}{3} \log x,$$

we have

$$\log \sqrt[3]{x} = \tfrac{1}{3}(0.5318 - 2).$$

We write the log as $1.5318 - 3$ instead of $0.5318 - 2$, so that we end up an integer characteristic after division.

$$\log \sqrt[3]{x} = \tfrac{1}{3}(1.5318 - 3)$$
$$= 0.5106 - 1$$

Similarly, it is sometimes convenient to write a logarithm like $0.5318 - 2$ in the form $8.5318 - 10$. This would be the case if subtraction were involved, so we would end up with a positive mantissa. These techniques are illustrated in the following examples.

Example 1 Find the following product:

$$x = (0.00587)(326500)$$

Solution. We find the product by finding the log of each factor, adding the logs and then finding the antilog. However, note that the second factor has four significant digits, and so you must interpolate to find log 326500.

Interpolation

log 327000 = 5.5145

log 326000 = 5.5132
difference = 0.0013

$\frac{5}{10}$ difference = 0.0007

log 326000 = 5.5132
log 326500 = 5.5139

Computation

log 326500 = 5 + .5139

log 0.00587 = −3 + .7686

log x = 2 + 1.2825

= 3.2825

Now we have to use interpolation again to find the antilog.

log 1910 = 3.2810 ⎤ 0.0015 ⎤
log x = 3.2825 ⎦   ⎬ 0.0023 $\frac{0.0015}{0.0023} = 0.7$
log 1920 = 3.2833   ⎦

From this we see that x is 0.7 of the way from 1910 to 1920, hence $x = 1917$.

Example 2 Find the following quotient:

$$x = \frac{0.00147}{762}$$

Solution. In a division problem, we subtract logarithms:

log 0.00147 = −3 + 0.1673

log 762 = 2 + 0.8820

In order to end up with a positive mantissa, we rewrite log 0.00147 in the following way:

log 0.00147 = −3 + 0.1673 = 7 + 0.1673 − 10 =   7.1673 − 10
log 762 =   2 + 0.8820   =   2.8820
                                        log x =   4.2853 − 10

                                                = −6 + 0.2853

222 *Exponential and Logarithmic Functions*

To find the antilog we have to interpolate

$$\left.\begin{array}{l}\log 0.00000192 = -6 + 0.2833 \\ \log x = -6 + 0.2853 \\ \log 0.00000193 = -6 + 0.2856\end{array}\right] 0.0023 \quad \dfrac{0.0020}{0.0023} = .9$$

with 0.0020 bracketed with the first two.

Hence $x = 0.000001929$.

Example 3 Find the fourth root of 0.0296. In other words, find x when

$$x = \sqrt[4]{0.0296}$$

Solution

$$\log x = \tfrac{1}{4}(-2 + 0.4713)$$
$$\log x = \tfrac{1}{4}(-4 + 2 + 0.4713)$$
$$= \tfrac{1}{4}(2.4713 - 4)$$
$$= 0.6178 - 1$$

Now interpolation again:

$$\left.\begin{array}{l}\log 0.414 = 0.6170 - 1 \\ \log x = 0.6178 - 1 \\ \log 0.415 = 0.6180 - 1\end{array}\right] 0.0010 \quad \dfrac{0.0008}{0.0010} = \dfrac{8}{10} = .8$$

Hence $x = 0.4148$.

SECTION 6-4, REVIEW QUESTIONS

1. We can replace multiplication by addition of logarithms according to the law: $\log(MN) = $ _____.   **log M + log N**
2. Division is replaced by subtraction according to the law $\log M/N = $ _____.   **log M − log N**
3. Powers and roots are found by using $\log N^k = $ _____.   **k log N**
4. To find antilogs from the table, the logarithm must be expressed as an integer plus a _____ mantissa.   **positive**
5. If we want to divide the logarithm $0.1234 - 7$ by 3 to find the log of a cube root, we would first write the log as $0.1234 - 7 = 2.1234 - $ _____.   **9**

6. If we want to subtract $\log M = 0.8796$ from $\log N = 0.1234$, we first write $\log N$ as $1.1234 - \underline{\hspace{1cm}}$.

SECTION 6-4, EXERCISES

Use logarithms to find the value of x in each of the following expressions.

1. $x = (53.8)(0.739)$
2. $x = (14.1)(0.299)$
3. $x = (98.43) \div (253.7)$
4. $x = (111) \div (0.000738)$
5. $x = (3.57 \times 10^{-4}) \div (9.75 \times 10^4)$
6. $x = [(640)(0.849)] \div (31.4)$
7. $x = (0.0315)^3$
8. $x = (0.008976)^4$
9. $x = (6.43)^3 \times (8.59)^4$
10. $x = 1 \div (1.23)^6$
11. $x = \sqrt[4]{0.0310}$
12. $x = \sqrt[7]{0.0956}$
13. $x = \sqrt[5]{(0.0654)^2}$
14. $x = \sqrt[4]{(0.796)^3}$
15. $x = \sqrt{\dfrac{(0.0732)(23.2)^2}{(0.895)^2 \times (73.5)^2}}$
16. $x = \dfrac{(6.38)^2 \times (8.43)^2}{\sqrt[3]{(0.635)^5}}$

6-5 EXPONENTIAL AND LOGARITHMIC EQUATIONS

We will now study some techniques that will allow us to solve certain types of conditional equations involving exponents or logarithms. These methods are illustrated in the following examples.

Example 1 Solve the equation $6^x = 12$.

Solution. Since $6 > 1$, we know that 6^x is a strictly increasing function whose range is the set of all positive numbers. Therefore, there will be exactly one root of the given equation. We find a formula for x by taking logarithms:

$$\log 6^x = \log 12$$

$$x \log 6 = \log 12$$

$$x = \log 12 / \log 6$$

From the Table of Common Logarithms we find $\log 12 = 1.0792$ and $\log 6 = 0.7782$. Therefore,

$$x = \frac{1.0792}{0.7782},$$

which can be calculated by using long division, logarithms, or a calculator. In any case $x = 1.39$.

Example 2 Solve $10^{x-3} = 2^{2x+1}$.

Solution. Taking logarithms of both sides gives an equivalent equation which can be solved for x.

$$\log 10^{x-3} = \log 2^{2x+1}$$

$$(x-3)\log 10 = (2x+1)\log 2$$

Since $\log 10 = 1$, we can simplify the equation in the following way.

$$x - 3 = (2x + 1)\log 2$$
$$x - 3 = 2x \log 2 + \log 2$$
$$x - 2x \log 2 = 3 + \log 2$$
$$x(1 - 2 \log 2) = 3 + \log 2$$
$$x = \frac{3 + \log 2}{1 - 2 \log 2}$$
$$= \frac{3 + \log 2}{1 - \log 4}$$
$$= \frac{3 + 0.3010}{1 - 0.6021}$$
$$= \frac{3.3010}{0.3979}$$
$$= 8.56$$

Equations involving logarithms sometimes can be solved by exponentiating both sides.

Example 3 Solve $\log(5x + 3) = \log x + 1$.

Solution. First consolidate the logarithm terms, and then exponentiate both sides.

$$\log(5x + 3) = \log x + 1$$
$$\log(5x + 3) - \log x = 1$$
$$\log\left(\frac{5x+3}{x}\right) = 1$$
$$\frac{5x+3}{x} = 10^1$$
$$5x + 3 = 10x$$

$$3 = 5x$$
$$x = \frac{3}{5}.$$

Example 4 Solve $\log x + \log (x + 3) = 1$

Solution.

$$\log x + \log (x + 3) = 1$$
$$\log x(x + 3) = 1$$
$$x(x + 3) = 10$$
$$x^2 + 3x - 10 = 0$$
$$(x + 5)(x - 2) = 0$$
$$x = -5 \text{ or } x = 2$$

Since $\log x$ and $\log (x + 3)$ are not defined for $x = -5$, the only solution is $x = 2$.

Example 5 Solve the equation $\ln x + 1 = \ln (x + 1)$.

Solution

$$\ln x + 1 = \ln (x + 1)$$
$$1 = \ln (x + 1) - \ln x$$
$$1 = \ln \left(\frac{x + 1}{x}\right)$$
$$e = \frac{x + 1}{x}$$
$$(e - 1)x = 1$$
$$x = \frac{1}{e - 1} = \frac{1}{1.72} = 0.583$$

Example 6 Solve the equation $\frac{1}{2}(e^x + e^{-x}) = 2$.

Solution.
$$\tfrac{1}{2}(e^x + e^{-x}) = 2$$
$$e^x + e^{-x} = 4$$

Now multiply both sides of the equation by e^x.
$$e^{2x} + 1 = 4e^x$$
$$(e^x)^2 - 4e^x + 1 = 0$$

This last equation is a quadratic equation in e^x. From the quadratic formula we get

$$e^x = \frac{-(-4) \pm \sqrt{16 - (4)(1)(1)}}{2}$$

$$= \frac{4 \pm \sqrt{12}}{2}$$

$$= 2 \pm \sqrt{3}$$

$$= 3.732 \text{ or } 0.268$$

Since $e^x = 3.732$ or $e^x = 0.268$, we have
$$x = \ln 3.732 \,,$$
or
$$x = \ln 0.268 \,.$$

From the Table of Natural Logarithms in the appendix, we find
$$x = 1.3169\,,$$
$$x = -1.3170\,.$$

Although we have lost some accuracy in the fourth decimal place in our work, the two roots are in fact negatives of each other, as can be seen from the fact that $\tfrac{1}{2}(e^x + e^{-x})$* is symmetric about the y-axis.

Problems involving compound interest frequently lead to logarithmic equations. For example, if a principal amount of money P is invested at an interest rate of r, then the amount A of money after n years is

$$A = P(1 + r)^n \,.$$

---

*$\tfrac{1}{2}(e^x + e^{-x})$ occurs frequently in applications and is called the hyperbolic cosine of x, denoted $\cosh x$; thus $\cosh x = \tfrac{1}{2}(e^x + e^{-x})$.

Example 7 How many years must you leave money invested at 6% before the amount doubles?

Solution. If P is the original amount, then the condition is

$$P(1 + 0.06)^n = 2P,$$

or

$$(1.06)^n = 2.$$

Taking logarithms, we get

$$n \log (1.06) = \log 2,$$

$$n = \frac{\log 2}{\log 1.06}$$

$$= \frac{0.3010}{0.0253}$$

$$= 11.9.$$

Therefore, your money will slightly more than double if invested at 6% for 12 years.

Another type of problem that leads to logarithms involves exponential decay. In many situations, the rate of growth or decay of a substance is proportional to the amount of the substance present at any given time. In such a case the amount A of substance present at a given time t is given by

$$A = A_0 e^{kt}$$

where A_0 is the amount at time $t = 0$, and k is a constant.

Example 8 A culture of bacteria contains 1000 bacteria initially, and 3000 bacteria after one hour. How many bacteria will the culture contain after 2 hours?

Solution. We have $A = A_0 e^{kt}$, where $A_0 = 1000$, and $A = 3000$ when $t = 1$. Thus

$$3000 = 1000 e^k$$

$$e^k = 3$$

When $t = 2$, we have

$$A = 1000 e^{k \cdot 2}$$

$$= 1000(e^k)^2$$
$$= 1000(3)^2$$
$$= 9000$$

SECTION 6-5, REVIEW QUESTIONS

1. If the unknown in a conditional equation occurs as an exponent, you can frequently solve the equation by first taking the _____ of both sides.   **logarithm**

2. If the unknown occurs in logarithmic expressions, you _____ both sides of the equation.   **exponentiate**

3. Before exponentiating both sides of the equation, it is convenient to combine all the _____ into one term.   **logarithms**

4. In solving a conditional equation involving logarithms, you must check that the roots you find give _____ numbers in the logarithmic expressions.   **positive**

5. The domain of $\log x$ is all positive numbers, so $\log x$ is not defined if x is zero or _____.   **negative**

6. The principal amount A that results from leaving P invested at an interest rate of r per year for n years is $A = $ _____. $P(1+r)^n$

7. The formula governing exponential growth or decay is $A = A_0 e^{kt}$, where k is a constant and A_0 is the amount at time $t = $ _____.   **0**

SECTION 6-5, EXERCISES

Solve the following equations.

1. $3^x = 10$
2. $6^x = 10$ Example 1,2
3. $5^x = 7$
4. $6^x = 3$
5. $3^x = 2^{x+1}$
6. $5^{x-2} = 7^x$
7. $10^{2x-3} = 43$
8. $e^{3x} = 16$
9. $e^{2x+1} = 3^x$
10. $10^x = 5^{x+2}$ Example 3,4
11. $\log(x+9) = \log x + 1$

12. $\log(3x+7) = \log x + 1$
13. $\log(2x+5) = \log(x+1) + 1$
14. $\log(3x+1) = \log(x-1) + 1$
15. $\log(x-8) = \log(x+1) - 1$
16. $\log(x-6) = \log(x+3) - 1$
17. $\log(3x-1) = \log(x+2) + 1$
18. $\log(2x-3) = \log(x+1) + 1$
19. $\log(10x+5) = 2 + \log(x-1)$
20. $\log(20x+7) = 2 + \log(x-2)$
21. $\log x + \log(x-9) = 1$
22. $\log(x+2) + \log(x-7) = 1$
23. $\log(2x+4) + \log(x-2) = 1$
24. $\log(3x-8) + \log(x+7) = 1$
25. $\log 5 + \log(1-x) = 1 + \log(x^2 + 2x - 1)$
26. $\log(2+2x) - 1 = \log(x^2 + 3x + 1) - \log 5$
27. $\log(x+1) + \log(2x+1) = \log(x^2 - 1)$
28. $\log(x+3) + \log(2x+1) = \log(x^2 + 3x)$

Example 6
29. $e^x + e^{-x} = 2$ 30. $e^x - e^{-x} = 0$
31. $e^x + 2e^{-x} = 3$ 32. $e^x - 3e^{-x} = 2$
33. $\frac{1}{2}(e^x + e^{-x}) = 3$ 34. $\frac{1}{2}(e^x - e^{-x}) = 1$

Example 7
35. $1000 is invested at 6% interest, compounded annually. What is the total amount A after 20 years?
36. A $100,000 house appreciates in value 5% each year. What is the house worth after 10 years?
37. How long does it take to double your money if you invest it at $4\frac{1}{2}\%$, compounded annually?
38. How long does it take to double your investment if you invest it at 3%, compounded annually?
39. If an amount of money triples in 30 years when interest is compounded annually, what is the interest rate?
40. Do exercise 39 if the money quadruples in 30 years.

Example 8
41. A culture contains 2000 bacteria initially and 8000 bacteria after one hour. How many bacteria will there be after (*a*) 2 hours? (*b*) 4 hours?
42. Radium decays exponentially, and half decays in 1600 years ($\frac{1}{2} A_o = A_o e^{k \cdot 1600}$). How much is left after 2000 years?

6-6 REVIEW FOR CHAPTER

1. Arrange the following in order of increasing magnitude.
   $2^{2/3}$, $(4^{5/2})(8^{-1})$, $(1/2)^{-4/3}$, 2^{-3}, $(2^{-2/9})^9$

Find the value of y if:

2. $8^y = (2^3)^2$ 3. $8^y = 2^{(3^2)}$ 4. $2^{(4^5)} = 16^y$ 5. $(2^4)^5 = 16^y$

Sketch the graph of each pair of functions on the same coordinate axis.

6. (a) $f(x) = 2^x$
   (b) $g(x) = \log_2 x$
   (c) Are f and g inverses of each other?
7. (a) $f(x) = 10^x$
   (b) $g(x) = \log x$
   (c) Are f and g inverses of each other?
8. (a) $f(x) = e^x$
   (b) $g(x) = \ln x$
   (c) Are f and g inverses of each other?

Express the following in exponential form:

9. $\log 35 = y$
10. $\log_2 25 = y$
11. $\log_e d = b$
12. $2 \log 5 = x$
13. $\ln 7 + \ln 6 = x$
14. $\frac{1}{2} \ln 25 - \ln 2 = x$
15. (a) What is the domain and range of f if $f(x) = 1^x$?
    (b) Does f have an inverse?
    (c) Can the number 1 be used as a base for logarithms?

Solve for x in each of the following:

16. $\log(x^2 - 1) - 2 \log(x - 1) = \log 3$
17. $\log_6(x + 9) + \log_6 x = 2$
18. $10^{\log 5} + 10^{\log 3} = 10^{\log x}$
19. $e^{\ln 4} + e^{\ln 8} = e^{\ln 3x}$
20. $4^{x+2} = 8^{3x-7}$
21. $2^{4x} = 3^{2x-5}$
22. $2^x = (\frac{1}{2})^{x-4}$
23. $(\log x)^2 - 5 \log x + 6 = 0$
24. $\log(4x) + \log x = 2$
25. $5^{x+3} - 5^{x+2} = 2^{3x}$

7 CIRCULAR FUNCTIONS

7-1 SINE AND COSINE FUNCTIONS

In this chapter we will discuss the functions sine, cosine, tangent, cosecant, secant, and cotangent. These are called the *circular* or *trigonometric* functions. In trigonometry these functions are associated with angle measurements. In calculus we also consider these as ordinary numerical functions defined on the whole real line. Our introduction of these functions will be independent of the idea of angle, but we will show their connection with angle measurement in Chapter 8.

To define the sine and cosine functions, we start with the unit circle $x^2 + y^2 = 1$. For any positive number s consider the point $P(s)$ on the circle, which you get to by going a distance s counterclockwise around the circle from the point $(1, 0)$.

We now define the cosine of s, written cos s, to be the x-coordinate of $P(s)$ and the sine of s, written sin s, to be the y-coordinate of $P(s)$, See Fig. 7-1.

The circumference of the unit circle is 2π units. Therefore the point $(0, 1)$ which is one-quarter the way around the circle is $\pi/2$ units counterclockwise from $(1, 0)$. Hence we have

$$P(\pi/2) = (0, 1),$$

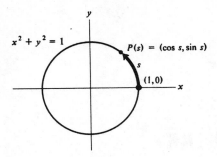

Figure 7-1

which means that

$$\cos \pi/2 = 0 \quad \text{and} \quad \sin \pi/2 = 1.$$

Similarly, the point $(-1, 0)$ is π units from $(1, 0)$, and the point $(0, -1)$ is $\frac{3}{2}\pi$ units counterclockwise from $(1, 0)$. We now have the following information:

$P(0) = (1, 0);$ $\quad\quad$ $\cos 0 = 1,$ $\quad\quad$ $\sin 0 = 0;$

$P(\pi/2) = (0, 1);$ $\quad\quad$ $\cos \pi/2 = 0,$ $\quad\quad$ $\sin \pi/2 = 1;$

$P(\pi) = (-1, 0);$ $\quad\quad$ $\cos \pi = -1,$ $\quad\quad$ $\sin \pi = 0;$

$P(3\pi/2) = (0, -1);$ $\quad\quad$ $\cos 3\pi/2 = 0,$ $\quad\quad$ $\sin 3\pi/2 = -1.$

For negative values of s, we let $P(s)$ be the point on the unit circle you get to by going $|s|$ units *clockwise* around the circle starting at $(1, 0)$. For negative values of s, we again define $\cos s$ and $\sin s$ so that $P(s) = (\cos s, \sin s)$. Hence, we also have the following facts:

$P(-\pi/2) = (0, -1);$ $\quad\quad$ $\cos (-\pi/2) = 0,$ $\quad\quad$ $\sin (-\pi/2) = -1;$

$P(-\pi) = (-1, 0);$ $\quad\quad$ $\cos (-\pi) = -1,$ $\quad\quad$ $\sin (-\pi) = 0;$

$P(-3\pi/2) = (0, 1);$ $\quad\quad$ $\cos (-3\pi/2) = 0,$ $\quad\quad$ $\sin (-3\pi/2) = 1.$

It should be clear that for $s = 2\pi$ we get back to the starting point, so that $P(2\pi) = (1, 0)$. Therefore, $\cos 2\pi = \cos 0 = 1$, and $\sin 2\pi = \sin 0 = 0$. For any value of s, $P(s) = P(s + 2\pi) = P(s + 4\pi) = \ldots$, and $P(s) = P(s - 2\pi) = P(s - 4\pi) = \ldots$. We now have our first two identities:

$$\cos s = \cos (s + 2k\pi),$$
$$\sin s = \sin (s + 2k\pi).$$
(k an integer)

This property of the sine and cosine functions is called *periodicity*. Any function f is *periodic* if there is a number p such that

$$f(x + p) = f(x)$$

for all values of x. The smallest value of p such that $f(x + p) = f(x)$ is called the *period* of the function. Hence the sine and cosine are periodic functions, and *they both have the period* 2π.

There are other properties of sine and cosine which are immediate from the definition. For example, Fig. 7-2 illustrates the fact that

$$\cos s = \cos (-s),$$
$$\sin s = -\sin (-s).$$

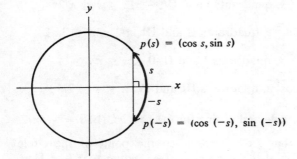

Figure 7-2

The following array of values also suggests a relationship between cosine and sine:

$\cos 0 = 1, \quad \cos \pi/2 = 0, \quad \cos \pi = -1, \quad \cos 3\pi/2 = 0, \quad \cos 2\pi = 1,$
$\sin \pi/2 = 1, \quad \sin \pi = 0, \quad \sin 3\pi/2 = -1, \quad \sin 2\pi = 0, \quad \sin 5\pi/2 = 1.$

The relation suggested by this table is

$$\cos s = \sin (s + \pi/2).$$

Geometrically this says that the x-coordinate of a point on the circle is the same as the y-coordinate of the point one-quarter circumference farther around in the counterclockwise direction. This fact is proved in Sec. 7-5.

Since $\cos s$ and $\sin s$ are the x- and y-coordinates of a point on the unit circle, they must satisfy the equation

$$x^2 + y^2 = 1.$$

That is, for every value of s,
$$(\cos s)^2 + (\sin s)^2 = 1.$$

Another consequence of $(\cos s, \sin s)$ being a point on the unit circle is the fact that
$$-1 \le \cos s \le 1 \quad \text{and} \quad -1 \le \sin s \le 1$$
for all values of s. To simplify the notation we write $\cos^2 s$ for $(\cos s)^2$, and $\sin^2 s$ for $(\sin s)^2$. Be careful to make the distinction between the notations $\cos^2 s$ and $\cos s^2$; $\cos^2 s = (\cos s)^2$ and $\cos s^2 = \cos(s^2)$. The identity above is usually written
$$\cos^2 s + \sin^2 s = 1.$$
From this identity, we also get immediately the following identities.

$\cos s = \sqrt{1 - \sin^2 s}$, quadrants I and IV, $-\pi/2 \le s \le \pi/2$

$\cos s = -\sqrt{1 - \sin^2 s}$, quadrants II and III, $\pi/2 \le s \le 3\pi/2$

$\sin s = \sqrt{1 - \cos^2 s}$, quadrants I and II, $0 \le s \le \pi$

$\sin s = -\sqrt{1 - \cos^2 s}$, quadrants III and IV, $\pi \le s \le 2\pi$

Example 1 Find the sine and cosine for (a) $s = 3\pi$, (b) $s = -5\pi/2$.

Solution. (a) Note that $3\pi = 2\pi + \pi$, so the point of the circle corresponding to 3π is also the point corresponding to $s = \pi$ [i.e., the point half a circumference from $(1, 0)$]. In Fig. 7-3(a) we see that $\cos 3\pi = \cos \pi = -1$, $\sin 3\pi = \sin \pi = 0$. Similarly, in Fig. 7-3(b)

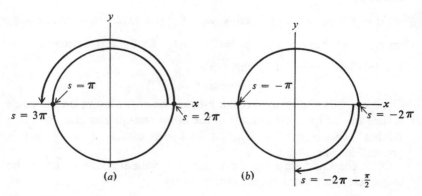

Figure 7-3

236 Circular Functions

we write $-5\pi/2 = -2\pi - \pi/2$, so the point of the circle corresponding to $-5\pi/2$ is one-quarter circumfereace clockwise from $(1, 0)$. Hence $\cos(-5\pi/2) = \cos(-\pi/2) = 0$, $\sin(-5\pi/2) = \sin(-\pi/2) = -1$.

Example 2 Find $\cos s$ if $\sin s = \sqrt{2}/2$, and s corresponds to a point of the circle in the second quadrant.

Solution. We know that $\cos^2 s + \sin^2 s = 1$, and that $\cos s$ is negative if s corresponds to a point in the second quadrant. Therefore,

$$\cos s = -\sqrt{1 - \sin^2 s}$$
$$= -\sqrt{1 - (\sqrt{2}/2)^2}$$
$$= -\sqrt{1 - 1/2}$$
$$= -\sqrt{2}/2.$$

Example 3 Given that $\cos(2.5) = -0.80$, find $\sin(2.5)$.

Solution. Since $\pi = 3.14$ and $\pi/2 = 1.57$, the point corresponding to $s = 2.5$ is in the second quadrant. Therefore $\sin 2.5 > 0$, and

$$\sin 2.5 = \sqrt{1 - \cos^2(2.5)}$$
$$= \sqrt{1 - (-0.80)^2}$$
$$= \sqrt{1 - 0.64}$$
$$= \sqrt{0.36}$$
$$= 0.60.$$

Example 4 Let $\pi/2 < s < \pi$, so that $P(s)$ is a point on the unit circle in the second quadrant. Find the number s_1 with $0 < s_1 < \pi/2$, so that $P(s_1)$ is in the first quadrant and symmetric to $P(s)$ about the y-axis. Write $\cos s$ and $\sin s$ in terms of $\cos s_1$, $\sin s_1$.

Solution. Observe from Fig. 7-4 that $P(\pi - s)$ and $P(s)$ are symmetric about the y-axis. So let $s_1 = \pi - s$. Two points symmetric about the y-axis have the same y-coordinate, and x-coordinates which are negatives of each other. Therefore,

$$\cos s = -\cos(\pi - s) = -\cos s_1,$$
$$\sin s = \sin(\pi - s) = \sin s_1.$$

The method of Example 4 shows how $\cos s$ and $\sin s$ can always

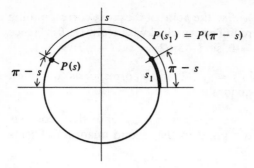

Figure 7-4

be expressed as $\pm\cos s_1$ and $\pm\sin s_1$ for some s_1 between 0 and $\pi/2$. If $P(s)$ is in the fourth quadrant, look at the value of s_1 so that $P(s)$ and $P(s_1)$ are symmetric about the x-axis. Then $s_1 = 2\pi - s$, $\cos s = \cos s_1$, and $\sin s = -\sin s_1$ (see Fig. 7-5). If $P(s)$ is in the third quadrant, then look at the value of s_1 so that $P(s)$ and $P(s_1)$ are

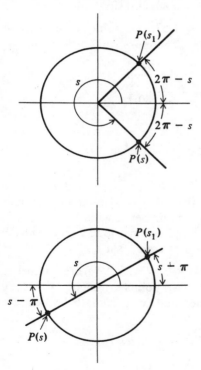

Figure 7-5

238 *Circular Functions*

symmetric about the origin. In this case $s_1 = s - \pi$, $\cos s = -\cos s_1$, ans $\sin s = -\sin s_1$.

SECTION 7-1, REVIEW QUESTIONS

1. The functions sine, cosine, etc., are called circular functions, or _____ functions.   **trigonometric**

2. In trigonometry these functions are associated with angular measurement, but here we consider them as ordinary functions defined on the real _____.   **line**

3. For any positive real number s, $\cos s$ is the x-coordinate of the point s units around the circumference of the unit circle from $(1, 0)$ in the _____ direction.   **counterclockwise**

4. Positive values of s are associated with points gotten to by going counterclockwise around the circle; negative values of s correspond to points gotten to by going _____ around the circle.   **clockwise**

5. $\cos s$ is the _____-coordinate of the point $P(s)$ on the unit circle,   **x**
   and $\sin s$ is its _____-coordinate.   **y**

6. Since $(\cos s, \sin s)$ is on the unit circle, for every number s we have $\cos^2 s + \sin^2 s =$ _____.   **1**

7. Since $\cos (s + 2\pi) = \cos s$ and $\sin (s + 2\pi) = \sin s$ for all real numbers s, cosine and sine are called _____ functions.   **periodic**

8. The period of both the functions cosine and sine is _____.   **2π**

9. If p is the period of a periodic function f, then $f(x + p) =$ _____ for all x.   **f(x)**

10. Using the symmetry properties of the unit circle, we can always write $\sin s$ or $\cos s$ in terms of $\sin s_1$ or $\cos s_1$ for some value of s_1 between 0 and _____.   **$\pi/2$**

SECTION 7-1, EXERCISES

Find the $\cos s$ and $\sin s$ of the following values of s:

1. 0   2. 2π 3. $-\dfrac{\pi}{2}$ 4. $-\pi$ **Example 1**

5. 3π 6. -5π 7. $-\dfrac{3\pi}{2}$ 8. $\dfrac{3\pi}{2}$

What quadrant is $P(s)$ in if:

9. $\sin s > 0, \cos s < 0$
10. $\sin s < 0, \cos s < 0$
11. $\sin s > 0, \cos s > 0$
12. $\sin s < 0, \cos s > 0$

Find $\cos s$ if:

Example 2

13. $\sin s = \dfrac{\sqrt{2}}{2}$, $P(s)$ in second quadrant.

14. $\sin s = -\dfrac{\sqrt{2}}{2}$, $P(s)$ in third quadrant.

15. $\sin s = \dfrac{1}{10}$, $P(s)$ in first quadrant.

16. $\sin s = \dfrac{1}{10}$, $P(s)$ in second quadrant.

Find $\sin s$ if:

17. $\cos s = -\dfrac{1}{2}$, $P(s)$ in second quadrant.

18. $\cos s = \dfrac{1}{2}$, $P(s)$ in fourth quadrant.

19. $\cos s = \dfrac{3}{4}$, $P(s)$ in first quadrant.

20. $\cos s = -\dfrac{3}{4}$, $P(s)$ in third quadrant.

Example 3

21. Find $\sin 1$, given $\cos 1 = .540$.
22. Find $\sin \dfrac{1}{2}$, given $\cos \dfrac{1}{2} = .877$.
23. Find $\sin 2$, given $\cos 2 = -.417$.
24. Find $\sin 3$, given $\cos 3 = -.990$.
25. Find $\cos 5$, given $\sin 5 = -.958$.
26. Find $\cos 4$, given $\sin 4 = -.757$.

Use the information given in exercises 21 through 26 to find the following values. Draw a figure showing $P(s)$ and an appropriate symmetric point $P(s_1)$ in the first quadrant.

Example 4

27. $\cos(\pi - 1)$
28. $\cos\left(\pi - \dfrac{1}{2}\right)$

240 *Circular Functions*

29. $\cos(-1)$
30. $\cos\left(-\dfrac{1}{2}\right)$
31. $\sin(2\pi - 5)$
32. $\sin(5 - \pi)$
33. $\cos(\pi - 3)$
34. $\cos(\pi + 3)$
35. $\sin(4 - \pi)$
36. $\sin(4 + \pi)$
37. $\sin(-4)$
38. $\sin(-5)$

Find $\cos s$ and $\sin s$ in terms of $\cos s_1$ and $\sin s_1$, where $P(s_1)$ is in the first quadrant, if:

39. $\dfrac{\pi}{2} < s < \pi$ and $P(s), P(s_1)$ are symmetric about the y-axis. **Example 4**

40. $\pi < s < \dfrac{3\pi}{2}$ and $P(s), P(s_1)$ are symmetric about the origin.

41. $\dfrac{3\pi}{2} < s < 2\pi$ and $P(s), P(s_1)$ are symmetric about the x-axis.

7-2 GRAPHS OF SIN x AND COS x

So far we have evaluated $\sin s$ only for values of s which are some multiple of $\pi/2$. These are the values of s that correspond to the points $(1, 0)$, $(0, 1)$, $(-1, 0)$, $(0, -1)$ on the unit circle. Now we will calculate the sine and cosine for multiples of $\pi/4$ and multiples of $\pi/6$. This will provide us with enough data to get a good idea of the graphs of $y = \sin x$ and $y = \cos x$.

The value $s = \pi/4$ corresponds to the intersection of the unit circle with the line $y = x$. See Fig. 7-6. We can find the x and y coordinates of this point by solving the two equations

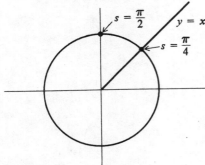

Figure 7-6

$$x^2 + y^2 = 1,$$
$$y = x.$$

Substituting x for y in the first equation gives us the following.

$$x^2 + x^2 = 1$$
$$2x^2 = 1$$
$$x^2 = 1/2$$
$$x = \sqrt{2}/2 = y$$

Therefore,
$$x = \cos \pi/4 = \sqrt{2}/2,$$
and
$$y = \sin \pi/4 = \sqrt{2}/2,$$

The values of $\cos s$, $\sin s$ for $s = 3\pi/4$, $5\pi/4$, $7\pi/4$ are obtained by symmetry. See Fig. 7-7. The values we read from Fig. 7-7 are

$$\cos \frac{\pi}{4} = \frac{\sqrt{2}}{2}, \qquad \sin \frac{\pi}{4} = \frac{\sqrt{2}}{2},$$

$$\cos \frac{3\pi}{4} = -\frac{\sqrt{2}}{2}, \qquad \sin \frac{3\pi}{4} = \frac{\sqrt{2}}{2},$$

$$\cos \frac{5\pi}{4} = -\frac{\sqrt{2}}{2}, \qquad \sin \frac{5\pi}{4} = -\frac{\sqrt{2}}{2},$$

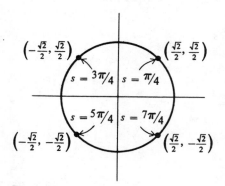

Figure 7-7

$$\cos\frac{7\pi}{4} = \frac{\sqrt{2}}{2}, \qquad \sin\frac{7\pi}{4} = -\frac{\sqrt{2}}{2}.$$

Of course, we now also know the values of sin s and cos s for negative multiples of $\pi/4$, and for values of the form $s = 2\pi + \pi/4, 2\pi + 3\pi/4$, etc.

Next consider the points on the unit circle corresponding to $s = \pi/3$ and $s = 2\pi/3$ (see Fig. 7-8). From symmetry considerations,

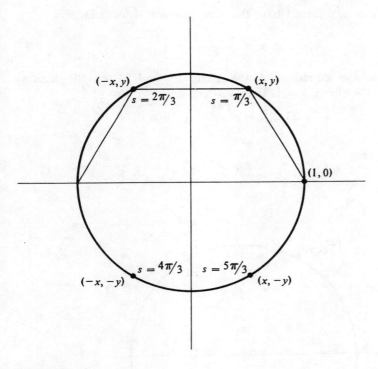

Figure 7-8

we see that if (x, y) is the point corresponding to $s = \pi/3$, then $(-x, y)$ corresponds to $s = 2\pi/3$. All three of the chords of Figure 7-8 have equal lengths, since they subtend equal arcs of length $\pi/3$. The horizontal chord has length $2x$. Therefore, the distance from (x, y) to $(1, 0)$ is also $2x$:

$$\sqrt{(x-1)^2 + (y-0)^2} = 2x,$$

or

$$x^2 - 2x + 1 + y^2 = 4x^2.$$

Graphs of Sin x and Cos x 243

Since (x, y) is a point of the unit circle, we can replace $x^2 + y^2$ in the last equation by 1, and get

$$-2x + 2 = 4x^2,$$
$$2x^2 + x - 1 = 0,$$
$$x = \frac{-1 \pm \sqrt{1 + 8}}{4}.$$

Since $x > 0$, we must have the plus sign in this formula, and

$$x = \frac{-1 + 3}{4} = \frac{1}{2}.$$

We use the equation of the circle to find y (or the identity $\cos^2 \pi/3 + \sin^2 \pi/3 = 1$, since $x = \cos \pi/3$).

$$(1/2)^2 + y^2 = 1 \quad (y \geq 0)$$
$$y^2 = 3/4$$
$$y = \sqrt{3}/2$$

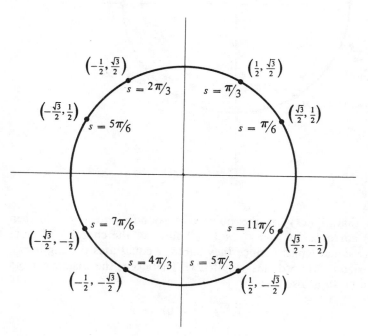

Figure 7-9

Finally, we have

$$\cos \pi/3 = 1/2, \quad \sin \pi/3 = \sqrt{3}/2.$$

This immediately also gives (see Fig. 7-9):

$$\cos 2\pi/3 = -1/2, \quad \sin 2\pi/3 = \sqrt{3}/2,$$
$$\cos 4\pi/3 = -1/2, \quad \sin 4\pi/3 = -\sqrt{3}/2,$$
$$\cos 5\pi/3 = 1/2, \quad \sin 5\pi/3 = -\sqrt{3}/2.$$

The point of the unit circle corresponding to $s = \pi/6$ is clearly symmetric about the line $y = x$ to the point $(1/2, \sqrt{3}/2)$ corresponding to $s = \pi/3$. Therefore, $(\sqrt{3}/2, 1/2)$ corresponds to $s = \pi/6$, and the values of $\cos s$, $\sin s$ for multiples of $\pi/6$ follow from this observation.

Collecting all these facts, we get the following table of values for $\cos s$ and $\sin s$.

s	$\cos s$	$\sin s$	s	$\cos s$	$\sin s$
0	1	0	π	-1	0
$\dfrac{\pi}{6}$	$\dfrac{\sqrt{3}}{2}$	$\dfrac{1}{2}$	$\dfrac{7\pi}{6}$	$-\dfrac{\sqrt{3}}{2}$	$-\dfrac{1}{2}$
$\dfrac{\pi}{4}$	$\dfrac{\sqrt{2}}{2}$	$\dfrac{\sqrt{2}}{2}$	$\dfrac{5\pi}{4}$	$-\dfrac{\sqrt{2}}{2}$	$-\dfrac{\sqrt{2}}{2}$
$\dfrac{\pi}{3}$	$\dfrac{1}{2}$	$\dfrac{\sqrt{3}}{2}$	$\dfrac{4\pi}{3}$	$-\dfrac{1}{2}$	$-\dfrac{\sqrt{3}}{2}$
$\dfrac{\pi}{2}$	0	1	$\dfrac{3\pi}{2}$	0	-1
$\dfrac{2\pi}{3}$	$-\dfrac{1}{2}$	$\dfrac{\sqrt{3}}{2}$	$\dfrac{5\pi}{3}$	$\dfrac{1}{2}$	$-\dfrac{\sqrt{3}}{2}$
$\dfrac{3\pi}{4}$	$-\dfrac{\sqrt{2}}{2}$	$\dfrac{\sqrt{2}}{2}$	$\dfrac{7\pi}{4}$	$\dfrac{\sqrt{2}}{2}$	$-\dfrac{\sqrt{2}}{2}$
$\dfrac{5\pi}{6}$	$-\dfrac{\sqrt{3}}{2}$	$\dfrac{1}{2}$	$\dfrac{11\pi}{6}$	$\dfrac{\sqrt{3}}{2}$	$-\dfrac{1}{2}$
π	-1	0	2π	1	0

We have used s to denote distance around the unit circle in the definition of sin s and cos s. To graph the sine and cosine functions, we now use x and y for the variables in the usual way.

The graphs of $y = \cos x$ and $y = \sin x$ are shown for $0 \le x \le 2\pi$ in Fig. 7-10 (below). By the periodicity of cosine and sine, both graphs repeat this pattern on every interval of length 2π.

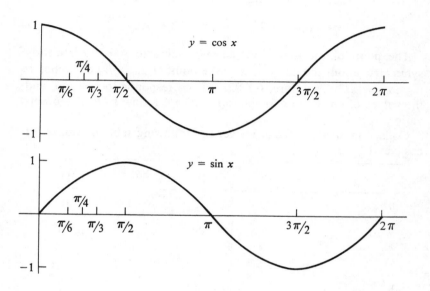

Figure 7-10

SECTION 7-2, REVIEW QUESTIONS

1. The part of the unit circle in the first quadrant has length $\pi/2$, and the point corresponding to $s = \pi/4$ lies on the line $y = $ _____.

2. If (x, y) is the point $P(s)$ corresponding to $s = \pi/4$, then $y = x$ and $x^2 + y^2 = $ _____.

3. Solving the equations $y = x$ and $x^2 + y^2 = 1$ gives $x = \cos \pi/4 = $ _____, and $y = \sin \pi/4 = $ _____.

4. From the values $\cos \pi/4 = \sqrt{2}/2$ and $\sin \pi/4 = \sqrt{2}/2$ we can compute from symmetry the values of $\cos s$, $\sin s$ when s is any multiple of _____.

246 *Circular Functions*

5. The points of the unit circle corresponding to $s = \pi/3$ and $s = 2\pi/3$ are symmetric about the _____-axis.

6. From this symmetry about the y-axis and the other geometric considerations we calculate that $\cos \pi/3 = 1/2$ and $\sin \pi/3 = $ _____.

7. Knowing that $\cos \pi/3 = 1/2$ allows us to calculate $\sin \pi/3$ from the identity $\cos^2 \pi/3 + $ _____ $ = 1$.

y

$\sqrt{3}/2$

$\sin^2 \dfrac{\pi}{3}$

SECTION 7-2, EXERCISES

1. Write decimal values for x and $\sin x$ for $x = 0, \dfrac{\pi}{6}, \dfrac{\pi}{4}, \dfrac{\pi}{3}, \dfrac{\pi}{2}$, and make a careful graph of $y = \sin x$ on $\left[0, \dfrac{\pi}{2}\right]$.

2. Proceed as in exercise 1 for $y = \cos x$.

Graph the point $P(s)$ on the unit circle and give its coordinates $(\cos s, \sin s)$ for each of the following values of s.

3. $\dfrac{13\pi}{6}$ 4. $-\dfrac{3\pi}{4}$ 5. $\dfrac{5\pi}{6}$ 6. 3π 7. $-\dfrac{2\pi}{3}$

8. $\dfrac{2\pi}{3}$ 9. -7π 10. $\dfrac{5\pi}{4}$ 11. $-\dfrac{3\pi}{4}$ 12. $-\dfrac{7\pi}{4}$

Graph the following functions for $0 \leq x \leq 2\pi$.

13. $y = -\sin x$
14. $y = -\cos x$
15. $y = \cos(-x)$
16. $y = \sin(-x)$
17. $y = 1 + \sin x$
18. $y = \cos x - 1$
19. $y = \sin(2\pi - x)$
20. $y = \cos(2\pi - x)$

7-3 GRAPHS OF
$y = A \ SIN \ (Bx+C) + D$,
$y = A \ COS \ (Bx+C) + D$

The graphs of $y = \sin x$ and $y = \cos x$ are shown in Fig. 7-11. Since $\sin(x + \pi/2) = \cos x$, the curves $y = \sin x$ and $y = \cos x$ have exactly the same shape, even though they are positioned differently on the plane. Any curve with this shape is called a sine curve. If we

Graphs of $y = A \ Sin \ (Bx + C) + D$, $y = A \ Cos \ (Bx + C) + D$

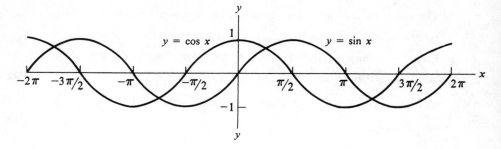

Figure 7-11

change the scale on either axis, or both axes, we will still call the resulting curve a sine curve, or sine wave. We will also refer to such a curve which has been translated up or down, or right or left, as a sine wave.

The graphs of equations of the form

$$y = A \sin (Bx + C) + D, \qquad y = A \cos (Bx + C) + D$$

all have this same basic sine wave shape. The graph of $y = \sin x$ between $x = 0$ and $x = 2\pi$ is called one cycle of the curve. In general, the corresponding part of the graph of any sine wave—over an interval equal to the period of the function—is called one *cycle* of the curve. The maximum values of the functions $A \cos (Bx + C)$ and $A \cos (Bx + C)$ is $|A|$. This maximum value is called the *amplitude* of the curve. For a translated curve $y = A \sin (Bx + C) + D$, or $y = A \cos (Bx + C) + D$, the amplitude is also $|A|$. For example, the maximum of $y = -2 \sin x + 3$ is 5, but the amplitude is 2. The cycle length and amplitude completely determine the shape of a sine wave, but not its position relative to the axes.

Let us consider first how each of the following graphs differs from the basic curve $y = \sin x$:

$$y = A \sin x,$$
$$y = \sin Bx,$$
$$y = \sin (x + C).$$

The curve $y = A \sin x$ is obtained from $y = \sin x$ by multiplying each y-value by A. Hence the length of a cycle (the period) remains 2π, and the amplitude becomes $|A|$ instead of 1. (See Fig. 7-12.) For

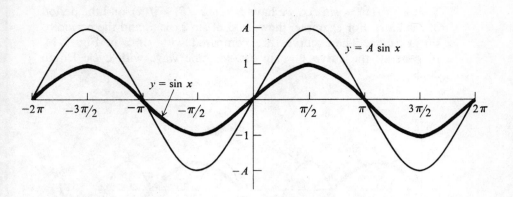

Figure 7-12

negative values of A, the sine wave would be reflected through the x-axis, and then multiplied by $|A|$. The example, $y = -2 \sin x$, is shown in Fig. 7-13.

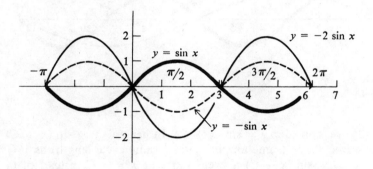

Figure 7-13

The graph of $y = \sin Bx$ looks like the graph of $y = \sin x$, but with the scale changed on the x-axis by a factor B. Since

$$B\left(x + \frac{2\pi}{B}\right) = Bx + 2\pi,$$

$$\sin B\left(x + \frac{2\pi}{B}\right) = \sin(Bx + 2\pi)$$

$$= \sin Bx.$$

Graphs of $y = A \sin(Bx + C) + D$, $y = A \cos(Bx + C) + D$ **249**

That is, if $f(x) = \sin Bx$, we have $f(x + 2\pi/B) = f(x)$, and the period of f is $2\pi/B$. For example, the period of $\sin 2x$ is π, and the period of $\sin \frac{1}{2}x$ is 4π. These curves are compared with $\sin x$ in Fig. 7-14. In general, the curve $y = \sin Bx$ is a sine wave with cycle length $2\pi/B$.

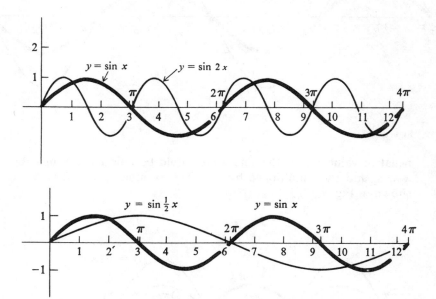

Figure 7-14

Finally, we consider $y = \sin(x + C)$. In general, $y = \sin(x + C)$ has the same shape (same amplitude and same cycle length) as $y = \sin x$. For $y = \sin(x + C)$ a cycle starts at $x = -C$ instead of 0, and ends at $2\pi - C$ instead of 2π. Hence the period is still 2π. If $C > 0$, the whole curve is just shoved to the left by C units. The graph of $y = \sin(x + 2)$ is compared with $y = \sin x$ in Fig. 7-15.

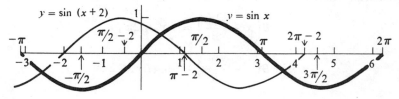

Figure 7-15

Now we can put these three effects together, and consider
$$y = A \sin (Bx + C).$$
The curve is a sine wave, with a cycle starting at $-C/B$, since $Bx + C = 0$ when $x = -C/B$. The length of a cycle is $2\pi/B$, because
$$B\left(x + \frac{2\pi}{B}\right) + C = Bx + C + 2\pi$$
Finally, the amplitude of the sine curve is $|A|$.

Example 1 Graph $y = 2 \sin (x - \pi)$.

Solution. This sine wave will have period 2π (since $B = 1$), and a cycle will start at $x = \pi$, or at $\pi - 2\pi = -\pi$. The amplitude is 2. The curves $y = \sin (x - \pi)$ and $y = 2 \sin (x - \pi)$ are shown in Fig. 7-16.

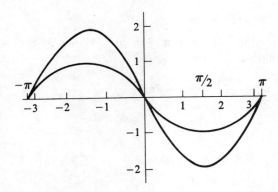

Figure 7-16

Example 2 Graph $y = \sin (2x + \pi/2)$.

Solution. This sine wave has period π, and a cycle will start where $2x + \pi/2 = 0$. Hence a cycle starts at $x = -\pi/4$ and ends at $-\pi/4 + \pi = 3\pi/4$. The amplitude of the curve is 1, as seen in Fig. 7-17.

The graphs of curves of the form
$$y = A \cos (Bx + C)$$
are obtained from the graph of $y = \cos x$ in the same way that the graphs of $y = A \sin (Bx + C)$ are obtained from $y = \sin x$. The curve

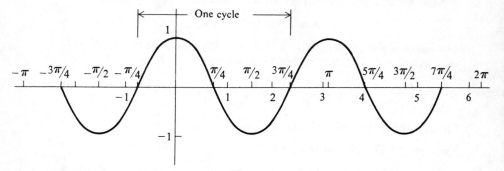

Figure 7-17

$y = A \cos(Bx + C)$ has amplitude $|A|$ and cycle length $2\pi/B$. A cosine-shaped cycle starts when $Bx + C = 0$; that is, when $x = -C/B$.

Example 3 Graph $y = \frac{1}{2} \cos(3x + \pi)$.

Solution. A cosine cycle starts at $3x + \pi = 0$, or $x = -\pi/3$. See Fig. 7-18. A cycle has length $2\pi/3$, since

$$3\left(x + \frac{2\pi}{3}\right) = 3x + 2\pi$$

The amplitude is 1/2.

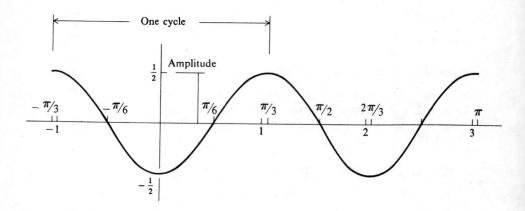

Figure 7-18

252 Circular Functions

SECTION 7-3, REVIEW QUESTIONS

1. Any curve with the same basic shape as $y = \sin x$ or $y = \cos x$ is called a _____ curve.   **sine**

2. Sine curves are also called sine _____.   **waves**

3. One cycle of a sine wave is the part of the graph over an interval equal to the _____ of the corresponding function.   **period**

4. The period of $\sin x$ is 2π, and the period of $\sin Bx$ is _____.   **2π/B**

5. $2\pi/B$ is the length of one _____ of $y = \sin Bx$.   **cycle**

6. The maximum value of $y = A \sin x$ is the _____ of the corresponding sine wave.   **amplitude**

7. The amplitude and period completely determine the _____ of any sine wave.   **shape**

8. The sine waves $y = A \sin Bx$ and $y = A \sin(Bx + C)$ have exactly the same shape, but their _____ start at different points.   **cycles**

9. In the sine curve, $y = A \sin(Bx + C)$, the amplitude is _____,   **|A|**
   and the period is _____.   **2π/B**

10. A determines amplitude ($|A|$), B determines the period ($2\pi/B$), and C determines the starting point of a _____.   **cycle**

11. A cosine-shaped cycle for $y = A \cos(Bx + C)$ starts where $Bx + C =$ _____.   **0**

12. A cycle for $y = A \cos(Bx + C)$ starts at $-C/B$ and has length _____.   **2π/B**

SECTION 7-3, EXERCISES

Graph the following functions. Give the amplitude and period, and indicate one cycle on the graph.

1. $y = 2 \sin x$
2. $y = 3 \cos x$
3. $y = -\frac{1}{2} \cos x$
4. $y = -\frac{3}{2} \sin x$
5. $y = \sin(x + \pi)$
6. $y = \cos\left(x - \frac{\pi}{2}\right)$
7. $y = \cos 2x$
8. $y = -\sin 3x$

9. $y = 2 \sin\left(\frac{1}{2}x\right)$ 10. $y = -3 \cos\left(\frac{1}{4}x\right)$

11. $y = \frac{3}{2} \cos(2x + \pi)$ 12. $y = -\frac{1}{2} \sin\left(\frac{1}{3}x - \frac{\pi}{2}\right)$

13. $y = \frac{1}{4} \sin\left(\frac{3}{2}x + \frac{\pi}{2}\right) + 2$ 14. $y = 2 \cos\left(\frac{2}{3}x - \pi\right) - 3$

15. $y = 2 \cos(x + 1)$ 16. $y = -\frac{1}{2} \sin(x - 2)$

17. $y = 1 - \sin(2x + 1)$ 18. $y = 2 + 2 \cos(3x - 2)$

19. $y = 1 + \sin \pi x$ 20. $y = \cos(\pi x + 1) - 2$

7-4 THE OTHER CIRCULAR FUNCTIONS

We now define the four remaining circular functions (or trigonometric functions), the tangent, secant, cosecant, and cotangent.

$$\text{tangent } s = \tan s = \frac{\sin s}{\cos s} \quad (\cos s \neq 0)$$

$$\text{secant } s = \sec s = \frac{1}{\cos s} \quad (\cos s \neq 0)$$

$$\text{cosecant } s = \csc s = \frac{1}{\sin s} \quad (\sin s \neq 0)$$

$$\text{cotangent } s = \cot s = \frac{\cos s}{\sin s} \quad (\sin s \neq 0)$$

These functions are defined on the whole real line, except for those points in which the denominator of the function is zero. Hence $\tan s$ and $\sec s$ are defined except for $s = \pm \pi/2, \pm 3\pi/2, \pm 5\pi/2$, etc. Similarly, $\csc s$ and $\cot s$ are defined except for $s = 0, \pm \pi, \pm 2\pi$, etc.

The graphs of $\tan s$, $\sec s$, $\sec s$, $\cot s$ all have vertical asymptotes, π units apart, corresponding to those points where the functions are not defined. The graphs of the cosecant, secant, and tangent functions are shown in Figs. 7-19, 7-20, and 7-21, along with the functions they are defined in terms of.

The values of $\tan s$, $\sec s$, and so on, can be calculated directly from the definition at points where $\sin s$ and $\cos s$ are known.

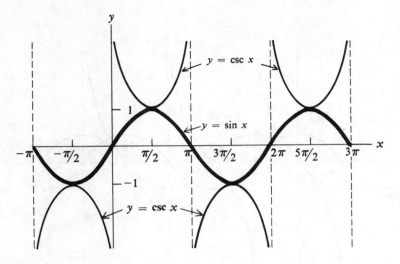

Figure 7-19

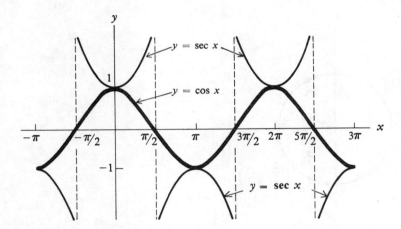

Figure 7-20

Example 1 Calculate $\tan s$, $\sec s$, and $\cot s$ for $s = 0$, $s = \pi/3$, $s = -\pi/4$, $s = 5\pi/6$.

Solution. The points of the unit circle corresponding to arc lengths of 0, $\pi/3$, $-\pi/4$, and $5\pi/6$ are shown in Fig. 7-22.
From these we read off the values below.

$$\tan 0 = \frac{\sin 0}{\cos 0} = \frac{0}{1} = 0$$

The Other Circular Functions 255

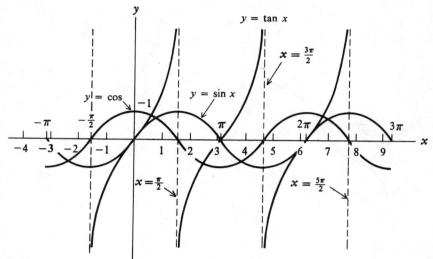

Figure 7-21

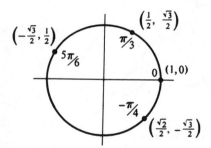

Figure 7-22

$$\tan \frac{\pi}{3} = \frac{\sin \pi/3}{\cos \pi/3} = \frac{\sqrt{3}/2}{1/2} = \sqrt{3}$$

$$\tan (-\pi/4) = \frac{\sin (-\pi/4)}{\cos (-\pi/4)} = \frac{-\sqrt{2}/2}{\sqrt{2}/2} = -1$$

$$\tan \frac{5\pi}{6} = \frac{\sin 5\pi/6}{\cos 5\pi/6} = \frac{1/2}{-\sqrt{3}/2} = -\frac{1}{\sqrt{3}}$$

Similarly, the values of sec s are the reciprocals of the corresponding values of cos s.

$$\sec 0 = \frac{1}{\cos 0} = \frac{1}{1} = 1$$

$$\sec \frac{\pi}{3} = \frac{1}{\cos \pi/3} = \frac{1}{1/2} = 2$$

$$\sec(-\pi/4) = \frac{1}{\cos(-\pi/4)} = \frac{1}{\sqrt{2}/2} = \sqrt{2}$$

$$\sec \frac{5\pi}{6} = \frac{1}{\cos 5\pi/6} = \frac{1}{-\sqrt{3}/2} = -\frac{2}{\sqrt{3}} = -\frac{2}{3}\sqrt{3}$$

The values of cot s are the reciprocals of the values of tan s.

$$\cot 0 = \frac{1}{0} \text{ is undefined.}$$

$$\cot \frac{\pi}{3} = \frac{1}{\sqrt{3}} = \frac{\sqrt{3}}{3}$$

$$\cot(-\pi/4) = \frac{1}{-1} = -1$$

$$\cot\left(\frac{5\pi}{6}\right) = -\sqrt{3}$$

The values of tan s, sec s, csc s, and cot s for $0 \le s \le \pi/2$ are collected in the table below. The values of the four functions in quadrants II, III, and IV can be found as just shown.

s	tan s	sec s	csc s	cot s
0	0	1	undefined	undefined
$\frac{\pi}{6}$	$\frac{\sqrt{3}}{3}$	$\frac{2\sqrt{3}}{3}$	2	$\sqrt{3}$
$\frac{\pi}{4}$	1	$\sqrt{2}$	$\sqrt{2}$	1
$\frac{\pi}{3}$	$\sqrt{3}$	2	$\frac{2\sqrt{3}}{3}$	$\frac{\sqrt{3}}{3}$
$\frac{\pi}{2}$	undefined	undefined	1	0

The Other Circular Functions

The identities for sin s and cos s also yield immediately similar identities for the remaining functions. For example, we know that

$$\sin(s + \pi/2) = \cos s.$$

Hence (see Figs. 7-19 and 7-20), we also have

$$\csc(s + \pi/2) = \sec s.$$

The important identity

$$\sin^2 s + \cos^2 s = 1$$

also yields immediate identities involving the other functions. Dividing these equations by $\sin^2 s$, we get

$$\frac{\sin^2 s}{\sin^2 s} + \frac{\cos^2 s}{\sin^2 s} = \frac{1}{\sin^2 s},$$

or

$$1 + \cot^2 s = \csc^2 s \ (s \neq 0, \pm \pi, \pm 3\pi \ldots).$$

Similarly, if we divide the same identity by $\cos^2 s$, we get

$$\frac{\sin^2 s}{\cos^2 s} + \frac{\cos^2 s}{\cos^2 s} = \frac{1}{\cos^2 s}$$

or

$$\tan^2 s + 1 = \sec^2 s \quad (s \neq \pm \frac{\pi}{2}, \pm \frac{\pi}{2}, \pm \frac{5\pi}{2} \ldots.)$$

The identities

$$\sin s = -\sin(-s), \quad \cos s = \cos(-s),$$

also yield corresponding identities for tan s, sec s, csc s, and cos s. For example,

$$\tan(-s) = \frac{\sin(-s)}{\cos(-s)} = \frac{-\sin s}{\cos s} = -\tan s,$$

$$\sec(-s) = \frac{1}{\cos(-s)} = \frac{1}{\cos s} = \sec s.$$

Of course all six trigonometric functions are periodic. For example,

$$\tan(s + 2\pi) = \frac{\sin(s + 2\pi)}{\cos(s + 2\pi)} = \frac{\sin s}{\cos s} = \tan s.$$

However, as Fig. 7-21 suggests, $\tan s$ (and its reciprocal, $\cot s$) actually have period π rather than 2π: $\tan(s + \pi) = \tan s$ for all s. The secant and cosecant have period 2π as is shown in Figs. 7-19 and 7-20.

Example 2 Find $\sin s$ and $\cos s$ given that $0 < s < \frac{\pi}{2}$, and $\tan s = \frac{4}{3}$.

Solution. Let $\sin s = A$. Since $\sin^2 s + \cos^2 s = 1$, and $\cos s$ is positive for $0 < s < \frac{\pi}{2}$, we then have $\cos s = \sqrt{1 - A^2}$. Hence

$$\tan s = \frac{\sin s}{\cos s} = \frac{A}{\sqrt{1 - A^2}} = \frac{4}{3},$$

$$3A = 4\sqrt{1 - A^2},$$

$$9A^2 = 16(1 - A^2),$$

$$25A^2 = 16,$$

$$A^2 = \frac{16}{25},$$

$$A = \frac{4}{5}.$$

Therefore $\sin s = 4/5$ and $\cos s = \sqrt{1 - (4/5)^2} = 3/5$.

Example 3 Find $\sec s$ given that $\pi < s < \frac{3\pi}{2}$, and $\tan s = 3$.

Solution 1. Let $\cos s = A$ and note that $A < 0$ since $\pi < s < \frac{3\pi}{2}$. Then $\sin s = -\sqrt{1 - A^2}$, since the sine is negative between π and $\frac{3\pi}{2}$. Hence

$$\tan s = 3 = -\frac{\sqrt{1 - A^2}}{A},$$

The Other Circular Functions 259

$$3A = -\sqrt{1-A^2}$$

$$9A^2 = 1 - A^2,$$

$$10A^2 = 1,$$

$$A = -\frac{1}{\sqrt{10}}.$$

Since $\cos s = -\frac{1}{\sqrt{10}}$,

$$\sec s = \frac{1}{\cos s} = -\sqrt{10}.$$

Solution 2. We can also find $\sec s$ directly from the identity

$$\tan^2 s + 1 = \sec^2 s.$$

We get

$$\sec^2 s = 3^2 + 1 = 10,$$

$$\sec s = \pm \sqrt{10}.$$

Since the cosine, and hence the secant, is negative between π and $\frac{3\pi}{2}$, $\sec s = -\sqrt{10}$.

Example 4 Find $\sin s$ given that $-\pi/2 < s < 0$ and $\cot s = -1/10$.

Solution. We use the identity $\csc^2 s = \cot^2 s + 1$ to first find $\csc s$.

$$\csc s = \pm\sqrt{\cot^2 s + 1} = \pm\sqrt{1/100 + 1} = \pm\tfrac{1}{10}\sqrt{101}$$

The given range of s values corresponds to points of the circle in the fourth quadrant, so $\sin s$ and $\csc s$ are negative. Hence,

$$\csc s = -\frac{\sqrt{101}}{10} \quad \text{and} \quad \sin s = -\frac{10}{\sqrt{101}}.$$

SECTION 7-4, REVIEW QUESTIONS

1. The six circular functions are cos s, sin s, tan s, and their reciprocals _____ , _____ , and _____ .   **sec s csc s**
   cot s
2. The circular functions are also called _____ functions.   **trigonometric**
3. Sin s is the y-coordinate of the point of the unit circle _____ units counterclockwise from $(1, 0)$ along the circle.   **s**
4. Sin s is the y-coordinate of the point s units around the unit circle, and cos s is the _____ -coordinate of this point.   **x**
5. The functions tan s, sec s, csc s, cot s are not defined everywhere, and the graphs have vertical _____ at the points where the functions are not defined.   **asymptotes**
6. Sec s and csc s are periodic of period 2π since sin s and cos s are, but tan s and cot s are periodic of period _____ .   **π**
7. The ratio of sin s and cos s has period π even though both functions have period _____ .   **2π**
8. The identity $\sin^2 s + \cos^2 s = 1$ yields the two new identities: $1 + \cot^2 s = \csc^2 s$, and _____ $+ 1 = \sec^2 s$.   **tan² s**
9. The fact that $\sin(-s) = -\sin s$ implies that $\csc(-s) = -\csc s$, and the fact that $\cos(-s) = \cos s$ implies that $\sec(-s) =$ _____ .   **sec s**

SECTION 7-4, EXERCISES

Give the domain and range of each of the following functions.

1. sin s
2. cos s
3. tan s
4. cot s
5. csc s
6. sec s

Calculate the following function values.

7. $\sin \dfrac{\pi}{6}$
8. $\cos \dfrac{\pi}{3}$
9. $\sec \dfrac{\pi}{4}$ **Example 1**
10. $\sec \dfrac{\pi}{6}$
11. $\tan \dfrac{2\pi}{3}$
12. $\cot \dfrac{5\pi}{6}$
13. $\csc \dfrac{3\pi}{2}$
14. $\sec \pi$
15. $\cot \dfrac{5\pi}{3}$

16. $\tan \frac{7\pi}{4}$ 17. $\tan \frac{3\pi}{2}$ 18. $\cot \pi$

19. $\sec \frac{\pi}{3}$ 20. $\sec \frac{\pi}{6}$ 21. $\sec \frac{3\pi}{2}$

22. $\csc \pi$

Find sin *s* and cos *s* if:

Example 2

23. $\tan s = \frac{3}{4}, 0 < s < \frac{\pi}{2}$

24. $\tan s = \frac{5}{12}, 0 < s < \frac{\pi}{2}$

25. $\tan s = -\frac{12}{5}, \frac{\pi}{2} < s < \pi$

26. $\tan s = -\frac{4}{3}, \frac{\pi}{2} < s < \pi$

27. $\tan s = 2, \pi < s < \frac{3\pi}{2}$

28. $\tan s = \frac{1}{2}, \pi < s < \frac{3\pi}{2}$

Find sec *s* if:

Example 3

29. $\tan s = 2, 0 < s < \frac{\pi}{2}$

30. $\tan s = 4, 0 < s < \frac{\pi}{2}$

31. $\tan s = -\frac{1}{3}, \frac{3\pi}{2} < s < 2\pi$

32. $\tan s = \frac{2}{3}, \pi < s < \frac{3\pi}{2}$

33. $\cot s = \frac{3}{4}, 0 < s < \frac{\pi}{2}$

34. $\cot s = \frac{2}{5}, 0 < s < \frac{\pi}{2}$

Find csc *s* if:

Example 4

35. $\cot s = -\frac{1}{3}, \frac{\pi}{2} < s < \pi$

36. $\cot s = -\frac{1}{5}, \frac{\pi}{2} < s < \pi$

37. $\tan s = -\frac{3}{5}, \frac{3\pi}{2} < s < 2\pi$

38. $\tan s = 6, \pi < s < \frac{3\pi}{2}$

Find csc s and sin s if:

39. $\cot s = 2, 0 < s < \frac{\pi}{2}$ Example 4

40. $\cot s = -3, \frac{\pi}{2} < s < \pi$

41. $\cot s = -\frac{1}{4}, \frac{3\pi}{2} < s < 2\pi$

42. $\cot s = -\frac{2}{3}, \frac{\pi}{2} < s < \pi$

Find tan s if:

43. $\sec s = 3, 0 < s < \frac{\pi}{2}$ Example 3

44. $\sec s = 4, 0 < s < \frac{\pi}{2}$

45. $\sec s = -\frac{5}{3}, \frac{\pi}{2} < s < \pi$

46. $\sec s = -\frac{13}{5}, \frac{\pi}{2} < s < \pi$

47. $\cos s = \frac{1}{2}, 0 < s < \frac{\pi}{2}$

48. $\cos s = \frac{1}{5}, 0 < s < \frac{\pi}{2}$

7-5 SUM AND REDUCTION FORMULAS FOR SINE AND COSINE

In this section we will show how $\sin(s_1 + s_2)$ and $\cos(s_1 + s_2)$ can be expressed as simple formulas involving $\sin s_1$, $\sin s_2$, $\cos s_1$, $\cos s_2$. Consider points $(\cos s_1, \sin s_1)$ and $(\cos s_2, \sin s_2)$ of the unit circle, corresponding to distances s_1 and s_2 from $(1, 0)$. See Fig. 7-23.

It is easy to see from the figure that the arc from $(1, 0)$ to the point $(\cos(s_1 + s_2), \sin(s_1 + s_2))$ has the same length (namely,

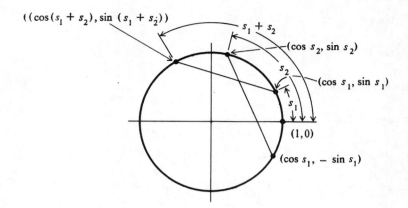

Figure 7-23

$s_1 + s_2$) as the arc from $(\cos s_1, -\sin s_1)$ to $(\cos s_2, \sin s_2)$. Therefore, the two chords will have the same length, which gives us the identity:

$$\sqrt{[\cos(s_1 + s_2) - 1]^2 + [\sin(s_1 + s_2) - 0]^2}$$
$$= \sqrt{(\cos s_2 - \cos s_1)^2 + (\sin s_2 + \sin s_1)^2}.$$

Squaring both sides and simplifying gives

$$\cos^2(s_1 + s_2) - 2\cos(s_1 + s_2) + 1 + \sin^2(s_1 + s_2) =$$
$$\cos^2 s_2 - 2\cos s_2 \cos s_1 + \cos^2 s_1 +$$
$$\sin^2 s_2 + 2\sin s_1 \sin s_2 + \sin^2 s_1.$$

Now we use the fact that

$$\cos^2(s_1 + s_2) + \sin^2(s_1 + s_2) = 1,$$
$$\cos^2 s_2 + \sin^2 s_2 = 1,$$
$$\cos^2 s_1 + \sin^2 s_1 = 1,$$

and we get

$$2 - 2\cos(s_1 + s_2) = 2 + 2\sin s_1 \sin s_2 - 2\cos s_1 \cos s_2,$$

or

$$\cos(s_1 + s_2) = \cos s_1 \cos s_2 - \sin s_1 \sin s_2.$$

If we replace s_2 by $-s_2$, we get the companion formula for $\cos(s_1 - s_2)$:

$$\begin{aligned}\cos(s_1 - s_2) &= \cos(s_1 + (-s_2)) \\ &= \cos s_1 \cos(-s_2) - \sin s_1 \sin(-s_2) \\ &= \cos s_1 \cos s_2 + \sin s_1 \sin s_2.\end{aligned}$$

Hence we have the difference formula for cosine:

$$\cos(s_1 - s_2) = \cos s_1 \cos s_2 + \sin s_1 \sin s_2.$$

The so-called reduction formulas we obtained previously from symmetry can be verified using these identities.

(a) $\cos(s + \pi) = -\cos s$ (a') $\cos s = -\cos(s - \pi)$

(b) $\cos(s + \pi/2) = -\sin s$ (b') $\cos s = -\sin(s - \pi/2)$

(c) $\cos(s - \pi/2) = \sin s$ (c') $\cos s = \sin(s + \pi/2)$

For an example, we will verify (a).

(a)
$$\begin{aligned}\cos(s + \pi) &= \cos s \cos \pi - \sin s \sin \pi \\ &= (\cos s)(-1) - (\sin s)(0) \\ &= -\cos s\end{aligned}$$

Formula (a') is obtained from (a) simply by replacing $s - \pi$, and formula (b') is obtained from (b) by replacing s by $s - \pi/2$.

Now we can use these formulas in turn to obtain a sum formula for sine. We use (c) as follows:

$$\begin{aligned}\sin(s_1 + s_2) &= \cos\left((s_1 + s_2) - \frac{\pi}{2}\right) \\ &= \cos\left(s_1 + \left(s_2 - \frac{\pi}{2}\right)\right) \\ &= \cos s_1 \cos\left(s_2 - \frac{\pi}{2}\right) - \sin s_1 \sin\left(s_2 - \frac{\pi}{2}\right)\end{aligned}$$

Now, we use (c) to replace $\cos(s_2 - \pi/2)$ by $\sin s_2$, and (b') to replace $-\sin(s_2 - \pi/2)$ by $\cos s_2$. We get $\sin(s_1 + s_2) = \cos s_1 \sin s_2 +$

$\sin s_1 \cos s_2$. We can rearrange this to the standard form:

$$\sin(s_1 + s_2) = \sin s_1 \cos s_2 + \cos s_1 \sin s_2.$$

If we replace s_2 by $-s_2$, and note that $\cos(-s_2) = \cos s_2$, and $\sin(-s_2) = -\sin s_2$, we get

$$\sin(s_1 - s_2) = \sin s_1 \cos s_2 - \cos s_1 \sin s_2.$$

The following reduction formulas for sine follow from this sum formula:

(a) $\sin(s + \pi) = -\sin s$ (a') $\sin s = -\sin(s - \pi)$
(b) $\sin(s + \pi/2) = \cos s$ (b') $\sin s = \cos(s - \pi/2)$
(c) $\sin(s - \pi/2) = -\cos s$ (c') $\sin s = -\cos(s + \pi/2)$

Observe that (a') is obtained from (a) by replacing s by $s - \pi$, (b') is obtained from (b) by replacing s by $s - \pi/2$, and (c') is obtained from (c) by replacing s by $s + \pi/2$. Formula (a), for example, is verified from the sum formula as follows:

$$\sin(s + \pi) = \sin s \cos \pi + \cos s \sin \pi$$
$$= \sin s(-1) + \cos s(0)$$
$$= -\sin s$$

Example 1 Evaluate $\sin 5\pi/12$.

Solution. We note that $\pi/6 + \pi/4 = 2\pi/12 + 3\pi/12 = 5\pi/12$. We know the values sine and cosine at $\pi/6$ and $\pi/4$, so we can use the sum formula:

$$\sin \frac{5\pi}{12} = \sin(\pi/6 + \pi/4)$$
$$= \sin \pi/6 \cos \pi/4 + \cos \pi/6 \sin \pi/4$$
$$= (1/2)(\sqrt{2}/2) + (\sqrt{3}/2)(\sqrt{2}/2)$$
$$= (1/4)(\sqrt{2} + \sqrt{6})$$

Example 2 Express $\sin(s + \pi/3)$ and $\cos(s + \pi/3)$ in terms of $\sin s$ and $\cos s$.

Solution.

$$\sin\left(s + \frac{\pi}{3}\right) = \sin s \cos \frac{\pi}{3} + \cos s \sin \frac{\pi}{3}$$

$$= \frac{1}{2} \sin s + \frac{\sqrt{3}}{2} \cos s$$

$$\cos\left(s + \frac{\pi}{3}\right) = \cos s \cos \frac{\pi}{3} - \sin s \sin \frac{\pi}{3}$$

$$= \frac{1}{2} \cos s - \frac{\sqrt{3}}{2} \sin s.$$

If a and b are any positive numbers such that $a^2 + b^2 = 1$, then $a = \cos \theta$ and $b = \sin \theta$ for some θ between 0 and $\pi/2$. In fact, since (a, b) and (b, a) are symmetric about the line $y = x$, we also have $a = \sin \theta'$ and $b = \cos \theta'$, where $\theta + \theta' = \pi/2$ (see Fig. 7-24). This observation allows us to write the following lines as combinations of $\sin s$ and $\cos s$ as a single sine or cosine function:

$$a \cos s - b \sin s = \cos s \cos \theta - \sin s \sin \theta = \cos(s + \theta)$$

$$a \sin s + b \cos s = \sin s \cos \theta + \cos s \sin \theta = \sin(s + \theta)$$

$$= \cos s \cos \theta' + \sin s \sin \theta' = \cos(s - \theta')$$

$$a \sin s - b \cos s = \sin s \cos \theta - \cos s \sin \theta = \sin(s - \theta)$$

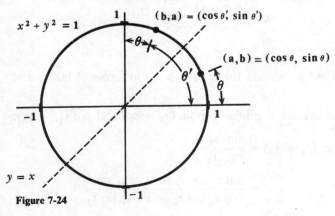

Figure 7-24

Example 3 Express the following as a single sine function of the form $\sin(s \pm \theta)$, or as a single function of the form $\cos(s \pm \theta)$, with $0 < \theta < \pi/2$:

(a) $\quad \dfrac{\sqrt{2}}{2} \cos s - \dfrac{\sqrt{2}}{2} \sin s;$

(b) $\quad \dfrac{\sqrt{3}}{2} \sin s + \dfrac{1}{2} \cos s.$

Solution. (a) Since $\dfrac{\sqrt{2}}{2} = \cos \dfrac{\pi}{4} = \sin \dfrac{\pi}{4}$, we have

$$\dfrac{\sqrt{2}}{2} \cos s - \dfrac{\sqrt{2}}{2} \sin s = \cos s \cos \dfrac{\pi}{4} - \sin s \sin \dfrac{\pi}{4}$$

$$= \cos\left(s + \dfrac{\pi}{4}\right).$$

(b) Here we have $\dfrac{\sqrt{3}}{2} = \cos \dfrac{\pi}{6}, \dfrac{1}{2} = \sin \dfrac{\pi}{6}$, or $\dfrac{\sqrt{3}}{2} = \sin \dfrac{\pi}{3}, \dfrac{1}{2} = \cos \dfrac{\pi}{3}$. This gives two possibilities:

$$\dfrac{\sqrt{3}}{2} \sin s + \dfrac{1}{2} \cos s = \sin s \cos \dfrac{\pi}{6} + \cos s \sin \dfrac{\pi}{6}$$

$$= \sin\left(s + \dfrac{\pi}{6}\right),$$

or

$$\dfrac{\sqrt{3}}{2} \sin s + \dfrac{1}{2} \cos s = \cos s \cos \dfrac{\pi}{3} + \sin s \sin \dfrac{\pi}{3}$$

$$= \cos\left(s - \dfrac{\pi}{3}\right).$$

Example 4 Find a formula for $\tan(s_1 + s_2)$ in terms of $\tan s_1$ and $\tan s_2$.

Solution. We use the formulas for $\sin(s_1 + s_2)$ and $\cos(s_1 + s_2)$:

$$\tan(s_1 + s_2) = \dfrac{\sin(s_1 + s_2)}{\cos(s_1 + s_2)}$$

$$= \dfrac{\sin s_1 \cos s_2 + \cos s_1 \sin s_2}{\cos s_1 \cos s_2 - \sin s_1 \sin s_2}$$

We can express this in terms of $\tan s_1$ and $\tan s_2$ by dividing numerator and denominator by $\cos s_1 \cos s_2$.

$$\tan(s_1 + s_2) = \frac{\dfrac{\sin s_1 \cos s_2}{\cos s_1 \cos s_2} + \dfrac{\cos s_1 \sin s_2}{\cos s_2 \cos s_1}}{\dfrac{\cos s_1 \cos s_2}{\cos s_1 \cos s_2} - \dfrac{\sin s_1 \sin s_2}{\cos s_1 \cos s_2}}$$

$$= \frac{\tan s_1 + \tan s_2}{1 - \tan s_1 \tan s_2}$$

SECTION 7-5, REVIEW QUESTIONS

1. $\text{Sin}(s_1 + s_2) = \sin s_1 \cos \underline{\hspace{1cm}} + \cos s_1 \sin \underline{\hspace{1cm}}$. s_2 s_2
2. $\text{Sin}(s_1 - s_2) = \sin s_1 \cos s_2 \underline{\hspace{1cm}} \cos s_1 \sin s_2$. $-$
3. $\text{Cos}(s_1 + s_2) = \cos s_1 \underline{\hspace{1cm}} - \sin s_1 \underline{\hspace{1cm}}$. $\cos s_2$ $\sin s_2$
4. $\text{Cos}(s_1 - s_2) = \cos s_1 \cos s_2 \underline{\hspace{1cm}} \sin s_1 \sin s_2$. $+$
5. If you replace s by $s - \pi$ in the identity $\sin(s + \pi) = -\sin s$, you get the identity $\sin \underline{\hspace{1cm}} = -\sin(s - \pi)$. s
6. The various reduction formulas, such as $\sin s = -\sin(s - \pi)$, allow us to write $\sin s$ and $\cos s$ in terms of $\sin s_1$ and $\cos s_1$ for some number s_1 between 0 and $\underline{\hspace{1cm}}$. $\pi/2$
7. If $\pi/2 < s < \pi$, then $s - \pi/2$ is between 0 and $\pi/2$; hence the reduction formula $\sin s = \cos(s - \pi/2)$ allows us to find $\sin s$ by looking up $\underline{\hspace{1cm}}$ in the table. $\cos(s - \pi/2)$

SECTION 7-5, EXERCISES

Evaluate

1. $\cos \dfrac{5\pi}{12}$ $\left(\dfrac{\pi}{6} + \dfrac{\pi}{4} = \dfrac{5\pi}{12}\right)$ 2. $\cos \dfrac{7\pi}{12}$ $\left(\dfrac{\pi}{3} + \dfrac{\pi}{4} = \dfrac{7\pi}{12}\right)$ Example 1

3. $\sin \dfrac{\pi}{12}$ $\left(\dfrac{\pi}{3} - \dfrac{\pi}{4} = \dfrac{\pi}{12}\right)$ 4. $\cos \dfrac{11\pi}{12}$ $\left(\dfrac{2\pi}{3} + \dfrac{\pi}{4} = \dfrac{11\pi}{12}\right)$

Verify the formulas for $\sin(s_1 + s_2)$ and $\cos(s_1 + s_2)$ for each of the following sums:

5. $\dfrac{2\pi}{3} = \dfrac{\pi}{2} + \dfrac{\pi}{6}$
6. $\dfrac{5\pi}{6} = \dfrac{\pi}{2} + \dfrac{\pi}{3}$
7. $\dfrac{3\pi}{4} = \dfrac{\pi}{2} + \dfrac{\pi}{4}$
8. $\dfrac{\pi}{2} = \dfrac{\pi}{3} + \dfrac{\pi}{6}$

Verify the formulas for $\sin(s_1 - s_2)$ and $\cos(s_1 - s_2)$ for each of the following differences:

9. $\dfrac{\pi}{6} = \dfrac{\pi}{3} - \dfrac{\pi}{6}$
10. $\dfrac{\pi}{4} = \dfrac{3\pi}{4} - \dfrac{\pi}{2}$
11. $\dfrac{\pi}{6} = \dfrac{\pi}{2} - \dfrac{\pi}{3}$
12. $\dfrac{\pi}{2} = \dfrac{5\pi}{6} - \dfrac{\pi}{3}$

Express the following in terms of $\sin s$ and $\cos s$.

Example 2
13. $\sin(s + \pi/6)$
14. $\cos(s + \pi/6)$
15. $\cos(s - \pi/4)$
16. $\sin(s + 3\pi/4)$
17. $\sin(s - \pi/3)$
18. $\cos(s - \pi/6)$
19. $\cos(s - \pi/2)$
20. $\sin(s - \pi/2)$
21. $\sin(s + \pi)$
22. $\cos(s - \pi)$
23. $\cos(s + 4\pi/3)$
24. $\sin(s + 7\pi/6)$

Write the following in the form $\sin(s \pm \theta)$ or $\cos(s \pm \theta)$ for some θ with $0 < \theta < \pi/2$.

Example 3
25. $\dfrac{1}{2}\sin s - \dfrac{\sqrt{3}}{2}\cos s$
26. $\dfrac{\sqrt{3}}{2}\cos s - \dfrac{1}{2}\sin s$
27. $\dfrac{1}{2}\sin s + \dfrac{\sqrt{3}}{2}\cos s$
28. $\dfrac{\sqrt{2}}{2}\cos s + \dfrac{\sqrt{2}}{2}\sin s$
29. $\dfrac{1}{2}\cos s + \dfrac{\sqrt{3}}{2}\sin s$
30. $\dfrac{\sqrt{2}}{2}\sin s - \dfrac{\sqrt{2}}{2}\cos s$

Express the following in terms of $\tan s$:

Example 4
31. $\tan(s + \pi/4)$
32. $\tan(s + \pi/6)$
33. $\tan(s + \pi/3)$
34. $\tan(s + \pi)$
35. Find a formula for $\tan(s_1 - s_2)$ in terms of $\tan s_1$ and $\tan s_2$.
36. Find a formula for $\cot(s_1 - s_2)$ in terms of $\cot s_1$ and $\cot s_2$.
37. Find a formula for $\cot(s_1 + s_2)$ in terms of $\cot s_1$ and $\cot s_2$.

7-6 MULTIPLE-ANGLE AND HALF-ANGLE FORMULAS

The basic sum formulas of the last section easily yield several valuable new identities. We start with the basic identities:

$$\sin(s_1 + s_2) = \sin s_1 \cos s_2 + \cos s_1 \sin s_2,$$
$$\cos(s_1 + s_2) = \cos s_1 \cos s_2 - \sin s_1 \sin s_2.$$

If we let $s_1 = s_2 = s$ in these formulas, we immediately get the so-called "double-angle" formulas:

$$\sin 2s = 2 \sin s \cos s,$$
$$\cos 2s = \cos^2 s - \sin^2 s.$$

Formulas for $\sin 3s$, $\cos 3s$, $\sin 4s$, $\cos 4s$, etc., can also be derived from the sum formulas, as we show in the next example.

Example 1 Find a formula for $\sin 3s$ in terms of $\sin s$ and $\cos s$, and then a formula for $\sin 3s$ in terms of $\sin s$ only.

Solution. We write $3s = 2s + s$, and use the sum formula and the formulas above for $\sin 2s$, $\cos 2s$.

$$\sin 3s = \sin(2s + s)$$
$$= \sin 2s \cos s + \cos 2s \sin s$$
$$= (2 \sin s \cos s) \cos s + (\cos^2 s - \sin^2 s) \sin s$$
$$= 2 \sin s \cos^2 s + \sin s \cos^2 s - \sin^3 s$$
$$= 3 \sin s \cos^2 s - \sin^3 s.$$

To express $\sin 3s$ in terms of $\sin s$ only we substitute $1 - \sin^2 s$ for $\cos^2 s$.

$$\sin 3s = 3 \sin s \cos^2 s - \sin^3 s$$
$$= 3 \sin s (1 - \sin^2 s) - \sin^3 s$$
$$= 3 \sin s - 4 \sin^3 s.$$

If we replace $\cos^2 s$ by $1 - \sin^2 s$ in the formula $\cos 2s = \cos^2 s - \sin^2 s$, we get
$$\cos 2s = 1 - 2\sin^2 s.$$

Similarly, we can replace $\sin^2 s$ by $1 - \cos^2 s$ to get
$$\cos 2s = 2\cos^2 s - 1.$$

These last formulas can be solved for $\sin^2 s$ or $\cos^2 s$. This allows us to express these second powers of circular functions in terms of first powers of $\cos 2s$.

$$\sin^2 s = \frac{1 - \cos 2s}{2}$$

$$\cos^2 s = \frac{1 + \cos 2s}{2}$$

These identities are important in calculus because they are used to develop formulas for the integration of trigonometric functions.

If we replace s by $s/2$ in the identities above, we get what are called the "half-angle" formulas.

$$\sin^2\left(\frac{s}{2}\right) = \frac{1 - \cos s}{2}$$

$$\cos^2\left(\frac{s}{2}\right) = \frac{1 + \cos s}{2}$$

Example 2 Find $\sin \pi/8$.

Solution. Since $\pi/8 = 1/2 \cdot \pi/4$, we can use the half-angle formulas to find $\sin^2 \pi/8$.

$$\sin^2 \frac{\pi}{8} = \frac{1 - \cos \pi/4}{2}$$

$$= \frac{1 - \sqrt{2}/2}{2}$$

$$= \frac{2 - \sqrt{2}}{4}$$

Because sine and cosine are positive for $0 < s < \pi/2$, we take the positive square root.

$$\sin \frac{\pi}{8} = \sqrt{\frac{2-\sqrt{2}}{4}}$$

The double-angle formulas for $\sin^2 s$ and $\cos^2 s$ give a corresponding double-angle formula for $\tan^2 s$.

$$\tan^2 s = \frac{\sin^2 s}{\cos^2 s}$$

$$= \frac{\frac{1-\cos 2s}{2}}{\frac{1+\cos 2s}{2}}$$

$$= \frac{1-\cos 2s}{1+\cos 2s}$$

We can replace s by $s/2$ in this identity and write it in the alternate form:

$$\tan^2 \frac{s}{2} = \frac{1-\cos s}{1+\cos s}.$$

Example 3 Find $\tan(-\pi/12)$.

Solution. We first note that $\tan(-\pi/12) < 0$, so we find $\tan^2(-\pi/12)$, and take the negative square root.

$$\tan^2\left(-\frac{\pi}{12}\right) = \frac{1-\cos(-\pi/6)}{1+\cos(-\pi/6)}$$

$$= \frac{1-\sqrt{3}/2}{1+\sqrt{3}/2}$$

$$= \frac{2-\sqrt{3}}{2+\sqrt{3}}$$

Therefore,

$$\tan\left(-\frac{\pi}{12}\right) = -\sqrt{\frac{2-\sqrt{3}}{2+\sqrt{3}}}.$$

SECTION 7-6, REVIEW QUESTIONS

2 sin s cos s

1. If we let $s_1 = s_2 = s$ in the formula $\sin(s_1 + s_2) = \sin s_1 \cos s_2 + \cos s_1 \sin s_2$, we get the double-angle formula $\sin 2s =$ _____.

$\cos^2 s - \sin^2 s$

2. The two double-angle formulas are $\sin 2s = 2 \sin s \cos s$ and $\cos 2s =$ _____.

$1 - \sin^2 s$

3. We can express $\cos 2s$ in terms of $\sin^2 s$ by starting with $\cos 2s = \cos^2 s - \sin^2 s$ and replacing $\cos^2 s$ by _____.

$2 \cos^2 s - 1$

4. We can write $\cos 2s$ in terms of $\sin^2 s$ ($\cos 2s = 1 - 2 \sin^2 s$), or in terms of $\cos^2 s$ ($\cos 2s =$ _____).

$\frac{1}{2}(1 - \cos 2s)$

5. Solving the second formula in question 4 above for $\cos^2 s$ gives $\cos^2 s = \frac{1}{2}(1 + \cos 2s)$, and solving the first formula for $\sin^2 s$ gives $\sin^2 s =$ _____.

SECTION 7-6, EXERCISES

Example 1

1. Find a formula for $\cos 3s$ in terms of $\sin s$ and $\cos s$.
2. Find a formula for $\cos 3s$ in terms of $\cos s$.
3. Find a formula for $\sin 4s = \sin(3s + s)$ in terms of $\sin s$ and $\cos s$.
4. Find a formula for $\cos 4s = \cos(3s + s)$ in terms of $\sin s$ and $\cos s$.

Use the half-angle formulas to evaluate the following.

Example 2

5. $\cos \dfrac{\pi}{8}$ 6. $\sin\left(-\dfrac{\pi}{8}\right)$ 7. $\cos\left(-\dfrac{\pi}{12}\right)$

8. $\sin \dfrac{\pi}{12}$ 9. $\sin \dfrac{5\pi}{8}$ 10. $\cos \dfrac{5\pi}{8}$

11. $\cos \dfrac{5\pi}{12}$ 12. $\sin \dfrac{5\pi}{12}$

Find $\tan^2 s$ for each of the following values of s.

Example 3

13. $\dfrac{\pi}{12}$ 14. $\dfrac{\pi}{8}$ 15. $\dfrac{5\pi}{8}$ 16. $\dfrac{5\pi}{12}$

17. $\dfrac{3\pi}{8}$ 18. $\dfrac{7\pi}{8}$

19. Use the formula for tan $(s_1 + s_2)$ to find a formula for tan $2s$.
20. Use the formula for cot $(s_1 + s_2)$ (Ex. 37, Sec. 7-5) to find a formula for cot $2s$.
21. Express $\sin^2 2s$ in terms of cos $4s$.
22. Express $\cos^2 2s$ in terms of cos $4s$.
23. Express $\sin^4 s$ in terms of cos $2s$ and cos $4s$. (Hint: $\sin^4 s = [½(1 - \cos 2s)]^2$.)
24. Express $\cos^4 s$ in terms of cos $2s$ and cos $4s$. (Hint: $\cos^4 s = [½(1 + \cos 2s)]^2$.)

7-7 FURTHER IDENTITIES

In this section we will practice using the kind of identities we have already developed to simplify expressions containing trigonometric functions. The table below contains the basic group of identities we will use to verify more complicated identities. These formulas below are fundamental and should be memorized.

BASIC IDENTITIES

$$\sin(-s) = -\sin s$$

$$\cos(-s) = \cos s$$

$$\tan(-s) = -\tan s$$

$$\tan s = \frac{\sin s}{\cos s}$$

$$\csc s = \frac{1}{\sin s}$$

$$\sec s = \frac{1}{\cos s}$$

$$\cot s = \frac{\cos s}{\sin s}$$

$$\sin(s_1 + s_2) = \sin s_1 \cos s_2 + \cos s_1 \sin s_2$$

$$\cos(s_1 + s_2) = \cos s_1 \cos s_2 - \sin s_1 \sin s_2$$

$$\tan(s_1 + s_2) = \frac{\tan s_1 + \tan s_2}{1 - \tan s_1 \tan s_2}$$

$$\sin 2s = 2 \sin s \cos s$$

$$\cos 2s = \cos^2 s - \sin^2 s$$

$$\sin^2 s = \frac{1 - \cos 2s}{2} \qquad \sin \frac{s}{2} = \pm\sqrt{\frac{1 - \cos s}{2}}$$

$$\cos^2 s = \frac{1 + \cos 2s}{2} \qquad \cos \frac{s}{2} = \pm\sqrt{\frac{1 + \cos s}{2}}$$

$$\sin^2 s + \cos^2 s = 1$$

$$\tan^2 s + 1 = \sec^2 s$$

$$\cot^2 s + 1 = \csc^2 s$$

Example 1 Verify $\dfrac{\sin s}{\csc s - \cot s} = 1 + \cos s$.

Solution. Note that the right side is defined for all values of s. The left side, however, is not defined when $\cot s$ or $\csc s$ is not defined, or when $\csc s = \cot s$. We make the same kind of agreement here as we do for algebraic identities. We consider the statement an identity provided the two sides are equal for all values of s for which both sides are defined. This is exactly the sense in which

$$\frac{x^2 - 1}{x - 1} = x + 1$$

is an algebraic identity.

If we express all the terms in the left formula in terms of $\sin s$ and $\cos s$ and simplify, we obtain the right side.

$$\frac{\sin s}{\csc s - \cot s} = \frac{\sin s}{\dfrac{1}{\sin s} - \dfrac{\cos s}{\sin s}}$$

$$= \frac{\sin s}{\dfrac{1 - \cos s}{\sin s}}$$

$$= \frac{\sin^2 s}{1 - \cos s}$$

$$= \frac{1 - \cos^2 s}{1 - \cos s}$$

$$= 1 + \cos s$$

Example 2 Verify $\dfrac{\sin 2s}{1 + \cos 2s} = \tan s$.

Solution. First express sin 2s and cos 2s in terms of sin s and cos s.

$$\frac{\sin 2s}{1 + \cos 2s} = \frac{2 \sin s \cos s}{1 + \cos^2 s - \sin^2 s}$$

$$= \frac{2 \sin s \cos s}{2 \cos^2 s}$$

$$= \frac{\sin s}{\cos s}$$

$$= \tan s$$

Example 3 Verify the identity

$$(\tan s - \sec s)^2 = \frac{1 - \sin s}{1 + \sin s} \qquad (1)$$

Solution. Since the right side contains only the sine function, we start by expressing the left side in terms of sin s and cos s and simplifying:

$$(\tan s - \sec s)^2 = \left(\frac{\sin s}{\cos s} - \frac{1}{\cos s}\right)^2$$

$$= \frac{(\sin s - 1)^2}{\cos^2 s} \qquad (2)$$

Comparing the right side of the given identity with Equation (2) above suggests that we multiply top and bottom of $(1 - \sin s)/(1 + \sin s)$ by $(1 - \sin s)$:

$$\frac{1 - \sin s}{1 + \sin s} = \frac{(1 - \sin s)^2}{(1 + \sin s)(1 - \sin s)}$$

$$= \frac{(1 - \sin s)^2}{1 - \sin^2 s}$$

$$= \frac{(1 - \sin s)^2}{\cos^2 s}$$

$$= \frac{(\sin s - 1)^2}{\cos^2 s}$$

We have now shown that both the left and right sides of (1) are identical to (2), and hence (1) is an identity.

There is no particular virtue in starting with one side of an identity and working until we have the other side. Because all steps in verifying an identity are reversible, we can work with both sides separately to try to put them in a common form as we did above.

Example 4 Verify $\dfrac{1 - \tan^2 s}{1 + \tan^2 s} = \cos 2s$.

Solution
$$\frac{1 - \tan^2 s}{1 + \tan^2 s} = \frac{1 - \dfrac{\sin^2 s}{\cos^2 s}}{1 + \dfrac{\sin^2 s}{\cos^2 s}}$$

$$= \frac{\cos^2 s - \sin^2 s}{\cos^2 s + \sin^2 s}$$

$$= \cos^2 s - \sin^2 s$$

$$= \cos 2s$$

Example 5 Verify $\dfrac{1 + \tan^2 s}{\csc^2 s} = \tan^2 s$.

Solution
$$\frac{1 + \tan^2 s}{\csc^2 s} = \frac{\sec^2 s}{\csc^2 s}$$

$$= \frac{\dfrac{1}{\cos^2 s}}{\dfrac{1}{\sin^2 s}}$$

$$= \frac{\sin^2 s}{\cos^2 s}$$

$$= \tan^2 s$$

SECTION 7-7, REVIEW QUESTIONS

1. $\sin(-s) = $ _____. $-\sin s$
   $\cos(-s) = $ _____. $\cos s$
   $\tan(-s) = $ _____. $-\tan s$
2. $\csc s = 1/$_____. $\sin s$
   $\sec s = 1/$_____. $\cos s$
3. $\sin(s_1 + s_2) = $ _____. $\sin s_1 \cos s_2 + \cos s_1 \sin s_2$
4. $\cos(s_1 + s_2) = $ _____. $\cos s_1 \cos s_2 - \sin s_1 \sin s_2$
5. $\sin 2s = $ _____. $2 \sin s \cos s$
   $\cos 2s = $ _____. $\cos^2 s - \sin^2 s$
6. $\sin^2 s + \cos^2 s = $ _____. 1
   $\tan^2 s + 1 = $ _____. $\sec^2 s$
   $\cot^2 s + 1 = $ _____. $\csc^2 s$

SECTION 7-7, EXERCISES

Show that each of the following is an identity.

1. $\cos s \tan s = \sin s$
2. $(1 - \cos s)(1 + \cos s) = \sin^2 s$
3. $\sec^2 s - \tan^2 s = 1$
4. $\dfrac{\cot s}{\csc s} = \cos s$
5. $\dfrac{\cos s}{1 + \sin s} = \dfrac{2 \cos s - \sin 2s}{1 + \cos 2s}$
6. $\tan s = \dfrac{\sin 2s}{1 + \cos 2s}$
7. $\dfrac{1}{\sec s - \tan s} - \dfrac{1}{\sec s + \tan s} = 2 \tan s$
8. $\sec s - \sin s \tan s = \cos s$
9. $\cot^2 s - \csc^2 s = -1$
10. $\dfrac{1}{\csc^2 2s} = 1 - \dfrac{1}{\sec^2 2s}$
11. $2 \csc 2s = \sec s \csc s$
12. $\dfrac{\csc^2 s - \cot^2 s}{\sec^2 s} = \cos^2 s$
13. $\tan s \sin 2s = 2 \sin^2 s$
14. $\cos^4 s = 1 - 2 \sin^2 s + \sin^4 s$
15. $\cot s \cos s = \csc s - \sin s$
16. $\dfrac{\csc s + 1}{\csc s - 1} = \dfrac{1 + \sin s}{1 - \sin s}$
17. $\dfrac{\cos s}{1 + \sin s} = \dfrac{1 - \sin s}{\cos s}$

Further Identities

18. $\sec^2 s + \tan^2 s + 1 = \dfrac{2}{\cos^2 s}$ 19. $\dfrac{\sin^2 s}{1 + \cos s} = 1 - \cos s$

20. $\dfrac{\cos s}{\csc s + 1} + \dfrac{\cos s}{\csc s - 1} = 2 \tan s$ 21. $\dfrac{\sin s}{\sec s + 1} + \dfrac{\sin s}{\sec s - 1} = 2 \cot s$

Give a counter example to show that each of the following is not an identity.

22. $\cos(s_1 - s_2) = \cos s_1 - \cos s_2$ 23. $\cos(s_1 + s_2) = \cos s_1 + \cos s_2$
24. $\cos 2s = 2 \cos s$ 25. $\sin 2s = 2 \sin s$

7-8 INVERSE CIRCULAR FUNCTIONS

Recall that every one-to-one function f has an inverse function, denoted f^{-1}. A pair (x, y) belongs to the function f if and only if the pair (y, x) belongs to the function f^{-1}. The fact expressed by the equation $y = f(x)$ can equally well be expressed by the equation $x = f^{-1}(y)$.

The exponential function e^x is strictly increasing, and hence is one-to-one. Its inverse function is given a special notation, $\ln x$. The function x^2 is not one-to-one on $(-\infty, \infty)$, but is one-to-one if we restrict the domain to $[0, \infty)$. The inverse to this function is denoted $\sqrt{x}$ (the non-negative number whose square is x).

It is convenient to introduce similar notations with respect to the circular functions. The sine is not one-to-one, but if we restrict the sine function to $[-\pi/2, \pi/2]$, we do get a one-to-one function. This restricted function is denoted $\operatorname{Sin} x$ (with a capital S). Thus (see Fig. 7-25),

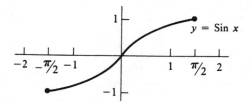

Figure 7-25

$$\operatorname{Sin} x = \sin x \text{ for } -\pi/2 \leq x \leq \pi/2.$$

The function $\operatorname{Sin} x$ has an inverse function, denoted $\operatorname{Sin}^{-1} x$, or Arcsin x. For any number x between -1 and 1, the number Arcsin x is the arc (between $-\pi/2$ and $\pi/2$) whose sine equals x. The graph of

$y = \mathrm{Sin}^{-1} x$ (or $y = \mathrm{Arcsin}\, x$) is exactly the same as the graph of $\mathrm{Sin}\, y = x$. This graph is the reflection of the graph $y = \mathrm{Sin}\, x$ in the line $y = x$ (Fig. 7-26).

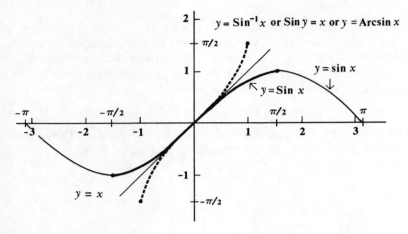

Figure 7-26

The cosine curve is periodic, like $y = \sin x$, and consequently is not one-to-one. However, if we restrict the domain of the cosine function to $[0, \pi]$ we do get a strictly decreasing function, and therefore a one-to-one function. We will denote this function $\mathrm{Cos}\, x$, with a capital C (see Fig. 7-27).

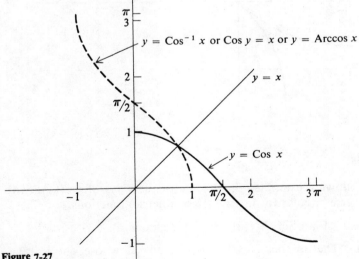

Figure 7-27

Inverse Circular Functions 281

$$\text{Cos } x = \cos x \text{ if } 0 \leq x \leq \pi$$

The inverse to Cos x is denoted $\text{Cos}^{-1} x$, or Arccos x. The equation $y = \text{Arccos } x$ means that y is the arc between 0 and π whose cosine is the number x:

$$y = \text{Cos}^{-1} x = \text{Arccos } x \quad \text{means} \quad \cos y = x \quad \text{and} \quad 0 \leq y \leq \pi.$$

We make a similar convention for the tangent function. In other words, for $-\pi/2 < x < \pi/2$, tan x is strictly increasing and therefore one-to-one. We denote by Tan x the tangent function restricted to the interval $(-\pi/2, \pi/2)$:

$$\text{Tan } x = \tan x \text{ for } -\pi/2 < x < \pi/2.$$

The inverse to Tan x is denoted $\text{Tan}^{-1} x$ or Arctan x. The curves $y = \text{Tan } x$ and $y = \text{Tan}^{-1} x$ are shown in Fig. 7-28.

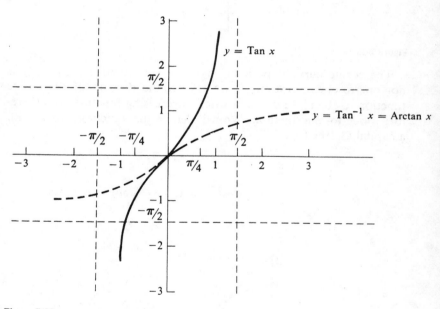

Figure 7-28

The following examples show some of the ways of working with these inverse circular (or inverse trigonometric) functions.

Example 1 Find $\text{Sin}^{-1}(1/2)$ and $\text{Sin}^{-1}(\sin 5\pi/6)$.

Solution. The number between $-\pi/2$ and $\pi/2$ whose sine is 1/2 is

$\pi/6$. Since $\sin \pi/6 = 1/2$ and $\pi/6$ is in the correct range $[-\pi/2, \pi/2]$, $\pi/6 = \mathrm{Sin}^{-1} 1/2$.

The number $5\pi/6$ also has sine equal to $1/2$. Since $\sin 5\pi/6 = 1/2$,

$$\mathrm{Sin}^{-1}(\sin 5\pi/6) = \mathrm{Sin}^{-1}(1/2) = \pi/6$$

Example 2 Find $\sin(\mathrm{Cos}^{-1}(-3/4))$.

Solution. The Arccos $(-3/4)$ is the length of the arc shown in Fig. 7-29. We know this arc corresponds to a point of the first or second quadrant by definition of $\mathrm{Cos}^{-1} x$. The arc is in the second quadrant since $-3/4$ is negative. We use the identity $\sin^2 x + \cos^2 x = 1$, and the fact that $\sin x > 0$ for x in the second quadrant, to find that

$$\sin(\mathrm{Cos}^{-1}(-3/4)) = +\sqrt{1 - \cos^2(\mathrm{Cos}^{-1}(-3/4))}$$
$$= +\sqrt{1 - (-3/4)^2}$$
$$= +\sqrt{1 - 9/16}$$
$$= \frac{\sqrt{7}}{4}.$$

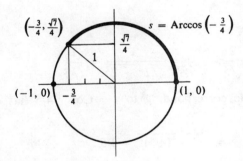

Figure 7-29

Example 3 Find $\sin(\mathrm{Tan}^{-1} 2/5)$ and $\cos(\mathrm{Tan}^{-1} 2/5)$.

Solution. If $s = \mathrm{Tan}^{-1} 2/5$, then $\tan s = 2/5$. If (x, y) is the point of the unit circle corresponding to arc s, then $x^2 + y^2 = 1$ and $y/x = \tan s = 2/5$. To get numbers in the ratio $2/5$ whose squares add to one, we divide both 2 and 5 by $\sqrt{2^2 + 5^2}$. The numbers are $2/\sqrt{2^2 + 5^2}$ and $5/\sqrt{2^2 + 5^2}$, since

Inverse Circular Functions

$$\left(\frac{2}{\sqrt{2^2+5^2}}\right)^2 + \left(\frac{5}{\sqrt{2^2+5^2}}\right)^2 = \frac{2^2}{2^2+5^2} + \frac{5^2}{2^2+5^2} = 1.$$

and

$$\frac{\frac{2}{\sqrt{2^2+5^2}}}{\frac{5}{\sqrt{2^2+5^2}}} = \frac{2}{5}.$$

Therefore the point (x, y) of the unit circle corresponding to the arc $\text{Tan}^{-1}\ 2/5$ is $(2/\sqrt{29}, 5/\sqrt{29})$. Hence $\sin(\text{Tan}^{-1}\ 2/5) = 5/\sqrt{29}$, and $\cos(\text{Tan}^{-1}\ 2/5) = 2/\sqrt{29}$.

Example 4 Evaluate $\cos(\text{Tan}^{-1}\ 1/2 - \text{Cos}^{-1}\ 4/5)$.

Solution. The point of the unit circle corresponding to arc length $s = \text{Tan}^{-1}\ 1/2$ has coordinates

$$x = \frac{2}{\sqrt{1^2+2^2}} = \frac{2}{\sqrt{5}}, \quad y = \frac{1}{\sqrt{1^2+2^2}} = \frac{1}{\sqrt{5}}.$$

The point of the unit circle corresponding to $t = \text{Cos}^{-1}\ 4/5$ has coordinates

$$x = 4/5, \quad y = \sqrt{1-(4/5)^2} = 3/5.$$

If $s = \text{Tan}^{-1}\ 1/2$, $t = \text{Cos}^{-1}\ 4/5$, then

$$\cos(s-t) = \cos s \cos t + \sin s \sin t.$$

From above we know that $\cos t = 4/5$, $\sin t = 3/5$, $\cos s = 2/\sqrt{5}$, $\sin s = 1/\sqrt{5}$. Hence,

$$\cos(\text{Tan}^{-1}\ 1/2 - \text{Cos}^{-1}\ 4/5) = \cos(s-t)$$

$$= \frac{2}{\sqrt{5}} \frac{4}{5} + \frac{1}{\sqrt{5}} \frac{3}{5}.$$

$$= \frac{8}{5\sqrt{5}} + \frac{3}{5\sqrt{5}}$$

$$= \frac{11}{5\sqrt{5}} = \frac{11\sqrt{5}}{25}$$

It is also possible to define inverse functions for suitably restricted versions of sec x, csc x, and cot x. However, we will not do this here. We remark that the notations $\sin^{-1} x$, $\cos^{-1} x$, and $\tan^{-1} x$ (without capital letters) are sometimes used to indicate *any* arc whose sine, cosine, or tangent is x. Thus one sometimes sees formulas such as

$$\sin^{-1} 1/2 = \pi/6,\ 5\pi/6,\ 13\pi/6,\ \ldots$$

$$\cos^{-1} \frac{\sqrt{2}}{2} = \pi/4,\ -\pi/4,\ 9\pi/4,\ \ldots$$

$$\tan^{-1} 1 = \pi/4,\ 5\pi/4,\ 9\pi/4,\ \ldots$$

The values $\text{Sin}^{-1} x$, $\text{Cos}^{-1} x$, and $\text{Tan}^{-1} x$, then, are called the *principal values*, which distinguishes them from the other possible values allowed by the notation $\sin^{-1} x$, $\cos^{-1} x$, $\tan^{-1} x$.

SECTION 7-8, REVIEW QUESTIONS

1. The equations $y = \text{Sin}^{-1} x$ and $y = \text{Arcsin } x$ both mean that $\sin y = x$, and y is in the closed interval between _____ and _____ . $-\pi/2 \quad \pi/2$
2. The equations $y = \text{Cos}^{-1} x$ and $y = \text{Arccos } x$ both mean that
   Cos _____ = _____ and $0 \leq y \leq \pi$. $y \quad x$
3. Sin^{-1} is the inverse of the _____ function restricted to the   sine
   interval $[-\pi/2, \pi/2]$, and Cos^{-1} is the inverse of the cosine function restricted to the interval _____ . $[0, \pi]$
4. If $y = \text{Tan}^{-1} x$, then _____ $< y <$ _____ . $-\pi/2 \quad \pi/2$
5. If $y = \text{Sin}^{-1} x$, then _____ $\leq y \leq$ _____ . $-\pi/2 \quad \pi/2$
6. If $y = \text{Cos}^{-1} x$, then _____ $\leq y \leq$ _____ . $0 \quad \pi$
7. For every value of x, $\tan(\text{Tan}^{-1} x) =$ _____ , $\sin(\text{Sin}^{-1} x)$ x
   = _____ , and $\cos(\text{Cos}^{-1} x) =$ _____ . $x \quad x$

x 8. If $-\pi/2 < x < \pi/2$, $\text{Tan}^{-1}(\tan x) =$ _____, and

x $\text{Sin}^{-1}(\sin x) =$ _____.

x 9. If $0 \leq x \leq \pi$, then $\text{Cos}^{-1}(\cos x) =$ _____.

SECTION 7-8, EXERCISES

Graph the following, and give their domain and range.

1. $y = 2 \text{ Arcsin } x$
2. $y = 2 \text{ Arccos } x$
3. $y = \text{Arccos}(2x)$
4. $y = \text{Arcsin}(2x)$

Evaluate the following.

Example 1
5. $\text{Sin}^{-1} \sqrt{2}/2$ 6. $\text{Cos}^{-1} 1/2$
7. $\text{Cos}^{-1} \sqrt{3}/2$ 8. $\text{Sin}^{-1} 1/2$
9. $\text{Sin}^{-1}(-\sqrt{3}/2)$ 10. $\text{Cos}^{-1}(-1/2)$
11. $\text{Cos}^{-1}(-\sqrt{2}/2)$ 12. $\text{Sin}^{-1}(-\sqrt{2}/2)$
13. $\text{Sin}^{-1}(\sin 3\pi/4)$ 14. $\text{Cos}^{-1}(\cos(-\pi/4))$
15. $\text{Cos}^{-1}(\cos 4\pi/3)$ 16. $\text{Sin}^{-1}(\sin 7\pi/6)$
17. $\text{Sin}^{-1}(\cos \pi/6)$ 18. $\text{Cos}^{-1}(\sin \pi/3)$
19. $\text{Cos}^{-1}(\sin(-\pi/4))$ 20. $\text{Sin}^{-1}(\cos(-\pi/4))$

Example 2
21. $\sin(\text{Cos}^{-1} 2/3)$ 22. $\cos(\text{Sin}^{-1} 1/3)$
23. $\cos(\text{Sin}^{-1}(-1/4))$ 24. $\sin(\text{Cos}^{-1}(-2/5))$

Example 3
25. $\sin(\text{Tan}^{-1} 1/3), \cos(\text{Tan}^{-1} 1/3)$
26. $\sin(\text{Tan}^{-1} 3/5), \cos(\text{Tan}^{-1} 3/5)$
27. $\sin(\text{Tan}^{-1}(-2/3)), \cos(\text{Tan}^{-1}(-2/3))$
28. $\sin(\text{Tan}^{-1}(-3/4)), \cos(\text{Tan}^{-1}(-3/4))$

Example 4
29. $\cos(\text{Tan}^{-1} 1 - \text{Cos}^{-1} 3/5)$
30. $\sin(\text{Tan}^{-1} 1/2 + \text{Sin}^{-1} 1/3)$
31. $\sin(\text{Sin}^{-1} 3/5 + \text{Cos}^{-1} 3/5)$
32. $\cos(\text{Sin}^{-1} 4/5 + \text{Cos}^{-1} 4/5)$

Show that each of the following is true.

33. $\text{Tan}^{-1} 1/5 + \text{Tan}^{-1} 2/3 = \pi/4$
34. $\text{Tan}^{-1} 3/7 + \text{Tan}^{-1} 2/5 = \pi/4$
35. $\text{Sin}^{-1} 3/5 + \text{Cos}^{-1} 12/13 = \text{Sin}^{-1} 56/65$
36. $\text{Sin}^{-1} 4/5 - \text{Sin}^{-1} 5/13 = \text{Cos}^{-1} 56/65$

7-9 TRIGONOMETRIC EQUATIONS

In previous sections we developed procedures for showing that certain trigonometric expressions were identities. In other words, we showed that these expressions were true for all values of the variable.

The goal of this section is to develop techniques to help us find *some* value of the variable that will make an expression a true statement. For example,

$$\sin x = 1$$

is true for $x = \pi/2$. Since $\sin x = \sin(x + 2N\pi)$, for every integer N, we also know that $\sin x = 1$ for $x = \pi/2, 5\pi/2, 9\pi/2...$. However if we were asked to solve $\sin x = 1$ for $0 \leq x \leq 2\pi$, the solution would be simply $\pi/2$.

Example 1 Solve $\cos x = 1/2$ for $0 \leq x \leq 2\pi$.

Solution. $\cos \pi/3 = 1/2$, hence $x = \pi/3$. However, we know that $\cos x = \cos(-x)$. Therefore $\cos(-\pi/3) = \cos 5\pi/3 = 1/2$ and $x = 5\pi/3$ is also a solution. Hence the solutions of $\cos x = 1/2$ in $[0, 2\pi]$ are $x = \pi/3$ and $5\pi/3$.

Example 2 Solve $\sin x + \cos x = 1$ for $0 \leq x < 2\pi$.

Solution.
$$\sin x + \cos x = 1$$
$$\sin x = 1 - \cos x$$
$$\sin^2 x = 1 - 2\cos x + \cos^2 x \text{ (square both sides)}$$
$$1 - \cos^2 x = 1 - 2\cos x + \cos^2 x$$
$$2\cos^2 x - 2\cos x = 0$$
$$2\cos x (\cos x - 1) = 0$$
$$2\cos x = 0 \quad \text{or} \quad \cos x - 1 = 0$$

Hence,
$$\cos x = 0 \quad \text{or} \quad \cos x = 1.$$

The possible values of x are

$$\frac{\pi}{2}, \frac{3\pi}{2}, \text{ and } 0.$$

We now must check each of the possible solutions to see if they are all solutions of the original equation:

$$\frac{\pi}{2}: \quad \sin\frac{\pi}{2} + \cos\frac{\pi}{2} = 1 + 0 = 1 \qquad \text{(check)}$$

$$\frac{3\pi}{2}: \quad \sin\frac{3\pi}{2} + \cos\frac{3\pi}{2} = -1 + 0 \neq 1 \qquad \text{(no)}$$

Trigonometric Equations

$0:$ $\sin 0 + \cos 0 = 0 + 1 = 1$ (check)

Therefore, $\pi/2$ and 0 are the solutions of $\sin x + \cos x = 1$, for $0 \leq x < 2\pi$.

Example 3 Solve $\cos^2 x - 2 \sin x = 1$ for $0 \leq x < 2\pi$.

Solution. Since $\cos^2 x = 1 - \sin^2 x$, we obtain

$$1 - \sin^2 x - 2 \sin x = 1,$$
$$\sin^2 x + 2 \sin x = 0,$$
$$\sin x(\sin x + 2) = 0.$$

Hence,

$$\sin x = 0 \quad \text{or} \quad \sin x = -2.$$

However, $-1 \leq \sin x \leq 1$, so $\sin x \neq -2$. Therefore $x = 0$ and $x = \pi$ are the only possible solutions in $[0, 2\pi)$.

$$0: \quad \cos^2 0 - 2 \sin 0 \stackrel{?}{=} 1$$
$$1 - 2(0) = 1 \quad \text{(check)}$$
$$\pi: \quad \cos^2 \pi - 2 \sin \pi \stackrel{?}{=} 1$$
$$(-1)^2 - 2(0) = 1 \quad \text{(check)}$$

Example 4 Find all values of x such that $\cos 2x - 4 \cos x + 3 = 0$.

Solution.
$$\cos 2x - 4 \cos x + 3 = 0$$
$$\cos^2 x - \sin^2 x - 4 \cos x + 3 = 0$$
$$\cos^2 x - (1 - \cos^2 x) - 4 \cos x + 3 = 0$$
$$2 \cos^2 x - 4 \cos x + 2 = 0$$
$$2(\cos x - 1)^2 = 0$$
$$\cos x = 1$$
$$x = 0, \pm 2\pi, \pm 4\pi, \pm 6\pi, \ldots, \pm 2n\pi, \ldots$$

Check:
$$\cos 2(2n\pi) - 4 \cos (2n\pi) + 3 \stackrel{?}{=} 0$$
$$\cos (4n\pi) - 4 \cos (2n\pi) + 3 \stackrel{?}{=} 0$$
$$1 - 4 + 3 = 0$$

SECTION 7-9, EXERCISES

Solve the following equations for $0 \le x < 2\pi$.

1. $2 \cos x - 1 = 0$
2. $4 \sin^2 x - 3 = 0$
3. $\sin^2 x - \cos^2 x + 1 = 0$
4. $2 \cos^2 x - \sqrt{3} \cos x = 0$
5. $3 \tan^2 x - 1 = 0$
6. $\sec^2 x - 4 \sec x + 4 = 0$
7. $3 \sec x + 2 = \cos x$
8. $\sin 2x - 4 \cos x = 0$
9. $2 \sin^2 x - 5 \sin x + 2 = 0$
10. $2 \sin x \cos x + \sin x = 0$
11. $\sqrt{3} \csc^2 x + 2 \csc x = 0$
12. $2 \sin^2 x + 3 \cos x - 3 = 0$
13. $\cos 2x = 0$
14. $\sin 2x = 0$
15. $\sin 2x + 2 \sin^2 \left(\dfrac{x}{2}\right) = 1$
16. $\cos 2x + 2 \cos^2 \dfrac{x}{2} = 1$
17. $\cos x = \dfrac{1 + \cos^2 x}{2}$
18. $\cot x + 2 \sin x = \csc x$
19. $\sec^2 x - 2 \tan x = 0$
20. $4 \tan^2 x - 3 \sec^2 x = 0$

7-10 REVIEW FOR CHAPTER

1. In what quadrants can $P(s)$ be if:
   (a) $\sin s > 0$
   (b) $\sin s < 0$
   (c) $\cos s > 0$
   (d) $\cos s < 0$
   (e) $\tan s > 0$
   (f) $\tan s < 0$
   (g) $\sec s > 0$
   (h) $\sec s < 0$
   (i) $\cot s > 0$
   (j) $\cot s < 0$
   (k) $\csc s > 0$
   (l) $\csc s < 0$

2. When will $\sin s = \sin s_1$ if $0 < s_1 < s < \pi$?
3. When will $\cos s = \cos s_1$ if $s \ne s_1$ and $0 < s_1 < \pi/2, 0 < s < 2\pi$?
4. Find $\sin s$, $\cos s$, $\sec s$, $\csc s$, and $\cot s$ if $\tan s = 2/3$ and $P(s)$ is in the third quadrant.

Graph the following functions; give the amplitude and the period.

5. $y = 2 \cos x$
6. $y = 3 \sin x$
7. $y = \sin 2x$
8. $y = \cos 3x$
9. $y = 2 \cos \tfrac{1}{2} x$
10. $y = 1 + \sin x$
11. $y = -1 + \cos x$
12. $y = \cos x + \sin x$
13. $y = 2 \sin(x + \pi/2)$
14. $y = 3 \cos(\pi - x)$

Find the value of $\sin s/2$, $\cos s/2$, and $\tan s/2$ for each of the following values of s.

15. $\dfrac{4\pi}{2}$
16. $\dfrac{7\pi}{6}$
17. $-\dfrac{2\pi}{3}$

Find the value of each of the following.

18. $\sin \dfrac{2\pi}{3}$

19. $\cos \dfrac{5\pi}{4}$

20. $\cos\left(\dfrac{\pi}{2} + \dfrac{\pi}{3}\right)$

21. $\cos \dfrac{5\pi}{12}$

22. $\sin \dfrac{5\pi}{12}$

23. $\tan \dfrac{5\pi}{12}$

Show that each of the following are true.

24. $\text{Tan}^{-1} \dfrac{1}{5} + \text{Tan}^{-1} \dfrac{2}{3} = \text{Tan}^{-1} 1$

25. $\text{Tan}^{-1} 3 + \text{Tan}^{-1} \dfrac{1}{3} = \dfrac{\pi}{2}$

26. $\text{Tan}^{-1} \dfrac{1}{3} + \text{Tan}^{-1} \dfrac{1}{5} = \text{Tan}^{-1} \dfrac{4}{7}$

27. $\cos(\pi/2 - s) = \sin s$

28. $\sin(2\pi - s) = -\sin s$

29. $\cos s \cos 2s - \sin s \sin 2s = \cos 3s$

30. $\cos 2s \cos s + \sin 2s \sin s = \cos s$

31. $2 \cos^2 s/2 - \cos s = 1$

32. $2 \sin s + \sin 2s = \dfrac{2 \sin^3 s}{1 - \cos s}$

33. $(\cos s - \sin s)^2 = 1 - \sin 2s$

34. $4 \sin^2 s \cos^2 s = 1 - \cos^2 2s$

35. $\cos s + \sin s = \dfrac{\cos 2s}{\cos s - \sin s}$

36. $\dfrac{2 \cos^2 s - \sin^2 s + 1}{\cos s} = 3 \cos s$

37. $\sin s \tan s + \cos s = \dfrac{1}{\cos s}$

38. $\dfrac{1}{\cos^2 s} + \tan^2 s + 1 = \dfrac{2}{\cos^2 s}$

39. $\sin^4 s - \sin^2 s \cos^2 s - 2 \cos^4 s = \sin^2 s - 2 \cos^2 s$

40. $\sec^2 s - \csc^2 s = (\tan s + \cot s)(\tan s - \cot s)$

41. $\tan s - \tan s_1 = \sec s \sec s_1 \sin(s - s_1)$

42. $\sin A = \sin(B + C)$ if $A + B + C = \pi$

43. $\cos A = -\cos(B + C)$ if $A + B + C = \pi$

290 *Circular Functions*

8 TRIGONOMETRY

8-1 ANGLES

An angle is the geometric configuration consisting of two rays (half-lines) originating at the same point. The meeting point of the two half-lines is called the *vertex*, and the two lines are called the *sides* of the angle.

An angle situated in a coordinate plane with one side along the positive x-axis, and the vertex at the origin, is said to be *in standard position*. (See Fig. 8-1.) The side along the x-axis is called the *initial side*, and the other side is called the *terminal side*. We can distinguish an initial side and a terminal side for any angle. Such an angle will be measured by the corresponding angle you get by sliding it into standard position. Figure 8-2 shows the two different standard angles you get for a given angle depending on which side is specified as the initial side. We will say that an angle is in the first, second, third, or fourth quadrant depending on whether the terminal side is in the first, second, third, or fourth quadrant. The angle of Fig. 8-2(*a*) is in the first quadrant, and the angle of Fig. 8-2(*b*) is in the fourth quadrant.

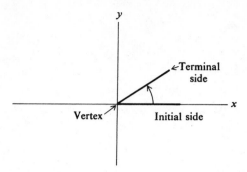

Figure 8-1

Figure 8-2

For the purpose of measuring angles, we can restrict our attention to angles in standard position by the agreement above. An angle in standard position is measured by the length of the arc of the unit circle which is cut off by the sides of the angle. Since any angle cuts the circle into *two* arcs, we stipulate that the angle is measured by the length of the arc from (1, 0) to the terminal side in the *counterclockwise* direction. The angle of Fig. 8-3 is measured by the length of arc s.

There are two different units for angle measurement: radians and degrees. The radian measure of an angle is simply the *length* of the arc of the unit circle the angle cuts off. Thus the angle θ of Fig. 8-3 would be an angle of $3\pi/4$ radians, since the arc s has length $3\pi/4$. A right angle in standard position (Fig. 8-4) would be an angle of $\pi/2$ radians, since the arc from (1, 0) to (0, 1) is one-quarter the total circumference of 2π.

The other unit used for angle measurement is the degree. If the unit circle is divided into 360 equal arcs, then the number of degrees

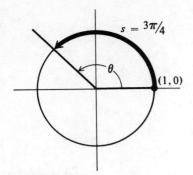

Figure 8-3

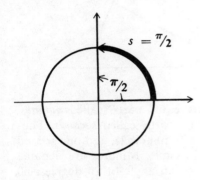

Figure 8-4

in an angle is the number of these arcs the angle cuts off. Again we consider the arc from the initial side to the terminal side *in the counterclockwise direction*. Hence an angle of 60 degrees, written 60°, is an angle that cuts off 60/360 of the whole unit circle. Since 60/360 (or 1/6) of the unit circle has length $\frac{1}{6}(2\pi) = \pi/3$, an angle of 60° is the same as an angle of $\pi/3$ radians. Similarly, the right angle ($\pi/2$ radians) cuts off one-quarter of the unit circle. Therefore a right angle has $1/4 \cdot 360 = 90°$. A 180°-angle is a *straight angle*; that is, an angle in which the two sides extend in opposite directions from the vertex along the same line.

Example 1 Express in degree measure the angles $\pi/6$, $\pi/4$, $\pi/3$, radians.

Solution. The angle π radians corresponds to a half circle, and therefore to 1/2 (360°) or 180°. Therefore (see Fig. 8-5)

$$\frac{\pi}{6} \text{ radians} = \frac{1}{6} \cdot 180° = 30°,$$

$$\frac{\pi}{4} \text{ radians} = \frac{1}{4} \cdot 180° = 45°,$$

$$\frac{\pi}{3} \text{ radians} = \frac{1}{3} \cdot 180° = 60°.$$

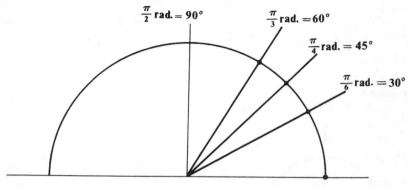

Figure 8-5

One degree is subdivided into sixty equal parts, called *minutes*. Each minute is divided into sixty equal parts, called *seconds*. Thus one minute is $1/60 \cdot 1/360 = 1/21600$ of a full circle, and one second is $1/60 \cdot 1/21600 = 1/1296000$ of a full circle. Minutes are denoted with the sign ′, and seconds with ″, so an angle of 20 degrees, 50 minutes, and 5 seconds would be written 20°50′5″. Fractions of a radian are expressed with decimals. For example, 90° is the same as $\pi/2$ radians, or 1.57 radians (approximately).

By our definition, the measure of any angle is some positive number of radians (between 0 and 2π), or some positive number of degrees (between 0° and 360°). It is also convenient to define what is meant by negative angles, and what is meant by angles of more than 2π radians (or 360°). By an angle of $-\pi/4$ radians we mean one congruent to the standard angle which cuts off an arc of $\pi/4$ from the unit circle in the *clockwise* direction from (1, 0). This convention is part of the standard mathematical agreement that counterclockwise is the positive direction, and clockwise is the negative direction.

The angle $-\pi/4$ could also be measured as $7\pi/4$ as shown in Fig. 8-6. This same angle could also be denoted by either $-45°$, or $315°$. Although it is convenient to introduce the concept of negative angles, notice that we no longer have a unique measure associated with a given angle.

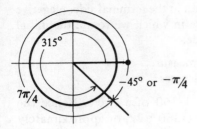

Figure 8-6

If θ_1 and θ_2 are two angles, we define their sum $\theta_1 + \theta_2$ as follows. If the initial side of θ_2 coincides with the terminal side of θ_1, and the vertices of θ_1 and θ_2 coincide, then $\theta_1 + \theta_2$ is the angle whose initial side is that of θ_1, and whose terminal side is that of θ_2. (See Fig. 8-7.) To permit this kind of definition, we must admit angles of more than 2π radians or $360°$. We therefore extend our definitions and agree that an angle of $2\pi + \theta$ radians is the same as an angle of θ radians.

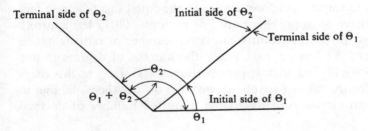

Figure 8-7

In terms of degree measurement, this means that an angle of $360° + \theta°$ is the same as an angle of $\theta°$. With this agreement, we have infinitely many measures of any given angle. For example, the angles of

$\pi/6$ rad., $13\pi/6$ rad., $25\pi/6$ rad., ..., $-11\pi/6$ rad., $-23\pi/6$ rad., ...

are all the same. In degrees, this angle could be described as

$$30°, 390°, 750°, \ldots, -330°, -690°, \ldots$$

We may think of the measure of an angle as the total number of radians (or degrees) through which we rotate the initial side to arrive at the terminal side. Angles of more than 2π radians correspond to the situation in which we rotate the initial side through more than one complete circle before coming to rest at the terminal side. Negative angle measures correspond to the case in which we rotate the initial side clockwise toward the terminal side.

Example 2 (a) Express 1° in radian measure.
(b) Express 1 radian in degrees.

Solution. (a) An angle of 1° is 1/360 of a circle. Hence the length of the arc on the unit circle is $1/360 \cdot 2\pi$, or approximately 3.1416/180. Hence

$$1° = 0.0175 \text{ radians}$$

(b) Similarly 1 radian corresponds to $1/2\pi$ of a whole circle, and therefore to $(1/2\pi \times 360°)$. Hence

$$1 \text{ radian} = 57.3°$$

Notice that

$$57.3 = \frac{1}{0.0175}$$

In the example above we have used the equal sign where in fact we only have an approximation. For example, 0.0174 is approximately (to three significant figures) the number of radians in one degree, and 57.3 is approximately the number of degrees in one radian. We will deal with approximations extensively in this chapter, and for the sake of simplicity we will now use the equal sign to mean approximate equality to the indicated number of decimal places.

Example 3 Find a negative measure in radians (between 0 and -2π) for each of the angles $\pi/2$, 4, $4\pi/3$.

Solution. The angle of $\pi/2$ radians divides the circle into arc of length $\pi/2$ and $2\pi - \pi/2$, or $3\pi/2$. Hence, $-3\pi/2$ radians is another measure for $\pi/2$ radians. Similarly, $-(2\pi - 4) = -(6.28 - 4) = -2.28$ radians corresponds to 4 radians, and $-(2\pi - 4\pi/3) = -2\pi/3$ radians corresponds to $4\pi/3$ radians.

Example 4 Give a measure in radians between 0 and 2π for angles of $5\pi/2$ and $-\pi/4$ radians, and a measure in degrees between 0° and 360° for angles of $-100°$ and 720°.

Solution. $5\pi/2$ radians is the same angle as $\pi/2$ radians, since $5\pi/2 = 2\pi + \pi/2$; $-\pi/4$ radians is the same angle as $7\pi/4$, since $2\pi - \pi/4 =$

$7\pi/4$. The angle $-100°$ is the same as $360° - 100° = 260°$; the angle $720°$ is the same as $0°$, since $360° + 360° = 720°$.

SECTION 8-1, REVIEW QUESTIONS

1. An angle consists of two _____ originating at the same point.   **rays**

2. The point where the two rays originate is called the _____.   **vertex**

3. The two rays (or segments) are called the _____ of the angle.   **sides**

4. We sometimes distinguish an initial side and a _____ side of an angle.   **terminal**

5. An angle is in standard position if the _____ side is along the positive x-axis and the vertex is at the origin.   **initial**

6. An angle in standard position is measured by the amount of arc of the unit circle which extends from the initial side to the terminal side in the _____ direction.   **counterclockwise**

7. In mathematics, counterclockwise is the positive direction, and clockwise is the _____ direction.   **negative**

8. The radian measure of an angle is the _____ of the arc of the unit circle from the x-axis counterclockwise to the terminal side.   **length**

9. An angle which cuts an arc 1 unit long off the unit circle is an angle of _____ radian(s).   **one**

10. An angle of 1° cuts off an arc which is one _____ th of the whole unit circle.   **360**

11. There are 360° in a complete circle and _____ radians in a complete circle.   **2π**

12. Negative angles correspond to arcs measured _____ from the x-axis.   **clockwise**

13. An angle of $2\pi + \theta$ radians is the same as an angle of _____ radians, and an angle of $360° + \theta°$ is the same as an angle of _____ degrees.   **θ**   **θ**

SECTION 8-1, EXERCISES

Find the degree measure between 0° and 360° for each of the following radian measures.

Example 1,
1. $\pi/3$
2. $\pi/6$
3. $\pi/2$
4. $\pi/4$
5. $2\pi/3$
6. $3\pi/4$
7. $5\pi/3$
8. $7\pi/6$
9. $5\pi/4$
10. $7\pi/4$
11. $\pi/8$
12. $\pi/12$
13. $\pi/10$
14. $\pi/5$
15. 2
16. 3
17. .5
18. 1.5

Find the radian measure between 0 and 2π for each of the following degree measures.

Example 2
19. 120°
20. 135°
21. 90°
22. 180°
23. 270°
24. 240°
25. 2°
26. 3°
27. 10°
28. 100°

Find a negative measure in radians between 0 and -2π and in degrees between 0 and $-360°$ for each of the following angles in radian measure.

Example 3
29. $3\pi/2$
30. π
31. $\pi/6$
32. $3\pi/4$
33. $11\pi/6$
34. $5\pi/3$
35. 1
36. 3

Find the radian measure between 0 and 2π and the degree measure between 0 and 360° for each of the following.

Example 4
37. 3π
38. 5π
39. $-\pi/2$
40. $-\pi$
41. 390°
42. 420°
43. $-45°$
44. $-30°$
45. $15\pi/2$
46. $9\pi/2$
47. 765°
48. 450°

8-2 TABLES OF TRIGONOMETRIC FUNCTIONS

With the definitions of the preceding section, it is easy to see how we can define the trigonometric functions on angles. We define the sine of an angle of s radians to be the sine of the number s; that is, sin s is the y-coordinate of the point where the terminal side of the angle intersects the unit circle. The cosine, tangent, and so on, are defined similarly. The cosine and sine of an angle of $\theta°$ are, respectively, cos s, sin s, where s is the corresponding radian measure. See Fig. 8-8. Since the radian measure of an angle of $\theta°$ is $(\theta/360) \cdot 2\pi$, or $(\theta/180)\pi$, we have

$$\sin \theta° = \sin\left(\frac{\theta\pi}{180}\right), \quad \cos \theta° = \cos\left(\frac{\theta\pi}{180}\right).$$

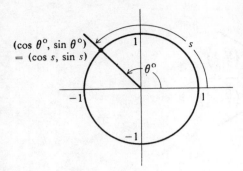

Figure 8-8

Similar agreements are made for the remaining functions: tangent, cotangent, secant, and cosecant. For example,

$$\sin 45° = \sin \frac{\pi}{4},$$

$$\cos 30° = \cos \frac{\pi}{6},$$

$$\tan 57.3° = \tan 1,$$

$$\sec(-60°) = \sec\left(-\frac{\pi}{3}\right),$$

$$\csc(130°) = \csc\left(\frac{130}{180}\pi\right),$$

$$\cot 1° = \cot(0.0174).$$

Earlier we saw that the trigonometric functions of any angle can be expressed in terms of some angle in the first quadrant. Figure 8-9 shows how each of the angles 150°, 210°, 330° has the same trigonometric functions, except for sign, as 30°. The angle of the first quadrant whose sine and cosine have the same numerical values as a given angle is called the *reference angle*. Hence 30° is the reference angle for each of the angles 150°, 210°, 330°.

Since degree measurement is commonly used in most applications, tables of values of trigonometric functions are usually given in terms of degrees. The Table of Trigonometric Functions in the Appendix gives angles for every 10′ of arc between 0° and 90°. To find a trigonometric function of an angle not in the first quadrant, we look up its reference angle, and then attach the appropriate sign.

Tables of Trigonometric Functions

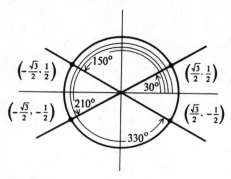

Figure 8-9

One other observations about the symmetry of the unit circle permits the simplification of tables of trigonometric functions. Observe (Fig. 8-10) that

$$\sin(45° + \theta°) = x_0 = \cos(45° - \theta°)$$
$$\cos(45° + \theta°) = y_0 = \sin(45° - \theta°)$$

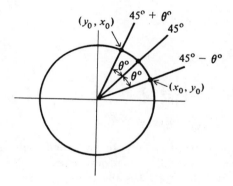

Figure 8-10

This means that the table need only contain sines and cosines of angles between 0° and 45°. The table is so arranged that when you reach 45°, you start to read backward. The entries for 46° are the same as those for 44°, only sin 46° = cos 44°, and cos 46° = sin 44°. Similarly, the entries for 50° are the same as those for 40°, and the entries for 80° are the same as those for 10°, only in each case the sine and cosine columns are reversed.

The portion of this table, for angles between 30° and 31° (and angles between 59° and 60°), is given below.

TABLE OF TRIGONOMETRIC FUNCTIONS

Angle θ									
Degrees	Radians	sin θ	csc θ	tan θ	cot θ	sec θ	cos θ		
30° 00′	.5236	.5000	2.000	.5774	1.732	1.155	.8660	1.0472	60° 00′
10	265	025	1.990	812	720	157	646	443	50
20	294	050	980	851	709	159	631	414	40
30	.5323	.5075	1.970	.5890	1.698	1.161	.8616	1.0385	30
40	352	100	961	930	686	163	601	356	20
50	381	125	951	969	675	165	587	327	10
31° 00′	.5411	.5150	1.942	.6009	1.664	1.167	.8572	1.0297	59° 00′
		cos θ	sec θ	cot θ	tan θ	csc θ	sin θ	Radians	Degrees
								Angle θ	

Example 1 Find sin 30°10′, cos 149°50′, tan (210°10′).

Solution. The reference angle for each of these angles is 30°10′, because

$$180° - 149°50' = 179°60' - 149°50'$$
$$= 30°10',$$

and
$$210°10' - 180° = 30°10'.$$

From the table we find

$$\sin 30°10' = 0.5025,$$
$$\cos 30°10' = 0.8646,$$
$$\tan 30°10' = 0.5812.$$

Since 149°10′ is in the second quadrant,

Tables of Trigonometric Functions

or
$$\cos(149°10') = -\cos(30°10')$$
$$\cos 149°10' = -0.8646.$$

Since $210°10'$ is in the third quadrant,
$$\tan(210°10') = \tan(30°10')$$
or
$$\tan(210°10') = 0.5812.$$

Example 2 Find $\cos 30°24'$.

Solution. The table has entries for $30°20'$ and $30°30'$, but not for $30°24'$. We use the entries we have to approximate $\cos 30°24'$ by linear interpolation. Since $24'$ is $4/10$ of the way from $20'$ to $30'$, we find the number that is $4/10$ of the way from $\cos 30°20'$ to $\cos 30°30'$.

$$\cos 30°20' = 0.8631$$
$$\cos 30°30' = \underline{0.8616}$$
$$\text{difference} = 0.0015$$

$$4/10 \text{ of difference} = 0.4 \times 0.0015 = 0.0006$$
$$\cos 30°24' = 0.8631 - 0.0006 = 0.8625$$

Notice that since the cosine decreases from $30°20'$ to $30°30'$, we *subtract* $4/10$ of the difference from the value for $\cos(30°20')$.

Example 3 Find $\sec(-120°43')$.

Solution. The reference angle for $-120°43'$ is $180° - (120°43')$, or $59°17'$, as shown in Fig. 8-11. Hence, $\sec(-120°43') = -\sec 59°17'$.

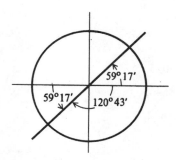

Figure 8-11

From the table, we find the values below.

$$\sec 59°20' = 1.961$$
$$\sec 59°10' = 1.951$$
$$\text{difference} = 0.010$$

Because 59°17' is 7/10 of the way from 59°10' to 59°20', we find 7/10 of 0.010, and add to 1.951:

$$0.7 \times 0.010 = 0.007,$$
$$\sec 59°17' = 1.951 + 0.007 = 1.958$$

We use interpolation in the same way to find the angle when we are given its sine, cosine, etc.

Example 4 Find the degree measure of angle θ in the first quadrant such that $\tan \theta = 0.5900$.

Solution. From the table we find

$$\tan 30°40' = 0.5930$$
$$\tan 30°30' = 0.5890$$
$$\text{difference} = 0.0040$$

The number 0.5900 is 0.0010 more than 0.5890, and hence is $0.0010/0.0040 = 1/4$ of the way from $\tan 30°30'$ to $\tan 30°40'$. The angle which is 1/4 of the way from 30°30' to 30°40' is 30°32.5', or 30°32'30". Usually we will not deal with seconds, and so ordinarily we would round this answer off to 30°33'.

SECTION 8-2, REVIEW QUESTIONS

1. The sine of an angle is the same as the sine of the number which is the angle's _____ measure.   **radian**

2. Angles are measured in radians or _____.   **degrees**

3. The cosine and sine of an angle are, respectively, the x- and y-coordinates of the point where the _____ side of the angle intersects the unit circle when the angle is in standard position.   **terminal**

4. The trigonometric functions of any angle are numerically the same as those of some angle in the _____ quadrant.   **first**

reference	5. The angle of the first quadrant you use to find sin θ, cos θ, etc., is called the _____ angle for θ.
90°	6. The reference angle is always between 0° and _____°.
180°	7. The reference angle for an angle θ° in the third quadrant is θ° − _____°.
180° − θ°	8. The reference angle for θ° in the third quadrant is θ° − 180°, and the reference angle for θ° in the second quadrant is _____.
linear	9. We obtain approximate values of the trigonometric functions for each minute of arc by _____ interpolation.
seven	10. Linear interpolation approximates sin 10°17′ by the number which is _____ tenths of the way from sin 10°10′ to sin 10°20′.

SECTION 8-2, EXERCISES

What is the reference angle for each of the following?

Example 1
1. −70°   2. 91°   3. 170°
4. −45°   5. −160°   6. 200°
7. 315°   8. −195°   9. 290°

10–18. Express sin x and cos x for each value of x in Exercise 1 through 9 in terms of the reference angle. [*Example:* sin (−70°) = −sin (70°) and cos (−70°) = cos (70°)].

19. (a) If $0 < x < \pi/2$ and $0 < y < \pi/2$, when will sin x = cos y?
    (b) If $0 < x < \pi/2$ and $0 < y < \pi/2$, when will tan x = cot y?

Use the Table of Trigonometric Functions to find the value of each of the following:

20. sin 237°20′   21. cos 125°30′   22. tan 185°40′
23. csc 322°10′   24. cot 110°10′   25. sec 127°50′

Use the table and linear interpolation to find the value of each of the following:

Example 2, 3
26. cos 175°55′   27. sin (−36°14′)   28. −sin 162°16′
29. −tan 154°35′   30. sec 315°45′   31. cot 326°25′

Find the degree measure of angle θ in the first quadrant determined by the following conditions.

Example 4
32. csc θ = 2.118   33. sin θ = 0.6841   34. cos θ = 0.8039
35. tan θ = 0.6766 36. sin θ = 0.8426   37. cos θ = 0.9576

8-3 RIGHT TRIANGLES

For any acute angle θ ($0 < \theta < \pi/2$), we consider the right triangle formed by dropping a perpendicular to the x-axis from the point where the terminal side intersects the unit circle. See Fig. 8-12. The right triangle formed this way has hypotenuse of length one. The side adjacent to the angle has length $\cos \theta$, and the side opposite the angle has length $\sin \theta$. From plane geometry we know that any right

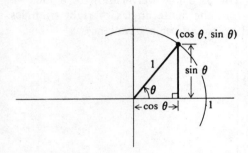

Figure 8-12

triangle with one acute angle equal to θ will be similar to this triangle.

For similar triangles, ratios of corresponding sides are equal. Therefore (see Fig. 8-13),

$$\sin \theta = \frac{\sin \theta}{1} = \frac{b}{c} = \frac{\text{opposite}}{\text{hypotenuse}}$$

$$\cos \theta = \frac{\cos \theta}{1} = \frac{a}{c} = \frac{\text{adjacent}}{\text{hypotenuse}}$$

$$\tan \theta = \frac{\sin \theta}{\cos \theta} = \frac{b}{a} = \frac{\text{opposite}}{\text{adjacent}}$$

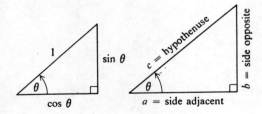

Figure 8-13

These formulas are sometimes used as definitions of the sine, cosine, and tangent of an acute angle. Since "opposite" and "adjacent" mean the same thing regardless of how the right triangle is oriented, the above formulas hold for any right triangle.

Consider a right triangle (Fig. 8-14) with acute angles α and β, sides a and b opposite these angles, and hypotenuse c. We know that $\cos \alpha = b/c$. The opposite-over-hypotenuse rule shows that $\sin \beta = b/c$. Similarly, $\cos \beta = a/c = \sin \alpha$. Since $\sin \alpha = \cos \beta$ and $\cos \beta = \sin \alpha$, and α and β are both in the first quadrant, we know that $\beta = (\pi/2) - \alpha$, or $\alpha + \beta = \pi/2$. This is a new verification of a fact we learned in plane geometry—that the acute angles of right triangles are complementary (i.e., add up to 90°, or $\pi/2$ radians).

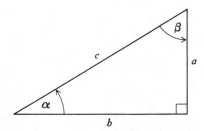

Figure 8-14

If we are given any side and angle of a right triangle or any two sides, we can then determine all the sides and angles. This is called *solving* the triangle. The following examples illustrate the method.

Example 1 Solve the right triangle that has one angle of 20° and hypotenuse equal to 10.

Solution. The triangle is shown in Fig. 8-15 with $\alpha = 20°$ and $c = 10$. We immediately know that $\beta = 70°$, because the two acute angles are complementary. We know that $\sin 20° = a/10$, and $\cos 20° = b/10$. Therefore,

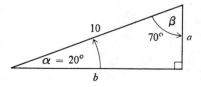

Figure 8-15

308 Trigonometry

$$a = 10 \sin 20° = 10 \times 0.3420 = 3.420,$$
$$b = 10 \cos 20° = 10 \times 0.9397 = 9.397.$$

Example 2 Let $a = 5$, $b = 7$ in a right triangle. Find α, β, and c.

Solution. We first calculate that
$$\tan \alpha = 5/7 = 0.7143.$$
From the Table of Trigonometric Functions we find
$$\tan 35°40' = 0.7177$$
$$\tan 35°30' = \underline{0.7133}$$
$$\text{difference} = 0.0044.$$
Since $\tan \alpha$ is $0.0010/0.0044 = 0.2$ of the way from 0.7133 to 0.7177, $\alpha = 35°32'$. Consequently $\beta = 54°28'$. We can find c from the formula $\sin \alpha = a/c = 5/c$:
$$c = 5/\sin \alpha$$
$$= 5/0.5812$$
$$= 8.603.$$
Notice that if we use the formula $\csc \alpha = c/a = c/5$, or
$$c = 5 \csc \alpha,$$
then the arithmetic involves multiplication instead of division. It is usually preferable to perform the calculation this way:
$$c = 5 \csc 35°32'$$
$$= 5 \times 1.721$$
$$= 8.605.$$

The ratios that can be read from a right triangle are very useful in determining one trigonometric function of an angle when another function is given.

Example 3 If θ is an acute angle such that $\sin \theta = 4/7$, find $\cos \theta$ and $\tan \theta$.

Solution. We draw a right triangle with one side 4 and hypotenuse 7. (See Fig. 8-16.) Then the angle opposite the side of length 4 will equal θ. The remaining side will have length $\sqrt{49 - 16} = \sqrt{33}$ by the Pythagorean theorem. Therefore, $\cos \theta = \sqrt{33}/7$, and $\tan \theta = 4/\sqrt{33}$.

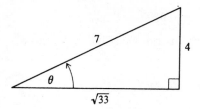

Figure 8-16

This method also works if $\sin \theta$ (or $\cos \theta$ or $\tan \theta$) is given in decimal form. For example, if we know that $\sin \theta = 0.3714$, then we can draw a right triangle with side 3174 and hypotenuse 10,000 and perform the same kind of calculation.

For angles not in the first quadrant, we can perform our right triangle computations with the reference angle. However, it is frequently easier to use a right triangle constructed on the given angle in standard position, with one or both legs labeled as negative.

Example 4 If $\cos \theta < 0$ and $\tan \theta = -\frac{1}{2}$, find $\sin \theta$ and $\cos \theta$.

Solution. The fact that $\cos \theta < 0$ and $\tan \theta < 0$ tells us that θ is in the second quadrant. We construct the right triangle shown in Fig. 8-17 from the information that $\tan \theta = -\frac{1}{2}$. Then the hypotenuse is $\sqrt{1 + 4} = \sqrt{5}$, and we can read off the values: $\sin \theta = 1/\sqrt{5}$, $\cos \theta = -2/\sqrt{5}$.

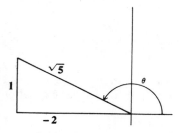

Figure 8-17

SECTION 8-3, REVIEW QUESTIONS

cos θ, sin θ

1. For an angle in standard position, the point where the terminal side of an angle θ intersects the unit circle has coordinates $x =$ _____, $y =$ _____.

2. If we drop a perpendicular from the point (cos θ, sin θ) to the x-axis, the resulting right triangle has hypotenuse of length _____. **1**

3. In a right triangle with hypotenuse 1, the side adjacent to angle θ has length cos θ, and the side opposite has length _____. **sin θ**

4. In any right triangle, sin θ is the ratio _____ side over the hypotenuse. **opposite**

5. In a right triangle, cos θ is the ratio _____ side over the _____. **adjacent hypotenuse**

6. The ratio of the side opposite angle θ over side adjacent to angle θ equals _____. **tan θ**

7. If α, β are the acute angles of a right triangle, a, b are the sides opposite them, and c is the hypotenuse, then sin α = _____, cos α = _____, tan α = _____. **a/c b/c a/b**

8. In solving a right triangle, it is better to use a trigonometric function for which the calculation is multiplication rather than _____. **division**

9. Instead of using $5/c = \sin 20°$, $c = 5/\sin 20°$, we would ordinarily use $c/5 = \csc 20°$, $c =$ _____. **5 csc 20°**

SECTION 8-3, EXERCISES

Let c be the hypotenuse of a right triangle with acute angles α and β, and sides a and b opposite α and β respectively. Solve the following triangles.

1. α = 20°, c = 5		2. β = 70°, c = 1	Example 1
3. α = 40°, c = 10		4. α = 38°, c = 10	
5. β = 25°, c = 3		6. β = 49°, c = 6	
7. a = 1, b = 4		8. a = 3, b = 2	Example 2
9. a = 10, b = 5		10. a = 15, b = 30	
11. a = 10,000, b = 4,452		12. a = 7,536, b = 10,000	

In 13 through 18, θ is an acute angle.

13. sin θ = 2/5, find cos θ, tan θ   Example 3
14. cos θ = 1/3, find sin θ, tan θ
15. tan θ = 1/2, find sin θ, cos θ
16. cot θ = 3, find sin θ, cos θ

17. $\csc \theta = 4$, find $\cos \theta$, $\tan \theta$
18. $\sec \theta = 1.5$, find $\sin \theta$, $\tan \theta$

Example 4
19. $\cos \theta = 1/2$, $\tan \theta < 0$, find $\sin \theta$ and $\tan \theta$
20. $\sin \theta = -2/3$, $\tan \theta < 0$, find $\cos \theta$, $\tan \theta$
21. $\cos \theta < 0$, $\tan \theta = 4/3$, find $\sin \theta$, $\cos \theta$
22. $\sin \theta > 0$, $\tan \theta = 3/4$, find $\sin \theta$, $\cos \theta$
23. $\cos \theta < 0$, $\sin \theta = 1/5$, find $\cos \theta$, $\tan \theta$
24. $\sin \theta < 0$, $\cos \theta = 2/3$, find $\sin \theta$, $\tan \theta$

25. A fish line 50 yards long makes an angle of 20° with the surface of the water. How deep is the bait?
26. The angle of elevation to the top of a 216-foot rock is 37° from a boat in the water. How far is the boat from a point directly below the highest point of the rock?
27. At a point 185 feet from the base of a tree the angle of elevation of the top is 55°. How tall is the tree?
28. How high is the clay pigeon when the shooter fires if he fires when the bird is 16 yards away and the shotgun makes an angle of 125° with his body? (See diagram below.)

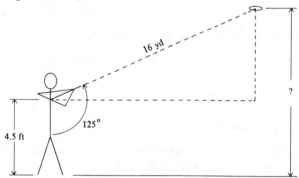

29. From an observation point the angles of depression of two people in a line with this point are 15° and 25°. Find the distance between the two people if the observation point is 400-feet high.
30. A tree grows on a horizontal plane. The angle of elevation to the top of the tree at rock A is 30° and at rock B, 100 feet nearer the tree, it is 45°. How high is the tree?
31. Find the angles of intersection of the diagonals of a rectangle 3 feet wide and 4 feet long.

8-4 LAW OF SINES

We turn now to the problem of solving triangles other than right triangles. We know that a triangle is uniquely determined if we are given:

(a) two angles and the included side (ASA), or
(b) two sides and the included angle (SAS), or
(c) three sides (SSS).

Suppose we are given angles α and β and included side c as shown in Fig. 8-18. We want to determine sides a and b, and angle γ. Of course $\gamma = 180° - (\alpha + \beta)$, so the real problem is finding sides a and b. If we drop the perpendicular of length h as shown, then we have

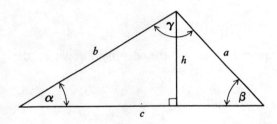

Figure 8-18

$$\sin \alpha = h/b, \quad h = b \sin \alpha ;$$
$$\sin \beta = h/a, \quad h = a \sin \beta .$$

Equating the two formulas for h, we get

$$b \sin \alpha = a \sin \beta .$$

$$\frac{a}{\sin \alpha} = \frac{b}{\sin \beta}$$

The same identity holds if one of the angles, α or β, is obtuse instead of acute as shown in Fig. 8-18. The case where β is obtuse is shown in Fig. 8-19. Since $\sin \beta' = \sin \beta$, we again have $h = a \sin \beta = b \sin \alpha$, or $a/\sin \alpha = b/\sin \beta$.

If we turn the triangle so that side b is horizontal, then the same argument shows that

$$a \sin \gamma = c \sin \alpha$$

Law of Sines 313

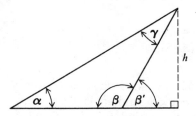

Figure 8-19

$$\frac{a}{\sin \alpha} = \frac{c}{\sin \gamma}$$

Combining these two equations, we get the identity known as the *Law of Sines*.

$$\frac{a}{\sin \alpha} = \frac{b}{\sin \beta} = \frac{c}{\sin \gamma}$$

If we are given any two angles of a triangle, we immediately know the third, since their sum is 180°. Hence if we are given any two angles and one side, we effectively have ASA, and the triangle is determined. Knowing one side, say a, and the three angles α, β, γ allows us to compute the remaining sides from the Law of Sines:

$$b = \frac{a \sin \beta}{\sin \alpha}, \qquad c = \frac{a \sin \gamma}{\sin \alpha}.$$

The other case in which we can use the Law of Sines is when we are given two of the sides and the angle opposite one of them. For example, if we are given a, α, and b, we can use

$$\frac{a}{\sin \alpha} = \frac{b}{\sin \beta}$$

to determine β. Knowing α and β, we determine γ, and then use

$$\frac{b}{\sin \beta} = \frac{c}{\sin \gamma}$$

to find c. The difficulty with this procedure is that two sides and the angle opposite one of them do not necessarily determine a unique triangle! In fact, there need not be *any* triangle that has a given set of numbers for a, α, and b.

Suppose we are given a, α, and b, and we want to know what possible triangles there are with sides a and b and angle α. Fig. 8-20 shows that if a is too small, there is no triangle with the given a, α, and b. The minimum possible value of a is $a = b \sin \alpha$, which yields

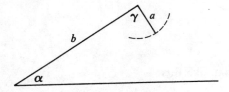

Figure 8-20

the right triangle of Fig. 8-21. Hence, if $a = b \sin \alpha$, there is a unique triangle formed with the given a, α, and b. If you use the Law of Sines to find β, here, you get

$$\frac{a}{\sin \alpha} = \frac{b}{\sin \beta}; \quad \sin \beta = \frac{b \sin \alpha}{a} = 1.$$

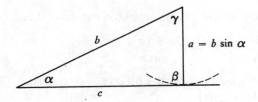

Figure 8-21

Hence (see Fig. 8-21),

$$\sin \beta = 1,$$
$$\beta = 90°,$$
$$\gamma = 90° - \alpha,$$
$$c = b \sin \gamma.$$

Figure 8-22 shows that there are two possible triangles with the given sides a, b, and angle α, if $a > b \sin \alpha$ and $a < b$.

Law of Sines 315

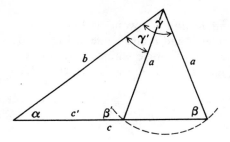

Figure 8-22

The arithmetic in this case looks like this:

$$b \sin \alpha < a < b, \quad \text{and} \quad \sin \beta = \frac{b \sin \alpha}{a} < 1.$$

The angle β can either be the acute angle shown, whose sine is $(b \sin \alpha)/a$, or its supplement, $\beta' = 180° - \beta$. The two lines give two different values for γ:

$$\gamma = 180° - (\alpha + \beta) \quad \text{or} \quad \gamma' = 180° - (\alpha + \beta')$$
$$= 180° - (\alpha + 180° - \beta)$$
$$= \beta - \alpha.$$

The different values γ, γ' give different values c, c' for the third side:

$$c = \frac{a \sin \gamma}{\sin \alpha}; \quad c' = \frac{a \sin \gamma'}{\sin \alpha}.$$

If $a > b$, there is a unique triangle formed with a, α, b (see Figure 8-23). Here, the angle β, determined by

$$\sin \beta = \frac{b \sin \alpha}{a},$$

must be the *acute* angle with this sine.

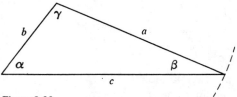

Figure 8-23

316 Trigonometry

Example 1 Solve the triangle with $\alpha = 20°$, $\beta = 60°$, $c = 8.3$.

Solution. With two angles given, the triangle is determined. First we find

$$\begin{aligned}\gamma &= 180° - (\alpha + \beta) \\ &= 180° - (20° + 60°) \\ &= 100°\end{aligned}$$

The reference angle for γ is 80°; $\sin 80° = 0.9848$, $\sin 20° = 0.3420$, and $\sin 60° = 0.8660$. We have (see Fig. 8-24):

$$\begin{aligned}a &= \frac{c \sin \alpha}{\sin \gamma} \\ &= \frac{(8.3)(0.3420)}{0.9848} \\ &= 2.88,\end{aligned}$$

$$\begin{aligned}b &= \frac{c \sin \beta}{\sin \gamma} \\ &= \frac{(8.3)(0.8660)}{0.9848} \\ &= 7.40.\end{aligned}$$

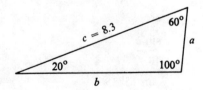

Figure 8-24

Example 2 Solve the triangle with $c = 2$, $b = 10$, $\gamma = 30°$.

Solution. $\sin 30° = 0.5000$, and since $c < b \sin \gamma = 10(1/2) = 5$, there is no triangle with this data. See Fig. 8-25.

Example 3 Solve the triangle with $b = 10$, $c = 7$, $\gamma = 30°$.

Solution. Here, b and γ are the same as in Example 2, only now c

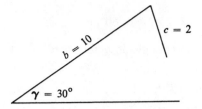

Figure 8-25

is between $b \sin \gamma = 5$ and $b = 10$. Therefore, there are two possible triangles.

$$\sin \beta = \frac{b \sin \gamma}{c}$$

$$= \frac{10(0.5000)}{7} = \frac{5}{7}$$

$$= 0.7143$$

We have the two cases

$$\beta = 45°35' \quad \text{or} \quad \beta' = 180° - \beta = 134°25';$$
$$\alpha = 180° - (45°35' + 30°)$$
$$= 104°25', \text{ or}$$
$$\alpha' = 180° - (30° + 134°25')$$
$$= 15°35'.$$

The corresponding values of the third side are

$$a = \frac{c \sin \alpha}{\sin \gamma} \qquad a' = \frac{c \sin \alpha'}{\sin \gamma}$$

$$= \frac{7 \sin 104°25'}{0.5000} \qquad = \frac{7 \sin 15°35'}{0.5000}$$

$$= \frac{7(0.9685)}{0.5} \qquad = \frac{7(0.2686)}{0.5}$$

$$= 13.5590 \qquad = 3.7604$$

$$= 13.56 \qquad = 3.76.$$

Example 4 Solve the triangle with $a = 3$, $b = 5$, $\beta = 45°$.

Solution. Since $b > a$, there is a unique triangle, and α is an acute angle.

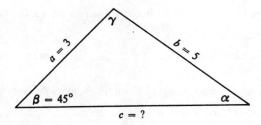

Figure 8-26

$$\sin \alpha = \frac{a \sin \beta}{b}$$

$$= \frac{3(0.707)}{5}$$

$$= 0.4242$$

$$\alpha = 25°6'$$

$$\gamma = 180° - (45° + 25°6')$$

$$= 109°54'$$

$$c = \frac{a \sin \gamma}{\sin \alpha}$$

$$= \frac{3(0.9357)}{0.425}$$

$$= 6.6$$

SECTION 8-4, REVIEW QUESTIONS

1. A triangle is uniquely determined if any two angles and one _____ are given.   **side**

2. If you know two angles of a triangle, you can determine the third because the sum of all three angles is _____ degrees.   **180**

3. A triangle is also determined by two sides and the _____ angle.   **included**

Law of Sines *319*

4. The Law of Sines states that if a, b, and c are the sides opposite angles α, β, and γ, then

$$\frac{a}{\sin \alpha} = \underline{\phantom{\frac{b}{\sin \beta}}} = \underline{\phantom{\frac{c}{\sin \gamma}}}.$$

$\dfrac{b}{\sin \beta}$ $\quad$ $\dfrac{c}{\sin \lambda}$

5. The Law of Sines allows you to solve a triangle if you know two sides and the angle _____ one of them.

opposite

6. If you know a, b, and α, you can determine _____ from the equation $a/\sin \alpha = b/\sin \beta$.

sin β

7. For given data a, α, b there might be no triangle formed, or there might be exactly one triangle possible, or there might be _____ possible triangles.

two

8. There are two possible triangles if $a < b$ but $a >$ _____ .

b sin α

9. If $a = b \sin \alpha$, there is one possible triangle, and it is a _____ triangle.

right

10. If you are given any two angles and one side, then there is exactly _____ possible triangle.

one

SECTION 8-4, EXERCISES

Solve the following triangles.

Example 1

1. $\alpha = 27°$, $\quad\quad\quad \gamma = 42°$, $\quad\quad\quad b = 24$
2. $\gamma = 29°30'$, $\quad\quad \beta = 48°30'$, $\quad\quad c = 8.4$
3. $\alpha = 132°$, $\quad\quad\quad \beta = 24°$, $\quad\quad\quad a = 135$
4. $\alpha = 50°$, $\quad\quad\quad\, \beta = 73°$, $\quad\quad\quad a = 5.8$
5. $\alpha = 102°$, $\quad\quad\quad \beta = 41°$, $\quad\quad\quad c = 52.8$
6. $\alpha = 48°30'$, $\quad\quad \gamma = 67°50'$, $\quad\quad b = 28.7$

In each of the following, determine the number of solutions without solving the triangles.

Example 2, 3, 4

7. $\alpha = 110°$, $\quad\quad\quad a = 5$, $\quad\quad\quad\quad b = 4$
8. $\beta = 60°$, $\quad\quad\quad\, b = 12$, $\quad\quad\quad c = 10$
9. $\gamma = 110°$, $\quad\quad\quad c = 36$, $\quad\quad\quad b = 36$
10. $\alpha = 30°$, $\quad\quad\quad a = 8$, $\quad\quad\quad\quad b = 7$
11. $\alpha = 45°$, $\quad\quad\quad a = 14$, $\quad\quad\quad b = 16$
12. $\alpha = 120°$, $\quad\quad\, a = 12$, $\quad\quad\quad b = 8$

13. If the information given in Exercises 7 through 12 determines one or more triangles, solve the triangles.

14. Use the diagram at the right to help prove that the area of $ABC = \tfrac{1}{2}ab \sin \gamma$.

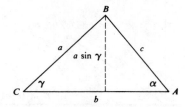

15. Prove that the area of triangle ABC above is also equal to

$$\tfrac{1}{2}bc \sin \alpha \quad \text{or} \quad \tfrac{1}{2}ac \sin \beta$$

Use the formulas in Exercise 15 to find the area of the following triangles:

16. $a = 8.6,$ $\qquad b = 7.9,$ $\qquad \gamma = 67°$
17. $a = 14.1,$ $\qquad c = 27.4,$ $\qquad \beta = 112°$
18. $c = 5.5,$ $\qquad b = 8,$ $\qquad \alpha = 103°30'$
19. One diagonal of a parallelogram is 24.8 and it makes an angle of 42°20' and 27°30' with the sides. Find the sides.
20. Two points A and B on a side of a road are 30 feet apart. A point C across the road is located so that angle CAB is 70° and angle ABC is 80°. How wide is the road?
21. A ship at sea simultaneously observed lights at A and B on either side of the harbor entrance. If the flash at A was 20° to the left of the ship's course, and at B was 15° to the right, how far is the ship from the harbor entrance if A and B are 200 yards apart?

8-5 LAW OF COSINES

The Law of Cosines is a generalized version of the Pythagorean theorem that allows us to solve a triangle when we are given two sides and the included angle—or three sides. Suppose we are given sides b and c, and included angle α. Draw the triangle with angle α in the standard position. The two possibilities, depending on whether α is acute or obtuse, are shown in Fig. 8-27. In either case the coordinates of the three vertices are $(0, 0)$, $(b \cos \alpha, b \sin \alpha)$, and $(c, 0)$.

Now we use the distance formula to compute the length a in terms of the coordinates of the two vertices $(b \cos \alpha, b \sin \alpha)$ and $(c, 0)$.

$$\begin{aligned} a^2 &= (b \cos \alpha - c)^2 + (b \sin \alpha - 0)^2 \\ &= b^2 \cos^2 \alpha - 2bc \cos \alpha + c^2 + b^2 \sin^2 \alpha \\ &= b^2 (\cos^2 \alpha + \sin^2 \alpha) + c^2 - 2bc \cos \alpha \end{aligned}$$

Hence $a^2 = b^2 + c^2 - 2bc \cos \alpha$. By the same argument applied to the other vertices, we obtain the three formulas that constitute the

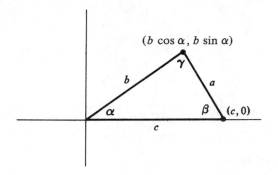

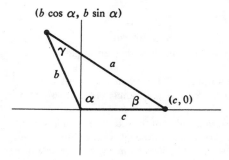

Figure 8-27

Law of Cosines.

$$a^2 = b^2 + c^2 - 2bc \cos \alpha$$
$$b^2 = a^2 + c^2 - 2ac \cos \beta$$
$$c^2 = a^2 + b^2 - 2ab \cos \gamma$$

Notice that if we know two sides and the included angle (e.g., b, c, and α), we can determine the other side:

$$a = \sqrt{b^2 + c^2 - 2bc \cos \alpha}\,.$$

Also, if we know three sides, the Law of Cosines allows us to determine the angles; for example,

$$\cos \alpha = \frac{a^2 - b^2 - c^2}{-2bc} = \frac{b^2 + c^2 - a^2}{2bc}\,.$$

Notice that there is no ambiguity over whether the angle is in the first or second quadrant if you know its cosine. The sine of θ and $180° - \theta$ are the same, but the cosine is positive in the first quadrant and negative in the second quadrant.

As was noted before, the Law of Cosines can be considered a generalization of the Pythagorean theorem. If angle γ is $90°$, the formula reduces to the familiar form $c^2 = a^2 + b^2$.

Example 1 Solve the triangle with $a = 4$, $b = 5$, $\gamma = 50°$.

Solution. First, we use the Law of Cosines to find c (Fig. 8-28).

$$c^2 = a^2 + b^2 - 2ab \cos \gamma$$
$$= 16 + 25 - 2(4)(5)(0.6428)$$
$$= 15.29$$
$$c = \sqrt{15.29}$$
$$= 3.91$$

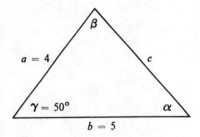

Figure 8-28

Now we can use either the Law of Sines or the Law of Cosines to find α and β. We find α by both methods:

$$a^2 = b^2 + c^2 - 2bc \cos \alpha ,$$
$$\cos \alpha = \frac{b^2 + c^2 - a^2}{2bc}$$
$$= \frac{25 + 15.29 - 16}{2(5)(3.91)}$$
$$= \frac{24.29}{39.1}$$

Law of Cosines 323

$$= 0.6212,$$
$$\alpha = 51°36'.$$

Using the Law of Sines we would have

$$\sin \alpha = \frac{a \sin \gamma}{c}$$
$$= \frac{4 \sin 50°}{3.91}$$
$$= \frac{4(0.7660)}{3.91}$$
$$= 0.7836,$$
$$\alpha = 51°36'.$$

Example 2 Solve the triangle with $a = 4$, $b = 5$, $c = 6$ (Fig. 8-29).

Solution. We find the first angle, say α, with the Law of Cosines:

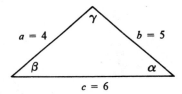

Figure 8-29

$$\cos \alpha = \frac{b^2 + c^2 - a^2}{2bc}$$
$$= \frac{25 + 36 - 16}{2(5)(6)}$$
$$= \frac{45}{60}$$
$$= 0.75,$$
$$\alpha = 41°25'.$$

The other angles can be found in the same way from the Law of Cosines, or we could use the Law of Sines as follows:

$$\sin \beta = \frac{b \sin \alpha}{a}$$

$$= \frac{5(0.6615)}{4}$$

$$= 0.8269,$$

$$\beta = 55°47'.$$

SECTION 8-5, REVIEW QUESTIONS

1. The distance formula is used to derive the Law of Cosines:
   $a^2 = b^2 + c^2 - 2bc \cos(\underline{\hspace{1cm}})$.   α

2. The other forms of the Law of Cosines are $b^2 = a^2 + c^2 - 2ac \cos \beta$ and $c^2 = \underline{\hspace{3cm}}$. $a^2 + b^2 - 2ab \cos\gamma$

3. If we take the angle in the Law of Cosines to be 90°, we get the $\underline{\hspace{3cm}}$ theorem.   Pythagorean

4. The Law of Cosines is used to solve a triangle when two sides and the $\underline{\hspace{3cm}}$ angle are given.   included

5. Triangles with SAS given are solved by the Law of Cosines, as are triangles with $\underline{\hspace{2cm}}$ sides given.   three

6. When one angle has been found by the Law of Cosines, the others can be found with the Law of $\underline{\hspace{3cm}}$.   Sines

SECTION 8-5, EXERCISES

Use the Law of Cosines to find the indicated part of the following triangles.

1. $\alpha = 60°$, $b = 10$, $c = 3$, $a = ?$
2. $a = 2\sqrt{61}$, $b = 8$, $c = 10$, $\gamma = ?$
3. $a = 4$, $b = 20$, $c = 18$,
   $\alpha = ?$ $\beta = ?$, $\gamma = ?$

Completely solve each of the following triangles.

4. $b = 8$, $c = 12$, $\alpha = 25°$
5. $a = 15$, $b = 16$, $c = 17$
6. $a = 2.5$, $b = 13$, $\gamma = 140°$
7. $a = 60$, $c = 30$, $\beta = 40°$

8. $a = 54$, $\quad c = 15$, $\quad \beta = 97°$
9. $a = 4.5$, $\quad b = 11$, $\quad c = 8.5$
10. Two sides and the included angle of a parallelogram are 3 feet, 4 feet, and 100°, respectively. Find the length of the longer diagonal.
11. Find the distance between points A and B on opposite sides of a river if you know the information given in the diagram below.

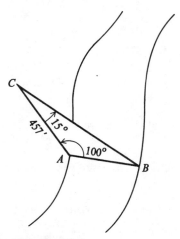

8-6 POLAR COORDINATES

The system of polar coordinates is an alternate way of associating pairs of numbers with the points of a plane. Recall that the cartesian coordinates of a point give the distances (right or left and up or down) from two given axes. Polar coordinates are based on a given ray or half-line. The ray is called the polar axis, and the endpoint is called the *pole*. The polar coordinates of any point in the plane are r (the distance from the point to the pole), and θ (the angle between the polar axis and the line from the pole to the point). See Fig. 8-30.

As usual, positive angles are measured counterclockwise, and negative angles are measured clockwise. Since there are infinitely many designations for a given angle, the polar coordinates of a point are not uniquely determined as are the cartesian coordinates. For example, $(1, 3\pi/4)$ and $(1, -5\pi/4)$ are both polar coordinates for the same point. (See Fig. 8-31.) In general, all the pairs $(r, \theta + 2k\pi)$, for k an integer, represent the same point. The pole is represented by any pair $(0, \theta)$.

It is also convenient to allow both degree and radian measure for angles. Thus $(2, 30°)$ and $(2, \pi/6)$ would represent the same point. However, a polar angle θ always means θ radians unless degrees are explicitly indicated.

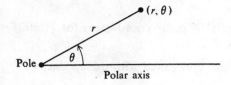

Figure 8-30

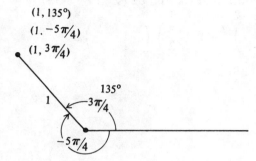

Figure 8-31

For polar coordinates we interpret negative values of r as follows: the point $(-r, \theta)$ is the same as the point $(r, \theta + \pi)$. For example, the point $(-2, \theta)$ is the point 2 units from the pole—but in the direction opposite to that of the terminal side of angle θ. Hence the point (r, θ) has all the possible polar designations (see Fig. 8-32):

$$(r, \theta + 2k\pi) \qquad k = 0, \pm 1, \pm 2, \pm 3, \ldots$$
$$(-r, \theta + (2k+1)\pi) \qquad k = 0, \pm 1, \pm 2, \pm 3, \ldots$$

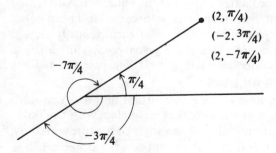

Figure 8-32

Polar Coordinates

Example 1 Find all possible sets of **polar coordinates** for $(1, \pi/2)$ with polar angle θ satisfying $|\theta| < 2\pi$.

Solution. The vertical line can be represented with these values of θ with $|\theta| < 2\pi$: $\quad \theta = \dfrac{\pi}{2}, -\dfrac{\pi}{2}, \dfrac{3\pi}{2}, -\dfrac{3\pi}{2}$.

The corresponding polar coordinates would be (see Fig. 8-33)

$$\left(1, \frac{\pi}{2}\right), \left(1, -\frac{3\pi}{2}\right), \left(-1, -\frac{\pi}{2}\right), \left(-1, \frac{3\pi}{2}\right).$$

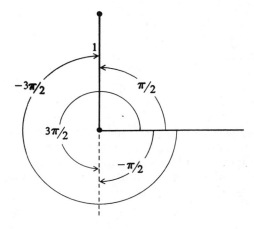

Figure 8-33

The graph of an equation in r and θ consists of all points that have a pair (r, θ) of polar coordinates satisfying the equation. Notice that the situation is different here than it is with cartesian coordinates, where there is only one pair of coordinates for each point. It may be that one pair of polar coordinates for a given point will satisfy a given equation, and another pair will not. If *any* pair satisfies the equation, the point is on the graph.

The simplest polar equations are $r = $ constant and $\theta = $ constant. The graph of $\theta = c$ is the line that makes an angle c with the polar axis. The points on the terminal side of the angle c are represented by coordinates (r, c) with $r > 0$, and the points on the other half-line by coordinates (r, c) with $r < 0$. See Fig. 8-34. for a graph of the line $\theta = \pi/3$.

If we have both polar and cartesian coordinates in a plane, we put

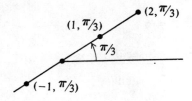

Figure 8-34

the pole at the origin and the polar axis along the positive x-axis. Then each point has cartesian coordinates (x, y) and polar coordinates (r, θ). The relationship between (x, y) and (r, θ) can be seen from Fig. 8-35.

$$x = r \cos \theta \qquad r = \sqrt{x^2 + y^2}$$
$$y = r \sin \theta \qquad \theta = \tan^{-1} \frac{y}{x}$$

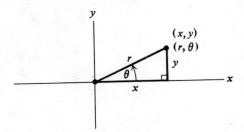

Figure 8-35

Notice that we do not want to restrict arctan to its principal value, which is between $-\pi/2$ and $\pi/2$. The formula $\theta = \tan^{-1} y/x$ should be interpreted to take into account the signs of both x and y. An exact determination of θ could be made with the two formulas:

$$\sin \theta = \frac{y}{r}, \qquad \cos \theta = \frac{x}{r}, \qquad (r > 0).$$

Any cartesian equation can be changed into a polar equation by substituting $r \cos \theta$ for x and $r \sin \theta$ for y. Similarly, any polar equation can be changed into a cartesian equation by expressing r and θ in terms of x and y.

Example 2 Write the polar equation of the line $2x - 3y = 6$ and the circle $x^2 - 2x + y^2 = 0$.

Polar Coordinates 329

Solution. The linear equation becomes

$$2r \cos \theta - 3r \sin \theta = 6$$

or

$$r = \frac{6}{2 \cos \theta - 3 \sin \theta} = \frac{2 \sec \theta}{\frac{2}{3} - \tan \theta}$$

See Fig. 8-36.

The equation of the circle (center $(1, 0)$, radius 1) becomes (see Fig. 8-36)

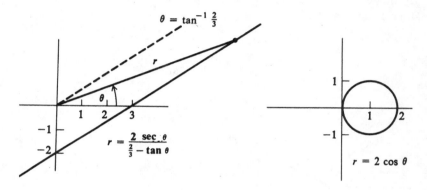

Figure 8-36

$$r^2 \cos^2 \theta - 2r \cos \theta + r^2 \sin^2 \theta = 0,$$
$$r^2 (\cos^2 \theta + \sin^2 \theta) - 2r \cos \theta = 0,$$
$$r^2 - 2r \cos \theta = 0,$$
$$r = 2 \cos \theta.$$

Example 3 Graph the cardioid $r = 1 + \cos \theta$.

Solution. We plot a table of values just as we would for a cartesian equation. Notice that since $\cos(\theta) = \cos(-\theta)$, the graph will be symmetric about the polar axis. Therefore, we need only let θ range from 0 to π. These points are shown in Fig. 8-37.

θ	0	$\frac{\pi}{6}$	$\frac{\pi}{4}$	$\frac{\pi}{3}$	$\frac{\pi}{2}$	$\frac{2\pi}{3}$	$\frac{3\pi}{4}$	$\frac{5\pi}{6}$	π
r	2	1.9	1.7	1.5	1	0.5	0.3	0.1	0

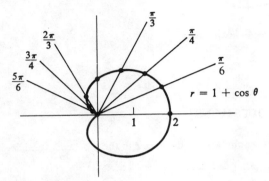

Figure 8-37

Example 4 Write the cartesian equation of the cardioid $r = 1 + \cos \theta$ of Example 3.

Solution. We substitute $\sqrt{x^2 + y^2}$ for r, and $\dfrac{x}{\sqrt{x^2 + y^2}}$ for $\cos \theta$, and simplify.

$$\sqrt{x^2 + y^2} = 1 + \frac{x}{\sqrt{x^2 + y^2}}$$

$$x^2 + y^2 = \sqrt{x^2 + y^2} + x$$

Example 5 Graph $r = \cos 2\theta$.

Solution. Since $\cos(-\theta) = \cos \theta$, we know that the curve will be symmetric about the polar axis. We see that r goes from 1 to 0 as θ goes from 0 to $\pi/4$ ($2\theta = \pi/2$). Therefore, the curve will loop into the origin (the pole) and will be tangent to the line $\theta = \pi/4$. For values of θ between $\pi/4$ and $\pi/2$, 2θ will be between $\pi/2$ and π, so r will be negative. The arc of the curve for θ between $\pi/4$ and $\pi/2$ will have the same shape as that for θ between 0 and $\pi/4$.

This kind of qualitative reasoning, plus a few plotted points, will give the shape of the curve. For example, when $\theta = \pi/6$, $r = \cos 2(\pi/6) = 1/2$. This curve is called a four leaved rose and is shown in Fig. 8-38.

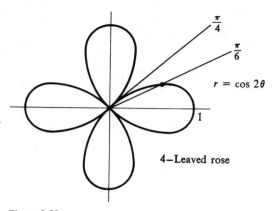

Figure 8-38

4–Leaved rose

SECTION 8-6, REVIEW QUESTIONS

pole
axis

1. The polar coordinates of a point are the distance r to the _____, and the angle θ from the polar _____.

counterclockwise

2. Positive angles are measured from the polar axis in the _____ direction.

3. A positive value of r corresponds to a distance r out the terminal side of the angle θ, and a negative value corresponds to a distance in the

opposite

_____ direction.

4. The points $(1, \theta)$ and $(-1, \theta)$ are on a line through the origin, and are

origin

symmetric about the _____.

infinitely

5. A point has a unique pair of cartesian coordinates but _____ many pairs of polar coordinates.

6. The graph of a polar equation consists of all points that have at least one

satisfy

pair of polar coordinates that _____ the equation.

7. The relation between polar and cartesian coordinates is given by

r cos θ   r sin θ

$x = $ _____, $y = $ _____.

8. Substituting $r \cos \theta$ for x and $r \sin \theta$ for y turns a cartesian equation into the

polar

_____ equation for the same curve.

9. Polar coordinates are found from cartesian coordinates by the formulas

$\sqrt{x^2+y^2}$, $\tan^{-1} \frac{y}{x}$

$r = $ _____, $\theta = $ _____.

332 *Trigonometry*

SECTION 8-6, EXERCISES

For each of the following points (expressed in polar coordinates) find all other possible polar coordinates (r, θ) with $|\theta| < 2\pi$.

1. $(2, 0)$
2. $(3, \pi)$
3. $(4, \pi/4)$ Example 1,
4. $(1, 2\pi/3)$
5. $(-3, -\pi/6)$
6. $(-10, -\pi/3)$

Give the polar coordinates (with $-\pi < \theta \leq \pi, r > 0$) of the points whose cartesian coordinates are given.

7. $(2, 2)$
8. $(0, 3)$
9. $(\sqrt{3}, 1)$
10. $(1, -\sqrt{3})$
11. $(-1, 0)$
12. $(0, -2)$

Give the cartesian coordinates of the points whose polar coordinates are given.

13. $(3, 2\pi/3)$
14. $(5, -\pi/6)$
15. $(7, 3\pi/4)$
16. $(4, -3\pi/4)$
17. $(8, \pi/2)$
18. $(3, \pi)$

Write the polar equation of the following cartesian equations.

19. $y = 3x + 5$ (line)
20. $x^2 + y^2 - 4y = 0$ (circle)   Example 2
21. $y = x^2$ (parabola)
22. $x^2/9 + y^2/4 = 1$ (ellipse)

Write the cartesian equation of the following polar equations.

23. $r = 3 \sin \theta$
24. $r = 5$ Example 4
15. $r = 2/(2 - \cos \theta)$
26. $r = -3 \cos \theta$

Sketch the graph of the following polar equations.

27. $r = \sin 5\theta$
28. $r = \cos 2\theta$ Example 3, 5
29. $r = 2 \cos \theta$
30. $r = \sin 3\theta$
31. $r = 3 \sin \theta$
32. $r = 2 \sin 3\theta$
33. $r = 3 \cos 2\theta$
34. $r \cos \theta = -3$
35. $r = 2 + 2 \cos \theta$
36. $r = (2 \cos \theta) - 1$
37. $r = 2 - 3 \cos \theta$
38. $r = 2 + \sin \theta$
39. $r^2 = \cos 2\theta$
40. $r^2 = 4 \sin 2\theta$

8-7 ROTATION OF AXES

In this section we will study the graphs of conic sections which are not in the standard orientation to the coordinate axes. Recall that to graph the parabola

$$y = x^2 - 4x + 3 \quad (1)$$

we complete the square, and write the equation in the form $y = (x - 2)^2 - 1$, or

$$y + 1 = (x - 2)^2 \quad (2)$$

However, if we introduce new variables $\bar{x}, \bar{y}$ by the equations

$$\bar{x} = x - 2, \quad \bar{y} = y + 1$$

then Equation (2) takes the particularly simple form

$$\bar{y} = \bar{x}^2 \quad (3)$$

We can think of $\bar{x}, \bar{y}$ as coordinates in a new coordinate system, with axes parallel to the given axes, and the new origin ($\bar{x} = \bar{y} = 0$) at $(2, -1)$. See Fig. 8-39.

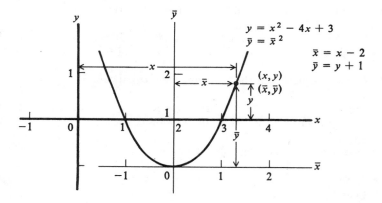

Figure 8-39

The system of $\bar{x}, \bar{y}$ coordinates is called a *translation* of the original system. Equation (1) takes the simplified form (3) in the translated system defined by (2). Equations of the form

$$Ax^2 + Cy^2 + Dx + Ey + F = 0 \quad (4)$$

can generally be put in one of the standard forms for a conic by translating the coordinates. In other words, by completing the square where necessary and introducing translated coordinates $\bar{x}, \bar{y}$, Equation

(4) will usually (see Problem 15 for exceptions) take one of these forms:

$$\bar{y} = a\bar{x}^2 \quad \text{or} \quad \bar{x} = a\bar{y}^2 \quad \text{(parabola)}$$

$$\frac{\bar{x}^2}{a^2} + \frac{\bar{y}^2}{b^2} = 1 \quad \text{(ellipse)} \tag{5}$$

$$\frac{\bar{x}^2}{a^2} - \frac{\bar{y}^2}{b^2} = \pm 1 \quad \text{(hyperbola)}$$

The most general second-degree equation in x and y is

$$Ax^2 + Bxy + Cy^2 + Dx + Ey + F = 0 \tag{6}$$

We now show that it is always possible to put Equation (6) in form (4) (i.e., remove the xy term) by introducing a *rotated* coordinate system. Therefore the graph of Equation (6) is always a conic section, but possibly not aligned with the axes in the usual way.

Consider a new pair of coordinate axes obtained by rotating the original axes through an angle α. Each point now has coordinates (x, y) in the original system and coordinates $(\bar{x}, \bar{y})$ in the rotated system (Fig. 8-40). The relationship between (x, y) and $(\bar{x}, \bar{y})$ is

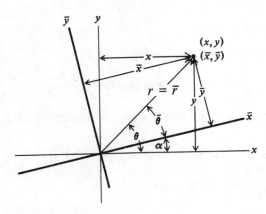

Figure 8-40

obtained by considering the corresponding polar coordinates, (r, θ) and $(\bar{r}, \bar{\theta})$. We know that $r = \bar{r}$, because the origins are the same, and that $\theta = \bar{\theta} + \alpha$ because of the rotation. Recall that

$$x = r \cos \theta \quad \text{and} \quad \bar{x} = \bar{r} \cos \bar{\theta} = r \cos \bar{\theta}$$
$$y = r \sin \theta \qquad\qquad \bar{y} = \bar{r} \sin \bar{\theta} = r \sin \bar{\theta}.$$

Therefore,
$$x = r \cos \theta$$
$$= r \cos (\bar{\theta} + \alpha)$$
$$= (r \cos \bar{\theta}) \cos \alpha - (r \sin \bar{\theta}) \sin \alpha$$
$$= \bar{x} \cos \alpha - \bar{y} \sin \alpha.$$

Similarly, we find that $y = \bar{x} \sin \alpha + \bar{y} \cos \alpha$, so the rotation of coordinates is effected by the equations

$$x = \bar{x} \cos \alpha - \bar{y} \sin \alpha,$$
$$y = \bar{x} \sin \alpha + \bar{y} \cos \alpha. \qquad (7)$$

If we make the substitution (7) in the general quadratic equation (6) we get another equation of the form

$$\bar{A}\bar{x}^2 + \bar{B}\bar{x}\bar{y} + \bar{C}\bar{y}^2 + \bar{D}\bar{x} + \bar{E}\bar{y} + \bar{F} = 0, \qquad (8)$$

where the coefficients $\bar{A}, \bar{B}, \bar{C}, \bar{D}, \bar{E}, \bar{F}$ depend on α. In particular

$$\bar{B} = -2A \sin \alpha \cos \alpha + B(\cos^2 \alpha - \sin^2 \alpha) + 2C \sin \alpha \cos \alpha$$
$$= (C - A) \sin 2\alpha + B \cos 2\alpha$$

The coefficient $\bar{B}$ will be zero if

$$A = C \text{ and } \alpha = \frac{\pi}{4}, \quad \text{or} \quad A \neq C \text{ and } \tan 2\alpha = \frac{B}{A - C}$$

Example 1 Eliminate the xy-term by rotation of coordinates, and graph the curve $xy = 1$.

Solution. Here $A = C = 0$, $B = 1$, $D = E = 0$, and $F = -1$. Since $A = C$, we know that a rotation through $45°$ will eliminate the xy-term. With $\alpha = 45°$, equations (7) become

$$x = \bar{x}\frac{\sqrt{2}}{2} - \bar{y}\frac{\sqrt{2}}{2}$$

$$y = \bar{x}\frac{\sqrt{2}}{2} + \bar{y}\frac{\sqrt{2}}{2}$$

Hence, the new equation is

$$\left(\frac{\bar{x} - \bar{y}}{\sqrt{2}}\right)\left(\frac{\bar{x} + \bar{y}}{\sqrt{2}}\right) = 1,$$

$$\frac{\bar{x}^2}{2} - \frac{\bar{y}^2}{2} = 1$$

We recognize this as a hyperbola within x-intercepts at $\pm\sqrt{2}$. See Fig. 8-41.

Example 2 Rotate the coordinates to eliminate the xy-term, and graph the curve $4x^2 - 24xy + 11y^2 = 500$ showing both sets of coordinates.

Solution. We find α from the formula

$$\tan 2\alpha = \frac{-24}{4 - 11} = \frac{24}{7}.$$

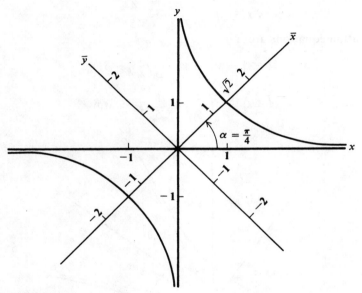

Figure 8-41

We first find $\cos 2\alpha$, and then find $\sin \alpha$ and $\cos \alpha$ from the identities

$$\sin \alpha = \sqrt{\frac{1 - \cos 2\alpha}{2}}, \quad \cos \alpha = \sqrt{\frac{1 + \cos 2\alpha}{2}}.$$

From the right triangle in Fig. 8-42. we see that $\sin 2\alpha = 24/25$, $\cos 2\alpha = 7/25$. Hence

$$\sin \alpha = \sqrt{\frac{1 - 7/25}{2}} = \frac{3}{5},$$

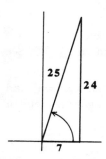

Figure 8-42

$$\cos \alpha = \sqrt{\frac{1 + 7/25}{2}} = \frac{4}{5}.$$

The rotation equations are

$$x = \bar{x}\left(\frac{4}{5}\right) - \bar{y}\left(\frac{3}{5}\right) = \frac{1}{5}(4\bar{x} - 3\bar{y}),$$

$$y = \bar{x}\left(\frac{3}{5}\right) + \bar{y}\left(\frac{4}{5}\right) = \frac{1}{5}(3\bar{x} + 4\bar{y}).$$

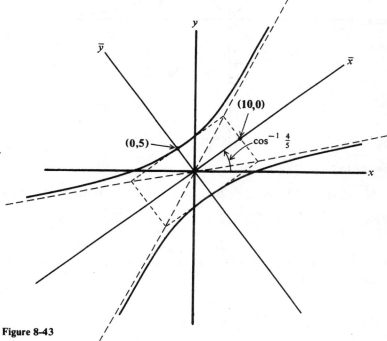

Figure 8-43

Substituting these expressions for x, y we get the new equation

$$4\tfrac{1}{25}(4\bar{x} - 3\bar{y})^2 - 24\tfrac{1}{25}(4\bar{x} - 3\bar{y})(3\bar{x} + 4\bar{y}) + 11\tfrac{1}{25}(3\bar{x} + 4\bar{y})^2 = 500,$$

$$4(16\bar{x}^2 - 24\overline{xy} + 9\bar{y}^2) - 24(12\bar{x}^2 + 7\overline{xy} - 12\bar{y}^2)$$
$$+ 11(9\bar{x}^2 + 24\overline{xy} + 16\bar{y}^2) = (500) \cdot (25)$$

$$-125\bar{x}^2 + 500\bar{y}^2 = (500) \cdot (25),$$

$$-\frac{\bar{x}^2}{100} + \frac{\bar{y}^2}{25} = 1.$$

The curve is a hyperbola with $\bar{y}$-intercepts ± 5, and asymptotes that are the diagonals of the box through $(0, \pm 5)$, $(\pm 10, 0)$ in the $\bar{x}, \bar{y}$-system. See Fig. 8-43.

SECTION 8-7, REVIEW QUESTIONS

1. Equations can sometimes be put in simpler form by introducing translated or _____ coordinates.
   rotated

2. If $\bar{x}, \bar{y}$ are the coordinates of the point (x, y) in the system rotated through an angle α, then $x = \bar{x}$ _____ $- \bar{y}$ _____
   cos α   sin α
   and $y = \bar{x}$ _____ $+ \bar{y}$ _____.
   sin α   cos α

3. Any second-degree equation in x and y represents some _____ section.
   conic

4. If a second degree equation has no xy-term, you ____ ____ ____ in the x and the y terms to tell what it is.
   complete the square

5. To eliminate the xy-term from a second-degree equation, you rotate through an angle α determined by $\tan 2\alpha =$ _____ if $A \neq C$.
   B/(A–C)

6. If $A \neq C$ you let $\alpha = \tfrac{1}{2} \tan^{-1} \dfrac{B}{A - C}$, and if $A = C$ you let $\alpha =$ _____.
   π/4

7. To find $\sin \alpha$ and $\cos \alpha$ from $\tan 2\alpha$, you first find the _____ of 2α.
   cosine

8. To find $\sin \alpha$ and $\cos \alpha$ from $\cos 2\alpha$ you use the identity $\cos \alpha = \sqrt{\dfrac{1 + \cos 2\alpha}{2}}$, $\sqrt{\dfrac{1 - \cos 2\alpha}{2}}$

   $\sin \alpha =$ _____.

SECTION 8-7, EXERCISES

Introduce a translated coordinate system to put the following in the standard form given in Equation (5). Graph the coordinate systems and the curves.

1. $x^2 + 4y^2 - 6x - 16y + 21 = 0$
2. $3y^2 - x + 12y + 13 = 0$
3. $4x^2 - 9y^2 + 24x - 18y - 9 = 0$
4. $-x^2 + y^2 - 2x - 2 = 0$

Rotate the coordinates to eliminate the xy-term, and graph the following curves.

Example 1, 2

5. $8x^2 - 4xy + 5y^2 = 36$
6. $x^2 - 2xy = 0$
7. $7x^2 - 6xy - y^2 = 0$
8. $xy = 4$
9. $x^2 + 4xy + 4y^2 = 9$
10. $x^2 + 2xy + 13y^2 = 884$

Rotate the coordinates to obtain an equation in $\bar{x}$ and $\bar{y}$ with no $\bar{x}\bar{y}$ term. Then introduce a translated system, $\bar{\bar{x}}, \bar{\bar{y}}$, to put the equation in the standard form of Equation (5). Graph the curve.

11. $5x^2 + 6xy + 5y^2 - 4x + 4y - 4 = 0$
12. $xy - 2y - 4x = 0$
13. $4x^2 - 4xy + y^2 - 8\sqrt{5}x - 16\sqrt{5}y = 0$
14. $9x^2 + 24xy + 16y^2 + 90x - 130y = 0$

15. Identify the graphs of these degenerate conics.
    (a) $x^2 + y^2 + 1 = 0$
    (b) $(x - 2)^2 + (y + 1)^2 = 0$
    (c) $(x - 2y)^2 = 0$
    (d) $x^2 - y^2 = 0$

16. Derive the second expression in Equation (7).
17. Find $\bar{A}, \bar{B}, \bar{C}, \bar{D}, \bar{E}$, and $\bar{F}$ of Equation (8) in terms of A, B, C, D, E, F and $\sin \alpha$ and $\cos \alpha$.

8-8 REVIEW FOR CHAPTER

Convert the following degree measures to radians.

1. $90°$
2. $78°$
3. $96°$
4. $\dfrac{12°}{\pi}$
5. $\dfrac{90°}{\pi}$
6. $\dfrac{\pi°}{180}$

Convert the following radian measures to degrees.

7. π
8. $\dfrac{\pi}{2}$
9. 10π
10. 10
11. 90
12. $\dfrac{180}{\pi}$

13. Write the sin, cos, and tan of the angle that the line passing through the origin and $(3, 4)$ makes with the positive x-axis.

Find the values of the other five trigonometric functions given the following information.

14. $\sin \theta = 4/5$ and $0 < \theta < \pi/2$
15. $\tan \theta = -1/2$ and $\pi/2 < \theta < \pi$
16. $\cos \theta = -3/7$ and $\pi < \theta < 3\pi/2$

Solve the following triangles.

17. $a = 3$, $b = 2$, $\gamma = 60°$
18. $a = 5$, $b = 6$, $c = 7$
19. $c = 160$, $\beta = 84°$, $\gamma = 54°$
20. $c = 11.7$, $\alpha = 46°30'$, $a = 6.35$
21. $a = 200$, $b = 210$, $\gamma = 43°30'$
22. $c = 16$, $\beta = 84°10'$, $\gamma = 53°50'$
23. $b = 512$, $\alpha = 74°40'$, $\beta = 30°20'$
24. $\alpha = 62°30'$, $a = 13.7$, $c = 15.2$
25. $a = 12.3$, $b = 34.9$, $c = 36.8$
26. $a = 2520$, $c = 1390$, $\beta = 54°24'$
27. $b = 472$, $c = 607$, $\alpha = 125°10'$
28. $a = 643$, $b = 778$, $c = 912$
29. $a = 189$, $b = 224$, $c = 355$

Write the polar equation of the following cartesian equations.

30. $y = -2x - 7$
31. $(x - 1)^2 + y^2 = 36$
32. $y = -4x^2 + 3$
33. $x^2/25 + y^2/9 = 1$

34–37. Graph the polar equations you found in Exercises 30 through 33.

Sketch the graph of the following polar equations:

38. $r = 3 \sin \theta$
39. $r = \sin 2\theta$
40. $r = 3 + \cos \theta$
41. $r = 5 + 2 \cos \theta$
42. $r^2 = 2 \cos 3\theta$
43. $r^2 = 4 \sin \theta$

Rotate the coordinates to eliminate the xy-term and, graph the curve for the following.

44. $9x^2 + 24xy + 16y^2 + 80x - 60y = 0$
45. $2xy - 3y^2 = 5$
46. $xy = 3$
47. $x^2 + 4xy + y^2 + 32 = 0$
48. $9x^2 + 4xy + 6y^2 + 12x + 36y + 44 = 0$
49. $x^2 - 10xy + y^2 + x + y + 1 = 0$
*50. $2x^2 + 3xy + 4y^2 + 2x - 3y + 5 = 0$

9 SYSTEMS OF EQUATIONS

9-1 TWO EQUATIONS IN TWO VARIABLES

In one common application of algebra, we have a single equation expressing some condition on a single physical quantity. The equation is then solved to find the value of the physical quantity. Frequently, however, there are several unknown quantities involved in an experiment, and several conditions which relate these quantities. In a mathematical formulation, this leads to several simultaneous equations (expressing the physical facts), with several unknowns in each equation. In this chapter we will develop methods for handling this kind of system of equations.

We know that a single equation in one variable typically has some finite number of solutions. On the other hand, a single equation in two variables, x and y, generally has an infinite number of ordered pairs (x, y) as solutions. The graph of a single equation in x and y is ordinarily a curve in the plane.

We frequently want to find the pairs (x, y) which simultaneously satisfy *two* given equations. Geometrically, solving two equations in x and y simultaneously amounts to finding the intersection of the graphs of the two separate equations.

Example 1 The area of a small rectangular rug is 12 square feet, and its perimeter is 16 feet. Find the dimensions of the rug.

Solution. Let x be one of the dimensions (length or width) and y be the other. Then the two conditions imposed on x and y are

$$xy = 12$$

and $$2x + 2y = 16.$$

The second equation is equivalent to $y = 8 - x$. If we replace y in the first equation by $8 - x$ we get

$$x(8 - x) = 12,$$
$$8x - x^2 = 12,$$
$$x^2 - 8x + 12 = 0,$$
$$(x - 6)(x - 2) = 0,$$
$$x = 6 \quad \text{or} \quad x = 2.$$

Since $y = 8 - x$, if $x = 6$, $y = 2$ and if $x = 2$, $y = 6$. Hence the rug must measure 6 by 2 feet. The two solutions of the system are (6, 2) and (2, 6). The graphs of the two equations are shown in Fig. 9-1.

Now we will formalize the ideas of the example above. We can always put all terms of an equation on one side and zero on the other ($xy = 12$ is equivalent to $xy - 12 = 0$). Therefore a *system of two equations in two variables* is a pair of equations of the form

$$P(x, y) = 0$$
$$Q(x, y) = 0 \tag{1}$$

A *solution* of the system (1) is a pair (x_0, y_0) which satisfies both equations. An *equivalent* system is any system which has the same set of solutions. To solve a system of equations we try to find an equivalent system whose solutions are obvious.

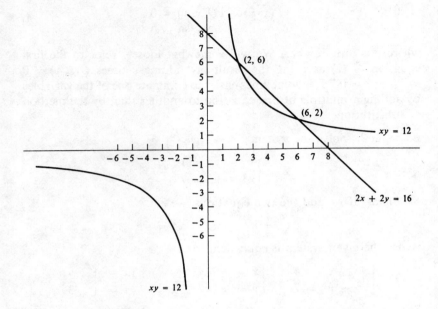

Figure 9-1

In any system we can replace any equation by one which is equivalent to it. Hence if we can solve $Q(x, y) = 0$ for y, and get an equivalent equation $y = R(x)$, then

$$\begin{cases} P(x, y) = 0 \\ Q(x, y) = 0 \end{cases} \text{ is equivalent to } \begin{cases} P(x, y) = 0 \\ y = R(x) \end{cases}$$

If we substitute $R(x)$ for y in the top equation, we get a single equation in x:

$$P(x, R(x)) = 0.$$

For each solution x_0 of this equation, the pair (x_0, y_0), where $y_0 = R(x_0)$ is a solution of the system. This technique, called the *method of substitution*, was the method used in Example 1.

Another system equivalent to (1) is

$$\begin{cases} P(x, y) + cQ(x, y) = 0 \\ Q(x, y) = 0 \end{cases} \quad (2)$$

where c is any number. We will somewhat loosely refer to the first equation of (2) as being the result of "adding c times $Q(x, y) = 0$ to $P(x, y) = 0$." It is sometimes easier to eliminate one of the variables by adding a multiple of one equation to another than by the method of substitution.

Example 2 Solve the system

$$\begin{cases} x^2 - y^2 = 1 \\ x^2 + y^2 = 7 \end{cases}$$

Solution. If we add the two equations, we get

$$2x^2 = 8 \quad \text{or} \quad x^2 = 4.$$

Hence the given system is equivalent to

$$\begin{cases} x^2 = 4 \\ x^2 + y^2 = 7 \end{cases} \quad \begin{cases} x = \pm 2 \\ y = \pm\sqrt{7 - x^2} \end{cases} \quad \begin{cases} x = \pm 2 \\ y = \pm\sqrt{3}. \end{cases}$$

The solutions are $(2, \sqrt{3})$, $(2, -\sqrt{3})$, $(-2, \sqrt{3})$, and $(-2, -\sqrt{3})$. The graphs of the equations of the original system are shown in Fig. 9-2.

The following examples illustrate other techniques for handling systems. It is easier to deal with a single equation, once we have eliminated one variable, than to always write both equations in the equivalent new system. We keep in mind that we always have all earlier equations at our disposal when solving a system.

Example 3 Solve

$$\begin{cases} x^2 y = 3 \\ \dfrac{y}{x^2 + 2} = 1 \end{cases}$$

Solution. We solve the second equation for y, substitute in the first, and solve for x.

$$y = x^2 + 2$$

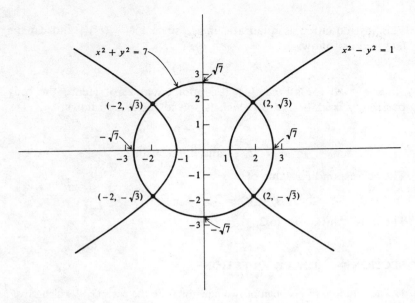

Figure 9-2

Now substitute for y in the first equation and solve.

$$x^2(x^2 + 2) = 3$$
$$x^4 + 2x^2 - 3 = 0$$
$$(x^2 + 3)(x^2 - 1) = 0$$
$$x^2 = 1$$
$$x = \pm 1$$

The only possible values for x are 1 and -1, and we find the corresponding y values from the second equation $y = x^2 + 2$:

$$x = 1, y = 3; x = -1, y = 3.$$

The solutions are $(1,3)$ and $(-1,3)$.

Example 4 Solve

$$\begin{cases} y = e^{2x} \\ y = e^x + 2 \end{cases}$$

Solution. Since the left sides are equal, we can equate the right sides. This gives

$$e^{2x} = e^x + 2,$$
$$e^{2x} - e^x - 2 = 0.$$

Two Equations in Two Variables 347

This last equation is quadratic in e^x, since $e^{2x} = (e^x)^2$, and can be factored as follows:

$$(e^x - 2)(e^x + 1) = 0.$$

Since $e^x > 0$ for all x, $e^x + 1$ cannot equal zero. Hence the only possibility is $e^x - 2 = 0$, which is solved for x as follows:

$$e^x - 2 = 0,$$
$$e^x = 2 \quad \text{and} \quad x = \ln 2.$$

The corresponding value of y is

$$y = e^x + 2 = 2 + 2 = 4.$$

The only solution is $(\ln 2, 4)$.

SECTION 9-1, REVIEW QUESTIONS

both

1. The solutions of a system of two equations are the pairs (x,y) which satisfy _____ equations.

intersect

2. The pairs (x,y) which satisfy both of the equations in a system are the coordinates of the points where the graphs of the two equations _____.

finite

3. Two curves generally intersect in some _____ number of points.

solutions

4. Two systems are equivalent if they have the same sets of _____.

substitution

5. If we can solve one equation for y in terms of x, we can solve the system by the method of _____.

equation

6. We can sometimes eliminate one variable by adding some multiple of one _____ to the other.

equivalent

7. The system of equations $P(x,y) + cQ(x,y) = 0$, $Q(x,y) = 0$ is _____ to the system $P(x,y) = 0$, $Q(x,y) = 0$.

SECTION 9-1, EXERCISES

Solve the following systems of equations, and sketch the graphs of both equations on the same coordinate axes.

Example 1, 4

1. $\begin{cases} y = x \\ y = 2x^2 \end{cases}$ 2. $\begin{cases} y = 2x \\ y = 3 - x^2 \end{cases}$

348 Systems of Equations

3. $\begin{cases} xy = 3 \\ y = x + 2 \end{cases}$
4. $\begin{cases} xy = 1 \\ 2x - y = 1 \end{cases}$

5. $\begin{cases} y = x^2 \\ y = 4 - x^2 \end{cases}$
6. $\begin{cases} y^2 = x \\ y^2 = 6 - x \end{cases}$

7. $\begin{cases} x^2 + y^2 = 34 \\ x^2 - y^2 = 16 \end{cases}$
8. $\begin{cases} x^2 + 4y^2 = 61 \\ 9x^2 - 25y^2 = 0 \end{cases}$ Example 2

9. $\begin{cases} x^2 + y^2 = 4 \\ 2x^2 + y^2 = 4 \end{cases}$
10. $\begin{cases} 5x^2 + 2y^2 = 18 \\ x^2 + 2y^2 = 18 \end{cases}$

11. $\begin{cases} y = x^2 - 4x \\ y = x - 4 \end{cases}$
12. $\begin{cases} y = 1 - x^2 - 2x \\ y = -x - 1 \end{cases}$ Example 3,4

13. $\begin{cases} x^2 + y^2 = 52 \\ 3x - 2y = 0 \end{cases}$
14. $\begin{cases} x^2 + y^2 = 20 \\ 2x + y = 0 \end{cases}$

Solve the following systems.

15. $\begin{cases} y = e^{2x} \\ y = 3e^x - 2 \end{cases}$
16. $\begin{cases} y = e^{2x} + 1 \\ y = 3 - e^x \end{cases}$ Example 4

17. $\begin{cases} y - 9 = e^{2x} \\ y = 6e^x \end{cases}$
18. $\begin{cases} y = e^{-2x} \\ y = 5e^{-x} - 6 \end{cases}$

19. $\begin{cases} y = e^{2x} + 1 \\ y = 1 - e^x \end{cases}$
20. $\begin{cases} y = e^{-2x} \\ y = -2e^{-x} - 1 \end{cases}$

21. What are the dimensions of a room with 108 square feet of floor space if the perimeter of the room is 42 feet? Example 1
22. Find the dimensions of a volleyball court whose area is 1800 square feet and whose perimeter is 180 feet.
23. Find two positive numbers whose difference is 4 such that the difference of the squares is 88.
24. Find two positive numbers whose sum is 10 and whose squares add up to 58.

9-2 TWO LINEAR EQUATIONS

Systems of linear equations arise in many different kinds of application, and are of very great importance. The simple form of such systems allows us to give an elegant and comprehensive description of their solutions. In this section we will treat the simplest linear system, which consists of two equations in two unknowns.

Recall that a *linear* equation in x and y is one of the form

$$ax + by = c. \tag{1}$$

We know that the graph of any such equation is a straight line. Hence we know what kind of solutions to expect from the *linear system*

$$\begin{cases} a_1 x + b_1 y = c_1 \\ a_2 x + b_2 y = c_2. \end{cases} \qquad (2)$$

The solutions of the system (2) will consist of the coordinates of the points where two lines intersect. Therefore *there will be exactly one solution if the lines are distinct but not parallel, no solutions if the lines are parallel, and infinitely many solutions if the two lines coincide.*

The two methods of Section 9-1 for obtaining an equivalent system certainly apply to linear systems in particular. Therefore we can use either the method of substitution, or the method of adding a multiple of one equation to the other. We illustrate both these methods in examples below.

Example 1 Solve
$$\begin{cases} 2x + 3y = 4 \\ x - 2y = -5 \end{cases}$$

Solution. We will use the method of substitution, and solve the second equation for x. The systems below are all equivalent to the given system.

$$\begin{cases} 2x + 3y = 4 \\ \quad x = 2y - 5 \end{cases} \quad \begin{cases} 2(2y - 5) + 3y = 4 \\ \quad x = 2y - 5 \end{cases}$$

$$\begin{cases} 7y - 10 = 4 \\ \quad x = 2y - 5 \end{cases} \quad \begin{cases} y = 2 \\ x = 2y - 5 \end{cases}$$

$$\begin{cases} y = 2 \\ x = (2)(2) - 5 = -1 \end{cases}$$

Hence the single solution is $(-1, 2)$. The graphs of the two lines are shown in Fig. 9-3.

In a system like that of Example 1—where there is just one solution—we say the equations are *independent*.

We need not always write equivalent new systems as we did above. It is simpler to deal with one variable at a time once we have eliminated one of the variables. The usual way of writing the work is illustrated in the next example.

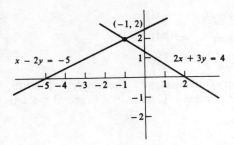

Figure 9-3

Example 2 Solve the system

$$\begin{cases} 3x - y = 5 \\ 2y - 5x = -8 \end{cases}$$

Solution. We solve the first equation for y, substitute in the second, and then solve for x.

$$y = 3x - 5$$
$$2(3x - 5) - 5x = -8$$
$$6x - 10 - 5x = -8$$
$$x = 2$$

Having found x, we can find y from the first equation above.

$$y = 3(2) - 5$$
$$y = 1$$

Hence the solution is $x = 2, y = 1$.

The next example illustrates what happens when the second equation does not really put any different conditions on x and y than the first equation.

Example 3 Solve the system

$$\begin{cases} 3x - y = 5 \\ 2y - 6x = -10 \end{cases}$$

Solution. We solve the first equation for y, substitute in the second, and then attempt to solve for x.

$$y = 3x - 5$$

Two Linear Equations

$$2(3x - 5) - 6x = -10$$
$$6x - 10 - 6x = -10$$
$$-10 = -10$$

This identity shows that every solution of the single equation $y = 3x - 5$ is a solution of the system. That is, the original system is equivalent to the system

$$\begin{cases} 3x - y = 5 \\ -10 = -10. \end{cases}$$

In this case we say the equations are *dependent*. The graphs of the two equations coincide (Fig. 9-4). The solutions are all pairs $(x, 3x - 5)$.

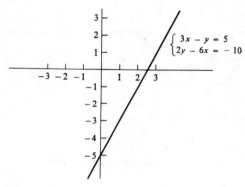

Figure 9-4

Example 4 Solve

$$\begin{cases} 3x - 2y = 6 \\ 5x + 3y = 10 \end{cases}$$

Solution. We could eliminate y by adding 2/3 of the second equation to the first. It is easier to keep integer coefficients so instead, we will multiply the first equation by 3, the second by 2, and then add. The first two steps give us

$$\begin{cases} 9x - 6y = 18 \\ 10x + 6y = 20. \end{cases}$$

Adding gives us
$$19x = 38,$$
$$x = 2.$$

Now substitute this value of x in the first equation to find y.

$$3(2) - 2y = 6$$
$$y = 0$$

The solution is thus $(2,0)$.

Example 5 Solve
$$\begin{cases} 7x + 3y = 5 \\ 21x + 9y = -18 \end{cases}$$

Solution. If we multiply the first equation by 3, we get the equivalent system
$$\begin{cases} 21x + 9y = 15 \\ 21x + 9y = -18. \end{cases}$$

This system is clearly *inconsistent*, that is, there are no solutions. The inconsistency would also be evident if we tried the method of substitution. We would get the following contradiction.

$$y = \frac{(5 - 7x)}{3}$$

$$21x + \frac{9(5 - 7x)}{3} = -18$$

$$21x + 15 - 21x = -18$$

$$15 = -18$$

The graphs of the original equations are parallel lines and have no point of intersection (Fig. 9-5).

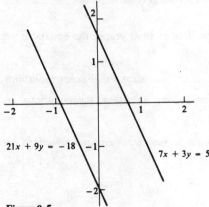

Figure 9-5

The three types of linear system of two equations in two unknowns are listed in the table below.

TYPE OF SYSTEM	NUMBER OF SOLUTIONS	GEOMETRIC INTERPRETATION
Independent equations	One	Lines intersect
Inconsistent equations	None	Lines are parallel
Dependent equations	Infinite number	Lines are coincident

SECTION 9-2, REVIEW QUESTIONS

linear   1. An equation of the form $ax + by = c$ is called a _____ equation in x and y.

line   2. The graph of a linear equation is a _____.

intersect   3. The solutions of a system of two linear equations will be the coordinates of the points where the two lines _____.

one   4. If the equations are independent (or consistent), the lines will intersect at _____ point.

parallel   5. If the equations are inconsistent, the two lines are _____.

no   6. Parallel lines do not intersect, and inconsistent equations have _____ simultaneous solutions.

dependent   7. If the two equations have the same line as their graph, the equations are called _____.

infinitely   8. Dependent equations have _____ many simultaneous solutions.

dependent   9. If you arrive at an identity while solving a system, the equations are _____.

inconsistent   10. If you arrive at a contradiction while solving a system, the equations are _____.

SECTION 9-2, EXERCISES

Solve the following systems of equations and tell if the system is independent, inconsistent, or dependent.

1. $2x + 4y = 10$
   $x - 3y = -5$

2. $3x - 2y = 7$
   $2x + y = 7$

3. $5x - y = -2$
   $3x + 5y = 10$

4. $x + 4y = 3$
   $-2x + 3y = 5$

5. $2x - y = 3$
   $-8x + 4y = -12$

6. $-x + 2y = 1$
   $3x - 6y = -3$

7. $-2x + y = 4$
   $6x - 3y = 10$

8. $4x - 2y = -1$
   $-6x + 3y = 0$

9. $2x - 3y = 0$
   $4x + 9y = 5$

10. $6x - 2y = 3$
    $3x + 4y = -1$

11. $3x - 9y = 15$
    $-4x + 12y = 8$

12. $4x - 6y = 10$
    $-6x + 9y = 15$

Solve the following by first finding $\dfrac{1}{x}$ and $\dfrac{1}{y}$.

13. $\dfrac{2}{x} + \dfrac{1}{y} = 2$
    $\dfrac{6}{x} - \dfrac{4}{y} = -1$

14. $\dfrac{2}{x} - \dfrac{3}{y} = 1$
    $\dfrac{4}{x} + \dfrac{9}{y} = 5$

15. $-\dfrac{3}{x} - \dfrac{9}{y} = 0$
    $\dfrac{6}{x} + \dfrac{6}{y} = -4$

16. $\dfrac{1}{x} + \dfrac{3}{y} = 5$
    $-\dfrac{2}{x} + \dfrac{1}{y} = -3$

17. A man can row 6 miles downstream in 1 hour, and return the 6 miles against the current in 2 hours. Find how fast he can row in still water and how fast the river is flowing.
18. The area of a field is increased by 1000 square yards if the length is increased by 10 yards and the width by 5. The same field decreases in area by 1100 square yards if the length is decreased by 5 yards and the width by 10 yards. Find the dimension of the field.
19. Jeff and Jim are 30 miles apart. If they start walking at the same time and travel in the same direction, Jeff will overtake Jim in 8 hours. If they walk toward each other, they will meet in 3 hours. How fast does each of them walk?
20. The sum of the acute angles of an obtuse triangle is 85°. If the difference of the acute angles is 19°, how large are the angles?

21. Find two numbers such that their sum is 12 and their difference is 3.
22. Find two numbers such that the sum of their reciprocals is 24 and the difference of their reciprocals is 4.

9-3 LINEAR SYSTEMS IN SEVERAL VARIABLES

In this section we will treat systems of linear equations in three or more variables. It is not necessary that the number of equations be the same as the number of unknowns. First, we will consider the following system of three linear equations in the three variables x, y, z.

Example 1 Solve
$$x - y + 4z = 1$$
$$y - 2z = 3$$
$$z = 1$$

Solution. From the third equation, we know immediately that $z = 1$. We can then find y from the second equation:
$$y = 3 + 2z$$
$$= 5$$

Now substitute 1 and 5 for z and y in the first equation to find x:
$$x = 1 - 4z + y$$
$$= 2$$

The solution is given by $x = 2$, $y = 5$, $z = 1$, so that the triple $(2,5,1)$ is the only solution.

In Example 1, the system is given in *echelon form*, or *triangular form*. The system is virtually solved as it stands: z is determined by the third equation, whereupon y is found from the second, and then x from the first. We can immediately determine the solutions of any system written in echelon form. The next example treats an echelon system of two equations in four unknowns.

Example 2 Solve
$$\begin{cases} x - y + z + 2w = 2 \\ \phantom{x - {}} y - 2z + w = 1 \end{cases}$$

356 Systems of Equations

Solution. This system is in echelon form, so it only remains to determine what the solutions are. For each value of z and w, we must have
$$y = 1 - w + 2z$$
From the first equation, we then have
$$\begin{aligned} x &= 2 - 2w - z + y \\ &= 2 - 2w - z + (1 - w + 2z) \\ &= 3 - 3w + z \end{aligned}$$
That is, for any given values of z and w, the quadruple $(3 - 3w + z, 1 - w + 2z, z, w)$ is a solution. Moreover, *all solutions must be of this form*. This infinite set of solutions is called a *two-parameter* family of solutions.

Any system of linear equations can be changed to an equivalent system in echelon form. The operations giving equivalent systems are the same as those of Section 9-2: You can multiply any equation by a constant, or add a constant multiple of one equation to another. The examples below illustrate the method for several different kinds of system.

Example 3 Solve the system
$$\begin{cases} x - y + 2z = 6 \\ 2x + y - z = 5 \\ x + 2y + z = 7 \end{cases}$$

Solution. We eliminate x from the second and third equations by subtracting twice the first equation from the second, and subtracting the first equation from the third. This gives
$$\begin{cases} x - y + 2z = 6 \\ 3y - 5z = -7 \\ 3y - z = 1 \end{cases}$$

Now we subtract the second equation from the third to eliminate y from the third equation.
$$\begin{cases} x - y + 2z = 6 \\ 3y - 5z = -7 \\ 4z = 8 \end{cases}$$

Now we can read off the solution from the last system:

Linear Systems in Several Variables

$$\begin{cases} z = 2; \\ 3y = -7 + 5z = 3; y = 1; \\ x = 6 - 2z + y; x = 3 \end{cases}$$

The triple (3,1,2) is the single solution.

Not every linear system has solutions. For example, the system of two equations: $x + y + z = 1$ and $x + y + z = 2$, is obviously inconsistent. The next example shows how inconsistency can appear when we reduce a system of three equations to echelon form.

Example 4 Solve the system

$$\begin{cases} x - 2y + z = 1 \\ x + 3y - 3z = 1 \\ 2x + y - 2z = 3 \end{cases}$$

Solution. We eliminate x from the second and third equations by subtracting the first from the second, and then subtracting twice the first from the third. This gives

$$\begin{cases} x - 2y + z = 1 \\ 5y - 4z = 0 \\ 5y - 4z = 1 \end{cases}$$

It is already obvious from the second and third equations that this last system is inconsistent. If we eliminate y by subtracting the second equation from the third, we get the obviously inconsistent echelon system

$$\begin{cases} x - 2y + z = 1 \\ 5y - 4z = 0 \\ 0 = 1 \end{cases}$$

Example 5 Solve the system

$$\begin{cases} x - 2y + z = 1 \\ x + 3y - 3z = 1 \\ 2x + y - 2z = 2 \end{cases}$$

Solution. This system differs from that of Example 4 only in the constant term, 2, in the third equation. If we eliminate x from the second and third equations as in Example 4, we now get

$$\begin{cases} x - 2y + z = 1 \\ 5y - 4z = 0 \\ 5y - 4z = 0 \end{cases}$$

Now we eliminate y from the third equation, and obtain the following echelon system:

$$\begin{cases} x - 2y + z = 1 \\ 5y - 4z = 0 \\ 0 = 0 \end{cases}$$

For any value of z, $y = (4/5)z$, and $x = 1 - z + 2y = 1 + (3/5)z$ gives a solution. The triples $(1 + (3/5)z, (4/5)z, z)$ are the one-parameter family of solutions.

In Example 2 we expressed the family of solutions of a system of two equations in x, y, z, w in terms of the two parameters z and w. In Example 5 we expressed the family of solutions in terms of the single parameter z. However, the unknowns which can be used as parameters to express the family of all solutions can not always be chosen arbitrarily. For example, suppose a system of two equations in x, y, z reduces to the following echelon form:

$$\begin{cases} x + y + 2z = 3 \\ z = 1 \end{cases}$$

Here z is obviously determined, but x can be expressed in terms of y, or y in terms of x. The family of all solutions can be given as all triples $(1 - y, y, 1)$, or as all triples $(x, 1 - x, 1)$.

Linear systems can have more equations than variables. In these cases, too, the solution set can be empty, or contain a single n-tuple, or be an infinite family.

Example 6 Solve the system

$$\begin{cases} x - y + z = 2 \\ x + 2y - z = -1 \\ 2x + y + 3z = 1 \\ 4x + 2y + 3z = 2 \end{cases}$$

Solution. First eliminate x from the second, third, and fourth equations by subtracting appropriate multiples of the first equation. This gives

$$\begin{cases} x - y + z = 2 \\ 3y - 2z = -3 \\ 3y + z = -3 \\ 6y - z = -6 \end{cases}$$

Now eliminate y from the third and fourth equations by subtracting the appropriate multiples of the second equation.

$$\begin{cases} x - y + z = 2 \\ 3y - 2z = -3 \\ 3z = 0 \\ 3z = 0 \end{cases}$$

We finally arrive at proper echelon form by eliminating z from the fourth equation by subtracting the third from it:

$$\begin{cases} x - y + z = 2 \\ 3y - 2z = -3 \\ 3z = 0 \\ 0 = 0 \end{cases}$$

The unique solution is $z = 0$, $y = -1$, $x = 1$.

SECTION 9-3, REVIEW QUESTIONS

echelon

1. We solve linear systems by reducing them to _____ form.

solutions

2. When a system is written in echelon form, it is simple to determine the _____.

infinite

3. A linear system can have no solutions, one solution, or an _____ family of solutions.

variables

4. An infinite family of solutions is called a one-parameter family, two-parameter family, etc., depending on how many of the _____ can have values assigned arbitrarily.

multiple

5. To obtain an equivalent echelon system, the first step is to eliminate the first variable from all equations but the first by adding some _____ of the first equation to each of the others.

second

6. When the first variable has been eliminated from all equations but the first, we eliminate the second variable from all equations but the _____.

by adding some multiple of the second equation.

7. If a system of four equations in three unknowns is reduced to echelon form, the fourth equation will contain _____ variables, and will therefore be simply a true or false statement.   **no**

8. If the fourth equation is a false statement, like $1 = 0$, the system is _____.   **inconsistent**

SECTION 9-3, EXERCISES

Solve the following systems of equations.

1. $\begin{cases} x + z = 8 \\ x + y + 2z = 17 \\ x + 2y + z = 16 \end{cases}$

2. $\begin{cases} x + 2y - z = 5 \\ x + y + 2z = 11 \\ x + y + 3z = 14 \end{cases}$ Example 1, 3, 4

3. $\begin{cases} x + 2y - z = -1 \\ 2x + 2y - 3z = -1 \\ 4x - y + 2z = 11 \end{cases}$

4. $\begin{cases} x - 2y + 3z = 6 \\ 2x + y - 2z = -1 \\ 3x - 3y - z = 5 \end{cases}$

5. $\begin{cases} 2x - y + z = 3 \\ x + 3y - 2z = 11 \\ 3x - 2y + 4z = 11 \end{cases}$

6. $\begin{cases} 2x - y + 2z = -8 \\ x + 2y - 3z = 9 \\ 3x - y - 4z = 3 \end{cases}$

7. $\begin{cases} x - y + 2z = 0 \\ x - 2y + 3z = -1 \\ 2x - 2y + z = -3 \end{cases}$

8. $\begin{cases} x + y = 2 + z \\ x - 3y = 1 - 2z \\ 3x + 3z = 4 + 5y \end{cases}$

9. $\begin{cases} 2x - y + z = 1 \\ x + 2y - 3z = -2 \\ 3x - 4y + 5z = 1 \end{cases}$

10. $\begin{cases} x + y + 2z = 3 \\ 3x - y + z = 1 \\ 2x + 3y - 4z = 8 \end{cases}$

11. $\begin{cases} x - 3y - z = 11 \\ x - 5y + z = 1 \end{cases}$

12. $\begin{cases} x + 2y - z = 8 \\ x + y + z = 0 \end{cases}$ Example 2

13. $\begin{cases} 2x + 6z - 18y = 6 \\ x - 3z - y = -3 \end{cases}$

14. $\begin{cases} 3x + 2z - 4y = 6 \\ 6x + 4z - 8y = 14 \end{cases}$

15. $\begin{cases} x + y - z + w = 7 \\ 2x - y + 3z + w = 1 \\ x - 5y + 9z - w = 3 \end{cases}$

16. $\begin{cases} x - y + z = -4 \\ y + z - w = 5 \\ 2x + z - 4w = 1 \end{cases}$

17. $\begin{cases} 2x + y + z + w = -3 \\ x - y + z - w = 2 \\ 3x + 2y + 2z + w = 3 \end{cases}$

18. $\begin{cases} x - 2z = -5 \\ -y + w = -5 \\ 2z + w = 2 \end{cases}$

Linear Systems in Several Variables

Example 6 19. $\begin{cases} 2x + 3y + 6z = 3 \\ -x + 2y + 2z = -1 \\ 3x - 4y - 5z = -6 \\ x + 2y + 3z = -2 \end{cases}$ 20. $\begin{cases} 5x - 4y + 7z = -4 \\ 9x - 5y + 5z = 8 \\ 4x + 3y + 3z = 2 \\ 2x + 5y - z = 6 \end{cases}$

9-4 MATRIX METHODS

If we think of a solution of a linear system as an ordered pair (or ordered triple, quadruple, etc.), then it is clear that it is immaterial which particular variables are used in a system. For example, the following systems are all essentially the same:

$$\begin{cases} x + y = 3 \\ y = 1 \end{cases} \quad \begin{cases} r + s = 3 \\ s = 1 \end{cases} \quad \begin{cases} x_1 + x_2 = 3 \\ x_2 = 1 \end{cases}$$

Each of these systems has the single solution (2,1).

In the systems above, it is not the choice of variables that matters, but the coefficients of the variables, and the constants on the right side. The array of coefficients and right-side constants of a linear system is called the *matrix of the system*. The matrix of each of the systems above is

$$\begin{bmatrix} 1 & 1 & 3 \\ 0 & 1 & 1 \end{bmatrix}. \tag{1}$$

The matrix

$$\begin{bmatrix} 1 & 1 \\ 0 & 1 \end{bmatrix} \tag{2}$$

is called the *matrix of the coefficients*, and (1) is sometimes also called the *augmented matrix* to distinguish it from (2).

In this section we will practice solving systems of equations by considering only the matrix of the system. It is clear that the matrix of the system contains all the necessary information, so there is no need to write the variables over and over again.

Any rectangular array of numbers is called a matrix. The horizontal lines of numbers in a matrix are its *rows*, and the vertical lines of numbers are its *columns*. The matrix (1) has two rows and three columns. The *main diagonal* of a matrix is the diagonal starting in the upper left corner (first row and first column). The main diagonal also contains the element in the second row and second column, the element in the third row and third column, etc.

Every matrix is the matrix of some linear system. For example, the matrix

$$\begin{bmatrix} 3 & 0 & 1 & 0 & 5 \\ 0 & 0 & 2 & -1 & 1 \\ -2 & 1 & 0 & 3 & 4 \end{bmatrix}$$

is the matrix of the system

$$\begin{aligned} 3x_1 \phantom{{}+x_2} + x_3 \phantom{{}-x_4} &= 5 \\ \phantom{3x_1 + x_2 +{}} 2x_3 - x_4 &= 1 \\ -2x_1 + x_2 \phantom{{}+x_3} + 3x_4 &= 4 \end{aligned} \qquad (3)$$

The fact that we use x_1, x_2, x_3, x_4 in (3) instead of x, y, z, w is not significant. In either case the solutions are the same set of ordered quadruples.

Two matrices are called equivalent if the corresponding linear systems are equivalent. We use the sign $\sim$ to indicate equivalence of matrices. Operations on rows which lead to equivalent matrices are the same as those operations on equations which lead to equivalent systems. The following three operators, called *elementary row operations*, clearly lead to equivalent matrices.

1. Interchange two rows (corresponds to writing the equations of a system in a different order).
2. Multiply a row by a nonzero constant (corresponds to multiplying both sides of one equation by a constant).
3. Add one row to another (corresponds to adding the corresponding sides of two equations).

Adding a multiple of one row to another is accomplished by performing two elementary row operations.

An *echelon matrix* is one which corresponds to a system in echelon form. Such a matrix has only zeros below the main diagonal. The following are echelon matrices (the main diagonal elements are in boldface):

$$\begin{bmatrix} \mathbf{3} & -1 & 0 & 4 \\ 0 & \mathbf{1} & 2 & 0 \\ 0 & 0 & \mathbf{3} & 1 \end{bmatrix} \quad (4) \qquad \begin{bmatrix} \mathbf{2} & 1 & 3 & 5 \\ 0 & \mathbf{0} & 1 & 2 \end{bmatrix} \quad (5)$$

$$\begin{bmatrix} 1 & 2 & 2 \\ 0 & 1 & 0 \\ 0 & 0 & 0 \\ 0 & 0 & 0 \\ 0 & 0 & 0 \end{bmatrix} \quad (6) \qquad \begin{bmatrix} 3 & 0 & 4 & 1 \\ 0 & 0 & 0 & 3 \end{bmatrix} \quad (7)$$

Matrix (4) corresponds to a system of three equations in x, y, z, and the system has a unique solution ($z = 1/3$, etc.) Matrix (5) corresponds to a system of two equations in x, y, z. In this system, z must be 2, y can be assigned arbitrarily, and x is given in terms of y. Matrix (6) corresponds to a system of five equations in x and y. The system has the single solution (2,0). Matrix (7) corresponds to an echelon system of two equations in x, y, z. The system is inconsistent since the second equation reads $0x + 0y + 0z = 3$.

To solve a linear system, we can write its matrix, and then obtain an equivalent triangular matrix by row operations. The solutions of the system are then found from the echelon system corresponding to the triangular matrix.

Example 1 Solve

$$\begin{cases} x + 2y - z + w = 6 \\ 2x - y + w = -1 \\ x + z - 3w = -4 \\ y - 2z + w = 5 \end{cases}$$

Solution. We write the matrix of the system, and reduce it to triangular form as follows:

$$\begin{bmatrix} 1 & 2 & -1 & 1 & 6 \\ 2 & -1 & 0 & 1 & -1 \\ 1 & 0 & 1 & -3 & -4 \\ 0 & 1 & -2 & 1 & 5 \end{bmatrix} \sim$$

Add -2 times the first row to the second, and -1 times the first row to the third.

$$\begin{bmatrix} 1 & 2 & -1 & 1 & 6 \\ 0 & -5 & 2 & -1 & -13 \\ 0 & -2 & 2 & -4 & -10 \\ 0 & 1 & -2 & 1 & 5 \end{bmatrix} \sim$$

Interchange the second and fourth rows; divide the third row by 2.

$$\begin{bmatrix} 1 & 2 & -1 & 1 & 6 \\ 0 & 1 & -2 & 1 & 5 \\ 0 & -1 & 1 & -2 & -5 \\ 0 & -5 & 2 & -1 & -13 \end{bmatrix} \sim$$

Add the second row to the third; add 5 times the second row to the fourth.

$$\begin{bmatrix} 1 & 2 & -1 & 1 & 6 \\ 0 & 1 & -2 & 1 & 5 \\ 0 & 0 & -1 & -1 & 0 \\ 0 & 0 & -8 & 4 & 12 \end{bmatrix} \sim$$

Divide the fourth row by 4.

$$\begin{bmatrix} 1 & 2 & -1 & 1 & 6 \\ 0 & 1 & -2 & 1 & 5 \\ 0 & 0 & -1 & -1 & 0 \\ 0 & 0 & -2 & 1 & 3 \end{bmatrix} \sim$$

Add -2 times the third row to the fourth.

$$\begin{bmatrix} 1 & 2 & -1 & 1 & 6 \\ 0 & 1 & -2 & 1 & 5 \\ 0 & 0 & -1 & -1 & 0 \\ 0 & 0 & 0 & 3 & 3 \end{bmatrix}$$

The last matrix is triangular, and corresponds to the echelon system

$$\begin{aligned} x + 2y - z + w &= 6 \\ y - 2z + w &= 5 \\ -z - w &= 0 \\ 3w &= 3 \end{aligned}$$

The solution is given by $w = 1$, $z = -1$, $y = 2$, $x = 0$; that is, the quadruple $(0, 2, -1, 1)$ is the only solution.

Example 2 Solve

$$\begin{cases} x + 2y - z = 3 \\ 2x - y - z = 5 \\ 5x - 3z = 13 \\ 4x + 3y - 3z = 11 \end{cases}$$

Solution. Write the matrix, and reduce to echelon form.

$$\begin{bmatrix} 1 & 2 & -1 & 3 \\ 2 & -1 & -1 & 5 \\ 5 & 0 & -3 & 13 \\ 4 & 3 & -3 & 11 \end{bmatrix} \sim$$

Add -2 times the first row to the second, -5 times the first row to the third, and -4 times the first row to the fourth.

$$\begin{bmatrix} 1 & 2 & -1 & 3 \\ 0 & -5 & 1 & -1 \\ 0 & -10 & 2 & -2 \\ 0 & -5 & 1 & -1 \end{bmatrix} \sim$$

Add -2 times the second row to the third, and -1 times the second row to the fourth.

$$\begin{bmatrix} 1 & 2 & -1 & 3 \\ 0 & -5 & 1 & 1 \\ 0 & 0 & 0 & 0 \\ 0 & 0 & 0 & 0 \end{bmatrix}$$

The system is equivalent to the echelon system

$$x + 2y - z = 3$$
$$-5y + z = 1$$
$$0 = 0$$
$$0 = 0$$

For any value of z, we get $y = \frac{1}{5}(z - 1)$, $x = \frac{1}{5}(3z + 17)$. The solutions are the one-parameter family $(\frac{1}{5}(3z + 17), \frac{1}{5}(z - 1), z)$.

SECTION 9-4, REVIEW QUESTIONS

1. The array of coefficients and constants in a linear system is called the _____ of the system. **matrix**
2. A matrix is any rectangular array of _____. **numbers**
3. The rows of a matrix are the horizontal lines of entries, and the vertical lines of entries are the _____. **columns**
4. The elements in the first row and first column, second row and second column, etc., make up the _____ diagonal. **main**
5. Every matrix is the matrix of some _____ system. **linear**
6. Two matrices are equivalent if the corresponding linear systems are _____. **equivalent**
7. The operations on equations which yield equivalent systems correspond to operations on rows which yield _____ matrices. **equivalent**
8. The matrix of an echelon system is called an _____ matrix. **echelon**
9. An echelon matrix has only _____ below the main diagonal. **zeros**
10. We solve any linear system by reducing its matrix to an _____ matrix which is equivalent to it. **echelon**

SECTION 9-4, EXERCISES

Solve the following systems by matrix methods.

1. $x + 2y = 1$
   $2x + 3y = 3$

2. $x - 3y = -1$
   $2x - 5y = -1$

3. $3x + 2y = 4$
   $x + y = 2$

4. $-2x + 3y = -2$
   $x - 2y = 0$

5. $2x - 4y = 10$
   $-3x + 6y = -15$

6. $3x - 9y = 15$
   $-4x + 12y = -20$

7. $-x + 3y = 4$
   $4x - 12y = 8$

8. $x + 5y = 10$
   $3x + 15y = 5$

9. $x + y - z = 5$
   $x + 2y + z = 0$

10. $x + 2y - 5z = 7$
    $2x + 4y - 5z = 14$

11. $x + 2y - 3z = 9$
    $2x - y + 2z = -8$
    $-x + 3y - 4z = 15$

12. $2x + 4y + z = 4$
    $x - y + 3z = 7$
    $3x + y - 2z = -1$

13. $3x - 2y + 4z = 19$
    $3y + z = -3$
    $2x + y - z = -3$
    $x + 2y + z = 0$

14. $2x - 3y + 5z = 6$
    $x + y + z = 4$
    $x - y = 1$
    $2x + z = 5$

15. $x + 2y - z + w = 2$
    $2x + 5y - 2z - w = -2$
    $x + 3y + z + 2w = 6$
    $x - y + 2z - w = 1$

16. $x - 2y + z - 2w = -1$
    $x - y + 2z + w = 4$
    $2x - 2y - z - w = 2$
    $x + z = 1$

17. $x + y + 2z + w = 1$
    $2x + y - z + w = 0$
    $x + 2y + z - w = -1$
    $4x + 4y + 2z + w = 1$

18. $x + 2y + 3z - w = 6$
    $2x + 5y + 5z - w = 12$
    $x + 3y + 5z + 2w = 9$
    $-x - y + z + w = -1$

19. $x + y - z + w = 1$
    $-x - 2y + z + 2w = -2$
    $x - z + 4w = 0$
    $2x + y - 2z + 6w = 1$

20. $x + 2y + w = 3$
    $x + 3y + z + 3w = 5$
    $x + y - z - w = 1$
    $y + z + 2w = 2$

9-5 SECOND AND THIRD ORDER DETERMINANTS

We have seen that the solutions of any linear system can be determined by reducing the system to echelon form. For most systems, this is the simplest and most efficient method.

It is clear from considering the echelon form of a system that there will not be a unique solution unless there are at least as many equations as unknowns. For a system of n equations in n unknowns, we have a simple test for telling whether the system has a unique solution. The test involves the idea of a *determinant*, which we will introduce in this section. If a system of n equations in n unknowns does have a unique solution (that is, a unique n-tuple $(x_1, \ldots, x_n)$), then the values $x_1, \ldots, x_n$ can also be simply expressed in terms of determinants. In addition to their theoretical importance for systems of linear equations, determinants arise in many other branches of mathematics.

Consider first the general system of two equations in two unknowns.

$$\begin{cases} a_1 x + b_1 y = c_1 \\ a_2 x + b_2 y = c_2 \end{cases} \tag{1}$$

The methods of Section 9-4 allow us to write a general solution for this system. The system has the following unique solution, provided $a_1 b_2 - a_2 b_1 \neq 0$.

$$x = \frac{c_1 b_2 - c_2 b_1}{a_1 b_2 - a_2 b_1} \qquad y = \frac{a_1 c_2 - a_2 c_1}{a_1 b_2 - a_2 b_1} \tag{2}$$

In the solution (2) notice that the denominators are the same, and involve only the coefficients of the unknowns. The numerator in the formula for x is obtained from the denominator by replacing the coefficients of x (a_1 and a_2) by the corresponding constant terms (c_1 and c_2). Similarly, the numerator in the formula for y is obtained by replacing the coefficients of y (b_1 and b_2) by c_1 and c_2, respectively. We define the common denominator of the formulas (2) to be the *determinant of the coefficients* of the system (1), and denote it as follows:

$$\begin{vmatrix} a_1 & b_1 \\ a_2 & b_2 \end{vmatrix} = a_1 b_2 - a_2 b_1. \tag{3}$$

With this definition, the formulas (2) can be written

$$x = \frac{\begin{vmatrix} c_1 & b_1 \\ c_2 & b_2 \end{vmatrix}}{\begin{vmatrix} a_1 & b_1 \\ a_2 & b_2 \end{vmatrix}}, \qquad y = \frac{\begin{vmatrix} a_1 & c_1 \\ a_2 & c_2 \end{vmatrix}}{\begin{vmatrix} a_1 & b_1 \\ a_2 & b_2 \end{vmatrix}}. \tag{4}$$

These formulas are sometimes called *Cramer's rule* for solving the system.

Notice that system (1) has the unique solution (4) if and only if the determinant of the coefficients is not zero. If $a_1 b_2 - a_2 b_1 = 0$, then the left sides of the two equations are proportional ($a_1/a_2 = b_1/b_2$). Therefore the equations are either dependent (if $c_1/c_2 = a_1/a_2 = b_1/b_2$), or inconsistent (if $c_1/c_2 \neq a_1/a_2 = b_1/b_2$).

Example 1 Solve by Cramer's rule:

$$\begin{cases} 3x - y = 5 \\ x + 2y = 7 \end{cases}$$

Solution.

$$x = \frac{\begin{vmatrix} 5 & -1 \\ 7 & 2 \end{vmatrix}}{\begin{vmatrix} 3 & -1 \\ 1 & 2 \end{vmatrix}} = \frac{10 + 7}{6 + 1} = \frac{17}{7} \qquad y = \frac{\begin{vmatrix} 3 & 5 \\ 1 & 7 \end{vmatrix}}{\begin{vmatrix} 3 & -1 \\ 1 & 2 \end{vmatrix}} = \frac{21 - 5}{7} = \frac{16}{7}$$

Example 2 Solve

$$\begin{cases} 2x - 3y = 5 \\ -6x + 9y = 10 \end{cases}$$

Solution. The determinant of the coefficients is zero:

$$\begin{vmatrix} 2 & -3 \\ -6 & 9 \end{vmatrix} = 18 - 18 = 0.$$

Therefore the left sides of the two equations are proportional. The equations are inconsistent, since the constant terms are not in the same ratio as the left sides.

The general system of three equations in three unknowns can be written:

$$a_1 x + b_1 y + c_1 z = d_1$$

$$a_2 x + b_2 y + c_2 z = d_2 \qquad (5)$$

$$a_3 x + b_3 y + c_3 z = d_3.$$

If we reduce this system to echelon form and solve it, we find that there is a unique solution, provided

$$a_1 b_2 c_3 - a_1 b_3 c_2 + a_2 b_3 c_1 - a_2 b_1 c_3 + a_3 b_1 c_2 - a_3 b_2 c_1 \neq 0.$$

The unique solution is given by

$$x = \frac{d_1 b_2 c_3 - d_1 b_3 c_2 + d_2 b_3 c_1 - d_2 b_1 c_3 + d_3 b_1 c_2 - d_3 b_2 c_1}{a_1 b_2 c_3 - a_1 b_3 c_2 + a_2 b_3 c_1 - a_2 b_1 c_3 + a_3 b_1 c_2 - a_3 b_2 c_1},$$

$$y = \frac{a_1 d_2 c_3 - a_1 d_3 c_2 + a_2 d_3 c_1 - a_2 d_1 c_3 + a_3 d_1 c_2 - a_3 d_2 c_1}{a_1 b_2 c_3 - a_1 b_3 c_2 + a_2 b_3 c_1 - a_2 b_1 c_3 + a_3 b_1 c_2 - a_3 b_2 c_1}, \quad (7)$$

$$z = \frac{a_1 b_2 d_3 - a_1 b_3 d_2 + a_2 b_3 d_1 - a_2 b_1 d_3 + a_3 b_1 d_2 - a_3 b_2 d_1}{a_1 b_2 c_3 - a_1 b_3 c_2 + a_2 b_3 c_1 - a_2 b_1 c_3 + a_3 b_1 c_2 - a_3 b_2 c_1}.$$

The formulas (7) follow the same pattern as the solutions (2) of the system of two equations in two unknowns. The denominators in (7) are all the same, and involve only the coefficients of the unknowns. The numerator in the x formula is obtained by replacing the a's in the denominator (the coefficients of x) by the corresponding d's (the constant terms). Similar statements hold for the y and z formulas. As before, we define the determinant of the coefficients of (5) to be the common denominator of the solution formulas:

$$\begin{vmatrix} a_1 & b_1 & c_1 \\ a_2 & b_2 & c_2 \\ a_3 & b_3 & c_3 \end{vmatrix} = a_1 b_2 c_3 - a_1 b_3 c_2 + a_2 b_3 c_1 - a_2 b_1 c_3 + a_3 b_1 c_2 - a_3 b_2 c_1. \quad (8)$$

With this definition, Cramer's rule for solving a third order system is

$$x = \frac{\begin{vmatrix} d_1 & b_1 & c_1 \\ d_2 & b_2 & c_2 \\ d_3 & b_3 & c_3 \end{vmatrix}}{\begin{vmatrix} a_1 & b_1 & c_1 \\ a_2 & b_2 & c_2 \\ a_3 & b_3 & c_3 \end{vmatrix}}, \quad y = \frac{\begin{vmatrix} a_1 & d_1 & c_1 \\ a_2 & d_2 & c_2 \\ a_3 & d_3 & c_3 \end{vmatrix}}{\begin{vmatrix} a_1 & b_1 & c_1 \\ a_2 & b_2 & c_2 \\ a_3 & b_3 & c_3 \end{vmatrix}},$$

$$z = \frac{\begin{vmatrix} a_1 & b_1 & d_1 \\ a_2 & b_2 & d_2 \\ a_3 & b_3 & d_3 \end{vmatrix}}{\begin{vmatrix} a_1 & b_1 & c_1 \\ a_2 & b_2 & c_2 \\ a_3 & b_3 & c_3 \end{vmatrix}}.$$

A system of two equations in two unknowns, or three equations in three unknowns, has a unique solution if and only if the determinant of the coefficients is different from zero. If the determinant of the coefficients is zero, then the equations are dependent or inconsistent. When the determinant of the coefficients is zero, we use matrix methods to reduce the system to echelon form as in the last section.

The definition (8) is too long to use conveniently, so we need a simple way of expanding a third order determinant. Observe that (8) can be written as follows:

$$\begin{vmatrix} a_1 & b_1 & c_1 \\ a_2 & b_2 & c_2 \\ a_3 & b_3 & c_3 \end{vmatrix} =$$

$$= a_1(b_2c_3 - b_3c_2) - a_2(b_1c_3 - b_3c_1) + a_3(b_1c_2 - b_2c_1)$$

$$= a_1 \begin{vmatrix} b_2 & c_2 \\ b_3 & c_3 \end{vmatrix} - a_2 \begin{vmatrix} b_1 & c_1 \\ b_3 & c_3 \end{vmatrix} + a_3 \begin{vmatrix} b_1 & c_1 \\ b_2 & c_2 \end{vmatrix} \qquad (10)$$

The 2 by 2 determinants in (10) are the respective *minors* of the elements a_1, a_2, a_3. In general, the minor of any element of a determinant is the determinant you get by deleting the row and column of the element in question. One can show that a 3 by 3 determinant can be evaluated by multiplying the elements of any row (or column) by their respective minors, with signs attached to the terms according to the following checkerboard pattern.

$$\begin{vmatrix} + & - & + \\ - & + & - \\ + & - & + \end{vmatrix}$$

The expansion (10) is the expansion of the determinant by the first column. The expansion by the second row would be

$$\begin{vmatrix} a_1 & b_1 & c_1 \\ a_2 & b_2 & c_2 \\ a_3 & b_3 & c_3 \end{vmatrix} = -a_2 \begin{vmatrix} b_1 & c_1 \\ b_3 & c_3 \end{vmatrix} + b_2 \begin{vmatrix} a_1 & c_1 \\ a_3 & c_3 \end{vmatrix} - c_2 \begin{vmatrix} a_1 & b_1 \\ a_3 & b_3 \end{vmatrix}.$$

Example 3 Solve by Cramer's rule:

$$2x - y + z = 1$$
$$x + 2z = 2$$
$$-x + y - z = 0$$

Solution. The determinant (D) of the coefficients is

$$D = \begin{vmatrix} 2 & -1 & 1 \\ 1 & 0 & 2 \\ -1 & 1 & -1 \end{vmatrix}.$$

We expand by the second row, which contains a zero:

$$D = -(1)\begin{vmatrix} -1 & 1 \\ 1 & -1 \end{vmatrix} + 0\begin{vmatrix} 2 & 1 \\ -1 & -1 \end{vmatrix} - (2)\begin{vmatrix} 2 & -1 \\ -1 & 1 \end{vmatrix}$$

$$= -(1-1) + 0(-2+1) - 2(2-1)$$

$$= -2.$$

Since $D \neq 0$, there is a unique solution. We expand each of the determinants below by the column of constants:

$$x = \frac{1}{D}\begin{vmatrix} 1 & -1 & 1 \\ 2 & 0 & 2 \\ 0 & 1 & -1 \end{vmatrix}$$

$$= -\frac{1}{2}\left[+(1)\begin{vmatrix} 0 & 2 \\ 1 & -1 \end{vmatrix} - (2)\begin{vmatrix} -1 & 1 \\ 1 & -1 \end{vmatrix} + 0\right]$$

$$= -\frac{1}{2}[(0-2) - 2(1-1)] = 1,$$

$$y = \frac{1}{D}\begin{vmatrix} 2 & 1 & 1 \\ 1 & 2 & 2 \\ -1 & 0 & -1 \end{vmatrix}$$

$$= -\frac{1}{2}\left[-(1)\begin{vmatrix} 1 & 2 \\ -1 & -1 \end{vmatrix} + (2)\begin{vmatrix} 2 & 1 \\ -1 & -1 \end{vmatrix} - 0\right]$$

$$= -\frac{1}{2}[-(-1+2) + 2(-2+1)] = \frac{3}{2},$$

$$z = \frac{1}{D}\begin{vmatrix} 2 & -1 & 1 \\ 1 & 0 & 2 \\ -1 & 1 & 0 \end{vmatrix}$$

$$= -\frac{1}{2}\left[+(1)\begin{vmatrix} 1 & 0 \\ -1 & 1 \end{vmatrix} - (2)\begin{vmatrix} 2 & -1 \\ -1 & 1 \end{vmatrix} + 0\right]$$

$$= -\frac{1}{2}[(1-0) - 2(2-1)] = \frac{1}{2}.$$

SECTION 9-5, REVIEW QUESTIONS

1. A linear system cannot have a unique solution unless there are at least as many equations as _____.

   unknowns

2. If there are as many equations as unknowns, the matrix of the coefficients is a _____ matrix.

   square

3. The determinant of the square 2 by 2 matrix is defined by
   $$\begin{vmatrix} a_1 & b_1 \\ a_2 & b_2 \end{vmatrix} = \text{_____}.$$

   $a_1 b_2 - a_2 b_1$

4. Cramer's rule gives the solution of a system of n equations in n unknowns in terms of quotients of _____.

   determinants

5. In the formulas of Cramer's rule, the denominator of the quotient is always the determinant of the _____.

   coefficients

6. A system of n equations in n unknowns has a unique solution if and only if the determinant of the coefficients is not _____.

   zero

7. To expand a third order determinant, you multiply each element of some row (or column) by its _____ with the appropriate sign attached.

   minor

8. The minor of an element is the determinant gotten by deleting the _____ and _____ of that element.

   row
   column

9. The sign of each term in an expansion by minors is determined by the checkerboard pattern starting with the _____ sign in the first row and column.

   plus

SECTION 9-5, EXERCISES

Solve the following systems by Cramer's rule.

Example 1

1. $x - 2y = 6$
   $x + y = 3$

2. $2x + y = 8$
   $x - 2y = -1$

3. $3x + y = 7$
   $x + 3y = 1$

4. $-2x + 5y = 1$
   $3x - 4y = 2$

5. $2x - ay = 3$
   $x + 2y = 1$

6. $ax + by = 2$
   $3x - 5y = 0$

7. $x - 2y + z = 4$
   $2x + y - z = 1$
   $-x + 2y + z = 4$

8. $2x + y - z = -2$
   $x + 2y + 2z = 5$
   $-2x + y + z = 0$ Example 3

9. $4x - y + z = -5$
   $3x + 2y - 4z = -5$
   $x + y + z = 4$

10. $3x + 2y + 2z = 4$
    $x - 5y + z = 1$
    $2x + y - 3z = 7$

11. $3x - y - 2z + 4 = 0$
    $2y - z - 3 = 0$
    $3x - 5y + 10 = 0$

12. $x - 3y + 2z + 1 = 0$
    $2x + y - z - 8 = 0$
    $-x + 4y + 3z + 9 = 0$

13. $\dfrac{3}{x} - \dfrac{1}{y} = 7$
    $\dfrac{3}{y} - \dfrac{1}{z} = 5$
    $\dfrac{3}{z} - \dfrac{1}{x} = 0$

14. $\dfrac{1}{x} + \dfrac{2}{y} - \dfrac{1}{z} = 7$
    $\dfrac{2}{x} - \dfrac{1}{y} + \dfrac{1}{z} = 7$
    $\dfrac{1}{x} - \dfrac{3}{y} - \dfrac{1}{z} = -3$

9-6   GENERAL PROPERTIES OF DETERMINANTS

Higher order determinants can also be evaluated by expanding them by any row or column. For example, if we expand the general 4 by 4 determinant by the first column, we get

$$\begin{vmatrix} a_1 & b_1 & c_1 & d_1 \\ a_2 & b_2 & c_2 & d_2 \\ a_3 & b_3 & c_3 & d_3 \\ a_4 & b_4 & c_4 & d_4 \end{vmatrix} = a_1 \begin{vmatrix} b_2 & c_2 & d_2 \\ b_3 & c_3 & d_3 \\ b_4 & c_4 & d_4 \end{vmatrix} - a_2 \begin{vmatrix} b_1 & c_1 & d_1 \\ b_3 & c_3 & d_3 \\ b_4 & c_4 & d_4 \end{vmatrix}$$

$$+ a_3 \begin{vmatrix} b_1 & c_1 & d_1 \\ b_2 & c_2 & d_2 \\ b_4 & c_4 & d_4 \end{vmatrix} - a_4 \begin{vmatrix} b_1 & c_1 & d_1 \\ b_2 & c_2 & d_2 \\ b_3 & c_3 & d_3 \end{vmatrix} \quad (1)$$

The signs in the expansion are determined by the familiar checkerboard pattern

$$\begin{vmatrix} + & - & + & - \\ - & + & - & + \\ + & - & + & - \\ - & + & - & + \end{vmatrix}$$

It is a fact that we will not verify here that we get the same result by expanding by any row or any column, using the signs of the checkerboard pattern. This principle holds for higher order determinants as well, and we will take this as our definition. A 4 by 4 determinant is a linear combination of 3 by 3 determinants, a 5 by 5 determinant is a linear combination of 4 by 4 determinants; and so on.

The utility of the above definition is based on the fact that Cramer's rule also holds for higher order linear systems. A system of n equations in n unknowns has a unique solution (an n-tuple of numbers) provided the n by n determinant of the coefficients is not zero. The values of the unknowns are quotients of determinants, and in each case the denominator is the determinant of the coefficients. The numerator in the formula for the jth unknown is the determinant gotten by replacing the jth column in the denominator by the column of constants.

Example 1 Solve for x by Cramer's rule:

$$2x + y - z + w = 3$$
$$y + z - w = 1$$
$$3x \phantom{{}+y+z} + 2w = 0$$
$$-2x + 4y + z + 2w = 1$$

Solution. There will be a unique solution provided the determinant of the coefficients is not zero, and x is given by

$$x = \frac{\begin{vmatrix} 3 & 1 & -1 & 1 \\ 1 & 1 & 1 & -1 \\ 0 & 0 & 0 & 2 \\ 1 & 4 & 1 & 2 \end{vmatrix}}{\begin{vmatrix} 2 & 1 & -1 & 1 \\ 0 & 1 & 1 & -1 \\ 3 & 0 & 0 & 2 \\ -2 & 4 & 1 & 2 \end{vmatrix}}.$$

We first evaluate the determinant of the coefficients, expanding by the third row since it contains two zeros.

$$\begin{vmatrix} 2 & 1 & -1 & 1 \\ 0 & 1 & 1 & -1 \\ 3 & 0 & 0 & 2 \\ -2 & 4 & 1 & 2 \end{vmatrix} = 3\begin{vmatrix} 1 & -1 & 1 \\ 1 & 1 & -1 \\ 4 & 1 & 2 \end{vmatrix} - 2\begin{vmatrix} 2 & 1 & -1 \\ 0 & 1 & 1 \\ -2 & 4 & 1 \end{vmatrix}$$

(then we expand both 3 by 3 determinants by the first column)

$$= 3[1(2 + 1) - 1(-2 - 1) + 4(1 - 1)]$$
$$\quad - 2[2(1 - 4) - 0 - 2(1 + 1)]$$
$$= 3[6] - 2[-10]$$
$$= 38$$

We next evaluate the numerator, expanding by the third row:

$$x = \frac{1}{38}\begin{vmatrix} 3 & 1 & -1 & 1 \\ 1 & 1 & 1 & -1 \\ 0 & 0 & 0 & 2 \\ 1 & 4 & 1 & 2 \end{vmatrix}$$

$$= \frac{1}{38}(-2)\begin{vmatrix} 3 & 1 & -1 \\ 1 & 1 & 1 \\ 1 & 4 & 1 \end{vmatrix}$$

(now expand by the first column)

$$= -\frac{2}{38}[3(1 - 4) - 1(1 + 4) + 1(1 + 1)]$$

$$= -\frac{2}{38}[-12]$$

$$= \frac{24}{38} = \frac{12}{19}.$$

The values of y, z, and w can be found in a similar way (problems 1, 2, and 3). If the four by four numerators for y and z are expanded by the third row, they will reduce to two three by three determinants

General Properties of Determinants

because of the two zeros in the row. The four by four numerator for w has three zeros in its third row, and so reduces to a single three by three determinant.

Each of the 3 by 3 determinants in the expansion (1) has six terms, so there are 24 terms in the expansion of a 4 by 4 determinant. A 5 by 5 determinant is a linear combination of five 4 by 4 determinants, so the expansion of a 5 by 5 determinant has $5 \times 24 = 120$ terms altogether. Since expansion by minors becomes cumbersome very rapidly as the size of the determinant increases, we need other methods to evaluate higher order determinants.

The following facts can be checked directly for 2 by 2 or 3 by 3 determinants. We can prove the same result for all higher order determinants by using the expansion by minors and mathematical induction. We will not carry out the details here.

ROW OPERATIONS ON DETERMINANTS

1. *If two rows of a determinant are interchanged, the determinant changes sign.*

$$\begin{vmatrix} 4 & 1 & 2 \\ -6 & 8 & 9 \\ 1 & 1 & 1 \end{vmatrix} = - \begin{vmatrix} 1 & 1 & 1 \\ -6 & 8 & 9 \\ 4 & 1 & 2 \end{vmatrix}$$

(Interchange first and third rows.)

2. *If each element of some row is multiplied by* k, *the value of the determinant is multiplied by* k.

$$\begin{vmatrix} 2 & 0 & 1 & 5 \\ 4 & 12 & 0 & -8 \\ 3 & 1 & -1 & 4 \\ 6 & 5 & 2 & 1 \end{vmatrix} = 4 \begin{vmatrix} 2 & 0 & 1 & 5 \\ 1 & 3 & 0 & -2 \\ 3 & 1 & -1 & 4 \\ 6 & 5 & 2 & 1 \end{vmatrix}$$

(Factor 4 from the second row.)

3. *If some multiple of one row is added to another, the determinant is unchanged.*

$$\begin{vmatrix} 1 & 2 & 3 \\ 3 & 5 & 7 \\ 4 & -3 & 6 \end{vmatrix} = \begin{vmatrix} 1 & 2 & 3 \\ 3-2 & 5-4 & 7-6 \\ 4 & -3 & 6 \end{vmatrix} = \begin{vmatrix} 1 & 2 & 3 \\ 1 & 1 & 1 \\ 4 & -3 & 6 \end{vmatrix}$$

(Add -2 times the first row to the second.)

Another basic property of determinants is the fact that if the rows and columns are interchanged, the determinant has the same value. For example,

$$\begin{vmatrix} 1 & 2 \\ 8 & 10 \end{vmatrix} = \begin{vmatrix} 1 & 8 \\ 2 & 10 \end{vmatrix},$$

$$\begin{vmatrix} 1 & 2 & 3 \\ 5 & 4 & 3 \\ 7 & 7 & 6 \end{vmatrix} = \begin{vmatrix} 1 & 5 & 7 \\ 2 & 4 & 7 \\ 3 & 3 & 6 \end{vmatrix}.$$

From this fact, it follows immediately that for each of the row properties above, there is a corresponding column property.

1. *If two columns are interchanged, the determinant changes sign.*
2. *If each element of some column is multiplied by* k, *the determinant is multiplied by* k.
3. *If some multiple of one column is added to another, the determinant is unchanged.*

The next example shows how the basic row (or column) properties can be used to simplify the expansion of a determinant. The procedure is similar to that for putting a matrix in echelon form.

Example 2 Evaluate

$$\begin{vmatrix} 1 & 4 & -3 & 5 \\ 2 & 3 & 0 & 1 \\ 4 & 1 & 5 & -1 \\ -1 & 1 & 2 & 0 \end{vmatrix}$$

Solution. By adding appropriate multiples of the first row to each of the other rows, we obtain three zeros in the first column.

$$\begin{vmatrix} 1 & 4 & -3 & 5 \\ 2 & 3 & 0 & 1 \\ 4 & 1 & 5 & -1 \\ -1 & 1 & 2 & 0 \end{vmatrix}$$

Add -2 times the first row to the second.

$$= \begin{vmatrix} 1 & 4 & -3 & 5 \\ 0 & -5 & 6 & -9 \\ 4 & 1 & 5 & -1 \\ -1 & 1 & 2 & 0 \end{vmatrix}$$

Add -4 times the first row to the third, and 1 times the first row to the fourth.

$$= \begin{vmatrix} 1 & 4 & -3 & 5 \\ 0 & -5 & 6 & -9 \\ 0 & -15 & 17 & -21 \\ 0 & 5 & -1 & 5 \end{vmatrix}$$

Expand by the first column.

$$= \begin{vmatrix} -5 & 6 & -9 \\ -15 & 17 & -21 \\ 5 & -1 & 5 \end{vmatrix}$$

Factor -5 from the first column.

$$= -5 \begin{vmatrix} 1 & 6 & -9 \\ 3 & 17 & -21 \\ -1 & -1 & 5 \end{vmatrix}$$

Add -3 times the first row to the second, and add 1 times the first row to the third.

$$= -5 \begin{vmatrix} 1 & 6 & -9 \\ 0 & -1 & 6 \\ 0 & 5 & -4 \end{vmatrix}$$

Expand by the first column.

$$= -5 \begin{vmatrix} 17 & -21 \\ -1 & 5 \end{vmatrix}$$

$$= -5(85 - 21) = (-5)(64) = -320$$

Example 3 Use row operations to convert the determinant to one with zeros below the main diagonal.

380 *Systems of Equations*

$$\begin{vmatrix} 4 & 3 & -4 \\ 1 & 0 & -2 \\ -2 & 1 & 2 \end{vmatrix}$$

Solution

$$= \begin{vmatrix} 4 & 3 & -4 \\ 1 & 0 & -2 \\ -2 & 1 & 2 \end{vmatrix}$$

Interchange the first and second rows.

$$= -\begin{vmatrix} 1 & 0 & -2 \\ 4 & 3 & -4 \\ -2 & 1 & 2 \end{vmatrix}$$

Multiply the first row by -4 and add to the second; multiply the first row by 2 and add to the third.

$$= -\begin{vmatrix} 1 & 0 & -2 \\ 0 & 3 & 4 \\ 0 & 1 & -2 \end{vmatrix}$$

Interchange the second and third rows.

$$= +\begin{vmatrix} 1 & 0 & -2 \\ 0 & 1 & -2 \\ 0 & 3 & 4 \end{vmatrix}$$

Multiply the second row by -3 and add to the third row.

$$= \begin{vmatrix} 1 & 0 & -2 \\ 0 & 1 & -2 \\ 0 & 0 & 10 \end{vmatrix}$$

Expand by the first column.

$$= \begin{vmatrix} 1 & -2 \\ 0 & 10 \end{vmatrix} = 10$$

It is clear that any triangular determinant (zeros below the main diagonal) is just the product of the main diagonal elements.

General Properties of Determinants

Example 4 Evaluate

$$D = \begin{vmatrix} 3 & 0 & -1 & 4 \\ -6 & 5 & 2 & 2 \\ 0 & 7 & 0 & 9 \\ -15 & 6 & 5 & -8 \end{vmatrix}$$

Solution. Multiply the third column by 3, and add to the first column. This does not change the value, so

$$D = \begin{vmatrix} 0 & 0 & -1 & 4 \\ 0 & 5 & 2 & 2 \\ 0 & 7 & 0 & 9 \\ 0 & 6 & 5 & -8 \end{vmatrix}.$$

Expanding by the first column shows that $D = 0$.

The argument of Example 4 shows that any determinant in which two columns (or rows) are proportional must be equal to zero.

SECTION 9-6, REVIEW QUESTIONS

1. A 4 by 4 determinant is defined to be a linear combination of four _____ by _____ determinants.   *[3 by 3]*

2. To expand an n by n determinant, you multiply each element of some row (or column) by its _____, and add the results with the appropriate signs from the checkerboard pattern.   *[minor]*

3. The minor of an element is the determinant you get by deleting the _____ and the _____ containing that element.   *[row, column]*

4. The minor of an element in an n by n determinant is an _____ by _____ determinant.   *[n–1 by n–1]*

5. A system of n linear equations in n unknowns has a unique solution provided the determinant of the _____ is not zero.   *[coefficients]*

6. If a determinant has value 10 and you interchange two rows, the new determinant has value _____.   *[–10]*

7. If two rows of a determinant are proportional, the value of the determinant

382 *Systems of Equations*

is _____ .   **zero**

8. To simplify determinants by row operations, you add multiples of the first row to the others to get _____ in the first column.   **zeros**

9. If the first element of the first column is a_1, and all the other elements of the first column are zeros, then the value of the determinant is a_1 times its _____ .   **minor**

SECTION 9-6, EXERCISES

1. Find y in Example 1.   Example 1
2. Find z in Example 1.
3. Find w in Example 1.
4. In Example 1, expand the numerator for x by the third column.

Solve the following systems by using determinants.

5. $2x + z = -4$ 6. $x + y = 1$ Example 1,2,3
   $x + y + w = -3$ $2x - z + w = 5$
   $y + z - w = 0$ $y + 2z = -2$
   $x - 3y = 0$ $-x + y = -1$

7. $3x - 2z + 5w = 2$ 8. $x - 2y + 3z = -3$
   $-2x - y + 4w = 0$ $2x + z - 2w = 3$
   $x + 2y - 3z - 5w = 1$ $x + y - z - w = 4$
   $3y + 5z + 2w = -3$ $-x + 2y + w = 0$

Write an equal determinant in which the first *column* has entries 1, 0, 0, 0.

9. $\begin{vmatrix} 1 & 2 & 1 & -1 \\ 2 & 1 & 1 & 3 \\ -1 & -1 & 2 & 3 \\ 3 & 6 & 1 & 4 \end{vmatrix}$ 10. $\begin{vmatrix} 1 & -1 & 3 & 2 \\ 3 & 1 & 5 & 8 \\ -2 & -1 & 6 & 3 \\ 5 & 2 & 10 & 7 \end{vmatrix}$ Example 2

Write an equal determinant in which the first *row* has entries 1, 0, 0, 0.

11. $\begin{vmatrix} 1 & 2 & -1 & 4 \\ 2 & 1 & 0 & 3 \\ -1 & 2 & 4 & 1 \\ 1 & 1 & -2 & 3 \end{vmatrix}$ 12. $\begin{vmatrix} 1 & -1 & 3 & 5 \\ -2 & 1 & 0 & -4 \\ 1 & 3 & -2 & 1 \\ 4 & -2 & 1 & 3 \end{vmatrix}$

Use row operations to convert the determinant to one with zeros below the main diagonal, and evaluate it.

Example 3

13. $\begin{vmatrix} 1 & 2 & -1 \\ -1 & -1 & 3 \\ 3 & 5 & 2 \end{vmatrix}$

14. $\begin{vmatrix} 1 & -2 & 2 \\ 2 & -3 & -1 \\ -3 & 8 & -3 \end{vmatrix}$

Tell how each of the right hand determinants below can be obtained by a row or column operations on the left hand determinant.

15. $\begin{vmatrix} 1 & 1 & 0 \\ 1 & 1 & 2 \\ 2 & 4 & 2 \end{vmatrix} = 2 \begin{vmatrix} 1 & 1 & 0 \\ 1 & 1 & 2 \\ 1 & 2 & 1 \end{vmatrix}$

16. $\begin{vmatrix} 1 & 2 & -1 \\ 1 & 1 & 2 \\ 1 & 1 & 3 \end{vmatrix} = \begin{vmatrix} 1 & 2 & -1 \\ 1 & 1 & 2 \\ 2 & 2 & 5 \end{vmatrix}$

17. $\begin{vmatrix} 2 & 4 & 1 \\ 0 & -1 & 3 \\ 3 & 1 & -1 \end{vmatrix} = \begin{vmatrix} 2 & 0 & 3 \\ 4 & -1 & 1 \\ 1 & 3 & -1 \end{vmatrix}$

18. $\begin{vmatrix} 2 & 3 & 7 \\ 3 & 2 & -5 \\ 5 & 7 & -5 \end{vmatrix} = - \begin{vmatrix} 2 & 3 & 7 \\ 5 & 7 & -5 \\ 3 & 2 & -5 \end{vmatrix}$

19. $\begin{vmatrix} 4 & -1 & 1 \\ 3 & 2 & -4 \\ 7 & 1 & -3 \end{vmatrix} = \frac{1}{3} \begin{vmatrix} 4 & -1 & 3 \\ 3 & 2 & -12 \\ 7 & 1 & -9 \end{vmatrix}$

20. $\begin{vmatrix} 2 & -1 & 5 \\ 5 & -4 & 1 \\ 1 & 1 & 3 \end{vmatrix} = \begin{vmatrix} 1 & -1 & 5 \\ 1 & -4 & 1 \\ 2 & 1 & 3 \end{vmatrix}$

9-7 REVIEW FOR CHAPTER

Solve the following systems of equations.

1. $\begin{cases} x^2 + y^2 = 25 \\ x - y = 0 \end{cases}$
2. $\begin{cases} x^2 - 4x + 3 = 0 \\ x - y + 1 = 0 \end{cases}$
3. $\begin{cases} x^2 - y^2 = 4 \\ x + y = 1 \end{cases}$
4. $\begin{cases} 3x^2 - y^2 = 3 \\ 2y - x = 8 \end{cases}$
5. $\begin{cases} xy = -12 \\ x + 14 = 2y \end{cases}$
6. $\begin{cases} xy = 6 \\ 2x - y = 1 \end{cases}$
7. $\begin{cases} x^2 - y^2 = 4 \\ x^2 + y^2 = 16 \end{cases}$
8. $\begin{cases} x^2 - 7y = 2 \\ x^2 - y^2 = 12 \end{cases}$
9. $\begin{cases} x^2 - y^2 = 11 \\ 2x^2 - 5y^2 = 7 \end{cases}$
10. $\begin{cases} x^2 - y^2 = 13 \\ 2x^2 + y^2 = -1 \end{cases}$
11. $\begin{cases} y = \left(\dfrac{1}{2}\right)^x \\ x + 2y = 2 \end{cases}$
12. $\begin{cases} y = \left(\dfrac{1}{3}\right)^x \\ x + 3y = 3 \end{cases}$
13. $\begin{cases} y = e^{2x} \\ y = 2e^x + 1 \end{cases}$
14. $\begin{cases} y = e^{4x} \\ y = 4e^{2x} + 4 \end{cases}$

Solve each of the following systems of equations by any method.

15. $\begin{cases} 3x = 1 - 2y \\ \dfrac{9}{2}x - 6y = 3 \end{cases}$
16. $\begin{cases} y = \dfrac{1}{3}x + 5 \\ y = \dfrac{1}{3}x - 5 \end{cases}$
17. $\begin{cases} y = -\dfrac{1}{4}x + 2 \\ x + 4y + 2 = 0 \end{cases}$
18. $\begin{cases} \dfrac{3x + 1}{5} = \dfrac{3y + 2}{4} \\ \dfrac{2x - 1}{5} + \dfrac{3y - 2}{4} = 2 \end{cases}$
19. $\begin{cases} \dfrac{x}{3} + \dfrac{y}{6} = \dfrac{2}{3} \\ \dfrac{2x}{5} + \dfrac{y}{4} = \dfrac{1}{5} \end{cases}$
20. $\begin{cases} \dfrac{x - 5}{4} = \dfrac{6y}{8} \\ 2x - 3y = 5 \end{cases}$

21. $\begin{cases} -2x + y + 3z = 0 \\ -4x + 2y + 6z = 0 \end{cases}$
22. $\begin{cases} -10x + 4y - 5z = 20 \\ 2x - \dfrac{4}{5}y + z = 4 \end{cases}$

23. $\begin{cases} 20x - 20y - 30z = 0 \\ 15x - 10y - 25z = 0 \\ 10x - 20y - 10z = 0 \end{cases}$
24. $\begin{cases} x + 2y + z = 3 \\ 2x - y + 3z = 7 \\ 3x + y + 4z = 10 \end{cases}$

25. $\begin{cases} 3x + 5y + 2z = 0 \\ 12x - 15y + 4z = 12 \\ 6x - 25y - 8z = 8 \end{cases}$
26. $\begin{cases} 2x - y + 4z = 3 \\ 3x + 2y - 2z = -1 \\ x - 4y + 10z = 7 \end{cases}$

27. $\begin{cases} \dfrac{1}{x} + \dfrac{2}{y} - \dfrac{1}{z} = 5 \\ \dfrac{2}{x} - \dfrac{1}{y} + \dfrac{1}{z} = 1 \\ \dfrac{1}{z} - \dfrac{3}{y} - \dfrac{1}{x} + 7 = 0 \end{cases}$
28. $\begin{cases} \dfrac{2}{x} + \dfrac{3}{y} + \dfrac{4}{z} = 3 \\ \dfrac{1}{x} - \dfrac{2}{z} = 0 \\ \dfrac{6}{x} - \dfrac{6}{y} = 8 \end{cases}$

29. $\begin{cases} 3x - 4y + 10z + w = -6 \\ 2x - 3y - z - w = -22 \\ x + y + z + 2w = 2 \\ -x - y + 2z - 2w = 4 \end{cases}$
30. $\begin{cases} 5x + 3y + 2z - w = 2 \\ 2x - y + z + 2w = -3 \\ -2x + 2y - z + w = 1 \\ 3x + 4y - 2z + 3w = -15 \end{cases}$

Evaluate the following determinants.

31. $\begin{vmatrix} -1 & 3 & 0 & 4 \\ 2 & 5 & -3 & 1 \\ 4 & -1 & 0 & 2 \\ 3 & 13 & -2 & 6 \end{vmatrix}$
32. $\begin{vmatrix} 3 & 0 & -2 & 5 \\ -2 & -1 & 4 & 0 \\ 1 & 2 & -3 & -5 \\ 0 & 3 & 5 & 2 \end{vmatrix}$

33. A newsboy has three times as many dimes as quarters. If he has $7.15 in dimes and quarters, how many of each coin does he have?

34. If you had $3.05 worth of nickels and dimes, and if the number of dimes you had was three more than twice the number of nickels, how many nickels and how many dimes would you have?

10 COMPLEX NUMBERS

10-1 ADDITION AND MULTIPLICATION

The history of numbers reflects a continuing extension of one number system into a larger one. In this chapter we will consider the *complex number system*, which is the "most natural" system from many points of view. The simple assumption that there is a new number whose square is -1 leads to many new results of profound importance.

The first number system was simply the set of counting numbers, or *natural numbers*:

$$0, 1, 2, 3, 4, 5, 6, \ldots$$

This system had two operations, *addition* and *multiplication*, which were relevant to the counting process. For example, two pomegranates plus three pomegranates is five pomegranates; three baskets of fish with five fish in each basket makes fifteen fish altogether.

The inverse operations, *subtraction* and *division*, are not always possible within the system of natural numbers. For example, $7 - 3$ is a natural number, but $3 - 7$ is not. To extend the natural numbers to a system in which subtraction is always possible, it is necessary to consider all the integers:

$$\ldots, -4, -3, -2, -1, 0, 1, 2, 3, 4, 5, \ldots$$

To further extend the system so that division is always possible, it is necessary to include all *rational* numbers; that is, all numbers m/n where m and n are integers and $n \neq 0$.

The system of rational numbers is closed under all four operations: $+, -, \times, \div$. However, we cannot solve many simple polynomial equations within this system. Since $\sqrt{2}, \sqrt{3}, \sqrt{5}$, etc., are not rational numbers, the equations

$$x^2 = 2, \quad x^2 = 3, \quad x^2 = 5, \tag{1}$$

etc., do not have solutions within this system. It is therefore natural to extend the system from the rationals to the set of all *real* numbers. This system is again closed under the four operations, and does allow us to solve the equations (1) within the system.

However, there are still simple polynomial equations which cannot be solved within the system of real numbers. For example, the equations

$$x^2 = -1, \quad x^2 = -2, \quad x^2 + x + 1 = 0,$$

have no real solutions. We therefore ask whether the real numbers can be further extended to a system in which every polynomial equation does have a solution.

First, we invent a new number, i, with the property that $i^2 = -1$. This is certainly necessary if the equation $x^2 = -1$ is to have a solution. Next, we see what other numbers we must include if we want a number system which contains all the real numbers and i.

If multiplication, with its usual properties, is to extend to this new system, then the number bi, for every real number b, must be in the system. Moreover, it must be the case that

$$0 \cdot i = 0 \quad \text{and} \quad 1 \cdot i = i$$

since the properties $0 \cdot a = 0$ and $1 \cdot a = a$ are basic properties of multiplication within the real number system. If addition is also to extend, we must have $a + bi$ in the system for all real numbers a and b.

We call this extended set, consisting of all numbers $a + bi$, where a and b are real, the system of *complex numbers*. Each complex number z has a unique representation in the *standard form* $z = a + bi$. The number a is the *real part* of z, denoted $Re\ z$, and b is the *imaginary* part of z, denoted $Im\ z$. That is,

$$Re\ (a + bi) = a, \quad Im\ (a + bi) = b.$$

Since the representation of a complex number in standard form is unique, two complex numbers are equal if and only if their real parts are equal and their imaginary parts are equal. *Each equation involving complex numbers is therefore equivalent to* two *equations involving real numbers*.

In the rest of this section and Section 10-2 we will develop the elementary algebraic properties of complex numbers. In Section 10-4 we will see that every polynomial equation with complex numbers as coefficients does have at least one complex solution.

The addition operation extends naturally to the complex numbers by the definition

$$(a + bi) + (c + di) = (a + c) + (b + d)i.$$

Hence, to add two complex numbers, you add the real parts and add the imaginary parts. The negative (additive inverse) of $a + bi$ is clearly $-a - bi$, since

$$\begin{aligned}(a + bi) + (-a - bi) &= (a - a) + (b - b)i \\ &= 0 + 0i \\ &= 0 + 0 \\ &= 0.\end{aligned}$$

With this definition, all the addition axioms for real numbers are also satisfied for complex numbers. It follows that any algebraic property of the real numbers which involves only addition or subtraction is also a property of complex numbers.

Example 1 Write in standard form

$$(2 - 3i) - [(5 + i) - (6 - 4i)]$$

Solution. By the comment just above, the laws of signs must hold for complex numbers just as for real numbers. We eliminate parentheses and then add the real parts and the imaginary parts.

$$\begin{aligned}(2 - 3i) - [(5 + i) - (6 - 4i)] &= (2 - 3i) - (5 + i) + (6 - 4i) \\ &= 2 - 3i - 5 - i + 6 - 4i \\ &= 3 - 8i\end{aligned}$$

Multiplication also extends naturally from the real to the complex numbers. We simply use the usual rule for expanding a product of two sums, and replace i^2 by -1:

$$(a + bi)(c + di) = ac + bdi^2 + bci + adi$$
$$= (ac - bd) + (bc + ad)i.$$

It is easy to check that this is a commutative and associative operation, and that 1 is still a multiplicative identity.

Example 2 Write in standard form

$$(2 + 5i)(7 - i) + (3 + i)^2$$

Solution.

$$(2 + 5i)(7 - i) + (3 + i)^2 = [(14 + 5) + (35 - 2)i] + (9 - 1 + 6i)$$
$$= 19 + 33i + 8 + 6i$$
$$= 27 + 39i$$

To see what the multiplicative inverse of a complex number must be, we first consider a specific example. We ask what is the multiplicative inverse of $2 - 3i$. If we let $x + yi$ be this inverse, then we must have

$$(2 - 3i)(x + yi) = 1$$
$$(2x + 3y) + (-3x + 2y)i = 1$$

This equation is equivalent to the two real equations:

$$2x + 3y = 1$$
$$-3x + 2y = 0$$

Multiplying the first by 3 and the second by 2 gives

$$6x + 9y = 3$$
$$-6x + 4y = 0$$

Hence $13y = 3$, and $y = 3/13$. The second original equation gives $3x = 2y$, so $3x = 6/13$, and $x = 2/13$. The multiplicative inverse of $2 - 3i$ is therefore $2/13 + 3/13i$.

We check the result by multiplying.

$$(2 - 3i)\left(\frac{2}{13} + \frac{3}{13}i\right) = \left(\frac{4}{13} + \frac{9}{13}\right) + \left(-\frac{6}{13} + \frac{6}{13}\right)i$$

$$= 1 + 0i = 1$$

The same process as above shows that the inverse of any nonzero complex number $a + bi$ (where $a \neq 0$ or $b \neq 0$) is given by:

$$(a + bi)^{-1} = \frac{1}{a + bi} = \frac{a}{a^2 + b^2} - \frac{b}{a^2 + b^2}i.$$

That is, if we solve the equation $(a + bi)(x + yi) = 1$ for x and y, we get

$$x = \frac{a}{a^2 + b^2} \quad \text{and} \quad y = \frac{-b}{a^2 + b^2}.$$

We check the result above by multiplication as follows:

$$(a + bi)\left(\frac{a}{a^2 + b^2} - \frac{b}{a^2 + b^2}i\right) = \frac{1}{a^2 + b^2}(a + bi)(a - bi)$$

$$= \frac{1}{a^2 + b^2}[(a^2 + b^2) + (ab - ab)i]$$

$$= 1.$$

Division of complex numbers, like division of real numbers, is just multiplication by the multiplicative inverse. That is:

$$\frac{a + bi}{c + di} = (a + bi)(c + di)^{-1}$$

$$= (a + bi)\left(\frac{c}{c^2 + d^2} - \frac{d}{c^2 + d^2}i\right).$$

Example 3 Perform the indicated operations and write the result in standard form.

(a) $\dfrac{3 - 7i}{8}$ (b) $\dfrac{2 + 5i}{3 + i}$ (c) $\dfrac{5}{1 - 2i}$

Solution. (a) The inverse of the real number $8 = 8 + 0i$ is just $1/8 = 1/8 + 0i$. Hence

$$\frac{3-7i}{8} = (3-7i)\left(\frac{1}{8} + 0i\right) = \frac{3}{8} - \frac{7}{8}i.$$

(b) The inverse of $3 + i$ is

$$(3+i)^{-1} = \frac{3}{9+1} - \frac{1}{9+1}i = \frac{3}{10} - \frac{1}{10}i.$$

Therefore

$$\frac{2+5i}{3+i} = (2+5i)\left(\frac{3}{10} - \frac{1}{10}i\right)$$

$$= \left(\frac{6}{10} + \frac{5}{10}\right) + \left(\frac{15}{10} - \frac{2}{10}\right)i$$

$$= \frac{11}{10} + \frac{13}{10}i.$$

(c) Since

$$(1-2i)^{-1} = \frac{1}{5} + \frac{2}{5}i,$$

$$\frac{5}{1-2i} = 5\left(\frac{1}{5} + \frac{2}{5}i\right) = 1 + 2i.$$

Example 4 Simplify $(2 - 3i)/(1 + 2i)$ by multiplying numerator and denominator by $1 - 2i$.

Solution.

$$\frac{2-3i}{1+2i} = \frac{(2-3i)(1-2i)}{(1+2i)(1-2i)}$$

$$= \frac{(2-6) + (-3-4)i}{(1+4) + (2-2)i}$$

$$= \frac{-4-7i}{5}$$

$$= -\frac{4}{5} - \frac{7}{5}i.$$

SECTION 10-1, REVIEW QUESTIONS

1. To solve the equation $x^2 + 1 = 0$, we need a number system in which the square of some number is _____.

−1

2. The number whose square is -1 is denoted _____.   i

3. Since $r^2 \geq 0$ for every real number r, i is not equal to any _____ number.   real

4. Each complex number has a unique representation in the _____ form $a + bi$.   standard

5. If $a + bi = c + di$, then $a = $ _____ and $b = $ _____.   c   d

6. Any equality of complex numbers is the same as two _____ equations.   real

7. The real part of $a + bi$ is a and the imaginary part is _____.   b

8. $Re(5 + 7i) = $ _____ and $Im(5 + 7i) = $ _____.   5   7

9. The additive inverse of $a + bi$ is _____, and the multiplicative inverse of $a + bi$ is _____. $-a\ -bi$ $\dfrac{a}{a^2+b^2} - \dfrac{b}{a^2+b^2}i$

10. The product $(c + di)\left[\dfrac{a}{(a^2 + b^2)} - \dfrac{b}{(a^2 + b^2)}i\right]$ is the same as the quotient $(c + di)/($ _____ $)$. $a + bi$

SECTION 10-1, EXERCISES

Simplify and write in standard form each of the following.

1. $(1 + 4i) + (3 + 5i)$
2. $(-2 + 6i) + (2 - 6i)$ **Example 1**
3. $(-1 + 5i) + 2i$
4. $(5 + 3i) + (7 + 2i) + (3 - 4i)$

5. $(2 + 3i)(4 + 7i)$
6. $(2 - 3i)(6 + 4i)$ **Example 2**
7. $2i(-3i)(1 - 6i)$
8. $i(3 + 5i)$
9. $(4 - 3i)^2$
10. $(2 + 3i)(3 - 2i)(6 - 4i)$
11. $(x - yi)(x + yi)$
12. $(-2x + yi)(2x - yi)$

Find the additive inverse of each of the following numbers.

13. $5 - 4i$
14. $-4 - 3i$
15. $1 + i$
16. $a - bi$
17. $a + bi$
18. $-2x - 3yi$

Write the multiplicative inverse of each of the following complex numbers in standard form, and check by multiplication.

19. i
20. $-i$
21. $1 + i$
22. $2 + 3i$
23. $4 - 3i$
24. $4 + 7i$

Find the following quotients.

25. $3 \div 2i$
26. $1 \div (2 + i)$
27. $(7 + 6i) \div (3 - 4i)$ **Example 3,4**
28. $(1 + i) \div (2 - i)$
29. $(4 + 3i) \div (2 + 5i)$
30. $-5i \div (3 + 5i)$

10-2 THE ALGEBRA OF COMPLEX NUMBERS

With the definitions we have made, all the axioms for addition and multiplication for real numbers also extend to the complex numbers. The rules of algebra for real numbers are simply the consequences of these axioms. Thus it follows that any rule of algebra involving only addition and multiplication (and the inverse operations, subtraction and division) also holds for complex numbers. A few of these elementary properties are listed below. We use single letters z, w, u, v, etc., to stand for arbitrary complex numbers.

$$-(z + w) = -z - w \qquad -(z - w) = w - z$$

$$(-z)w = -(zw) \qquad (-z)(-w) = zw$$

$$\frac{1}{-z} = -\frac{1}{z} \qquad \frac{-z}{-w} = \frac{z}{w}$$

$$\frac{z}{w} + \frac{u}{w} = \frac{z + u}{w} \qquad \frac{z}{w} + \frac{u}{v} = \frac{zv + wu}{wv}$$

$$\frac{z}{w} \cdot \frac{u}{v} = \frac{zu}{wv} \qquad \frac{z}{w} \div \frac{u}{v} = \frac{z}{w} \cdot \frac{v}{u}$$

$$zw = 0 \text{ if and only if } z = 0 \text{ or } w = 0$$

$$\frac{z}{w} = \frac{u}{v} \text{ if and only if } zv = wu$$

Example 1 Show that the fractions are equal.

$$\frac{1 - 2i}{5 - 5i} \quad \text{and} \quad \frac{2 - 3i}{9 - 7i}$$

Solution. The fractions are equal provided

$$(1 - 2i)(9 - 7i) = (5 - 5i)(2 - 3i)$$

We evaluate both products:

$$(1 - 2i)(9 - 7i) = (9 - 14) + (-18 - 7)i = -5 - 25i$$
$$(5 - 5i)(2 - 3i) = (10 - 15) + (-10 - 15)i = -5 - 25i$$

Since the cross products are equal, the fractions are equal.

Example 2 Write in standard form

$$\frac{1 + i}{1 - i} \div \frac{2 + i}{1 + 2i}$$

Solution.

$$\frac{1+i}{1-i} \div \frac{2+i}{1+2i} = \frac{1+i}{1-i} \cdot \frac{1+2i}{2+i}$$

$$= \frac{(1+i)(1+2i)}{(1-i)(2+i)}$$

$$= \frac{-1+3i}{3-i}$$

$$= (-1+3i)\left(\frac{3}{10} + \frac{1}{10}i\right)$$

$$= \left(-\frac{3}{10} - \frac{3}{10}\right) + \left(\frac{9}{10} - \frac{1}{10}\right)i$$

$$= -\frac{3}{5} + \frac{4}{5}i$$

Since the standard algebraic rules hold for complex numbers, equations with complex coefficients can be treated in the usual way.

Example 3 Solve $(1 + 3i)z + (2 - 4i) = 7 - 3i$.

Solution. The following are all equivalent equations:

$$(1 + 3i)z + (2 - 4i) = 7 - 3i,$$

$$(1 + 3i)z = (7 - 3i) - (2 - 4i)$$

$$= 7 - 3i - 2 + 4i$$

$$= 5 + i,$$

$$z = \frac{5+i}{1+3i}$$

$$= (5+i)\left(\frac{1}{10} - \frac{3}{10}i\right)$$

$$= \left(\frac{5}{10} + \frac{3}{10}\right) + \left(\frac{1}{10} - \frac{15}{10}\right)i$$

$$= \frac{4}{5} - \frac{7}{5}i.$$

Although all the properties of addition and multiplication extend to the complex number system, *the properties of the order relation do not.* One of the basic properties of order is that $x^2 \geq 0$ for every real number x. Since $i^2 = -1$, it follows that the order relation $<$ can *not* be extended to the complex system. The inequality signs, $<$ and $>$, are *never* used between complex numbers with non-zero imaginary parts.

From the defining property of i, it is clear that every negative real number has two complex square roots. For example, $2i$ and $-2i$ are square roots of -4, $\sqrt{5}i$ and $-\sqrt{5}i$ are square roots of -5, etc. We now extend the use of the radical sign to indicate these complex roots for negative real numbers:

$$\sqrt{-A} = \sqrt{A}\,i \quad \text{if} \quad A \geq 0.$$

Note that $\sqrt{A}$ still stands for the *positive* root of A. Thus

$$\sqrt{-4} = 2i, \quad \sqrt{-25} = 5i, \quad \sqrt{-\frac{1}{8}} = \frac{1}{2\sqrt{2}}i = \frac{\sqrt{2}}{4}i.$$

The rules for combining radicals do *not* hold for negative radicands. For example,

$$\sqrt{-4}\,\sqrt{-9} = (2i)(3i) = 6i^2 = -6,$$

but

$$\sqrt{(-4)(-9)} = \sqrt{36} = 6.$$

That is, the rule $\sqrt{A}\,\sqrt{B} = \sqrt{AB}$ only holds if $A \geq 0$ and $B \geq 0$. To combine radical expressions with negative radicands, you must first express them in terms of i.

Example 4 Express in standard form $(3 + 2\sqrt{-5})(4 - \sqrt{-7})$.

Solution.

$$(3 + 2\sqrt{-5})(4 - \sqrt{-7}) = (3 + 2\sqrt{5}i)(4 - \sqrt{7}i)$$
$$= (12 + 2\sqrt{35}) + (8\sqrt{5} - 3\sqrt{7})i$$

With our definition of the square root of a negative number, the standard quadratic formula gives two complex roots for every quadratic equation with real coefficients.

Example 5 Solve $x^2 + 2x + 5 = 0$, and check the roots by substituting in the equation.

Solution. The quadratic formula is

$$x = \frac{-b \pm \sqrt{b^2 - 4ac}}{2a}$$

398 *Complex Numbers*

and here, $a = 1$, $b = 2$, and $c = 5$. Therefore

$$x = \frac{-2 \pm \sqrt{4 - 20}}{2}$$

$$= \frac{-2 \pm \sqrt{-16}}{2}$$

$$= \frac{-2 \pm 4i}{2}$$

$$= -1 \pm 2i.$$

That is, $-1 + 2i$ and $-1 - 2i$ are the roots. We check the root $-1 + 2i$ in the equation.

$$(-1 + 2i)^2 + 2(-1 + 2i) + 5 \stackrel{?}{=} 0$$

$$[(1 - 4) - 4i] - 2 + 4i + 5 \stackrel{?}{=} 0$$

$$-3 - 4i - 2 + 4i + 5 \stackrel{?}{=} 0$$

$$0 \stackrel{\checkmark}{=} 0$$

The root $-1 - 2i$ can be checked similarly.

In the complex number system, not only negative numbers, but indeed *all complex numbers have square roots.* Since $(-z)^2 = z^2$, each complex number has two square roots, which are negatives of each other. The next example shows how to find the square roots of a complex number.

Example 6 Find the square roots of $-3 + 4i$.

Solution. Let $x + yi$ be such a square root. Then

$$(x + yi)^2 = -3 + 4i,$$

$$x^2 - y^2 + (2xy)i = -3 + 4i.$$

Hence we have the two real equations

$$x^2 - y^2 = -3 \quad \text{and} \quad 2xy = 4.$$

We substitute $2/x$ for y in the first equation.

$$x^2 - \left(\frac{2}{x}\right)^2 = -3$$

$$x^2 - \frac{4}{x^2} + 3 = 0$$

$$x^4 + 3x^2 - 4 = 0$$

$$(x^2 + 4)(x^2 - 1) = 0$$

Since x and y are real, $x^2 + 4 > 0$ and $x = \pm 1$ are the only solutions. If $x = 1$, $y = 2/x = 2$; and if $x = -1$, $y = 2/x = -2$. Hence the two roots are $1 + 2i$ and $-1 - 2i$.

SECTION 10-2, REVIEW QUESTIONS

complex	1. The addition and multiplication axioms for real numbers are also valid for _____ numbers.
real	2. Although the addition and multiplication axioms extend to the complex numbers, the order relation, $<$, is only defined on _____ numbers.
axioms	3. Any property of addition or multiplication which holds for real numbers must also hold for the complex numbers, since these properties are all consequences of the _____ for addition and multiplication.
real	4. Equations with complex coefficients can be solved the same way as equations with _____ coefficients.
$\sqrt{A}\,i$	5. If A is a positive real number, we define $\sqrt{-A}$ to be _____.
positive	6. In the definition $\sqrt{-A} = \sqrt{A}\,i$, $\sqrt{A}$ still stands for the _____ square root of A.
negative	7. The rule $\sqrt{a}\,\sqrt{b} = \sqrt{ab}$ does not hold if a and b are _____.
i	8. To combine radical expressions with negative radicands, you must first write them in terms of _____.
two	9. Zero has one square root, and every other complex number has _____ complex square roots.
real	10. To solve an equation like $z^2 = a$, where z and a are complex numbers, you solve two _____ equations.

SECTION 10-2, EXERCISES

Check whether or not the following fractions are equal.

1. $\dfrac{1+i}{1-i}, \dfrac{3+1}{1-3i}$
2. $\dfrac{2+i}{1-3i}, \dfrac{1+3i}{4-2i}$ — Example 1
3. $\dfrac{2+2i}{3-2i}, \dfrac{4+8i}{5+i}$
4. $\dfrac{-2+5i}{1+4i}, \dfrac{3+2i}{5-i}$

Write in standard form.

5. $\left(\dfrac{1+i}{2+i}\right)\left(\dfrac{1-2i}{3-i}\right)$
6. $\left(\dfrac{-1+2i}{-2-i}\right) \cdot \left(\dfrac{1+i}{1+3i}\right)$ — Example 2
7. $\left(\dfrac{2+3i}{-1+4i}\right) \div \left(\dfrac{1-i}{1+3i}\right)$
8. $\left(\dfrac{1-3i}{-1+i}\right) \div \left(\dfrac{4-i}{2-3i}\right)$
9. $\left(\dfrac{4+2i}{1+2i}\right) \div \left(\dfrac{2-4i}{2-i}\right)$
10. $\left(\dfrac{6+3i}{2+i}\right) \div \left(\dfrac{6-3i}{2i}\right)$
11. $\left(\dfrac{6-12i}{3i}\right) \div \left(\dfrac{2}{5i}\right)$
12. $\left(\dfrac{2i}{1+i}\right) \div \left(\dfrac{i}{3}\right)$

Solve the following equations for z.

13. $2z + (2+i) = 4 - 3i$
14. $3z - 2i = 3 + 4i$ — Example 3
15. $2iz + (1+i) = 1 - i$
16. $iz + (3-i) = 1 + 2i$
17. $(2+i)z + (1-i) = 3 + 2i$
18. $(1-2i)z + (3-4i) = 4 - 7i$

Express the following products in standard form.

19. $5(2 + \sqrt{-1})$
20. $3(\sqrt{-4} + 5)$ — Example 4
21. $(2 - \sqrt{-9})(-4 + \sqrt{-1})$
22. $(1 - \sqrt{-4})(2 + \sqrt{-25})$
23. $(1 + \sqrt{-5})(2 - \sqrt{-3})$
24. $(2 - \sqrt{-7})(1 + \sqrt{-8})$

Solve the following quadratic equations and check one root by substituting in the equation.

25. $x^2 - x - 2 = 0$
26. $x^2 - 5x + 6 = 0$ — Example 5
27. $x^2 - 2x + 2 = 0$
28. $x^2 + 2x + 2 = 0$
29. $x^2 - x + 1 = 0$
30. $x^2 + 2x + 3 = 0$
31. $2x^2 - x + 1 = 0$
32. $3x^2 + 3x + 1 = 0$

Find the square roots of the following.

33. $2i$
34. $-8i$
35. $8 - 6i$ — Example 6
36. $-5 + 12i$
37. $-3 - 4i$
38. $24 + 10i$

The Algebra of Complex Numbers

10-3 CONJUGATE AND MODULUS

Each complex number $z = x + yi$ is uniquely determined by the ordered pair of real numbers (x, y). Therefore we can identify the complex numbers with the points of the plane, letting $x + yi$ correspond to the point with coordinates (x, y). We will then speak interchangeably of "the number z" and "the point z". The complex number zero corresponds to the origin. The points of the x axis, also called the *real axis*, correspond to the real numbers $x + 0i$. The y axis, also called the *imaginary axis*, consists of the pure imaginary numbers $0 + yi$. With this identification, we speak of the plane as the z plane, or the *complex plane*.

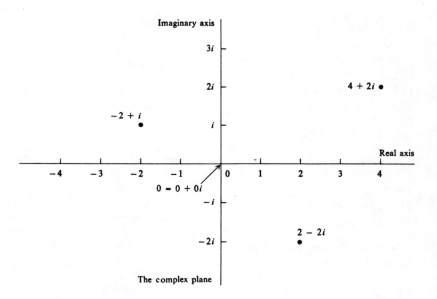

Figure 10-1

Addition of two complex numbers has a simple geometric interpretation in the complex plane. The point $z_1 + z_2$ is the fourth vertex of the parallelogram with $z_1, 0,$ and z_2 as contiguous vertices (Fig. 10-2).

The *absolute value* of a real number is its distance from the origin. We use the same notation for complex numbers, so that

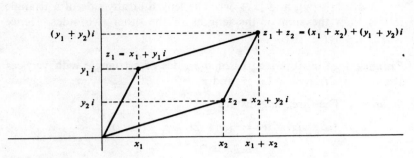

Figure 10-2

$$|x + yi| = \sqrt{x^2 + y^2}. \tag{1}$$

The absolute value of a complex number is also called its *modulus*. As is the case for real numbers, the distance between two complex numbers is the absolute value of their difference: If $z_1 = x_1 + yi_1$ and $z_2 = x_2 + yi_2$, then

$$|z_1 - z_2| = |(x_1 - x_2) + i(y_1 - y_2)|$$
$$= \sqrt{(x_1 - x_2)^2 + (y_1 - y_2)^2}.$$

As for real numbers, $|-z| = |z|$, so that $|z_1 - z_2| = |z_2 - z_1|$.

The following inequality, called the *triangle inequality*, is of basic importance in more advanced courses:

$$|z + w| \leq |z| + |w|.$$

The origin of the name "triangle inequality" can be seen in Fig. 10-3. The lengths of the sides of the triangle with vertices at 0, z, and

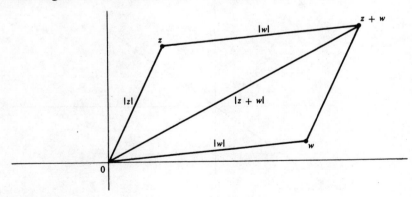

Figure 10-3

$z + w$ are $|z|$, $|w|$, and $|z + w|$. The length of any side of a triangle is less than the sum of the lengths of the other two sides. Hence $|z + w| \leq |z| + |w|$.

Example 1 Find the lengths of the sides of a triangle with vertices at $z = 3 + 2i$, $w = -1 - i$, and $u = 2 - i$.

Solution. The three lengths are

$$|z - w| = |(3 + 2i) - (-1 - i)|$$
$$= |4 + 3i|$$
$$= \sqrt{25} = 5,$$
$$|z - u| = |(3 + 2i) - (2 - i)|$$
$$= |1 + 3i|$$
$$= \sqrt{10},$$
$$|w - u| = |(-1 - i) - (2 - i)|$$
$$= |-3| = 3.$$

Notice that $5 + \sqrt{10} > 3$, $5 + 3 > \sqrt{10}$, and $3 + \sqrt{10} > 5$.

The *conjugate* of the complex number $z = x + yi$ is the number $x - yi$, which is denoted $\bar{z}$. Thus if $z = x + yi$,

$$\bar{z} = x - yi. \qquad (2)$$

From Fig. 10-4 we see that z and $\bar{z}$ are symmetric about the real axis. It is immediate from the definition (2) that

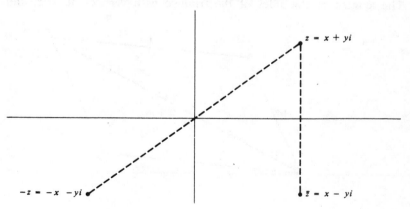

Figure 10-4

$$\bar{\bar{z}} = z \quad \text{for all } z,$$

and

$$z = \bar{z} \quad \text{if and only if } z \text{ is real}.$$

The importance of conjugation depends on the fact that it preserves sums and products. That is, the conjugate of a sum is the sum of the conjugates, and the conjugate of a product is the product of the conjugates. Similar statements hold for subtraction and division.

$$\overline{z_1 + z_2} = \bar{z}_1 + \bar{z}_2 \quad (3) \qquad \overline{z_1 - z_2} = \bar{z}_1 - \bar{z}_2 \quad (3')$$

$$\overline{z_1 z_2} = \bar{z}_1 \bar{z}_2 \quad (4) \qquad \overline{(z_1/z_2)} = \bar{z}_1/\bar{z}_2 \quad (4')$$

We will verify (4) and show how (4') follows from it. Let $z_1 = x_1 + y_1 i$ and $z_2 = x_2 + y_2 i$, then

$$z_1 z_2 = (x_1 + y_1 i)(x_2 + y_2 i) = (x_1 x_2 - y_1 y_2) + (x_1 y_2 + x_2 y_1)i \quad (5)$$

$$\bar{z}_1 \bar{z}_2 = (x_1 - y_1 i)(x_2 - y_2 i) = (x_1 x_2 - y_1 y_2) - (x_1 y_2 + x_2 y_1)i \quad (6)$$

From (5) and (6) we see that $\overline{z_1 z_2} = \bar{z}_1 \bar{z}_2$. Now let $z_1/z_2 = z_3$, so that

$$z_1 = z_2 z_3.$$

Then we have from (4) that

$$\bar{z}_1 = \overline{z_2 z_3} = \bar{z}_2 \bar{z}_3.$$

Therefore
$$\bar{z}_1/\bar{z}_2 = \bar{z}_3 = \overline{(z_1/z_2)}.$$

The verification of (3) is simple and (3') follows from (3) like (4') follows from (4) (problem 29).

Example 2 Compute both $\overline{(z/w)}$ and $\bar{z}/\bar{w}$ for $z = -3 + i$ and $w = 1 + 2i$.

Solution.
$$\frac{z}{w} = \frac{-3+i}{1+2i}$$
$$= (-3+i)\left(\frac{1}{5} - \frac{2}{5}i\right)$$
$$= \left(-\frac{3}{5} + \frac{2}{5}\right) + \left(\frac{1}{5} + \frac{6}{5}\right)i$$
$$= -\frac{1}{5} + \frac{7}{5}i$$

Hence
$$\overline{(z/w)} = -\frac{1}{5} - \frac{7}{5}i$$

Computing $\bar{z}/\bar{w}$, we get
$$\frac{\bar{z}}{\bar{w}} = \frac{-3-i}{1-2i}$$
$$= (-3-i)\left(\frac{1}{5} + \frac{2}{5}i\right)$$
$$= \left(-\frac{3}{5} + \frac{2}{5}\right) + \left(-\frac{1}{5} - \frac{6}{5}\right)i$$
$$= -\frac{1}{5} - \frac{7}{5}i.$$

One of the useful properties of conjugation is the fact that $z\bar{z}$ is a *real* number:

$$z\bar{z} = (x+iy)(x-iy) = x^2 - (iy)^2 = x^2 + y^2 = |z|^2. \qquad (7)$$

Since it is easy to divide a complex number by a real number, we can effectively perform a division of complex numbers by multiplying numerator and denominator by the conjugate of the denominator:

$$\frac{z}{w} = \frac{z\bar{w}}{w\bar{w}} = \frac{z\bar{w}}{|w|^2}.$$

Example 3 Write in standard form $(3 - 2i)/(5 + 3i)$.

Solution. We proceed by multiplying numerator and denominator by the conjugate $\overline{5 + 3i} = 5 - 3i$.

$$\frac{3 - 2i}{5 + 3i} = \frac{(3 - 2i)(5 - 3i)}{(5 + 3i)(5 - 3i)}$$

$$= \frac{(15 - 6) + (-10 - 9)i}{25 + 9}$$

$$= \frac{9}{34} - \frac{19}{34}i$$

From (7) we can easily show that the modulus of a product (or quotient) is the product (or quotient) of the moduli. That is,

$$|z_1 z_2| = |z_1||z_2|, \qquad (8)$$

and

$$\left|\frac{z_1}{z_2}\right| = \frac{|z_1|}{|z_2|}. \qquad (9)$$

To verify (8), we write

$$|z_1 z_2|^2 = (z_1 z_2)(\overline{z_1 z_2})$$

$$= (z_1 z_2)(\bar{z}_1 \bar{z}_2)$$

$$= (z_1 \bar{z}_1)(z_2 \bar{z}_2)$$

$$= |z_1|^2 |z_2|^2. \qquad (10)$$

Since $|z_1|$, $|z_2|$, and $|z_1 z_2|$ are nonnegative, (8) follows from (10) by taking square roots. The quotient rule (9) can be proved similarly.

Example 4 Find the modulus of

$$\frac{(3 - 2i)(7 + i)}{(5 + i)}$$

Solution. From (9) and (10) we have

$$\left|\frac{(3 - 2i)(7 + i)}{(5 + i)}\right| = \frac{|3 - 2i||7 + i|}{|5 + i|}$$

$$= \frac{\sqrt{13}\sqrt{50}}{\sqrt{26}}$$

$$= \frac{\sqrt{13} \cdot 5\sqrt{2}}{\sqrt{26}}$$
$$= 5.$$

SECTION 10-3, REVIEW QUESTIONS

1. We identify the complex number $x + yi$ with the point with coordinates _____.

    (x, y)

2. The x axis is also called the _____ axis, and the y axis is also called the _____ axis.

    real
    imaginary

3. The real axis consists of the real numbers $x + 0i$, and the imaginary axis consists of the pure _____ numbers $0 + yi$.

    imaginary

4. The absolute value of z is its distance from the _____.

    origin

5. The distance from z to the origin, 0, is $|z|$, and the distance between two complex numbers z and w is _____.

    |z − w|

6. The inequality $|z + w| \leq |z| + |w|$ is called the _____ inequality.

    triangle

7. The conjugate of z is denoted _____.

    $\bar{z}$

8. The points z and $\bar{z}$ are symmetric about the _____ axis.

    real

9. The equality $z = \bar{z}$ holds only if z is a _____ number.

    real

10. Conjugation preserves addition, subtraction, multiplication, and division, so that $\overline{z + w} =$ _____, $\overline{z - w} =$ _____, $\overline{zw} =$ _____, and $\overline{(z/w)} =$ _____.

    $\bar{z} + \bar{w}$ $\bar{z} - \bar{w}$
    $\bar{z}\,\bar{w}$ $\bar{z}/\bar{w}$

11. Division by a complex number z is accomplished easily by multiplying numerator and denominator by _____.

    $\bar{z}$

12. For any complex number z, $z\bar{z}$ is the real number _____.

    $|z|^2$

13. $|zw| =$ _____ and $|z/w| =$ _____.

    |z| |w| |z|/|w|

SECTION 10-3, EXERCISES

Find the following absolute values.

1. $|3 - 4i|$
2. $|1 - 2i|$

408 *Complex Numbers*

3. $|-5 + i|$
4. $|7 - 3i|$
5. $|-5 - 3i|$
6. $|5 + 12i|$

Find the distance between the two given complex numbers.

7. $-3 + i, 2 - 4i$
8. $1 + 4i, -2 + i$ Example 1
9. $3 - 5i, 4 - 3i$
10. $-1 + 7i, 1 + 4i$
11. $2 - 5i, 5$
12. $1 + 4i, 6i$

Write in standard form.

13. $\dfrac{-4 + 3i}{2 + i}$
14. $\dfrac{4 + 2i}{3 - i}$ Example 3

15. $\dfrac{8 + i}{2 - i}$
16. $\dfrac{-5 + 14i}{-4 + i}$

17. $\dfrac{3 - 4i}{5 + 2i}$
18. $\dfrac{-2 + 5i}{-4 - i}$

For each of the following, compute $\overline{zw}$, $\bar{z}\,\bar{w}$, $\overline{(z/w)}$, and $\bar{z}/\bar{w}$.

19. $z = 2 + i, w = 1 + i$
20. $z = 1 - 3i, w = 2 + i$ Example 2
21. $z = -3 + i, w = 4 - i$
22. $z = 2 - 2i, w = -1 - i$

Find the modulus of each of the following.

23. $\dfrac{(1 - i)(2 + 3i)}{4 + 2i}$
24. $\dfrac{(1 + 5i)(5 - 2i)}{2 - 3i}$ Example 4

25. $\dfrac{(3 - 2i)(5 - i)}{-2 + 2i}$
26. $\dfrac{(4 + 3i)(1 + i)}{-2 + i}$

27. $\dfrac{5(4 - i)}{3 - i}$
28. $\dfrac{(3 + i)}{(6 - 2i)(1 - 3i)}$

29. Verify that $\overline{z_1 + z_2} = \bar{z}_1 + \bar{z}_2$, and show how it follows from this that $\overline{z_1 - z_2} = \bar{z}_1 - \bar{z}_2$. (cf. the verification of (4) and (4′).)

10-4 COMPLEX ROOTS OF POLYNOMIAL EQUATIONS

We remarked in an earlier chapter that every polynomial can be factored into linear factors and irreducible (unfactorable) quadratic factors. At that time we considered only real numbers, and hence only polynomials with real coefficients. Now let us consider polynomials of the form

$$P(x) = a_n x^n + a_{n-1} x^{n-1} + \cdots + a_1 x + a_0 \qquad (1)$$

Complex Roots of Polynomial Equations 409

where $a_0, a_1, \ldots, a_n$ are *complex* numbers, and x is allowed to take on complex values. If $a_n \neq 0$, the polynomial (1) is of *degree n*.

A polynomial with coefficients in a certain number system is called *irreducible* if it cannot be written as a product of two polynomials of smaller degree with coefficients in the given system. Hence the concept of irreducibility depends on what number system you consider. For example, the quadratic polynomial

$$x^2 - 3$$

cannot be factored into polynomials with rational coefficients, but can be factored into linear polynomials with *real* coefficients:

$$x^2 - 3 = (x - \sqrt{3})(x + \sqrt{3}).$$

Similarly, the polynomial $x^2 + 1$ cannot be factored into factors with real coefficients, but can be factored if we allow complex coefficients:

$$x^2 + 1 = (x + i)(x - i).$$

Hence the polynomial $x^2 + 1$ is irreducible with respect to the real numbers, but not with respect to the complex numbers. In fact, no non-constant polynomial is irreducible with respect to the complex numbers! This follows from the following fact, called the *fundamental theorem of algebra*:

Every complex polynomial of degree 1 or greater has a complex zero.

That is, if P is the polynomial (1), and $n \geq 1$, then there is some complex number z_1 such that $P(z_1) = 0$.

The fundamental theorem of algebra is just what its name implies— one of the most basic and beautiful theorems in all of mathematics. Our initial assumption was simply to add to our number system a solution of the equation $x^2 = -1$. As a consequence, we obtain the far-reaching result that *every* polynomial equation has a solution. The proof of the fundamental theorem of algebra is by no means simple, and all known proofs depend on deep results from complex function theory.

The division algorithm holds for complex polynomials just as for real polynomials. That is, given polynomials $A(x)$ and $B(x)$, there are unique polynomials $Q(x)$ and $R(x)$ such that the degree of $R(x)$ is less than the degree of $B(x)$, and

$$A(x) = B(x)Q(x) + R(x). \qquad (2)$$

$Q(x)$ is called the *quotient* and $R(x)$ is called the *remainder*. If $B(x)$

is a linear polynomial $x - z_1$, then $R(x)$ must be a constant c (a zero degree polynomial), and (2) becomes

$$A(x) = (x - z_1)Q(x) + c. \qquad (3)$$

From (3) we see, just as in the case of real polynomials, that $A(z_1) = 0$ if and only if $c = 0$; that is,

$A(z_1) = 0$ if and only if $x - z_1$ divides $A(x)$.

Now we combine (4) with the fundamental theorem of algebra to show that every complex polynomial of degree n can be factored into n complex linear factors. Let $n \geq 1$, and

$$P(x) = a_n x^n + \cdots + a_0.$$

From the fundamental theorem, we know that $P(z_1) = 0$ for some complex number z_1, and hence that $x - z_1$ divides $P(x)$:

$$P(x) = (x - z_1)Q_1(x).$$

If $Q_1(x)$ is a constant, then $P(x)$ has degree one (i.e., $n = 1$), and we are done. Otherwise, $Q_1(x)$ has some zero z_2, and $x - z_2$ divides $Q_1(x)$, so

$$P(x) = (x - z_1)Q_1(x)$$
$$= (x - z_1)(x - z_2)Q_2(x).$$

We continue in this way until we arrive at a quotient $Q_n(x)$ of degree zero ($Q_n(x) = c$) and get

$$P(z) = (z - z_1)(z - z_2) \cdots (z - z_n)c.$$

It is easy to see that c is the coefficient of z^n, so $c = a_n$, and

$$P(x) = a_n(x - z_1) \cdots (x - z_n).$$

If the numbers $z_1, \ldots, z_n$ are distinct, the equation $P(x) = 0$ has exactly n complex roots. In any case, an nth degree polynomial equation has at least one complex root, and at most n roots.

The results above of course also hold for polynomials with real coefficients. That is, a real nth degree polynomial can always be factored into n linear factors, some or all of which may have complex coefficients.

Although the factorization of a polynomial is always possible in theory, it may be impossible to find the exact factors for a particular example. For instance, there is no known technique which will allow us to express exactly the linear factors of $x^5 + 3x^2 - 3$. There are, however, advanced techniques that can be used to approximate the coefficients in the linear factors to any desired accuracy.

Example 1 Factor into complex linear factors $4x^2 - 4x + 5$.

Solution. We find the roots of the equation $4x^2 - 4x + 5 = 0$, which will give us the linear factors.

$$x = \frac{4 \pm \sqrt{16 - 4 \cdot 4 \cdot 5}}{8}$$

$$= \frac{4 \pm \sqrt{16 - 16 \cdot 5}}{8}$$

$$= \frac{4 \pm 4 \cdot 2i}{8}$$

$$= \frac{1}{2} \pm i$$

Therefore $[x - (1/2 + i)]$ and $[x - (1/2 - i)]$ are linear factors of $4x^2 - 4x + 5$. The leading coefficient is 4, so

$$4x^2 - 4x + 5 = 4\left[x - \left(\frac{1}{2} + i\right)\right]\left[x - \left(\frac{1}{2} - i\right)\right]$$

$$= [2x - (1 + 2i)][2x - (1 - 2i)].$$

If the roots we get for the quadratic equation $P(x) = 0$ turn out to be real numbers, then of course the "complex" linear factors of $P(x)$ are in fact real linear factors.

If a real quadratic polynomial has complex roots, then they are always conjugates of each other, as we can see from the quadratic formula. In general, if the complex number z_1 is a root of any *real* polynomial equation, then so is its conjugate. To show this, let

$$P(x) = a_0 + a_1 x + \cdots + a_n x^n$$

where $a_0, a_1, \ldots, a_n$ are real numbers so that $\bar{a}_0 = a_0$, $\bar{a}_1 = a_1$, etc.

Then

$$\overline{P(x)} = \overline{a_0 + a_1 x + \cdots + a_n x^n}$$

$$= \overline{a_0} + \overline{a_1 x} + \cdots + \overline{a_n x^n}$$

$$= \bar{a}_0 + \bar{a}_1 \bar{x} + \cdots + \bar{a}_n(\bar{x})^n$$
$$= a_0 + a_1 \bar{x} + \cdots + a_n(\bar{x})^n$$
$$= P(\bar{x}).$$

This shows that if $P(x)$ is a polynomial with real coefficients, and x is any complex number, then $P(\bar{x}) = \overline{P(x)}$. In particular, if $P(z_1) = 0$, then $P(\bar{z}_1) = \overline{P(z_1)} = \bar{0} = 0$. That is, if z_1 is a root of the equation $P(x) = 0$, then so is $\bar{z}_1$, and the roots of a real polynomial equation occur in conjugate pairs.

Example 2 Find a real quadratic polynomial $P(x)$ with the zero $5 + 2i$.

Solution. If $5 + 2i$ is a zero of the real polynomial $P(x)$, then $\overline{5 + 2i} = 5 - 2i$ must be the other zero. Hence $[x - (5 + 2i)]$ and $[x - (5 - 2i)]$ are the linear factors of $P(x)$, and

$$P(x) = [x - (5 + 2i)][x - (5 - 2i)]$$
$$= x^2 - 10x + 29.$$

Notice that the *only* possible real polynomials with $5 + 2i$ as one zero are multiples of $x^2 - 10x + 29$.

Example 3 Find a real polynomial of smallest degree which has the zeros 3, i, and $1 - i$. Write it as a product of real linear and quadratic factors which are irreducible over the reals.

Solution. Since the polynomial is to have real coefficients, the conjugates $\bar{i} = -i$ and $\overline{1-i} = 1+i$ must also be roots. Therefore the polynomial must have the factors $x - 3$, $x - i$, $x + i$, $x - (1 - i)$, and $x - (1 + i)$. Hence

$$P(x) = (x - 3)(x - i)(x + i)(x - (1 - i))(x - (1 + i))$$
$$= (x - 3)(x^2 + 1)(x^2 - 2x + 2).$$

The polynomial above is the only real fifth degree polynomial with the given zeros, except for constant multiples, $cP(x)$. If we specify that the leading coefficient is one, then the polynomial is unique.

Example 4 Solve the equation

$$P(x) = x^4 - 4x^3 + 9x^2 - 16x + 20 = 0$$

given that $2 + i$ is one root, and factor $P(x)$ into linear factors.

Solution. Since $2 + i$ is a root, and the coefficients are real, $2 - i$ must also be a root. Therefore $x - (2 + i)$ and $x - (2 - i)$ are factors, and their product must divide $P(x)$:

$$[x - (2 + i)][x - (2 - i)] = x^2 - 4x + 5 .$$

Dividing $P(x)$ by $x^2 - 4x + 5$, we get

$$
\begin{array}{r}
x^2 + 4 \\
x^2 - 4x + 5 \overline{\smash{)}x^4 - 4x^3 + 9x^2 - 16x + 20} \\
\underline{x^4 - 4x^3 + 5x^2} \\
4x^2 - 16x + 20 \\
\underline{4x^2 - 16x + 20}
\end{array}
$$

The roots of $x^2 + 4 = 0$ are $2i$ and $-2i$. Hence the four roots of $P(x) = 0$ are $2 + i$, $2 - i$, $2i$, and $-2i$, and the factorization is

$$P(x) = [x - (2 + i)][x - (2 - i)][x - 2i][x + 2i] .$$

SECTION 10-4, REVIEW QUESTIONS

n   1. If $a_n \neq 0$, the polynomial $a_n x^n + \cdots + a_1 x + a_0$ is of degree _____.

2. The fundamental theorem of algebra says that every non-constant poly-

zero   nomial has at least one complex _____.

0   3. A number z_1 is called a zero of polynomial P provided $P(z_1) =$ _____.

$x - z_1$ 4. For any polynomial $P(x)$, $P(z_1) = 0$ if and only if _____ divides $P(x)$.

linear   5. Every nth degree polynomial can be factored into n _____ factors with complex coefficients.

6. An nth degree polynomial equation has at least one root, and at most

n   _____ distinct roots.

7. If z_1 is a root of the polynomial equation $P(x) = 0$, where P has real co-

$\bar{z}_1$ efficients, then _____ is also a root.

8. The complex roots of a real polynomial equation occur in pairs which are

conjugates   _____ of each other.

SECTION 10-4, EXERCISES

Factor into complex linear factors.

1. $x^2 + 9$
2. $x^2 + 100$
3. $x^2 - 2x + 5$
4. $x^2 - 4x + 5$
5. $x^2 - 4x - 4$
6. $x^2 - 2x - 1$
7. $2x^2 - 12x + 20$
8. $2x^2 - 2x + 1$

Example 1

Write the real quadratic equation which has the given complex root, if the coefficient of x^2 is one.

9. $3 + i$
10. $2 + 3i$
11. $4 - i$
12. $3 + 4i$
13. $-2 + 2i$
14. $-5 - i$
15. $-\frac{1}{2} - \frac{\sqrt{3}}{2}i$
16. $\frac{1}{2} + \frac{\sqrt{2}}{2}i$

Example 2

Find the real polynomial of smallest degree, with leading coefficient one, which has the following zeros. Write the polynomial as a product of real factors which are irreducible over the reals.

17. $1, 2i, 1 + i$
18. $3, -4i, 1 - 2i$
19. $-2, 7, -3 + 4i$
20. $5, -4, -7 + 2i$
21. $3 - 5i, -2 + 4i$
22. $-5 + i, 2 + 3i$

Example 3

Solve the following equations, given the one indicated root.

23. $x^3 - x^2 + x - 1 = 0, x = 1$
24. $x^3 + 2x^2 + 9x + 18 = 0, x = -2$
25. $x^4 - 2x^3 + 3x^2 - 2x + 2 = 0, x = 1 + i$
26. $x^4 - 6x^3 + 14x^2 - 24x + 4 = 0, x = 3 - i$
27. $x^4 - 8x^3 + 24x^2 - 32x + 20 = 0, x = 1 - i$
28. $x^4 - 12x^3 + 57x^2 - 132x + 136 = 0, x = 4 + i$

Example 4

10-5 POLAR FORM OF COMPLEX NUMBERS

The polar representation for points in the plane has a useful application to complex numbers. Let r and θ be the polar coordinates of the point corresponding to the complex number z. That is, r is the distance from z to the origin, and θ is the angle between the positive real axis and the ray from 0 to z (Fig. 10-5). The distance r is the *modulus* of z, $r = |z|$, and the angle θ is called the *argument* of z. These terms are frequently abbreviated mod z, and arg z:

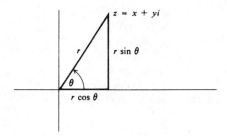

Figure 10-5

$$r = |z| = \operatorname{mod} z,$$
$$\theta = \arg z.$$

For example, mod $(1 + i) = \sqrt{2}$ and arg $(1 + i) = \pi/4$; mod $(-3) = 3$ and arg $(-3) = \pi$; mod $(1 - \sqrt{3}i) = 2$ and arg $(1 - \sqrt{3}i) = -\pi/3$; mod $7 = 7$ and arg $7 = 0$.

From Fig. 10-5, we see that if $r = \operatorname{mod} z$ and $\theta = \arg z$, then we have the following *trigonometric representation*, or *polar representation*:

$$z = r \cos \theta + ir \sin \theta$$
$$= r(\cos \theta + i \sin \theta). \qquad (1)$$

In particular, a complex number lies on the unit circle ($r = 1$) if and only if it has the polar form $z = \cos \theta + i \sin \theta$.

Any complex number z can be written in polar form simply by factoring out $|z|$ from its standard representation: If $z = x + yi$, then

$$z = \sqrt{x^2 + y^2}\left(\frac{x}{\sqrt{x^2 + y^2}} + i\frac{y}{\sqrt{x^2 + y^2}}\right).$$

Here
$$r = \sqrt{x^2 + y^2},$$
$$\cos \theta = \frac{x}{\sqrt{x^2 + y^2}}, \qquad (2)$$
$$\sin \theta = \frac{y}{\sqrt{x^2 + y^2}}.$$

Any two numbers whose squares add up to 1 are the cosine and sine of some angle, so the last two equations of (2) determine θ, or arg z.

Example 1 Find mod z and arg z for each of the following:

(a) $z = -2 - 2i$ (b) $z = \sqrt{3} + i$ (c) $z = 2 - 2\sqrt{3}i$

Solution.

(a) $|-2 - 2i| = \sqrt{8} = 2\sqrt{2} = \text{mod}(-2 - 2i)$

$$-2 - 2i = 2\sqrt{2}\left(\frac{-2}{2\sqrt{2}} - \frac{2}{2\sqrt{2}}i\right) = 2\sqrt{2}\left(\frac{-1}{\sqrt{2}} - \frac{1}{\sqrt{2}}i\right)$$

Hence $\cos\theta = -1/\sqrt{2}$, $\sin\theta = -1/\sqrt{2}$, and
$\theta = -3\pi/4 = \arg(-1 - 2i)$.

(b) $|\sqrt{3} + i| = \sqrt{4} = 2 = \text{mod}(\sqrt{3} + i)$

$$\sqrt{3} + i = 2\left(\frac{\sqrt{3}}{2} + \frac{1}{2}i\right)$$

Hence $\cos\theta = \sqrt{3}/2$, $\sin\theta = 1/2$, and $\theta = \pi/6 = \arg(\sqrt{3} + i)$.

(c) $|2 - 2\sqrt{3}i| = \sqrt{16} = 4 = \text{mod}(2 - 2\sqrt{3}i)$

$$2 - 2\sqrt{3}i = 4\left(\frac{2}{4} - \frac{2\sqrt{3}}{4}i\right) = 4\left(\frac{1}{2} - \frac{\sqrt{3}}{2}i\right)$$

Hence $\cos\theta = 1/2$, $\sin\theta = -\sqrt{3}/2$, and
$\theta = -\pi/3 = \arg(2 - 2\sqrt{3}i)$.

Multiplication and division of complex numbers are especially simple when the numbers are written in polar form. Let $z_1 = r_1(\cos\theta_1 + i\sin\theta_1)$ and $z_2 = r_2(\cos\theta_2 + i\sin\theta_2)$. Then

$$z_1 z_2 = r_1 r_2 (\cos\theta_1 + i\sin\theta_1)(\cos\theta_2 + i\sin\theta_2)$$
$$= r_1 r_2 [(\cos\theta_1 \cos\theta_2 - \sin\theta_1 \sin\theta_2)$$
$$+ i(\sin\theta_1 \cos\theta_2 + \cos\theta_1 \sin\theta_2)]$$
$$= r_1 r_2 [\cos(\theta_1 + \theta_2) + i\sin(\theta_1 + \theta_2)]. \qquad (3)$$

From (3) we see again that the modulus of the product is the product of the moduli:

$$|z_1 z_2| = |z_1| |z_2| = r_1 r_2 .$$

We get the additional fact from (3) that the argument of a product is the sum of the arguments. The rule for quotients also is immediate from (3). If $z_1/z_2 = z_3$, then $z_1 = z_2 z_3$, so

$$r_1 = r_2 r_3 \quad \text{or} \quad r_3 = \frac{r_1}{r_2}$$

and
$$\theta_1 = \theta_2 + \theta_3 \quad \text{or} \quad \theta_3 = \theta_1 - \theta_2 .$$

That is, the modulus of the quotient is the quotient of the moduli, and the argument of a quotient is the difference of the arguments. This also gives us the polar representation for z^{-1}.

$$\text{mod}(z^{-1}) = \text{mod}\left(\frac{1}{z}\right) = \frac{\text{mod } 1}{\text{mod } z} = \frac{1}{|z|}$$

$$\arg(z^{-1}) = \arg\left(\frac{1}{z}\right) = \arg 1 - \arg z = -\arg z$$

Hence if $z = r(\cos \theta + i \sin \theta)$,

$$z^{-1} = \frac{1}{r}(\cos(-\theta) + i \sin(-\theta))$$

$$= \frac{1}{r}(\cos \theta - i \sin \theta) .$$

Example 2 Find the modulus and argument of z and w, and express them in polar form.

(a) $z = (3 + 3i)(1 - \sqrt{3}i)$ (b) $w = \dfrac{2i}{3 + 3i}$

Solution.

(a) $\quad \text{mod } z = |3 + 3i| |1 - \sqrt{3}i| = (3\sqrt{2})(2) = 6\sqrt{2}$

$\arg z = \arg(3 + 3i) + \arg(1 - \sqrt{3}i)$

$$= \frac{\pi}{4} - \frac{\pi}{3} = \frac{-\pi}{12}$$

$$z = 6\sqrt{2}\left[\cos\left(\frac{-\pi}{12}\right) + i \sin\left(\frac{-\pi}{12}\right)\right]$$

(b) $$\text{mod } w = \frac{|2i|}{|3+3i|} = \frac{2}{3\sqrt{2}} = \frac{\sqrt{2}}{3}$$

$$\arg w = \arg(2i) - \arg(3+3i)$$

$$= \frac{\pi}{2} - \frac{\pi}{4} = \frac{\pi}{4}$$

$$w = \frac{\sqrt{2}}{3}\left(\cos\frac{\pi}{4} + i\sin\frac{\pi}{4}\right)$$

From the product rule, we get a simple rule for obtaining the powers of number in polar form. If

$$z = r(\cos\theta + i\sin\theta),$$

then
$$z^2 = r^2[\cos(\theta+\theta) + i\sin(\theta+\theta)]$$
$$= r^2(\cos 2\theta + i\sin 2\theta).$$

Repeating the multiplication by z gives

$$z^3 = r^3(\cos 3\theta + i\sin 3\theta),$$
$$z^4 = r^4(\cos 4\theta + i\sin 4\theta),$$

and so on. In general, for any positive integer n,

$$z^n = r^n[\cos(n\theta) + i\sin(n\theta)]. \qquad (4)$$

This formula is called *De Moivre's Theorem*. Since

$$z^{-1} = \frac{1}{r}[\cos(-\theta) + i\sin(-\theta)],$$

$$z^{-n} = \frac{1}{r^n}[\cos(-n\theta) + i\sin(-n\theta)]$$

$$= \frac{1}{r^n}[\cos(n\theta) - i\sin(n\theta)].$$

Example 3 Write in standard form.

(a) $(1+i)^{10}$ (b) $(1+i)^{-3}$

Solution. First write $1+i$ in polar form:

$$1+i = \sqrt{2}\left(\frac{1}{\sqrt{2}} + \frac{1}{\sqrt{2}}i\right) = \sqrt{2}\left(\cos\frac{\pi}{4} + i\sin\frac{\pi}{4}\right)$$

From DeMoivre's Theorem (4), we have

(a) $(1 + i)^{10} = (\sqrt{2})^{10} \left(\cos \frac{10\pi}{4} + i \sin \frac{10\pi}{4} \right)$

$= 2^5 \left(\cos \frac{5\pi}{2} + i \sin \frac{5\pi}{2} \right)$

$= 32(0 + i)$

$= 32i$

(b) $(1 + i)^{-3} = (\sqrt{2})^{-3} \left[\cos \left(\frac{-3\pi}{4} \right) + i \sin \left(\frac{-3\pi}{4} \right) \right]$

$= \frac{1}{2\sqrt{2}} \left(\frac{-\sqrt{2}}{2} - \frac{i\sqrt{2}}{2} \right)$

$= -\frac{1}{4} - \frac{1}{4} i$

We know that for any complex number a, the equation $x^n - a = 0$ has at least one complex root, and at most n distinct complex roots. In fact, every non-zero complex number has n distinct nth roots, and DeMoivre's Theorem provides a simple way of finding them. We start with $a = 1$, and ask for the complex nth roots of 1. If $z^n = 1$, and $z = r(\cos \theta + i \sin \theta)$, then

$$z^n = r^n(\cos n\theta + i \sin n\theta) = 1$$

Therefore r must be 1, and $n\theta$ must be zero or some multiple of 2π. Hence θ must be one of the n distinct numbers

$$0, \frac{2\pi}{n}, \frac{4\pi}{n}, \ldots, \frac{(n-1)2\pi}{n}.$$

For example, for $n = 5$,

$$\theta = 0, \frac{2\pi}{5}, 2 \cdot \frac{2\pi}{5}, 3 \cdot \frac{2\pi}{5}, \quad \text{or} \quad 4 \cdot \frac{2\pi}{5}.$$

The five fifth roots of 1 are 1 itself, and four other complex numbers equally spaced around the unit circle (see Fig. 10-6).

Example 4 Find the cube roots of 1.

Solution. If $\cos 3\theta + i \sin 3\theta = 1$, then $3\theta = 0$, or $3\theta = 2\pi$, or $3\theta = 4\pi$; that is, $\theta = 0$, or $2\pi/3$, or $4\pi/3$. Notice that the equations

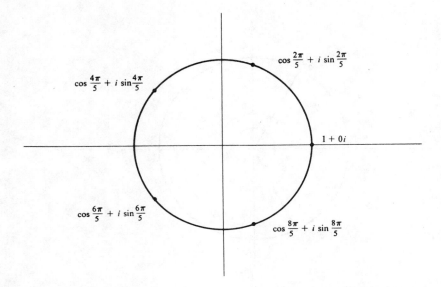

Figure 10-6

$3\theta = 6\pi$, $3\theta = 8\pi$, $3\theta = 10\pi$, etc., need not be considered, since $6\pi/3 = 2\pi$ is the same angle as 0, $8\pi/3 = 2\pi + 2\pi/3$ is the same angle as $2\pi/3$, and $10\pi/3 = 2\pi + 4\pi/3$ is the same angle as $4\pi/3$. The three roots are (see Fig. 10-7):

$$\cos 0 + i \sin 0 = 1,$$

$$\cos \frac{2\pi}{3} + i \sin \frac{2\pi}{3} = -\frac{1}{2} + \frac{\sqrt{3}}{2} i,$$

$$\cos \frac{4\pi}{3} + i \sin \frac{4\pi}{3} = -\frac{1}{2} - \frac{\sqrt{3}}{2} i.$$

The nth roots of any complex number $z_0 = r_0(\cos \theta_0 + i \sin \theta_0)$ are found in much the same way. If $z = r(\cos \theta + i \sin \theta)$ and $z^n = z_0$, then

$$r^n = r_0 \quad \text{so} \quad r = \sqrt[n]{r_0}$$

Since $\cos n\theta = \cos \theta_0$ and $\sin n\theta = \sin \theta_0$, we must have $n\theta = \theta_0$, or $n\theta = \theta_0 + 2\pi, \ldots$, or $n\theta = \theta_0 + (n-1)2\pi$. The n possibilities are

Polar Form of Complex Numbers

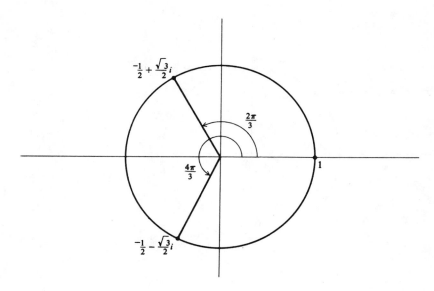

Figure 10-7

$$\frac{\theta_0}{n}, \frac{\theta_0}{n} + \frac{2\pi}{n}, \frac{\theta_0}{n} + \frac{4\pi}{n}, \ldots, \frac{\theta_0}{n} + \frac{(n-1)2\pi}{n}.$$

Example 5 Find the fourth roots of -16.

Solution.
$$-16 = 16(\cos \pi + i \sin \pi)$$
If $z^4 = -16$, then mod $z = \sqrt[4]{16} = 2$, and arg z is one of the numbers

$$\frac{\pi}{4}, \frac{\pi}{4} + \frac{2\pi}{4}, \frac{\pi}{4} + 2 \cdot \frac{2\pi}{4}, \frac{\pi}{4} + 3 \cdot \frac{2\pi}{4}.$$

The numbers are

$$2\left(\frac{\sqrt{2}}{2} + i\frac{\sqrt{2}}{2}\right) = \sqrt{2} + i\sqrt{2},$$

$$2\left(\frac{-\sqrt{2}}{2} + i\frac{\sqrt{2}}{2}\right) = -\sqrt{2} + i\sqrt{2},$$

$$2\left(\frac{-\sqrt{2}}{2} - i\frac{\sqrt{2}}{2}\right) = -\sqrt{2} - i\sqrt{2},$$

$$\left(\frac{\sqrt{2}}{2} - i\frac{\sqrt{2}}{2}\right) = \sqrt{2} - i\sqrt{2}.$$

Again the roots are equally spaced points, but now on the circle of radius $\sqrt[4]{16}$. The argument of the first point is 1/4 the argument of the given number -16, or $\pi/4$. (See Fig. 10-8).

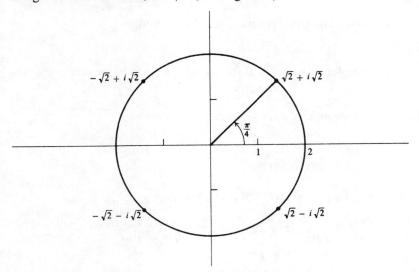

Figure 10-8

SECTION 10-5, REVIEW QUESTIONS

1. The modulus of z is written $|z|$ or _____, and the argument of z is written _____. **mod z**

   arg z

2. Mod $z = \sqrt{x^2 + y^2}$ if $z = x + yi$, and arg z is the angle between the real axis and the ray from the origin to _____. **z**

3. If $|z| = r$ and arg $z = \theta$, then $r(\cos \theta + i \sin \theta)$ is called the _____ representation of z. **polar**

4. The polar representation of any number z can be written by factoring _____ from the standard form. **|z|**

Polar Form of Complex Numbers 423

5. If $z = \sqrt{x^2 + y^2}\left(\dfrac{x}{\sqrt{x^2+y^2}} + \dfrac{y}{\sqrt{x^2+y^2}}i\right)$ and arg $z = \theta$, then $\cos\theta =$ _____ $\dfrac{x}{\sqrt{x^2+y^2}}$ and $\sin\theta =$ _____ $\dfrac{y}{\sqrt{x^2+y^2}}$.

6. The modulus of a product is the product of the moduli, and the argument of a product is the _____ **sum** _____ of the arguments.

7. If $z = r(\cos\theta + i\sin\theta)$, then the polar form of z^{-1} is $z^{-1} =$ _____ $\dfrac{1}{r}[\cos(-\theta) + i\sin(-\theta)]$.

8. DeMoivre's Theorem states that $[r(\cos\theta + i\sin\theta)]^n =$ _____ $r^n[\cos(n\theta) + i\sin(n\theta)]$.

9. The n complex nth roots of 1 are equally spaced points on the unit **circle** _____.

10. The complex roots of z_0 all lie on the circle of radius $\sqrt[n]{|z_0|}$ _____.

SECTION 10-5, EXERCISES

Find mod z, arg z, mod $(1/z)$ and arg $(1/z)$.

Example 1

1. $z = 3 - 3i$
2. $z = -3 + \sqrt{3}i$
3. $z = -3 - 3\sqrt{3}i$
4. $z = -4 + 4i$

Write z, w, zw, and z/w in polar form.

Example 2

5. $z = -3 + 3i$, $w = \sqrt{3} - i$
6. $z = -\sqrt{3} - i$, $w = -1 + \sqrt{3}i$
7. $z = 2 - 2i$, $w = -4 + \dfrac{4\sqrt{3}}{3}i$
8. $z = -5 + \dfrac{5\sqrt{3}}{3}i$, $w = 7 + 7i$

Find the modulus of each of the following numbers.

9. $(3 + 2i)(1 - i)$
10. $(4 - i)(3 + 4i)$
11. $(7 - i)/(3 + i)$
12. $(4 + 4i)/(6 + 3i)$
13. $\dfrac{(2 - 4i)(3 + i)}{(5 - i)}$
14. $\dfrac{(-1 + 5i)(4 + 3i)}{(-1 - 3i)}$

Write the following in standard form.

Example 3

15. $(1 + i)^3$
16. $(1 + i)^{-2}$

17. $\left(\dfrac{1}{2}+\dfrac{\sqrt{3}}{2}i\right)^{-4}$

18. $\left(\dfrac{\sqrt{3}}{2}+\dfrac{1}{2}i\right)^{5}$

19. $(1-\sqrt{3}i)^3$

20. $(-2-2i)^4$

Write the indicated roots in standard form.

21. fourth roots of 1
22. sixth roots of 1
23. cube roots of 27
24. fourth roots of -16
25. square roots of $-4i$
26. square roots of $9i$
27. cube roots of -8
28. fourth roots of 16

Example 4,5

10-6 REVIEW FOR CHAPTER

If $z = a + bi$ and $w = c + di$, with $b \neq 0$ and $d \neq 0$, verify each of the following.

1. If $z + w$ is a real number, then $b + d = 0$ or $d = -b$.
2. If zw is a real number, then $bc + ad = 0$.
3. If $z + w$ and zw are both real numbers, then z and w are conjugate complex numbers.

Perform the indicated operation and write in standard form.

4. $\dfrac{1+i}{1-i}$

5. $\dfrac{-3+2i}{7+4i}$

6. $\dfrac{6-3i}{2+i}$

7. $\left(\dfrac{1-i}{1+3i}\right)\left(\dfrac{2+3i}{-1+4i}\right)$

8. $\left(\dfrac{2+i}{3-i}\right)\left(\dfrac{1-4i}{1+3i}\right)$

9. $\left(\dfrac{2}{2-5i}\right) \div \left(\dfrac{2i}{-2-5i}\right)$

10. $\left(\dfrac{1-i}{1+3i}\right) \div \left(\dfrac{2+3i}{-1+4i}\right)$

Solve the following equations.

11. $(2+i)x + 3i = 1 - i$
12. $(1-2i) + 3x = 4 - 3i$
13. $x + yi = -5 + 4i$
14. $2x + 3yi - 6 + 9i = 0$
15. $-x + 4yi = (2 + 6i) - (7 - 2i)$
16. $x + yi = (2 - i)(2 + i)$

Find the modulus of each of the following.

17. $\dfrac{(2 - i)(3 + i)}{4 - 2i}$

18. $\dfrac{(4 - 3i)(2 + 3i)}{3 + 4i}$

Solve each of the following polynomial equations given the indicated root and factor each polynomial into linear factors.

19. $P(x) = x^3 + x \qquad x = i$

20. $P(x) = 2x^3 - 19x^2 + 42x + 26 \qquad x = -\dfrac{1}{2}$

21. $P(x) = x^3 - 7x^2 + 31x - 25 \qquad x = 3 + 4i$

22. $P(x) = x^4 - 2x^3 + 2x^2 - 8x - 8 \qquad x = 2i$

Find the modulus and argument of z, w, zw, and z/w, and write all four numbers in polar form.

23. $z = 1 - i \qquad w = \sqrt{2} - \sqrt{2}i$

24. $z = 3 + 3i \qquad w = 2 + 2\sqrt{3}i$

25. $z = 4 - 4\sqrt{3}i \qquad w = -1 - i$

Express each of the following complex numbers in standard form.

26. $2\left(\cos \dfrac{\pi}{4} + i \sin \dfrac{\pi}{4}\right)$

27. $3\left(\cos \dfrac{3\pi}{2} + i \sin \dfrac{3\pi}{2}\right)$

28. $4\left(\cos \dfrac{5\pi}{6} + i \sin \dfrac{5\pi}{6}\right)$

29. $2\left(\cos \dfrac{2\pi}{3} + i \sin \dfrac{2\pi}{3}\right)$

Write each of the following in standard form.

30. $(1 - i)^6$

31. $\left(\dfrac{\sqrt{3}}{2} + \dfrac{1}{2}i\right)^{10}$

Find the indicated roots of each of the following numbers.

32. Cube roots of 8
33. Cube roots of -125
34. Fourth roots of 81
35. Fourth roots of 1

11 PERMUTATIONS, COMBINATIONS, AND PROBABILITY

11-1 PERMUTATIONS

The theory of probability had its origins in practical problems arising in games of chance. These problems involve counting the number of possible outcomes of some operation. For example, we might want to know how many ways a pair of dice can fall, and how many of these give a total of seven spots. Or we might want to know how many different poker hands there are, and how many of these are straight flushes.

Many counting problems depend on knowing how many different arrangements there are of the elements of some set. For example, we might ask how many ways the 26 letters of the alphabet can be arranged in a row. The number is enormous (approximately 4×10^{26}), and listing all possibilities is simply not feasible. Consequently, we need to develop special techniques to handle this kind of counting problem.

One of the basic principles of counting is illustrated by the following simple problem:

Suppose there are four roads from town X to town Y, and three roads from town Y to town Z. How many routes are there from X to Z, passing through Y? The answer is almost obvious: Each of the four paths from X to Y could be combined with any of the three paths from Y to Z. There are 4 × 3 = 12 possibilities altogether.

In general, suppose you have two operations, O_1 and O_2 (for example, O_1 could be the operation of choosing a road from X to Y). If there are m possibilities for O_1, and n possibilities for O_2, then there are $m \cdot n$ possible ways to perform O_1 and then O_2. This principle extends to three or more operations. If there are respectively m, n, and p possibilities for O_1, O_2, and O_3, then there are altogether $m \cdot n \cdot p$ possible ways of performing O_1, then O_2, then O_3.

Example 1 How many even three-digit numbers larger than 500 can be formed using the digits 4, 5, and 6?

Solution. The problem is how many ways can you perform the three operations—O_1: choosing the first (hundreds) digit; O_2: choosing the second (tens) digit; and O_3: choosing the third (units) digit. There are two possibilities for O_1, since the first digit must be 5 or 6 if the number is to exceed 500. There are three possibilities for O_2, since any of the digits 4, 5, or 6 can be used. There are two possibilities for O_3 (4 or 6), since the number is to be even. Therefore there are 2 × 3 × 2 = 12 possible numbers.

Example 2 How many different sequences of heads and tails can you get in four flips of a coin? What are the chances of getting all heads?

Solution. There are four operations, and two possible outcomes of each. Therefore there are 2 × 2 × 2 × 2 = 16 possible sequences. Only one of these sequences consists of all heads. The chances of getting four heads is therefore 1/16.

Any ordered arrangement of the elements of a set is called a *permutation* of the set. If a set has n elements, a permutation of the set is an ordered n-tuple. For example, the permutations of the set $\{a, b, c\}$ are the following six ordered triples:

$$(a,b,c), (a,c,b), (b,a,c), (b,c,a), (c,a,b), (c,b,a)$$

We think of the permutations of an n-element set as the outcome of n successive operations: O_1: choosing the first element; O_2: choosing the second element; ... ; O_n: choosing the nth element. There are n possibilities for O_1, since any element can be listed first. There are only $n - 1$ possibilities for O_2, since there are only $n - 1$

elements left to choose from after a first element has been selected. Similarly, there are $n - 2$ possibilities for O_3, $n - 3$ for O_4, and so on, until finally there is only one possibility for O_n (choosing the last element). According to our fundamental principle of counting, we have the following result.

There are $n \cdot (n - 1) \cdot (n - 2) \ldots 3 \cdot 2 \cdot 1$ permutations of an n-element set.

Example 3 A scrabble player has four squares, each with a different letter. How many ways can the squares be arranged to form possible words?

Solution. There are four choices for the first square, then three choices for the second square, then two choices for the third square, and finally one choice for the fourth square. Hence there are $4 \cdot 3 \cdot 2 \cdot 1 = 24$ possible permutations of the four squares.

The product $1 \cdot 2 \cdot 3 \cdot \cdots \cdot n$ occurs frequently in other branches of mathematics, as well as in counting processes, so we introduce a notation for it. The expression $n!$ is read *n-factorial*, and defined by

$$n! = 1 \cdot 2 \cdot 3 \cdot \cdots \cdot n$$

We saw above that there are $n!$ permutations of any n-element set. Hence there are $26!$ permutations of the alphabet, and $4!$ permutations of four scrabble tiles.

It is convenient to define $0! = 1$ in much the same way as we define $x^0 = 1$. The first few factorials are: $0! = 1$, $1! = 1$, $2! = 2$, $3! = 6$, $4! = 24$, $5! = 120$, $6! = 720$, $7! = 5,040$, $8! = 40,320$, $9! = 362,880$, and $10! = 3,628,800$.

Example 4 (a) How many numbers with no more than 10 digits are there? (b) How many permutations of the 10 digits $0, 1, \ldots, 9$ are there?

Solutions. The two questions (a) and (b) are very different. In (a) we are allowed to use any digit $0, 1, \ldots, 9$ in any place in the number. There are 10 operations, each with 10 possibilities, so the total number is $10^{10} = 10,000,000,000$. The numbers go from 0 (which is counted in the form $0,000,000,000$) to $9,999,999,999$. In (b) we are asked for permutations of $0, 1, \ldots, 9$, so that each digit must be used just once in each number. Hence there are $10! = 3,628,800$ possibilities.

Since $n!$ increases very rapidly as n increases, we develop some special ways of dealing with expressions containing factorials. The product of the numbers from 1 to n can be written in any of the following ways:

$$n! = n \cdot (n-1)! = n \cdot (n-1) \cdot (n-2)! = \cdots$$

For example, $10! = 10 \cdot 9 \cdot 8 \cdot 7!$. Some techniques for dealing with these expressions are illustrated in the next example.

Example 5   Evaluate (a) $10!/7!$ and (b) $(7! - 5!)/6!$.

Solution.

(a)
$$\frac{10!}{7!} = \frac{10 \cdot 9 \cdot 8 \cdot 7!}{7!}$$
$$= 10 \cdot 9 \cdot 8$$
$$= 720$$

(b)
$$\frac{7! - 5!}{6!} = \frac{7 \cdot 6 \cdot 5! - 5!}{6 \cdot 5!}$$
$$= \frac{5!(7 \cdot 6 - 1)}{6 \cdot 5!}$$
$$= \frac{41}{6}$$

Notice how much simpler the computations above are than computing directly $3{,}628{,}800/5{,}040$, and $(5{,}040 - 120)/720$.

SECTION 11-1, REVIEW QUESTIONS

mn

1. If there are m possible ways to perform one operation O_1, and n ways to perform a second operation O_2, then there are _____ ways to perform O_1 and O_2.

3·5·7

2. If there are three paths from A to B, five paths from B to C, and seven paths from C to D, then there are _____ paths from A to B to C to D.

permutation

3. An ordered arrangement of the elements of a set is called a _____ of the set.

4. A permutation of an *n*-element set is an _____ *n*-tuple. **ordered**

5. In forming an ordered *n*-tuple with the *n* elements of some set, there are _____ choices for the first element, _____ choices for the second, and so on. **n   n−1**

6. An *n*-element set has _____ permutations. **n!**

7. *n*! is read *n*-_____, and denotes the product $n(n-1)(n-2)\cdots 2\cdot 1$. **factorial**

SECTION 11-1, EXERCISES

1. How many three-digit numbers can be formed using the digits 2, 3, 4?   **Example 1, 2**
2. How many two-digit numbers can be written using only odd digits 1, 3, 5, 7, 9?
3. How many even three-digit numbers larger than 500 can be written using the digits 2, 4, 7?
4. How many odd three-digit numbers less than 700 can be written using the digits 3, 4, 9?
5. How many different sequences of heads and tails can you get in five flips of a coin? What are the chances of getting all tails?
6. A true-false test has six questions. How many different ways can you make out the answer sheet? What are the chances of guessing the correct answer to all questions?
7. A multiple choice quiz has five questions, and there are three possible answers to each. How many possible sequences of answers are there?
8. A single die is thrown three times. What are the chances of guessing the sequence of three numbers which are thrown?
9. How many three-letter "words" can be formed using the letters A, B, C, D, E if the middle letter must be a vowel?
10. How many four-letter "words" can be formed using the letters A, B, C, D, E if both middle letters must be vowels?
11. The Hawaiian alphabet has seven consonants and five vowels. How many possible two-letter words can be formed if the second letter must be a vowel?
12. How many possible two-letter Hawaiian words could be written using first a consonant and then a vowel? Using two vowels?
13. A salesman's route takes him from town A, to town B, to town C, and then back to town A. There are three roads from A to B, two from B to C, and four from C to A. How many different trips can he make?
14. If the salesman of Exercise 13 goes from A to B to C to B to A, how many different trips can he make?
15. If there are four roads from X to Y, how many ways can you go from X to Y and back without using the same road twice.
16. How many trips can the salesman of Exercise 14 make if he does not use the same road twice?

Permutations   431

Example 3,4

17. How many license plates can you make using two letters (O and Q are excluded) followed by four non-zero digits? (Write answer as a product.)
18. How many license plates can you make using any three letters followed by any three non-zero digits? (Write answer as a product.)
19. How many ways can five scrabble squares be arranged to form possible five-letter words?
20. How many possible three-letter words can a player form from five different scrabble squares?
21. How many different four-digit numbers can be written using each of the digits 2, 4, 6, 8 exactly once?
22. How many five-letter "words" can be formed using each of the letters A, B, C, D, E exactly once?
23. How many permutations of the letters X, Y, Z, W are there?
24. How many ways can four people be lined up for a photograph?
25. How many permutations of the odd digits are there?
26. How many permutations of the non-zero even digits are there?
27. How many ways can a club with 10 members fill three officer's chairs, if a person can hold only one office?
28. How many ways can an eleven-man team choose two different members to be captain and manager?
29. Three women and four men are to pose for a photograph, with the women in the first row and the men in the second. How many arrangements are possible?
30. Three boys and two girls are lined up for a relay race team with boy and girl alternating. How many arrangements are possible?

Example 5

Evaluate each of the following.

31. $\dfrac{8!}{5!}$

32. $\dfrac{12!}{10!}$

33. $\dfrac{14!}{11!}$

34. $\dfrac{8!}{4!}$

35. $\dfrac{42!}{40!2!}$

36. $\dfrac{8!}{5!3!}$

37. $\dfrac{7!}{5!2!}$

38. $\dfrac{9!}{5!4!}$

39. $\dfrac{5!!}{49!2!}$

40. $\dfrac{100!}{98!5!}$

41. $\dfrac{(n+2)!}{n!}$

42. $\dfrac{(n+3)!}{(n-1)!}$

43. $\dfrac{(n-1)!}{(n+1)!}$

44. $\dfrac{(n-1)!}{n!}$

432 Permutations, Combinations, and Probability

II-2 COMBINATIONS

A permutation of a set is an ordering of all the elements of the set. That is, a permutation is an ordered n-tuple whose entries are all distinct elements of the set. We also frequently want to count the ordered k-tuples (with $k < n$) formed with distinct elements from some n-element set. The example below treats a typical question of this sort.

Example 1 How many code words can be formed using three distinct letters?

Solution. We think of the process of forming a code word as three successive operations: picking the first letter, picking the second letter, and picking the third letter. There are 26 choices for the first letter. Since the letters are to be distinct, there are 25 choices for the second letter, and 24 choices for the third. Therefore there are $26 \times 25 \times 24$ possible "words" formed from three distinct letters.

An ordering of k elements chosen from an n-element set is called a *permutation of* n-*elements*, k *at a time*. For example, a permutation of 10 things, 3 at a time is simply an ordered triple whose entries are distinct choices from the 10 elements.

We use the notation $P(n;k)$ for the *number* of permutations of n things, k at a time. There are n choices for the first element, $(n-1)$ choices for the second element, and so on. When we have chosen $(k-1)$ elements in the ordered k-tuple, then there are $n - (k-1)$, or $n - k + 1$ possible choices for the last entry. The basic counting principle shows that

$$P(n;k) = n(n-1)(n-2)\cdots(n-k+1)$$
$$= \frac{n!}{(n-k)!}. \qquad (1)$$

We use $P(n;n)$ for the number of permutations of n things, and then (1) gives us

$$P(n;n) = \frac{n!}{0!} = n!$$

which agrees with our earlier result.

Example 2 How many ways can you make up a relay swimming team composed of a freestyler, a backstroker, and a breaststroker, choosing from seven swimmers all of whom can swim all strokes?

Solution. This team can be thought of as an ordered triple, or a permutation of the seven swimmers, three at a time. Therefore there are $P(7;3)$ possibilities:

$$P(7;3) = \frac{7!}{(7-3)!} = \frac{7!}{4!} = 7 \cdot 6 \cdot 5 = 210.$$

Example 3 How many ways can a president, vice-president, secretary, and choir director be elected in a club with 11 members?

Solution. Each possibility can be thought of as an ordered quadruple. Therefore the answer is the number of permutations of 11 things, 4 at a time:

$$P(11;4) = \frac{11!}{(11-4)!} = \frac{11!}{7!} = 11 \cdot 10 \cdot 9 \cdot 8 = 7,920.$$

In addition to *ordered* k-tuples formed from the elements of some set, we also sometimes want to count *unordered* k-tuples. An unordered k-tuple is simply a k-element subset of some given set. To illustrate the difference, consider the three-element set $\{a,b,c\}$. The *permutations* of this set, two at a time, are the following six ordered pairs:

$$(a,b), (a,c), (b,a), (b,c), (c,a), (c,b).$$

The *subsets* of $\{a,b,c\}$ having two elements are

$$\{a,b\}, \{a,c\}, \{b,c\}.$$

The subsets $\{a,b\}$ and $\{b,a\}$ are the same, but the permutations (a,b) and (b,a) are different. Since each two-element subset has two permutations, there are half as many two-element subsets of $\{a,b,c\}$ as there are permutations of $\{a,b,c\}$, two at a time.

A k-element subset of an n-element set is called a *combination of the* n *elements,* k *at a time.* The number of combinations of n things k at a time is denoted $C(n;k)$. Each k-element subset has $k!$ permutations. That is, each combination of n things, k at a time, corresponds to $k!$ different permutations of n things, k at a time. Therefore there are $k!$ times as many permutations of n things k at a time as there are combinations of n things k at a time:

$$k!\, C(n;k) = P(n;k).$$

This gives us the following formula for $C(n;k)$:

$$C(n;k) = \frac{P(n;k)}{k!},$$

$$C(n;k) = \frac{n!}{(n-k)!\,k!}.\qquad(2)$$

Example 4 How many different five-man basketball teams can be formed from a group of eight boys?

Solution. The question is how many five-element combinations (subsets) are there in a set of eight. The answer is

$$C(8;5) = \frac{8!}{(8-5)!\,5!}$$
$$= \frac{8 \cdot 7 \cdot 6}{3!} = 8 \cdot 7$$
$$= 56.$$

Notice that the question above could also have been phrased: How many ways can you pick three boys *not* to be on the team? In other words, $C(8;5) = C(8;8-5)$.

In general, it is easy to see from (2) that

$$C(n;k) = C(n;n-k).\qquad(3)$$

Example 5 You are required to answer any three questions on a five-question quiz. How many choices do you have?

Solution. The number of three-question subsets of the set of five questions is

$$C(5;3) = \frac{5!}{(5-3)!\,3!} = \frac{5 \cdot 4}{2!}$$
$$= 10.$$

If you would rather think of how many choices you have for the set of two questions you need not answer, the answer is $C(5;2)$, which of course also equals 10.

As an application of the formula for counting combinations, we consider the problem of determining the coefficients in the binomial expansion $(a + b)^n$. The terms in the expansion of

$$(a+b)(a+b)(a+b)\cdots(a+b)$$

will consist of all products you get by choosing an *a* or *b* from each pair of parentheses. If you choose k *a*'s (and hence $n-k$ *b*'s), you get the term $a^k b^{n-k}$. The coefficient of $a^k b^{n-k}$ in the expansion is the number of different ways k *a*'s can be chosen from the n sets of parentheses. That is, the coefficient of $a^k b^{n-k}$ is $C(n;k)$. The notation $\binom{n}{k}$ is also used for these *binomial coefficients*, so we have

$$(a + b)^n = \binom{n}{n} a^n b^0 + \binom{n}{n-1} a^{n-1} b^1 + \binom{n}{n-2} a^{n-2} b^2$$
$$+ \cdots + \binom{n}{k} a^k b^{n-k} + \cdots + \binom{n}{1} a^1 b^{n-1} + \binom{n}{0} a^0 b^n. \quad (4)$$

From (3) we see that $\binom{n}{k} = \binom{n}{n-k}$, so the coefficients in (4) are symmetric. For example,

$$(a + b)^6 = a^6 + 6a^5 b + 15a^4 b^2 + 20a^3 b^3 + 15a^2 b^4 + 6ab^5 + b^6.$$

Writing out the factorial expression for $\binom{n}{k}$, we have

$$\binom{n}{k} = \frac{n!}{(n-k)! \, k!} = \frac{n(n-1) \cdots (n-k+1)}{k!}. \quad (5)$$

Hence the binomial expansion (4) can also be written

$$(a + b)^n = a^n + na^{n-1} b + \frac{n(n-1)}{2!} a^{n-2} b^2$$
$$+ \cdots + \frac{n(n-1) \cdots (n-k+1)}{k!} a^k b^{n-k}$$
$$+ \cdots + nab^{n-1} + b^n. \quad (6)$$

SECTION 11-2. REVIEW QUESTIONS

k 1. A permutation of n things, k at a time, is on ordered _____ -tuple.

2. The number of ordered k-tuples formed from n things is the same as the

permutations number of _____ of n things, k at a time.

3. The number of permutations of n things, k at a time, denoted $P(n;k)$ is

n!/(n-k)! given by the formula _____.

4. A combination of k things chosen from a set of n things is just a k-element

subset _____ of the given set.

5. The number of k-element subsets of an n-element set, denoted $C(n;k)$, is

n!/(n–k)!k! given by the formula _____.

6. The number $C(n;k)$ of combinations of n things k at a time is also called a _____ coefficient. **binomial**
7. The binomial coefficients are denoted _____. $\binom{n}{k}$
8. The binomial coefficient $\binom{n}{k}$ is the coefficient of the term _____ in the expansion of $(a+b)^n$. $a^k b^{n-k}$
9. The coefficient $\binom{n}{k}$ of the term $a^k b^{n-k}$ is the same as the coefficient $\binom{n}{n-k}$ of the term _____. $a^{n-k} b^k$

SECTION 11-2, EXERCISES

Evaluate.

1. $P(7;2)$
2. $P(5;3)$
3. $P(10;4)$
4. $P(9;3)$
5. $C(6;2)$
6. $C(5;4)$
7. $C(10;8)$
8. $C(8;5)$
9. How many code words can be formed using four distinct letters? **Example 1**
10. How many four-digit numbers can be written using four different odd digits?
11. How many three-digit numbers can be written using distinct even non-zero digits?
12. How many three-letter "words" can be formed using three different letters chosen from T, X, Y, Z, W?
13. How many ways can you pick the first four hitters from a nine-man baseball team? **Example 2,3**
14. How many ways can you pick a first, second and third baseman from a group of seven infielders?
15. How many ways can three class offices be filled by different students in a class of 20 people?
16. A kindergarten class with six students appoints a new class president and vice-president each week. How many weeks are required to exhaust all possibilities?
17. How many four-letter radio station call-letter combinations are there ending in A, if the first letter must be K and there are no repetitions?
18. If the first call-letter for a radio station must be K or W, and no repetitions are allowed, how many four-letter possibilities are there ending in A?
19. How many five-man basketball teams can be formed from a group of nine boys? **Example 4,5**
20. How many nine-man baseball teams can be formed from twelve boys?
21. How many ways can you choose ten questions to answer from a fifteen-problem quiz?
22. How many ways can you choose four questions to omit from a nine-question exam?
23. How many four-element subsets does an eight-element set have?
24. How many three-element subsets does a nine-element set have?

25. How many two-element sets does a ten-element set have?
26. How many five-element sets does a ten-element set have?

Evaluate the binomial coefficients.

27. $\binom{6}{4}$ 28. $\binom{5}{3}$

29. $\binom{10}{3}$ 30. $\binom{8}{4}$

31. Find the coefficient of a^3b^2 in $(a+b)^5$.
32. Find the coefficient of a^4b^2 in $(a+b)^6$.
33. Find the coefficient of a^8b^4 in $(a+b)^{12}$.
34. Find the coefficient of a^3b^7 in $(a+b)^{10}$.

II-3 DISTINGUISHABLE PERMUTATIONS; PARTITIONS

It sometimes happens that we do not wish to distinguish (or cannot distinguish) one permutation of several objects from another. Consider, for example, a collection of three red blocks and two white blocks. We ask how many different color patterns can be made by arranging these blocks in a row. We label the three red blocks R_1, R_2, R_3, and the two white blocks W_1 and W_2. One of the permutations of the five blocks is

$$R_1 R_3 W_1 R_2 W_2$$

This permutation gives the color pattern red, red, white, red, white. We get this same color pattern if we permute the red blocks among themselves, or the white blocks among themselves. For example,

$$R_2 R_3 W_2 R_1 W_1$$

is a different permutation, but it gives the same color pattern. We will say these two permutations are *indistinguishable*. By contrast, the two permutations

$$R_2 R_3 W_1 R_1 W_2 \quad \text{and} \quad R_3 W_2 R_1 R_2 W_1$$

are *distinguishable*, since they give different color patterns.

For any given color pattern of the five blocks, there are 3! different permutations of the red blocks among themselves which give the same color pattern. With each of these 3! permutations of the red blocks, there are 2! permutations of the white blocks which also give the same color pattern. Therefore there are $3! \cdot 2!$ permutations for each color pattern. Since there are a total of 5! permutations of the blocks, there are 5!/3!2! distinguishable permutations, or different color patterns.

We will use the notation $\binom{5}{3,2}$ for the number of distinguishable permutations of five things, if three of them are mutually indistinguishable, and the remaining two are mutually indistinguishable. As we saw above, $\binom{5}{3,2} = \frac{5!}{3!2!}$.

More generally, suppose we have n objects which fall into p groups, so that the objects in each group are indistinguishable from each other. Let there be k_1 elements in the first group, k_2 in the second, etc., and k_p in the pth group. For example, a collection of pool balls with k_1 one-balls, k_2 two-balls, etc., would fit this pattern. We denote the number of distinguishable permutations of the n elements by

$$\binom{n}{k_1, \ldots, k_p}.$$

The same reasoning used for the blocks shows that there are $k_1! k_2! \ldots k_p!$ indistinguishable permutations for each distinguishable permutation. Therefore

$$\binom{n}{k_1, \ldots, k_p} = \frac{n!}{k_1! k_2! \ldots k_p!}.$$

Example 1 How many distinguishable permutations are there of a collection of three quarters, two dimes, and four nickels?

Solution. There are $3 + 2 + 4 = 9$ elements altogether, and the number of distinguishable permutations is

$$\binom{9}{3,2,4} = \frac{9!}{3!\,2!\,4!} = \frac{9 \cdot 8 \cdot 7 \cdot 6 \cdot 5}{3 \cdot 2 \cdot 2}$$

$$= 3 \cdot 2 \cdot 7 \cdot 6 \cdot 5$$

$$= 1{,}260.$$

Example 2 Six dice are rolled, and the upfaces show three 2s, two 5s, and a 6. How many different six-digit numbers can be formed with the dice?

Solution. Different numbers will correspond to distinguishable permutations of the six dice. The three sets of indistinguishable elements have three, two, and one elements, respectively. Therefore the number of different six-digit numbers is

$$\binom{6}{3,2,1} = \frac{6!}{3!\,2!\,1!} = \frac{6\cdot 5\cdot 4}{2}$$
$$= 60.$$

In the examples above we considered sets which break up into various subsets of indistinguishable elements. Now let us consider a set divided into subsets in a perfectly arbitrary way. Any such decomposition of a set into mutually exclusive subsets is called a *partition* of the set. That is, a partition of a set S is a collection of subsets of S, $\{S_1, S_2, \ldots\}$, such that every element of S is in exactly one of the subsets S_i.

The set $\{a,b,c,d,e\}$ can be partitioned into two subsets, one with three elements and one with two elements, in many different ways. The following are just a few of the possible 3,2-partitions:

$$\{\{a,b,c\},\ \{d,e\}\},$$
$$\{\{a,d,e\},\ \{b,c\}\},$$
$$\{\{b,c,d\},\ \{a,e\}\}.$$

The partition $\{\{a, b, c\}, \{d, e\}\}$ is of course the same as the partition $\{\{a, c, b\}, \{e, d\}\}$, since $\{a, b, c\}$ and $\{a, c, b\}$ denote the same set, and $\{d, e\}$, $\{e, d\}$ denote the same set. In other words, a permutation of the elements of a subset does not change the subset, and so does not give a new partition. The elements of each subset are indistinguishable from each other insofar as partitioning is concerned.

Example 3 How many partitions of a five-element set are there into two subsets, one with three elements, and one with two elements.

Solution. We regard the elements of the three-element set as mutually indistinguishable, since permuting these elements among themselves gives the same subset. A similar statement holds for the elements of the two element set. The number of 3,2-partitions is then the same as the number of distinguishable permutations:

$$\binom{5}{3,2} = \frac{5!}{3!\,2!}.$$

The same kind of reasoning used to count distinguishable permutations gives us the number of partitions of an n-element set into subsets with prescribed numbers of elements. If there are to be p subsets in a partition, and the subsets have $k_1, k_2, \ldots, k_p$ elements, respectively (so that $k_1 + \cdots + k_p = n$), then the number of possible partitions is

$$\binom{n}{k_1, \ldots, k_p} = \frac{n!}{k_1! \ldots k_p!}.$$

Example 4 A group of 12 students is to divide into three committees to plan three different aspects of a group project. How many ways can these committees be formed if they are to contain three, four, and five people, respectively?

Solution. The number of 3,4,5-partitions of a 12-element set is

$$\binom{12}{3,4,5} = \frac{12!}{3!\,4!\,5!} = \frac{12 \cdot 11 \cdot 10 \cdot 9 \cdot 8 \cdot 7 \cdot 6}{3 \cdot 2 \cdot 4 \cdot 3 \cdot 2}$$

$$= 12 \cdot 11 \cdot 10 \cdot 3 \cdot 7$$

$$= 27{,}720.$$

SECTION 11-3, REVIEW QUESTIONS

1. If two permutations are to be counted separately, they are called _____.   **distinguishable**

2. Distinguishable permutations are counted separately, but two _____ permutations are considered to be the same for counting purposes.   **indistinguishable**

3. In a set of n objects, k of them are indistinguishable. By permuting these k objects among themselves, you get _____ different permutations of the n objects which are indistinguishable.   **k!**

4. By grouping the indistinguishable elements of a set into subsets, we form a _____ of the set.   **partition**

5. A partition of a set is a collection of subsets such that each element of the given set belongs to exactly _____ subset in the collection.   **one**

6. The number of ways of partitioning an n-element set into subsets with $k_1, \ldots, k_p$ elements is denoted _____. $\binom{n}{k_1, \ldots, k_p}$

7. The number $\binom{n}{k_1, \ldots, k_p}$ is expressed in terms of factorials as _____. $\frac{n!}{k_1! \ldots k_p!}$

SECTION 11-3, EXERCISES

How many distinguishable permutations are there of:

Example 1,2
1. Three red blocks and four white blocks?
2. Four red blocks and two white blocks?
3. Two quarters, two dimes, and three pennies?
4. Three dimes, two nickels, and five pennies?
5. The letters in "Mississippi"? (Mississippi is a state on the mainland.)
6. The letters in "Kaunakakai"? (Kaunakakai is a town on Molokai.)
7. How many four-digit numbers can be written with two 4's and two 8's? Write them all.
8. How many five-digit numbers can be written with three 2's and two 7's? Write them all.
9. How many five-symbol code words can be formed using three a's and two b's?
10. How many six-symbol code words can be formed using two x's and four y's?

Example 3
11. How many partitions of a four-element set {a, b, c, d} into two subsets with two elements each are there? Write them.
12. How many partitions of a seven-element set are there into two subsets, one with five elements and one with two elements?
13. How many partitions of a seven-element set are there into three sets, having two, two, and three elements respectively?
14. How many partitions of a nine-element set are there into three sets, having four, three, and two elements respectively?

Example 4
15. How many ways can eight men be assigned to four double rooms?
16. How many ways can six men be assigned to two triple rooms?
17. How many ways can a club with nine members be split into three committees with two, three, and four members?
18. How many ways can a hiking group of seven people be split up into three groups of three, two, and two to try three different trails?
19. How many ways can eight coins be arranged so that three heads and five tails are showing?
20. How many ways are there of flipping a coin five times and getting three heads and two tails?

11-4 PROBABILITY

The simplest part of the theory of probability treats questions like the following: What are the chances of rolling a seven with two dice? What are the chances of getting three heads in a row with three flips of a coin? What are the chances of being dealt a flush in five-card stud?

To treat such questions, we start with the idea of the set of possible *outcomes* of some operation. For example, there are 52 outcomes in the operation of drawing a card from a deck, two possible outcomes in flipping a coin, and $C(52;5)$ outcomes in dealing a five-card poker hand. We will be concerned here with the case where all the possible outcomes are equally likely. For example, a head and a tail are equally likely when a coin is flipped, and any particular set of five cards is as likely as any other hand in one deal.

A subset of the set of all possible outcomes of an operation is called an *event*. For example, the event of rolling an even number on one die is the subset $\{2,4,6\}$ of the set of all possible outcomes $\{1,2,3,4,5,6\}$. The event of drawing an ace from a deck of cards is the subset $\{$A-clubs, A-diamonds, A-hearts, A-spades$\}$ of the set of all 52 possible outcomes.

We now define the *probability* of an event in this situation of an operation with a finite number of equally likely outcomes. If there are n outcomes, the probability of any one of them is $1/n$. If there are k outcomes in an event, the probability of the event is k/n. We will use $\mathbf{P}(E)$ to denote the probability of the event E.

For the operation of flipping a coin, there are two outcomes, which we will denote H and T. There are four possible events for this operation, which are the four subsets of $\{H,T\}$: $\emptyset$ (the empty set), $\{H\}$, $\{T\}$, and $\{H,T\}$ (the set of all outcomes is considered a subset of itself). The empty set corresponds to the *impossible event* of getting neither a head nor tail, and this event has probability 0. The set of all outcomes, $\{H,T\}$, corresponds to the *certain event* of getting either a head or tail, and this event has probability 1.

Operation: Flipping a coin
Outcomes: $\{H,T\}$
Events: $\{\emptyset, \{H\}, \{T\}, \{H,T\}\}$

Probabilities: $\mathbf{P}(\emptyset) = 0$; $\mathbf{P}(\{H\}) = \dfrac{1}{2}$; $\mathbf{P}(\{T\}) = \dfrac{1}{2}$;

$\mathbf{P}(\{H,T\}) = 1$

Example 1 What is the probability of throwing a sum of 3 on one roll of two dice?

Solution. We indicate each outcome by an ordered pair (m,n), where m is the number on the first die and n is the number on the second die. There are 36 possible outcomes: $(1,1), (1,2), \ldots, (1,6),$ $(2,1), \ldots,(6,6)$. The event of rolling a 3 consists of the outcomes

(1,2) and (2,1). Hence the probability of this event is $2/36 = 1/18$.

Example 2 What is the probability of dealing a five-card poker hand containing only aces and kings?

Solution. The total number of five-card hands which can be dealt from a deck of 52 cards is

$$C(52;5) = \frac{52!}{47!\ 5!}.$$

There are eight aces and kings in the deck, so the number of five-card hands consisting only of aces and kings is

$$C(8;5) = \frac{8!}{3!\ 5!}$$

Since the total number of hands (outcomes) is $C(52;5)$, and the total number of hands in the event is $C(8;5)$, the probability of this event is

$$\frac{C(8;5)}{C(52;5)} = \frac{8!\ 47!\ 5!}{3!\ 5!\ 52!} = \frac{8!\ 47!}{3!\ 52!}$$

$$= \frac{8 \cdot 7 \cdot 6 \cdot 5 \cdot 4}{52 \cdot 51 \cdot 50 \cdot 49 \cdot 48}$$

$$= \frac{1}{46{,}410}.$$

If E_1 and E_2 are two events for an operation, then the event "E_1 or E_2" consists of all the outcomes in either set. That is, the union $E_1 \cup E_2$ is the event "E_1 or E_2". If E_1 and E_2 have no common elements, they are said to be *disjoint*, and then $P(E_1 \cup E_2) = P(E_1) + P(E_2)$; but if E_1 and E_2 are not disjoint, then $P(E_1 \cup E_2) < P(E_1) + P(E_2)$. Disjoint and non-disjoint events are illustrated in the next example.

Example 3 (a) What is the probability of rolling a 7 or 11 in one roll of the dice? (b) What is the probability of drawing a face card or a heart from a deck of cards?

Solution. (a) The event of rolling a seven consists of the six outcomes (1,6), (2,5), (3,4), (4,3), (5,2), (6,1), and the event of rolling an 11 consists of the two outcomes (6,5), (5,6). These events are disjoint, so the event "7 or 11" has 8 outcomes out of the 36 possible outcomes. The probability of the event "7 or 11" is therefore $8/36 = 0.222$.

(b) The event of drawing a face card (J, Q, K, A) has 16 outcomes, and the event of drawing a heart has 13 outcomes. However, 4 of the 13 hearts are face cards, so these events are not disjoint. The total number of outcomes which are "face cards or hearts" is $16 + 9 = 25$. The probability of this event is $25/52 = 0.481$.

Let O_1 be an operation with m outcomes, and O_2 a second operation with n outcomes. Assume that the outcomes of O_2 do not depend on those of O_1. Then we can form a third operation, O_3, which consists of performing first O_1, and then O_2. The operation O_3 has $m \cdot n$ outcomes. Let E_1 be an event for O_1 containing k of the m possible outcomes of O_1, and E_2 be an event containing l of the n possible outcomes of O_2. The event "first E_1, then E_2" is an event for O_3, which has $k \cdot l$ of the possible $m \cdot n$ outcomes. Therefore the probability of "E_1, then E_2", which we denote (E_1, E_2), is

$$P(E_1, E_2) = \frac{kl}{mn} = \frac{k}{m} \cdot \frac{l}{n} = P(E_1)P(E_2).$$

Example 4 What is the probability of rolling a five on the first roll of a pair of dice, and then a 7 or 11 on the second roll?

Solution. The first event consists of the outcomes (4,1), (3,2), (2,3), (1,4), and has probability $4/36 = 1/9$. The second event, 7 or 11, has probability $2/9$ as we saw in Example 3. Therefore the probability of "first 5, then 7 or 11" is

$$\frac{1}{9} \cdot \frac{2}{9} = \frac{2}{81}.$$

If an event E contains k of the n outcomes of an operation, then the event E' containing the other $n - k$ outcomes is called the *complement* of E. Since $P(E) = k/n$, and $P(E') = (n - k)/n = 1 - k/n$, we have the important relationship

$$P(E') = 1 - P(E).$$

This principle is frequently useful in computing probabilities for repeated operations.

Example 5 What is the probability of rolling at least one sum of 6 in three rolls of a pair of dice?

Solution. We compute first the probability of the complementary event: no sum of 6 appears in the three rolls. The event of getting a 6 consists of the outcomes (5,1), (4,2), (3,3), (2,4), (1,5), and hence has probability 5/36. The probability of not getting a six on any roll is

Probability 445

therefore $1 - 5/36 = 31/36$. The probability of not getting a 6 on three consecutive rolls of the dice is

$$\frac{31}{36} \cdot \frac{31}{36} \cdot \frac{31}{36} = 0.639.$$

Hence the probability that at least one 6 appears in the three rolls is $1 - 0.639 = 0.361$.

SECTION 11-4, REVIEW QUESTIONS

outcomes 1. The basic idea of elementary probability is the set of _____ of some operation.

36 2. The operation of flipping a coin has two outcomes, and rolling a pair of dice has _____ outcomes.

event 3. Any subset of the set of all outcomes is called an _____.

certain 4. The empty set is called the impossible event, and the set of all outcomes is the _____ event.

0
1 5. The impossible event has probability _____, and the certain event has probability _____.

0
1 6. The probability of any event is a number between _____ and _____.

k/n 7. If an event E contains k of the n possible outcomes of an operation, then $P(E) = $ _____.

P(E₁)+P(E₂) 8. If E_1 and E_2 are disjoint events, then the probability of E_1 or E_2 is given by $P(E_1 \cup E_2) = $ _____.

P(E₁)P(E₂) 9. If an operation is performed twice, then the probability of getting event E_1 the first time and event E_2 the second time is _____.

complement 10. The event E' containing all the outcomes not in E is called the _____ of E.

1 11. Since E and E' are disjoint events, and their union is the certain event, $P(E) + P(E') = $ _____.

SECTION 11-4, EXERCISES

If a pair of dice is thrown once, what is the probability that the sum thrown is:

1. 8
2. 10
3. 4
4. 11
5. 6
6. 5
7. an even number
8. less than 5
9. more than 7
10. an odd number
11. How many ways can you deal a five-card hand containing four aces? What is the probability of being dealt four aces at poker?
12. How many five-card hands consisting of only aces, kings, and queens are there? What is the probability of being dealt such a hand?
13. How many poker hands are there containing all hearts? What is the probability of being dealt a heart flush?
14. How many poker hands contain no card higher than a 6 (i.e. 2, 3, 4, 5, 6 only)? What is the probability of being dealt such a hand?
15. In one particularly revolting poker game, 3's and 9's are wild. What is the probability of being dealt only 3's and 9's in the first two cards?
16. What is the probability that the first two cards you are dealt in seven-card stud are both aces?
17. If the two one-eyed jacks and four deuces are wild cards, what is the probability that the first two cards dealt are wild cards?
18. What is the probability that all three down cards in seven-card stud will be aces?

Example 1

Example 2

In a single roll of a pair of dice, what is the probability that the sum will be:

19. 8 or 4
20. 10 or 11
21. 3 or 12
22. 2 or 6
23. 7 or 11 or 12
24. 6 or 7 or 8
25. 6 or a pair
26. 10 or a pair
27. odd or a multiple of 3
28. even or a multiple of 3

Example 3

If a card is drawn from a deck, what is the probability that:

29. It is an ace or king?
30. It is a 10 or lower (10, 9, 8, 7, 6, 5, 4, 3, 2)?
31. It is a heart or a spade face card?
32. It is a spade or even (2, 4, 6, 8, 10) diamond?
33. It is an ace or a club?
34. It is a red card or a face card?

A pair of dice are rolled twice. What is the probability of getting:

35. A five, then a seven?
36. A six, then a three?
37. A seven, then a seven?
38. An eight, then a seven?
39. At least one seven?
40. At least one nine?
41. At least one even number?
42. At least one number divisible by 4?

Example 4

What is the probability of:

Example 5
43. Getting at least one head in three flips of a coin?
44. Drawing at least one heart in two draws from full decks?
45. Drawing at least one face card in four draws from full decks?
46. Drawing at least one red card in three draws from full decks?

II-5 REVIEW FOR CHAPTER

Evaluate each of the following.

1. $\dfrac{63!}{60!\,3!}$ 2. $\dfrac{37!}{30!\,7!}$ 3. $\dfrac{n!}{(n-3)!\,3!}$

4. $\dfrac{n!\,(n-2)!}{(n+2)!}$ 5. $\dfrac{n!\,n!}{(n+1)!\,(n-1)!}$ 6. $\dfrac{(n+7)!}{n!\,7!}$

7. How many four-digit numbers can be formed using the digits 5, 6, 7, and 8?
8. How many ways can you go from A to B to C if there are two paths from A to B, and five paths from B to C?
9. How many five-letter "words" may be formed using the letters a, b, c, d, and e, if no letter can be used twice?
10. How many three-letter "words" may be formed using the letters a, b, c, d, and e, if you can use any letter as often as you wish?

Evaluate each of the following.
11. $P(8;3)$ 12. $P(10;4)$ 13. $P(27;4)$
14. $C(9;4)$ 15. $C(11;3)$ 16. $C(32;6)$
17. How many basketball teams can you pick from 11 players, all of whom can play all positions?
18. How many basketball teams can you pick from 11 players if 5 of them are guards, 5 forwards, and 1 center?
19. How many baseball batting orders can a manager form if he only changes the first five positions?
20. What is the probability of throwing (a) a 10 on one roll of a pair of dice? (b) A 4 on one roll of a pair of dice? (c) A 5 on one roll of a pair of dice?
21. A student is told he can answer 8 questions of an 11-question examination. How many different ways are there for him to choose the questions he will answer?
22. How many ways can nine men be assigned to three triple rooms?
23. How many seven-digit numbers can be written using the digit 6 three times, the digit 4 twice, and the digit 3 twice?

24. How many partitions of a seven-element set are there into two subsets, one with four elements and one with three elements?

Write out the expansion for each of the following.

25. $(2x + 2y)^5$

26. $\left(\dfrac{1}{2}x - 3y\right)^6$

27. Find the fourth term in the expansion of $(x - 3)^7$.
28. Find the seventh term in the expansion of $(2x - 1/3y)^{15}$.
29. Four cards are drawn from a deck of cards. Find the probability that (a) two are 10s and two are 7s. (b) All four are aces.
30. A committee of three is to be chosen at random from a club having nine men and eight women members. Find the probability that all three are men.
31. What is the probability that the committee of exercise 30 will consist of one man and two women?
32. If seven coins are tossed, find the probability that (a) all seven will be heads. (b) Three will be tails and four heads. (c) Five will be heads and two tails.
33. Determine the probability that a random family of five children has at least three girls and at least one boy. Assume that half of all children are boys and half girls.

12 SEQUENCES AND SERIES

12-1 INDUCTION

In this chapter we will deal with *sequences*, which are arrays of numbers like $x_1, x_2, x_3, x_4, \ldots$, and with *series*, which are indicated sums like $x_1 + x_2 + x_3 + \cdots$. Both the concepts "sequence" and "series" depend on the idea that we have a particular number x_n associated with each natural number n. Before we treat general sequences and series, we therefore need to know more about the set of natural numbers itself. In this section we will discuss the so-called inductive properties of the natural numbers, and what a proof by mathematical induction is.

We have agreed that the set of natural numbers consists of $1, 2, 3, 4, \ldots$, and so on. This is a suitable agreement for most purposes. However, if we want to prove something about the set of all natural numbers, the phrase "and so on" is a little too vague. Therefore we now introduce a definition for the set **N** of natural numbers which is exact enough so we can base formal arguments on it.

The basic idea of natural numbers is simple: you start with the multiplicative identity, 1; then you add 1, and get the number $1 + 1$, or 2; then you add 1 again, and get the number $1 + 1 + 1$, or 3; etc. Now what does "*etc.*" mean? Simply that for each number k in **N**, the number $k + 1$ is also in **N**. We incorporate this idea in the following definition.

> The set N of natural numbers is the smallest set of numbers S with the two properties: (a) $1 \in S$ and (b) $k + 1 \in S$ whenever $k \in S$.

In other words, if S has properties (a) and (b), then S contains all natural numbers. Moreover, **N** is the smallest set which satisfies (a) and (b). A set which satisfies (a) and (b) has the *inductive properties*. Hence by definition any set with the inductive properties contains all natural numbers.

To see how arguments are based on this definition, we consider the following problem: Is the number $2^{2^k} - 1$ always divisible by 3 if k is a natural number? The first thing a sensible person would do would be to check some special cases like $k = 1, 2, 3, 4$. We let a_k denote $2^{2^k} - 1$, and compute.

$k = 1 \qquad a_1 = 2^{2^1} - 1 = 2^2 - 1 = 3 = 3 \cdot 1$

$k = 2 \qquad a_2 = 2^{2^2} - 1 = 2^4 - 1 = 15 = 3 \cdot 5$

$k = 3 \qquad a_3 = 2^{2^3} - 1 = 2^8 - 1 = 255 = 3 \cdot 85$

$k = 4 \qquad a_4 = 2^{2^4} - 1 = 2^{16} - 1 = 65{,}535 = 3 \cdot 21{,}845$

We see that a_k is divisible by 3 for $k = 1, k = 2, k = 3$, and $k = 4$. Of course we can not show that 3 divides $2^{2^k} - 1$ for *every* natural number k just by checking a few special cases, but at least we know that the theorem might be true.

To show that the statement is true for *all* natural numbers k, we give what is called a *proof by induction*. We let S be the set of all natural numbers k such that 3 divides $2^{2^k} - 1$. If we can show that S satisfies the inductive properties (a) and (b), then it will follow that S contains every natural number. We have already shown that $1 \in S$ since $a_1 = 2^2 - 1 = 3$ is divisible by 3. Hence S satisfies property (a). Now we try to find a pattern which relates the $(k + 1)$st to the kth case in order to show that S also has property (b).

The first four cases again are

$$a_1 = 2^2 - 1,$$
$$a_2 = 2^4 - 1 = (2^2 - 1)(2^2 + 1),$$
$$a_3 = 2^8 - 1 = (2^4 - 1)(2^4 + 1),$$
$$a_4 = 2^{16} - 1 = (2^8 - 1)(2^8 + 1).$$

We observe that a_k is a factor of a_{k+1}, at least for $k = 1,2,3$, so that if 3 divides a_k, 3 also divides a_{k+1}. If we show that a_k divides a_{k+1} for all k, then we will have verified that S also satisfies the inductive property (b), which completes the proof by induction. All the details of the proof are carried out in the following example.

Example 1 Prove by induction that 3 divides $2^{2^k} - 1$ for all natural numbers k.

Solution. Let $a_k = 2^{2^k} - 1$ as above, and let S be the set of natural numbers for which 3 divides a_k. As we saw above, $1 \in S$ since $a_1 = 3$, so S satisfies property (a). Let k be any natural number in S; that is, assume that 3 divides $a_k = 2^{2^k} - 1$. Now write a_{k+1} as follows:

$$\begin{aligned} a_{k+1} &= 2^{2^{k+1}} - 1 \\ &= 2^{2^k \cdot 2} - 1 \\ &= (2^{2^k})^2 - 1 \\ &= (2^{2^k} - 1)(2^{2^k} + 1) \\ &= a_k \cdot (2^{2^k} + 1). \end{aligned}$$

Since 3 divides a_k, 3 divides a_{k+1}. Hence $k + 1$ is in S whenever k is in S. Therefore S also satisfies property (b), and so $S = N$.

Example 2 Prove by induction that $2k \geq k + 1$ for all $k \in \mathbf{N}$.

Solution. For $k = 1$, $2k = 2$, and $k + 1 = 2$, so the statement is true for $k = 1$. Now let k be any natural number for which the statement is true; that is, assume that

$$2k \geq k + 1. \tag{1}$$

For $k + 1$, the statement is

$$2(k + 1) \geq (k + 1) + 1. \tag{2}$$

This is a consequence of (1) by the following argument:

$$\begin{aligned} 2(k + 1) &= 2k + 2 \\ &\geq k + 1 + 2 \qquad \text{using (1)} \\ &\geq (k + 1) + 1. \end{aligned}$$

Hence (2) follows from (1). The set of numbers for which the statement $2k \geq k + 1$ is true has both inductive properties, and hence contains all natural numbers.

Example 3 Show that the sum of the first n odd integers equals n^2, for every $n \in N$.

Solution. The odd integers are $1, 3, 5, \ldots$, and the nth odd integer is $2n - 1$. Hence we want to show that for every $n \in N$,

$$1 + 3 + 5 + \cdots + (2n - 1) = n^2. \tag{3}$$

The formula holds for $n = 1$. Let n be any positive integer for which (3) is true. The sum of the first $n + 1$ odd integers is then

$$[1 + 3 + 5 + \cdots + (2n - 1)] + (2n + 1)$$
$$= n^2 + (2n + 1) \qquad \text{using (3)}$$
$$= (n + 1)^2.$$

That is, the sum of the first $n + 1$ odd integers is $(n + 1)^2$ provided the sum of the first n odd integers is n^2. Inductive property (b) holds, as well as (a), so the statement is true for all natural numbers.

It is important to notice that inductive proherty (b) is not sufficient by itself. For example, let us try to show that the sum of the first n even integers is $n^2 + n + 2$; that is, that

$$2 + 4 + 6 + \cdots + 2n = n^2 + n + 2. \tag{4}$$

If n is a natural number for which (4) is a true statement, then the sum of the first $n + 1$ even integers is

$$[2 + 4 + 6 + \cdots + 2n] + 2(n + 1)$$
$$= [n^2 + n + 2] + 2(n + 1) \qquad \text{using (4)}$$
$$= (n^2 + 2n + 1) + n + 3$$
$$= (n + 1)^2 + (n + 1) + 2.$$

Hence the formula holds for $n + 1$ whenever it holds for n. However, the formula does *not* hold for $n = 1$, since $2 \neq (1^2) + (1) + 2$. In fact, the formula does not hold for any n since the sum of the first n even integers is always strictly less than $n^2 + n + 2$ (see Exercise 17).

The discussion above shows that you cannot make an inductive proof by verifying only inductive property (b), which is the "k to $k + 1$" part of the argument. You must have a starting place for your trip up the natural numbers. If you are going to prove that something is true for *all* natural numbers, then your starting place must be 1. However, if you can show that some natural number N is in a set S, and that $k + 1 \in S$ whenever $k \in S$ (and $k \geq N$), then S contains all $n \geq N$. In other words, your starting place in an inductive proof can be any integer N, and then your assertion holds for all $n \geq N$.

Example 4. Show that $2n + 1 < 2^n$ if $n \geq 3$.

Solution. For $N = 3$ we verify the inequality: $2 \cdot 3 + 1 = 7 < 2^3 = 8$. Now suppose that $k \geq 3$ and $2k + 1 < 2^k$. Then

$$
\begin{aligned}
2(k + 1) + 1 &= 2k + 3 \\
&= (2k + 1) + 2 \\
&< 2^k + 2 \qquad &\text{by assumption} \\
&< 2^k + 2^k \qquad &2 < 2^k \text{ if } k \geq 3 \\
&= 2 \cdot 2^k \\
&= 2^{k+1}.
\end{aligned}
$$

SECTION 12-1, REVIEW QUESTIONS

natural   1. The symbol N denotes the set of _____ numbers.

2. The natural numbers form the smallest set which contains 1, and contains

k + 1   _____ whenever it contains k.

3. The conditions (a) $k \in S$ and (b) $k + 1 \in S$ whenever $k \in S$ are called

inductive   _____ properties.

contains   4. Any set with both inductive properties _____ the set **N**.

5. To prove a statement by induction, you show that the set of numbers for

true   which the statement is _____ has the inductive properties.

6. A proof by induction shows that a given statement holds for the set of all

natural   _____ numbers.

SECTION 12-1, EXERCISES

Example 1   1. Show that $n^2 + n$ is even (divisible by 2) for all n.
2. Show that $n^2 + 3n$ is even for all n.
3. Show that $n^3 + 2n$ is divisible by 3 for all n.

4. Show that $n^3 + 5n$ is divisible by 3 for all n.
5. Show that 4 divides $3^{2n} - 1$ for all n.
6. Show that 8 divides $3^{2n} - 1$ for all n.
7. Show that 9 divides $2^{3n} + 1$ for all n.
   (Hint: $a^3 + 1 = (a + 1)(a^2 - a + 1)$.)
8. Show that 8 divides $2^{3n} - 1$ for all n.
   (Hint: $a^3 - 1 = (a - 1)(a^2 + a + 1)$.)
9. Show that $3n \geq n + 2$ for all n.
10. Show that $4n \geq n + 3$ for all n.
11. Show that $n^2 + 1 \geq 2n$ for all n.
12. Show that $n^3 + 2 \geq 3n$ for all n.
13. Show that $2n \leq 2^n$ for all n.
14. Show that $3n \leq 3^n$ for all n. (Hint: $3 \leq 2 \cdot 3^k$.)
15. Show that $2^{n-1} \leq n!$ for all n. (Hint: $2 \leq k + 1$.)
16. Show that if $a > 0$, then $(a + 1)^n \geq 1 + na$ for all n.

Example 2

Show that each of the following is true for every natural number n.

17. $2 + 4 + 6 + \cdots + 2n = n(n + 1)$

Example 3

18. $1 + 2 + 3 + \cdots + n = \dfrac{1}{2}n(n + 1)$

19. $1 + 5 + 9 + \cdots + (4n - 3) = n(2n - 1)$

20. $1 + 4 + 7 + \cdots + (3n - 2) = \dfrac{n}{2}(3n - 1)$

21. $1 + 2 + 4 + \cdots + 2^{n-1} = 2^n - 1$

22. $\dfrac{1}{2} + \dfrac{1}{4} + \dfrac{1}{8} + \cdots + \dfrac{1}{2^n} = 1 - \dfrac{1}{2^n}$

23. $1^2 + 2^2 + 3^2 + \cdots + n^2 = \dfrac{n(n + 1)(2n + 1)}{6}$

24. $1^2 + 3^2 + 5^2 + \cdots + (2n - 1)^2 = \dfrac{n(4n^2 - 1)}{3}$

25. Show that $2n + 1 \leq 2^n$ if $n \geq 3$.

Example 4

26. Show that $6n + 6 \leq 3^n$ if $n \geq 3$.
27. Show that $n^2 \leq 2^n$ if $n \geq 4$.
    (Hint: use Exercise 25.)
28. Show that $3n^2 + 3n + 1 \leq 2 \cdot 3^n$ if $n \geq 3$.
    (Hint: use Exercise 26.)
29. Show that $n^3 \leq 3^n$ if $n \geq 3$.
    (Hint: use Exercise 28.)
30. Show that $2n + 1 < n!$ if $n \geq 4$.
31. Show that $n^2 \leq n!$ if $n \geq 5$.
    (Hint: use Exercise 30.)

12-2 SEQUENCES

A *sequence* is an infinite array of numbers of the form

$$a_1, a_2, a_3, a_4, \ldots, a_n, \ldots$$

For example, we would indicate the sequence of odd integers by

$$1, 3, 5, 7, 9, \ldots, 2n - 1, \ldots \qquad (1)$$

Similarly, the array

$$1, \frac{1}{4}, \frac{1}{9}, \frac{1}{16}, \ldots, \frac{1}{n^2}, \ldots \qquad (2)$$

indicates the sequence whose nth term is $1/n^2$. The important idea is that in a sequence you have a definite well-defined nth term for every natural number n. This can be formalized by saying that *a sequence is a function defined on the set of all natural numbers.*

If a is such a function, then we write a_n instead of $a(n)$ for the nth *term*, which is just the value of the function at the integer n. We also use the notation $\{a_n\}$ to indicate the sequence whose terms are $a_1, a_2, a_3, \ldots$. One way of defining a sequence, like any other function, is to give a formula for the value at any n; that is, to give a formula for a_n. The sequences (1) and (2) are defined, respectively, by the formulas $a_n = 2n - 1$, and $a_n = 1/n^2$.

Example 1 Write the first five terms, and the tenth term of the sequences $\{a_n\}$, $\{b_n\}$, and $\{c_n\}$ defined by (a) $a_n = (2n + 1)^2$; (b) $b_n = 1 + 1/10^n$; and (c) $c_n = 7$.

Solution. (a) $a_1 = 9$; $a_2 = 25$; $a_3 = 49$; $a_4 = 81$; $a_5 = 121$; and $a_{10} = (2 \cdot 10 + 1)^2 = 21^2 = 441$.

(b) $b_1 = 1.1$; $b_2 = 1.01$; $b_3 = 1.001$; $b_4 = 1.0001$; $b_5 = 1.00001$; and $b_{10} = 1.0000000001$.

(c) $c_1 = 7$; $c_2 = 7$; $c_3 = 7$; $c_4 = 7$; $c_5 = 7$; and $c_{10} = 7$.

We frequently also indicate a sequence by listing the first few terms. In this case the presumption is that the first few terms establish an obvious pattern, which makes it clear what the rest of the terms are, and allows us to determine the nth term for any n.

Example 2 Find a formula for the nth term of the sequence $a = \{1, 9, 25, 49, 81, \ldots\}$.

Solution. The terms of the sequence are perfect squares, and can be written $\{1^2, 3^2, 5^2, 7^2, 9^2, \ldots\}$. Hence the nth term is the square of the nth odd integer. The nth odd integer is $2n - 1$, and therefore $a_n = (2n - 1)^2$.

Another way of defining a sequence is to use the inductive properties of the natural numbers. Suppose we give a value for a_1, and for each k tell how to find a_{k+1} in terms of a_k. Then we can find a_n for any n. This is called a *recursive definition*, or an *inductive definition*.

Example 3 Let $a_1 = 1$ and $a_{k+1} = 2a_k + 1$ for each $k \in \mathbf{N}$. Write the first five terms of this inductively defined sequence.

Solution. We simply follow the prescription given in the inductive definition—each new term is found by multiplying the last term by 2, and then adding 1. We get

$$a_1 = 1,$$
$$a_2 = 2 \cdot 1 + 1 = 3,$$
$$a_3 = 2 \cdot 3 + 1 = 7,$$
$$a_4 = 2 \cdot 7 + 1 = 15,$$
$$a_5 = 2 \cdot 15 + 1 = 31.$$

One of the important applications of the inductive definition is the formal treatment of the exponential notation x^n. We have already discussed the properties of x^n based on the informal idea of "x used as a factor n times". Now let us see how a formal definition allows us to make inductive proofs of these properties.

For any number x, we define a sequence of numbers $x^1, x^2, x^3, \ldots$. That is, we define what is meant by x^n for any $n \in \mathbf{N}$. The inductive definition is:

$$x^1 = x, \tag{3}$$
$$x^{k+1} = x \cdot x^k. \tag{4}$$

Example 4 Prove by induction that $(xy)^n = x^n y^n$ for all $n \in \mathbf{N}$.

Proof. For $n = 1$, the statement is $(xy)^1 = x^1 y^1$. By definition (4), $(xy)^1 = xy$, $x^1 = x$, and $y^1 = y$, so the identity holds for $n = 1$. Now let k be any integer for which the statement is true; that is, assume that

$$(xy)^k = x^k y^k. \tag{5}$$

Then,

$$(xy)^{k+1} = (xy)(xy)^k \quad \text{by the inductive definition (4)}$$
$$= (xy)(x^k y^k) \quad \text{by assumption (5)}$$
$$= (x \cdot x^k)(y \cdot y^k) \quad \text{commutative and associative laws}$$
$$= x^{k+1} y^{k+1}. \quad \text{by definition (4)}$$

Hence the law holds for $k + 1$ whenever it holds for k. The law also holds for $n = 1$, and therefore for all $n \in \mathbf{N}$.

SECTION 12-2, REVIEW QUESTIONS

natural

1. A sequence is a function defined on the set of _____ numbers.

2. The nth term of the sequence is the value of the function at the natural number _____.

n

a_n

3. The nth term of the sequence a is written _____.

4. A sequence can be defined inductively by specifying a_1, and defining a_{k+1} in terms of _____.

a_k

5. When a sequence is indicated by listing several terms within braces, these terms must indicate an obvious pattern from which we can determine the value of a_n for every _____.

n

SECTION 12-2, EXERCISES

Write the first four terms and the indicated term of each sequence below.

Example 1

1. $a_n = 3^{n-1}$ seventh term   2. $a_n = n^2(n + 1)$ eighth term

3. $a_n = 2 + \dfrac{1}{3^n}$ sixth term   4. $a_n = n + \dfrac{1}{2^n}$ sixth term

Find a formula for the nth term of each of the following sequences.

Example 2

5. $a = \{4, 8, 12, 16, \ldots\}$ 6. $a = \{1, 3, 5, 7, \ldots\}$
7. $a = \{-2, 4, -8, 16, \ldots\}$ 8. $a = \{1, 3, 6, 10, \ldots\}$
9. $a = \{-4, -3, -2, -1, 0, \ldots\}$ 10. $a = \{-3, 5, -7, 9, \ldots\}$
11. $a = \left\{\dfrac{1}{2}, \dfrac{2}{3}, \dfrac{3}{4}, \dfrac{4}{5}, \ldots\right\}$ 12. $a = \left\{\dfrac{1}{3}, \dfrac{1}{4}, \dfrac{1}{5}, \dfrac{1}{6}, \ldots\right\}$
13. $a = \{3, -7, 11, -15, \ldots\}$ 14. $a = \{-1, 1, -1, 1, \ldots\}$

15. $a = \left\{1, 1, \dfrac{3}{4}, \dfrac{1}{2}, \dfrac{5}{16}, \ldots\right\}$ 16. $a = \left\{\dfrac{1}{3}, \dfrac{-1}{5}, \dfrac{1}{9}, \dfrac{-1}{17}, \ldots\right\}$

Write the first five terms of the sequences defined inductively:

17. $a_1 = -1$ $a_{k+1} = a_k + 2$ 18. $a_1 = 3$ $a_{k+1} = a_k - 1$ *Example 3*

19. $a_1 = 2$ $a_{k+1} = -3a_k$ 20. $a_1 = 4$ $a_{k+1} = \dfrac{1}{10}a_k$

21. $a_1 = 2$ $a_{k+1} = a_k^2$ 22. $a_1 = 3$ $a_{k+1} = a_k(a_k + 2)$

Prove the following laws by mathematical induction.

23. Prove that $(x/y)^n = x^n/y^n$ for all x, all $y \neq 0$, and all $n \in \mathbf{N}$. *Example 4*
24. Let m be any fixed natural number. Show that $(x^m)^n = x^{(mn)}$ for all $n \in \mathbf{N}$.
    [Hint: By definition, $(x^m)^{n+1} = x^m \cdot (x^m)^n$.]

12-3 ARITHMETIC AND GEOMETRIC SEQUENCES

Any formula giving a_{k+1} in terms of a_k will define some sequence, once the first term is also specified. In this section we will look at the sequences defined by two particularly simple recursive formulas.

The first type of sequence, called an *arithmetic sequence*, is characterized by a recursive formula of the form

$$a_{k+1} = a_k + d$$

where d is some fixed number. That is, *an arithmetic sequence is one in which each term is found by adding a constant, d, to the preceding term*. Another way of saying the same thing is that an arithmetic sequence, sometimes also called an *arithmetic progression*, is one in which the difference between any two successive terms is a constant.

$$a_{k+1} - a_k = d.$$

Example 1 Find the seventh term of the arithmetic sequence $\{4, 9, 14, 19, \ldots\}$, and find a formula for the nth term.

Solution. The common difference d is 5 since

$9 - 4 = 14 - 9 = 19 - 14 = 5$. Hence the succeeding terms would be

$$4, 9, 14, 19, 24, 29, 34, 39, \ldots$$

and the seventh term is 34. In general, the terms can be written

Sequences 459

$$4$$
$$4 + 5$$
$$4 + 2 \cdot 5$$
$$4 + 3 \cdot 5$$

etc. Hence the nth term is

$$a_n = 4 + (n - 1)5.$$

The general arithmetic sequence with initial term a_1 and common difference d is defined inductively by

$$a_1 = a_1 \quad \text{and} \quad a_{k+1} = a_k + d.$$

The nth term is clearly

$$a_n = a_1 + (n - 1)d.$$

A formal proof by induction that this is correct is left as an exercise.

Example 2 The third term of an arithmetic sequence is 1 and the sixth term is -8. Find the initial term a_1, the common difference d, and a formula for a_n. Write the first six terms.

Solution. We have

$$a_3 = a_1 + 2d = 1,$$
$$a_6 = a_1 + 5d = -8.$$

We solve these two simultaneous linear equations for a_1 and d. Subtracting gives

$$-3d = 9,$$
$$d = -3.$$

Hence
$$a_1 + 2(-3) = 1,$$
$$a_1 = 7.$$

Now we have

$$a_n = 7 + (n - 1) \cdot (-3).$$

The first six terms are 7, 4, 1, -2, -5, -8.

The second kind of sequence we will look at is characterized by the fact that there is a common ratio between successive terms. That is, each term is found by multiplying the preceding term by some number r. A *geometric sequence*, or *geometric progression*, is one defined inductively as follows:

$$a_1 = a_1 \quad \text{and} \quad a_{k+1} = ra_k$$

where r is some fixed number.

Example 3 Find the eighth term of the geometric sequence $\{1, -2, 4, -8, \ldots\}$, and find a formula for a_n.

Solution. The common ratio is -2, since $-2/1 = 4/-2 = -8/4 = -2$. Hence the first eight terms are:

$$1, -2, 4, -8, 16, -32, 64, -128.$$

The nth term is

$$a_n = 1 \cdot (-2)^{n-1}.$$

In general, if the initial term is a_1, and the common ratio is r, the successive terms of a geometric sequence are

$$a_1, a_1 r, a_1 r^2, a_1 r^3, a_1 r^4, \ldots.$$

Hence the nth term is given by

$$a_n = a_1 r^{n-1}.$$

Example 4 A man plans to win at roulette by tripling his bet each time he loses. If he starts with a $2.00 bet, and loses five times in a row, what must the sixth bet be.

Solution. The successive bets will form the geometric sequence $2, 2 \cdot 3, 2 \cdot 3^2, 2 \cdot 3^3, \ldots$. The sixth bet (the sixth term in the sequence) will be

$$a_6 = 2 \cdot 3^5 = 2 \cdot 243 = 486 \quad \text{dollars.}$$

The *arithmetic mean*, or *average*, of two numbers a and b is $(a + b)/2$. The numbers

$$a, \frac{(a+b)}{2}, b$$

are terms in an arithmetic sequence, with

$$d = \frac{b - a}{2}.$$

More generally, we say we have *inserted* n *arithmetic means between* a *and* b provided we have found n numbers $m_1, m_2, \ldots, m_n$ so that

$$a, m_1, m_2, \ldots, m_n, b$$

are terms in an arithmetic sequence. The same terminology applies to geometric sequences. Thus the geometric mean of a and b is

$\sqrt{ab}$, since a, $a\sqrt{b/a} = \sqrt{ab}$, $a(\sqrt{b/a})^2 = b$ are in a geometric sequence.

Example 5 (a) Insert four arithmetic means between 5 and 20.
(b) Insert three geometric means between 16 and 1.

Solution. (a) The arithmetic sequence must be

$$5, 5 + d, 5 + 2d, 5 + 3d, 5 + 4d, 20 = 5 + 5d.$$

Hence
$$20 = 5 + 5d,$$
$$d = 3,$$

and the four arithmetic means are 8, 11, 14, 17.

(b) The geometric sequence must be

$$16, 16r, 16r^2, 16r^3, 1 = 16r^4.$$

Hence
$$r^4 = \frac{1}{16},$$

$$r = \pm \frac{1}{2}.$$

There are two sets of geometric means, and the corresponding sequences are

$$16, -8, 4, -2, 1 \quad \text{and} \quad 16, 8, 4, 2, 1.$$

SECTION 12-3, REVIEW QUESTIONS

1. Any sequence defined inductively by a formula of the form $a_{k+1} = a_k + d$ is called an _____ sequence.

    arithmetic

2. In an arithmetic sequence the difference between any two successive terms is the constant _____.

    d

3. The nth term of an arithmetic sequence is given by $a_n = $ _____.

    $a_1 + (n-1)d$

4. A geometric sequence is one defined by an inductive formula of the form $a_{k+1} = $ _____.

    ra_k

5. The ratio of any two successive terms is a constant in a _____ sequence.

    geometric

6. The nth term of an arithmetic sequence is $a_n = a_1 + (n-1)d$, and the nth term of a geometric sequence is given by $a_n = $ _____.

    $a_1 r^{n-1}$

7. To insert three arithmetic means between a and b means to find three numbers m_1, m_2, and m_3 so that a, m_1, m_2, m_3, b are terms in an _____ sequence.   arithmetic

8. If a, m, b are terms in a geometric sequence, then m is the _____ mean of a and b.   geometric

9. The geometric mean of a and b is $\sqrt{ab}$, and the arithmetic mean is _____. $\dfrac{a+b}{2}$

SECTION 12-3, EXERCISES

Find the indicated term and a formula for the nth term of each of the following arithmetic sequences.

1. $a = \{3, 8, 13, \ldots\}$ sixth term   Example 1
2. $a = \{-2, -5, -8, \ldots\}$ twelfth term
3. $a = \{\frac{1}{2}, \frac{3}{4}, 1, \ldots\}$ seventh term
4. $a = \{1, \frac{-1}{2}, -2, \ldots\}$ fifth term
5. $a = \{8, -2, -12, \ldots\}$ sixth term
6. $a = \{1, 1.1, 1.2, \ldots\}$ eighth term
7. The second term of an arithmetic sequence is 6, and the fourth term is 16. Find a_1, the common difference, and a formula for a_n. Write the first five terms.   Example 2
8. The first term of an arithmetic sequence is 1/3, and the fifth term is 3. Find the common difference and a formula for a_n. Write the first six terms.
9. The third term of an arithmetic sequence is 15, and the seventh term is -27. Find a_1, the common difference, and a formula for a_n. Write the first six terms.

Find the indicated term and a formula for the nth term of each of the following geometric sequences.

10. $a = \{2, 6, 18, \ldots\}$ sixth term   Example 3
11. $a = \{5, 10, 20, \ldots\}$ eighth term
12. $a = \{25, -5, 1, \ldots\}$ seventh term
13. $a = \{1, \frac{-1}{2}, \frac{1}{4}, \frac{-1}{8}, \ldots\}$ sixth term
14. $a = \{i, -1, -i, \ldots\}$ fifth term
15. $a = \{\sqrt{2}, \sqrt{6}, 3\sqrt{2}, \ldots\}$ seventh term
16. A sum of $200 is invested today at 5% per year. What will be the total   Example 4

amount in 6 years if the interest is compounded annually?

17. A man agrees to work at the rate of 1 cent the first day, 2 cents the second day, 4 cents the third day, and so on. How much will the man earn on the tenth day? On the fifteenth day?

Example 5 18. Insert four arithmetic means between -2 and -8.

19. Insert ten arithmetic means between -3 and -36.
20. Insert seven arithmetic means between 13 and 29.
21. Insert three geometric means between 15 and 5/27.
22. Insert three geometric means between 1 and 256.
23. Insert two geometric means between $\sqrt{5}$ and 5.
24. Prove that if an arithmetic sequence $\{a_n\}$ is defined inductively by $a_1 = a$, $a_{k+1} = a_k + d$, then the nth term is given by $a_n = a + (n-1)d$.
25. Prove that if a geometric sequence $\{a_n\}$ is defined inductively by $a_1 = a$, $a_{k+1} = ra_k$, then the nth term is given by $a_n = a_{n-1}r = ar^{n-1}$.

12-4 SERIES

The indicated sum of the terms of a sequence is called a *series*. Hence a series is an array of the form

$$a_1 + a_2 + a_3 + \cdots + a_n + \cdots. \tag{1}$$

For the moment we will regard such an infinite sum merely as a formal expression. Later we will see how it is sometimes reasonable to interpret such an infinite sum as a real number.

For any natural number n, the nth *partial sum* of the series (1) is the sum of the first n terms. If we let S_n denote the nth partial sum, then

$$S_n = a_1 + a_2 + \cdots + a_n.$$

With each series $a_1 + a_2 + \cdots + a_n + \cdots$ we have now associated a new sequence $\{S_n\}$, called the *sequence of partial sums*. Notice that

$$S_{k+1} = (a_1 + a_2 + \cdots + a_k) + a_{k+1}$$
$$= S_k + a_{k+1}.$$

Hence we can make a formal inductive definition of the sequence of partial sums as follows:

$$\begin{aligned} S_1 &= a_1, \\ S_{k+1} &= S_k + a_{k+1}. \end{aligned} \tag{2}$$

Example 1 Write the partial sums S_1, S_2, S_3, S_4, and give a general formula for S_n for each of the series

$$1 + 2 + 3 + \cdots + k + \cdots, \tag{3}$$

$$1 + 4 + 9 + \cdots + k^2 + \cdots . \tag{4}$$

Solution. For series (3) we have $S_1 = 1$, $S_2 = 1 + 2 = 3$, $S_3 = 1 + 2 + 3 = 6$, and $S_4 = 1 + 2 + 3 + 4 = 10$. From Exercise 18 in Section 12-1, we know that $1 + 2 + \cdots + n = \frac{1}{2}n(n+1)$, or or

$$S_n = \frac{1}{2} n(n+1). \tag{5}$$

For series (4) we have $S_1 = 1$, $S_2 = 1 + 4 = 5$, $S_3 = 1 + 4 + 9 = 14$, and $S_4 = 1 + 4 + 9 + 16 = 30$. From Exercise 7 in Section 12-1, we have
$1 + 4 + 9 + \cdots + n^2 = (1/6)n(n+1)(2n+1)$. Hence

$$S_n = \frac{1}{6} n(n+1)(2n+1). \tag{6}$$

The symbol Σ (Greek letter sigma) is frequently used to abbreviate sums. We write $\sum_{k=1}^{n} a_k$ (this is read as the sum of a_k as k runs from 1 to n) for the nth partial sum S_n; thus

$$\sum_{k=1}^{n} a_k = a_1 + a_2 + \cdots + a_n. \tag{7}$$

For example, the kth term of the series (3) is just k, so formula (5) can be written

$$\sum_{k=1}^{n} k = \frac{1}{2} n(n+1). \tag{8}$$

Similarly, formula (6) can be written

$$\sum_{k=1}^{n} k^2 = \frac{1}{6} n(n+1)(2n+1). \tag{9}$$

The letter k in the definition (7) is called a dummy variable, which means that any other letter could be used instead. For example,

$$\sum_{i=1}^{n} a_i = \sum_{j=1}^{n} a_j = \sum_{k=1}^{n} a_k.$$

Example 2 If $a_k = k(k-1)$, find $\sum_{k=1}^{5} a_k$.

Solution.

$$\sum_{k=1}^{5} a_k = a_1 + a_2 + a_3 + a_4 + a_5$$

$$= 1(1-1) + 2(2-1) + 3(3-1) + 4(4-1) + 5(5-1)$$
$$= 0 + 2 + 6 + 12 + 20$$
$$= 40.$$

The following formulas use Σ notation to state some simple facts about grouping and factoring in sums.

$$\sum_{k=1}^{n} (a_k + b_k) = \sum_{k=1}^{n} a_k + \sum_{k=1}^{n} b_k \qquad (10)$$

$$\sum_{k=1}^{n} ca_k = c \sum_{k=1}^{n} a_k \qquad (11)$$

$$\sum_{k=1}^{n} c = nc \qquad (12)$$

Written out completely, these statements would read:

$$(a_1 + b_1) + (a_2 + b_2) + \cdots + (a_n + b_n)$$
$$= (a_1 + \cdots + a_n) + (b_1 + \cdots + b_n) \quad (10')$$
$$ca_1 + ca_2 + \cdots + ca_n = c(a_1 + a_2 + a_3 + \cdots + a_n) \quad (11')$$
$$\underbrace{c + c + \cdots + c}_{n \text{ terms}} = nc \quad (12')$$

Example 3 Find a formula for $\sum_{k=1}^{n} k(k-1)$ and check it against Example 2 for $n = 5$.

Solution. We use formulas (10), (11), and (12), together with the earlier formulas (8) and (9).

$$\sum_{k=1}^{n} k(k-1) = \sum_{k=1}^{n} (k^2 - k)$$

$$= \sum_{k=1}^{n} k^2 - \sum_{k=1}^{n} k \qquad \text{from (10)}$$

$$= \frac{1}{6} n(n+1)(2n+1) - \frac{1}{2} n(n+1)$$

from (8) and (9)

$$= \frac{1}{6} n(n+1)(2n+1-3)$$

$$= \frac{1}{3}n(n+1)(n-1)$$

For $n = 5$, we get

$$\sum_{k=1}^{5} k(k-1) = \frac{1}{3}(5)(6)(4) = 40,$$

which is in agreement with the direct computation of Example 2.

The sum of the terms of an arithmetic sequence is called an *arithmetic series*, and the sum of the terms of a geometric sequence is a *geometric series*. Thus (13) is an arithmetic series, and (14) a geometric series.

$$a_1 + (a_1 + d) + (a_1 + 2d) + \cdots + [a_1 + (n-1)d] + \cdots \quad (13)$$

$$a_1 + a_1 r + a_1 r^2 + \cdots + a_1 r^{n-1} + \cdots \quad (14)$$

Since the kth term of an arithmetic series is $a_1 + (k-1)d$, we can find a formula for the nth partial sum S_n as follows:

$$S_n = \sum_{k=1}^{n} [a_1 + (k-1)d]$$

$$= \sum_{k=1}^{n} a_1 + d \sum_{k=1}^{n} (k-1)$$

$$= na_1 + d\left[\frac{1}{2}(n-1)n\right] \quad \text{(this is (8), with } n-1 \text{ instead of } n\text{)}$$

$$= \frac{1}{2}n[2a_1 + (n-1)d]. \quad (15)$$

Since $a_1 + (n-1)d = a_n$, we can also write

$$S_n = \frac{1}{2}n(a_1 + a_n), \quad (16)$$

so S_n is n times the average of the first and nth terms.

Example 4 Find the sum of the positive odd integers less than 100.

Solution. The sum we want is the arithmetic series

$$1 + 3 + 5 + \cdots + 97 + 99.$$

The kth odd integer is $2k - 1$, so that $99 = 2 \cdot 50 - 1$ is the fiftieth odd integer. Hence

$$S_{50} = \frac{1}{2} 50(1 + 99) = 2{,}500.$$

A formula for the sum of n terms of a geometric series can be found as follows:

$$S_n = a_1 + a_1 r + a_1 r^2 + \cdots + a_1 r^{n-1},$$
$$rS_n = a_1 r + a_1 r^2 + \cdots + a_1 r^{n-1} + a_1 r^n.$$

Subtracting the second equation from the first gives

$$(1 - r)S_n = a_1 - a_1 r^n,$$
$$S_n = a_1 \left(\frac{1 - r^n}{1 - r} \right). \tag{17}$$

Example 5 A ball bounces 3/4 of the height from which it is dropped. If the ball is dropped from a height of 5 feet, how far has it traveled when it hits the ground the tenth time?

Solution. The ball travels 5 feet before it hits the ground the first time, $2 \cdot 5 \cdot 3/4$ feet more before it hits the ground the second time, and so on. Hence the total distance D is

$$D = 5 + 2 \cdot 5 \cdot \frac{3}{4} + 2 \cdot 5 \cdot \left(\frac{3}{4}\right)^2 + \cdots + 2 \cdot 5 \cdot \left(\frac{3}{4}\right)^9.$$

Writing the first term as $-5 + 10$, we get -5 plus a geometric series:

$$D = -5 + \left[10 + 10 \cdot \frac{3}{4} + 10 \cdot \left(\frac{3}{4}\right)^2 + \cdots + 10 \cdot \left(\frac{3}{4}\right)^9 \right]$$

$$= -5 + 10 \frac{\left(1 - \left(\frac{3}{4}\right)^{10}\right)}{\left(1 - \frac{3}{4}\right)} \qquad \text{using (17)}$$

$$= -5 + 40[1 - (3/4)^{10}].$$

Using logarithms, we find that $(3/4)^{10} = 0.0563$, and $D = 32.75$ feet.

SECTION 12-4 REVIEW QUESTIONS

1. An indicated sum $a_1 + a_2 + \cdots$ is called a _____ .   series
2. The sum of the first n terms of a series is called the nth

    _____   _____ .   partial   sum

3. If S_n is the nth partial sum of a series, then $S_{n+1} = S_n +$ _____ . a_{n+1}
4. The partial sum S_n can be written in Σ notation as $S_n =$ _____ . $\sum_{k=1}^{n} a_k$
5. The symbols $\sum_{k=1}^{n} a_k$ are read as the sum of _____ as k runs a_k

    from _____ to _____ .   1   n

6. The following formulas are simple grouping and factoring properties written in Σ notation:

    $\sum_{k=1}^{n} (a_k + b_k) =$ _____ + _____ $\sum_{k=1}^{n} a_k + \sum_{k=1}^{n} b_k$

    $\sum_{k=1}^{n} ca_k = c($ _____ $)$ $\sum_{k=1}^{n} a_k$

    $\sum_{k=1}^{n} c =$ _____ . nc

7. The sum of the terms of an arithmetic sequence is called an _____ series.   arithmetic

8. The sum of n terms of an arithmetic series is given as

    $S_n = 1/2\, n[2a_1 + (n-1)d]$, or $S_n = 1/2\, n($ _____ + _____ $)$. $a_1 + a_n$

9. For an arithmetic series, S_n is n times the _____ of the first and nth terms.   average

10. The sum of n terms of a geometric series is given by $S_n = a_1 \cdot ($ _____ $)$. $\dfrac{1-r^n}{1-r}$

SECTION 12-4, EXERCISES

Find S_1, S_2, S_3 and S_4 for each of the following series. Write a formula for S_n (cf. Exercises 17-24, Sec. 12-1), and check it for $n = 4$.   Example 1

1. $2 + 4 + 6 + \cdots + 2n + \cdots$
2. $1 + 5 + 9 + \cdots + (4n - 3) + \cdots$
3. $1 + 4 + 7 + \cdots + (3n - 2) + \cdots$
4. $1 + 2 + 4 + \cdots + 2^{n-1} + \cdots$
5. $\dfrac{1}{2} + \dfrac{1}{4} + \dfrac{1}{8} + \cdots + \dfrac{1}{2^n} + \cdots$

6. $1^2 + 3^2 + 5^2 + \cdots + (2n - 1)^2 + \cdots$

Perform the indicated addition.

Example 2

7. $\sum_{k=1}^{5} (3k + 1)$ 8. $\sum_{k=1}^{5} (2k + 1)$

9. $\sum_{k=1}^{4} (k^2 + k)$ 10. $\sum_{k=1}^{4} (2k^2 + 1)$

11. $\sum_{k=1}^{5} (2^{k-1} + 2k)$ 12. $\sum_{k=1}^{5} (2^{k-1} + k^2)$

13. $\sum_{k=1}^{6} (3 \cdot 2^{k-1} - k)$ 14. $\sum_{k=1}^{6} (2^{k-1} - 5)$

Find a formula for each of the following and check your answer against the corresponding answer in Exercises 7-14.

Example 3

15. $\sum_{k=1}^{n} (3k + 1)$ 16. $\sum_{k=1}^{n} (2k + 1)$

17. $\sum_{k=1}^{n} (k^2 + k)$ 18. $\sum_{k=1}^{n} (2k^2 + 1)$

19. $\sum_{k=1}^{n} (2^{k-1} + 2k)$ 20. $\sum_{k=1}^{n} (2^k + k^2)$

21. $\sum_{k=1}^{n} (3 \cdot 2^{k-1} - k)$ 22. $\sum_{k=1}^{n} (2^{k-1} - 5)$

Find the sum of the indicated number of terms of the following series.

Example 4

23. $1 + 7 + 13 + \ldots ; n = 10$
24. $2 + 6 + 10 + \ldots ; n = 21$
25. $10 + 13 + 16 + \ldots ; n = 7$
26. $15 + 20 + 25 + \ldots ; n = 10$
27. Find the sum of the even integers between 15 and 97.
28. Find the sum of the odd integers between 50 and 120.
29. If one cent is saved on the first of the month, two cents on the second, three cents on the third, etc., what is the total amount saved after 30 days?
30. A man starts to work at $10,000 per year and gets a $1,000 raise each year for 20 years. How much has he earned all together at the end of his 21st year?

Find the sum of the indicated number of terms of the geometric series.

Example 5

31. $1 + 3 + 9 + \ldots ; n = 5$
32. $\frac{1}{4} + \frac{1}{8} + \frac{1}{16} + \ldots ; n = 5$
33. $2 + 3 + \frac{9}{2} + \ldots ; n = 6$
34. $8 + 4 + 2 + \ldots ; n = 7$
35. A man earns $1.00 the first day he works, $2.00 the second, $4.00 the third, etc. How much has he earned after 30 days?

36. If a car depreciates 30% each year, what is the value at the end of five years of a car which cost $4,000?
37. A ball rebounds $\frac{9}{10}$ of the distance it falls. If you drop the ball from 100 feet, how far will it travel between the first time it hits the ground and the fifth time?
38. How far will the ball of Exercise 37 travel between the first time it hits the ground and the tenth time?

Prove the following formulas by induction.

39. $a_1 + (a_1 + d) + \cdots + [a_1 + (n-1)d] = \frac{1}{2}n[2a_1 + (n-1)d]$

40. $a + ar + ar^2 + \cdots + ar^{n-1} = \dfrac{a(r^n - 1)}{r - 1}$ if $r \neq 1$.

12-5 CONVERGENCE; INFINITE SERIES

Much of our everyday work with numbers deals with approximations rather than exact quantities. For example, we use the approximation 1.414 for $\sqrt{2}$, and the approximation 3.1416 for π. In fact, the successive decimal expressions for a number like π (3, 3.1, 3.14, 3.141, 3.1415, 3.14159, ...) really form a sequence of better and better approximations.

It is frequently important to know whether the terms of a sequence approach some limiting value as you go farther and farther out in the sequence. For example, consider the sequence $\{n/(n + 1)\}$:

$$\frac{1}{2}, \frac{2}{3}, \frac{3}{4}, \frac{4}{5}, \ldots, \frac{n}{n+1}, \ldots \quad (1)$$

It is clear that the terms of (1) get closer and closer to the number 1 as n increases. We express this by saying that the *limit of the sequence is* 1, or the *sequence converges to* 1. We also indicate this fact with the symbolism

$$\lim_{n \to \infty} \frac{n}{n+1} = 1,$$

which is read as "the limit as n tends to infinity of $n/(n + 1)$ is 1".

In general, we say that a sequence $\{a_n\}$ converges to a limit L if the difference between a_n and L becomes arbitrarily small in magnitude for all sufficiently large values of n. For example, all the terms beyond some term must be within 0.01 of L, all the terms beyond some other term must be within 0.001 of L, and so on. In this case we write

$$\lim_{n \to \infty} a_n = L.$$

Example 1 Find the following limits.

(a) $\lim\limits_{n \to \infty} \dfrac{n+1}{2n}$ (b) $\lim\limits_{n \to \infty} \dfrac{3n^3 - n}{2n^3 + 7n^2 + 1}$

Solution. (a) If we divide numerator and denominator by n, it becomes clear how the terms behave for large n.

$$\lim_{n \to \infty} \frac{n+1}{2n} = \lim_{n \to \infty} \frac{1 + \dfrac{1}{n}}{2}$$

$$= \frac{1}{2}$$

(b) Here we divide numerator and denominator by n^3.

$$\lim_{n \to \infty} \frac{3n^3 - n}{2n^3 + 7n^2 + 1} = \lim_{n \to \infty} \frac{3 - \dfrac{1}{n^2}}{2 + \dfrac{7}{n} + \dfrac{1}{n^3}}$$

$$= \frac{3}{2}$$

Example 2. Find the limit

$$\lim_{n \to \infty} \frac{n}{2^n}$$

Solution. From Exercise 27, Sec. 12-1, we know that $n^2 \leq 2^n$ if $n \geq 4$. Therefore, if $n \geq 4$,

$$\frac{n}{2^n} = \frac{n^2}{2^n} \cdot \frac{1}{n} \leq \frac{1}{n}.$$

Since $\lim\limits_{n \to \infty} \dfrac{1}{n} = 0$, and $0 \leq \dfrac{n}{2^n} \leq \dfrac{1}{n}$ for all $n \geq 4$,

$$\lim_{n \to \infty} \frac{n}{2^n} = 0.$$

Example 3. Find the limit

$$\lim_{n\to\infty} \frac{2^n + n}{2^{n-1} + 1}$$

Solution. Since $\dfrac{n}{2^n}$ approaches 0 as n increases, it follows that

$$\frac{n}{2^{n-1}} = \frac{1}{2} \cdot \frac{n}{2^n}$$

also approaches 0. We divide numerator and denominator of the given expression by 2^{n-1} and get

$$\lim_{n\to\infty} \frac{2^n + n}{2^{n-1} + 1} = \lim_{n\to\infty} \frac{2 + \dfrac{n}{2^{n-1}}}{1 + \dfrac{1}{2^{n-1}}}$$

$$= 2.$$

The most important application we will make of the idea of the limit of a sequence is to infinite series:

$$\sum_{k=1}^{\infty} a_k = a_1 + a_2 + \cdots + a_k + \cdots \qquad (2)$$

Let $\{S_n\}$ be the sequence of partial sums of the series (2), so that

$$S_1 = a_1,$$
$$S_2 = a_1 + a_2,$$
$$S_3 = a_1 + a_2 + a_3,$$
$$\cdots$$
$$S_n = a_1 + a_2 + a_3 + \cdots + a_n.$$

If the partial sums S_n approach some limit S, then we say that the series (2) *converges to* S, or that *the sum of the series is* S. We write

$$\sum_{k=1}^{\infty} a_k = S,$$

or
$$a_1 + a_2 + a_3 + \cdots + a_k + \cdots = S.$$

Example 4 Show that

$$\frac{1}{2} + \frac{1}{4} + \frac{1}{8} + \cdots + \frac{1}{2^n} + \cdots = 1.$$

Solution. The series is a geometric series with $a_1 = 1/2$, and $r = 1/2$. Hence the nth partial sum is

$$S_n = \frac{1}{2}\left[\frac{1 - \left(\frac{1}{2}\right)^n}{1 - \frac{1}{2}}\right]$$

$$= 1 - \left(\frac{1}{2}\right)^n.$$

By definition, the sum of the series is the limit of S_n. Since $(1/2)^n$ approaches 0 as n increases, we have

$$\sum_{k=1}^{\infty} \left(\frac{1}{2}\right)^k = \lim_{n \to \infty}\left[1 - \left(\frac{1}{2}\right)^n\right] = 1.$$

For any geometric series, we can write the formula for S_n as follows:

$$S_n = \sum_{k=1}^{n} a_1 r^{k-1} = a_1 \left(\frac{1 - r^n}{1 - r}\right)$$

$$= \frac{a_1}{1 - r} - \frac{a_1}{1 - r} r^n.$$

Since $a_1/(1 - r)$ does not depend on n, the limit depends only on the factor r^n in the second term. If $-1 < r < 1$, then r^n approaches zero as n increases. Hence any geometric series with $|r| < 1$ converges, and the sum is $a_1/(1 - r)$; that is,

$$\sum_{k=1}^{\infty} a_1 r^{k-1} = \frac{a_1}{1 - r} \quad \text{if } |r| < 1. \tag{3}$$

If $|r| \geq 1$, then it is easy to see that the partial sums do not converge. If $|r| > 1$, then r^n becomes numerically larger and larger, which prevents $\{S_n\}$ from converging. If $r = 1$ or -1, the series becomes, respectively,

$$a + a + a + \cdots + a + \cdots, \tag{4}$$

or
$$a - a + a - \cdots + (-1)^{n+1}a + \cdots. \tag{5}$$

The nth partial sum of (4) is na, which clearly does not converge except in the trivial case $a = 0$. The partial sums of (5) are

$$a, 0, a, 0, a, 0, \ldots$$

and this sequence also does not converge unless $a = 0$.

Example 5 Show that $0.3 + 0.03 + 0.003 + \cdots = 1/3$.

Solution. The series is the same as the geometric series

$$\frac{3}{10} + \frac{3}{10} \cdot \frac{1}{10} + \frac{3}{10} \cdot \left(\frac{1}{10}\right)^2 + \cdots + \frac{3}{10} \cdot \left(\frac{1}{10}\right)^{n-1} + \cdots.$$

Since $|r| = 1/10 < 1$, the series converges, and from (3) we have

$$S = \frac{\frac{3}{10}}{1 - \frac{1}{10}} = \frac{\frac{3}{10}}{\frac{9}{10}} = \frac{1}{3}.$$

The repeating decimal $0.33333\ldots$ is interpreted to be the geometric series of Example 4. The same kind of argument shows that every repeating decimal is a rational number.

Example 6 Write the repeating decimal $0.212121\ldots$ as a fraction.

Solution. The decimal indicates the following series:

$$0.21 + 0.0021 + 0.000021 + \cdots =$$

$$\frac{21}{100} + \frac{21}{100} \cdot \frac{1}{100} + \frac{21}{100} \cdot \left(\frac{1}{100}\right)^2 + \cdots.$$

This is a geometric series with $a_1 = 21/100$, and $r = 1/100$. Since $|r| < 1$, the series converges, and from (3) we have

$$S = \frac{\frac{21}{100}}{1 - \frac{1}{100}} = \frac{\frac{21}{100}}{\frac{99}{100}} = \frac{21}{99} = \frac{7}{33}.$$

The next example involves a type of series which is not a geometric series, but for which we can nevertheless find a formula for S_n.

Example 7 Find a formula for the nth partial sum, and find the sum of the series

$$\sum_{k=1}^{\infty} \left(\frac{1}{k} - \frac{1}{k+1}\right).$$

Solution. We write a few terms of S_n to see if we can establish a pattern.

$$S_1 = \frac{1}{1} - \frac{1}{2} = 1 - \frac{1}{2}$$

$$S_2 = \left(\frac{1}{1} - \frac{1}{2}\right) + \left(\frac{1}{2} - \frac{1}{3}\right) = 1 - \frac{1}{3}$$

$$S_3 = \left(\frac{1}{1} - \frac{1}{2}\right) + \left(\frac{1}{2} - \frac{1}{3}\right) + \left(\frac{1}{3} - \frac{1}{4}\right) = 1 - \frac{1}{4}$$

$$S_4 = \left(\frac{1}{1} - \frac{1}{2}\right) + \left(\frac{1}{2} - \frac{1}{3}\right) + \left(\frac{1}{3} - \frac{1}{4}\right) + \left(\frac{1}{4} - \frac{1}{5}\right) = 1 - \frac{1}{5}$$

It is clear from these examples, and is easy to prove by induction, that

$$S_n = 1 - \frac{1}{n+1}.$$

It follows that

$$\sum_{k=1}^{\infty} \left(\frac{1}{k} - \frac{1}{k+1}\right) = \lim_{n \to \infty} \left(1 - \frac{1}{n+1}\right) = 1. \tag{6}$$

Notice that (6) can also be written

$$\sum_{k=1}^{\infty} \frac{1}{k(k+1)} = 1. \tag{7}$$

SECTION 12-5, REVIEW QUESTIONS

1. If the terms of a sequence get closer and closer to a limiting number L, we say the sequence _____ to L.   *converges*

2. If $\{a_n\}$ converges to L, we write _____ = L.   *$\lim_{n \to \infty} a_n$*

3. $\lim_{n \to \infty} \frac{1}{n} = $ _____ and $\lim_{n \to \infty} \frac{n}{2^n} = $ _____.   *0   0*

4. A series $\sum_{k=1}^{\infty} a_k$ converges to S provided the sequence of partial sums has S as its _____.   *limit*

5. $S_n = \sum_{k=1}^{n} a_k$, and $\lim_{n \to \infty} S_n = S$, then the series $\sum_{k=1}^{\infty} a_k$ _____ to S.   *converges*

6. The nth partial sum of the geometric series $\sum_{k=1}^{\infty} a_1 r^{k-1}$ is _____.

476 *Sequences and Series*

$$S_n = \frac{a_1}{(1-r)} - \underline{}. \qquad \frac{a_1}{1-r} \cdot r^n$$

7. If $|r| < 1$, then $\dfrac{a_1}{(1-r)} \cdot r^n$ converges to _____.   0

8. If $|r| < 1$, then the sum of the geometric series $\sum_{k=1}^{\infty} a_1 r^{k-1}$ is _____. $\dfrac{a_1}{1-r}$

9. The repeating decimal $0.6666\ldots$ is the geometric series with $a_1 = $ _____   .6

     and $r = $ _____.   1/10

10. Any repeating decimal represents a _____ number.   rational

SECTION 12-5, EXERCISES

Find the following limits.

1. $\lim\limits_{n \to \infty} \dfrac{n+3}{n+2}$ 2. $\lim\limits_{n \to \infty} \dfrac{5n-1}{n+2}$ Example 1

3. $\lim\limits_{n \to \infty} \dfrac{n^2+n}{2n^2-10}$ 4. $\lim\limits_{n \to \infty} \dfrac{3n^2+5n}{2n^2+1}$

5. $\lim\limits_{n \to \infty} \dfrac{2n^3-5n^2+100}{n^3+100n^2+1000n}$ 6. $\lim\limits_{n \to \infty} \dfrac{n^4-n^2+1}{3n^4+n^2+1}$

The examples and exercises of Sec. 12-1 may be useful in the exercises below.

7. $\lim\limits_{n \to \infty} \dfrac{2n}{2^n}$ 8. $\lim\limits_{n \to \infty} \dfrac{3n}{2^{n-1}}$ Example 2,3

9. $\lim\limits_{n \to \infty} \dfrac{5n}{3^n}$ 10. $\lim\limits_{n \to \infty} \dfrac{10n+1}{2 \cdot 3^n}$

11. $\lim\limits_{n \to \infty} \dfrac{2^{n-1}+1}{2^n+n}$ 12. $\lim\limits_{n \to \infty} \dfrac{5 \cdot 2^n - n}{2^{n-1}+n}$

13. $\lim\limits_{n \to \infty} \dfrac{2^n+3^n}{3^n}$ 14. $\lim\limits_{n \to \infty} \dfrac{3 \cdot 4^n + 2^n}{2 \cdot 4^{n-1}}$

15. $\lim\limits_{n \to \infty} \dfrac{n^3}{3^{n-1}}$ 16. $\lim\limits_{n \to \infty} \dfrac{3^n+4n^3}{3^{n-1}}$

17. $\lim\limits_{n \to \infty} \dfrac{n+1}{n!}$ 18. $\lim\limits_{n \to \infty} \dfrac{n^2}{n!}$

Find the sums of the following geometric series.

19. $\dfrac{1}{4} + \dfrac{1}{8} + \dfrac{1}{16} + \cdots + \dfrac{1}{2^{n+1}} + \cdots$ Example 4

20. $1 + \dfrac{1}{2} + \dfrac{1}{4} + \cdots + \dfrac{1}{2^{n-1}} + \cdots$

21. $\dfrac{1}{3} - \dfrac{1}{9} + \dfrac{1}{27} - \cdots + (-1)^{n+1}\dfrac{1}{3^n} + \cdots$

22. $1 - \dfrac{1}{2} + \dfrac{1}{4} - \cdots + (-1)^{n+1}\dfrac{1}{2^n} + \cdots$

23. $-10 + 1 - .01 + \cdots + (-1)^n \dfrac{1}{10^n} + \cdots$

24. $10 - 2 + \dfrac{2}{5} - \cdots + 10\left(-\dfrac{1}{5}\right)^n + \cdots$

25. $1 + x + x^2 + \cdots + x^{n-1} + \cdots \ (|x| < 1)$
26. $1 - x + x^2 - \cdots + (-x)^{n-1} + \cdots \ (|x| < 1)$

Write the repeating decimals as fractions.

Example 5, 6

27. .555 ...
28. .777 ...
29. .131313 ...
30. .787878 ...
31. 12.212121 ...
32. 878.787878 ...

Find a formula for the nth partial sum, and find the sum of the series.

Example 7

33. $\displaystyle\sum_{k=1}^{\infty}\left(\dfrac{3}{k} - \dfrac{3}{k+1}\right)$
34. $\displaystyle\sum_{k=1}^{\infty}\left(\dfrac{1}{k+1} - \dfrac{1}{k+2}\right)$

35. $\displaystyle\sum_{k=1}^{\infty}\left(\dfrac{1}{k+1} - \dfrac{1}{k}\right)$
36. $\displaystyle\sum_{k=1}^{\infty}\left(\dfrac{1}{2k} - \dfrac{1}{2k+2}\right)$

37. $\displaystyle\sum_{k=1}^{\infty}\left(\dfrac{2}{3k} - \dfrac{2}{3k+3}\right)$
38. $\displaystyle\sum_{k=1}^{\infty}\left(\dfrac{5}{k^2} - \dfrac{5}{(k+1)^2}\right)$

39. $\displaystyle\sum_{k=1}^{\infty}\left(\dfrac{1}{(k+1)^3} - \dfrac{1}{k^3}\right)$
40. $\displaystyle\sum_{k=1}^{\infty}\left(\dfrac{2k+3}{k} - \dfrac{2k+5}{k+1}\right)$

12-6 REVIEW FOR CHAPTER

Show by mathematical induction that each of the following statements is true for every positive integer n.

1. $\dfrac{1}{2} + \dfrac{1}{6} + \dfrac{1}{12} + \cdots + \dfrac{1}{n(n+1)} = \dfrac{n}{n+1}$

2. $1 + 3 + 9 + \cdots + 3^{n-1} = \dfrac{3^n - 1}{2}$

3. $1 + \dfrac{1}{3} + \dfrac{1}{9} + \cdots + \dfrac{1}{3^{n-1}} = \dfrac{3}{2}\left(1 - \dfrac{1}{3^n}\right)$

4. $3n \le n^2 + 2$ (Hint: $n^2 + 2 + 3 \le n^2 + 2n + 3$)
5. $2^n > n$ (Hint: recall that $2n \ge n + 1$)

Write the first five terms and the indicated term of each sequence shown below.

6. $a_n = \dfrac{3}{n^2 + 1}$; seventh term

7. $a_n = \dfrac{5}{n(n+1)}$; eighth term

8. $a_n = \dfrac{(-1)^n(n-2)}{n}$; seventh term

9. $a_n = (-1)^{n-1} 3^{n+1}$; ninth term

Find a formula for the nth term of each of the following sequences.

10. $\{-1, 1, 3, 5, \ldots\}$

11. $\{-1, 1, -1, 1, -1, \ldots\}$

12. $\left\{1, \dfrac{2}{3}, \dfrac{3}{5}, \dfrac{4}{7}, \ldots\right\}$

13. $\left\{\dfrac{5}{2}, \dfrac{5}{6}, \dfrac{5}{12}, \dfrac{5}{20}, \ldots\right\}$

14. Insert five arithmetic means between 86 and 14.
15. Insert two geometric means between 686 and 2.
16. A sum of $1,000 is invested today at the rate of 6% per year. What will the total amount be in 10 years if the interest is compounded annually?
17. The fifth term of an arithmetic sequence is -16, and the twentieth term is -46. Find the common difference and a formula for a_n. Write out the first six terms and find the sum of the first ten terms.
18. The first term of an arithmetic sequence is $3\frac{1}{2}$ and the eighth term is $8\frac{1}{2}$. Find the common difference and a formula for a_n. Write the first seven terms and find the sum of the first six terms.

19. If $a_k = \dfrac{(-1)^k}{2^k}$, find $\sum\limits_{k=1}^{5} a_k$.

20. Find a formula for $\sum\limits_{k=1}^{n} 2k(1-k)$.

21. Find a formula for $\sum\limits_{k=1}^{n} k^2\left(1 - \dfrac{1}{k}\right)$.

Find the sum of the indicated number of terms for each of the following arithmetic or geometric series.

22. $-10 - 20 - 30 - \cdots$; $n = 623$

23. $\dfrac{2}{3} + \dfrac{4}{3} + \dfrac{8}{3} + \cdots$; $n = 8$

24. $\dfrac{4}{x} + \dfrac{8}{x^2} + \dfrac{16}{x^3} + \cdots$; $n = 725$

25. $(x + 6) + (x + 36) + (x + 66) + \cdots$; $n = 5$

Find each of the following limits.

26. $\lim\limits_{n \to \infty} \dfrac{n + 1}{n}$
27. $\lim\limits_{n \to \infty} \dfrac{n^2 - 2}{n^2 - 1}$

28. $\lim\limits_{n \to \infty} \dfrac{2n^2 + 3n - 2}{n^2 + 4}$
29. $\lim\limits_{n \to \infty} \dfrac{4n^3 - 3}{5n^3 - 3n^2 + 1}$

Find a common fraction for each repeating decimal.

30. $0.818181\ldots$
31. $0.138512512\ldots$

13 VECTORS

13-1 VECTORS AND SCALARS

Some of the most important quantities of physics are determined by a direction as well as a magnitude. For example, force, velocity, and acceleration all involve a direction. Such a quantity is called a *vector*. Numbers or quantities that are determined purely numerically are called *scalars* to distinguish them from vectors. Temperature, mass, and electric charge are examples of scalar quantities.

We will represent vectors geometrically by directed line segments. The length of the segment represents the magnitude of the vector, and the vector's direction is that of the segment. We write $\overrightarrow{PQ}$ to indicate the line segment joining P and Q and directed from P to Q. Since a vector represents a magnitude and a direction, but not a position, we agree that two directed segments $\overrightarrow{PQ}$ and $\overrightarrow{P'Q'}$ represent the same vector if they have the same length and direction. Thus a given vector **V** can be represented geometrically by many different directed segments as shown in Fig. 13-1. Any one of these directed segments will be called a *representative* of the vector **V**. We will write $\overrightarrow{PQ} = \overrightarrow{RS}$ to mean that the segments have the same length and direction, and we write $\mathbf{V} = \overrightarrow{PQ}$ to mean that $\overrightarrow{PQ}$ is a representative of **V**.

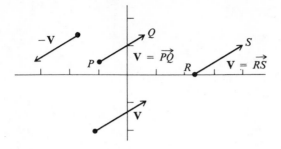

Figure 13-1

In print we use a bold face type, **V**, to represent a vector. In writing, we indicate a vector by an ordinary letter with a bar or arrow over it: $\bar{V}$ or $\vec{V}$. The magnitude of a vector is indicated with the absolute value sign:

$$|\mathbf{V}| = \text{magnitude of } \mathbf{V}.$$

The magnitude of **V** is the length of any representative $\vec{PQ}$ of **V**.

The zero vector, denoted $\vec{0}$, or **0** is the vector with zero magnitude. This vector can be represented by a single point, thought of as a *degenerate segment* $\vec{PP}$, and no direction is associated with it.

The zero vector acts as an identity for the vector addition, which we will define next. The sum of vectors **A** and **B**, denoted **A** + **B**, is the vector defined as follows.

If we pick any representative $\vec{PQ}$ of **A**, and let $\vec{QR}$ be the representative of **B** which originates at Q, then (Fig. 13-2),

$$\mathbf{A} + \mathbf{B} = \vec{PR}$$

It is easy to see that the definition of **A** + **B** does not depend on which segment $\vec{PQ}$ we chose to represent **A**.

We also see from Fig. 13-2 that $\vec{QS}$, which represents **B** + **A**, has the same length and direction as $\vec{PR}$, which represents **A** + **B**. Therefore we have the commutative law for vector addition.

Commutative Law of Addition

$$\mathbf{A} + \mathbf{B} = \mathbf{B} + \mathbf{A}$$

If $\mathbf{A} = \vec{PQ}$ and $\mathbf{0} = \vec{PP} = \vec{QQ}$, then $\mathbf{A} + \mathbf{0} = \vec{PQ} = \mathbf{0} + \mathbf{A}$, so that **0** is an identity element for vector addition.

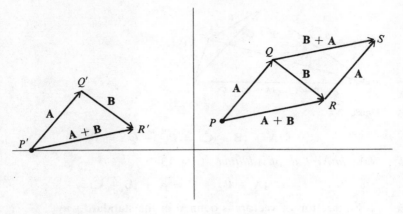

Figure 13-2

Identity Element for Addition

$$0 + A = A + 0 = A$$

Each vector **A** has an inverse $-\mathbf{A}$, with respect to addition. If $\mathbf{A} = \overrightarrow{PQ}$, then $-\mathbf{A} = \overrightarrow{QP}$, since $\overrightarrow{PQ} + \overrightarrow{QP} = \overrightarrow{PP} = \mathbf{0}$. That is, $-\mathbf{A}$ is the vector with the same magnitude as **A**, but opposite direction.

Inverse for Addition   (Fig. 13-3).

$$\mathbf{A} + (-\mathbf{A}) = \mathbf{0}$$

The operation of vector addition is also associative, as illustrated in Fig. 13-4. We let $\mathbf{A} = \overrightarrow{PQ}, \mathbf{B} = \overrightarrow{QR}, \mathbf{C} = \overrightarrow{RS}$, so that $\mathbf{A} + \mathbf{B} = \overrightarrow{PR}$ and $\mathbf{B} + \mathbf{C} = \overrightarrow{QS}$. Consequently,

$$(\mathbf{A} + \mathbf{B}) + \mathbf{C} = \overrightarrow{PR} + \overrightarrow{RS} = \overrightarrow{PS},$$

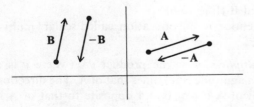

Figure 13-3

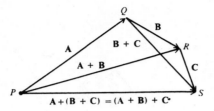

Figure 13-4

$$A + (B + C) = \vec{PQ} + \vec{QS} = \vec{PS},$$

Associative Law of Addition (Fig. 13-4).

$$(A + B) + C = A + (B + C)$$

Subtraction of vectors is defined in the standard way.

Definition of Subtraction

$$A - B = A + (-B)$$

Since $-B$ is the additive inverse of B, this means that $A - B$ is the vector you add to B to get A. In terms of directed segments, if $\vec{PQ} = A$, $\vec{PR} = B$, then $\vec{RQ} = A - B$. See Fig. 13-5.

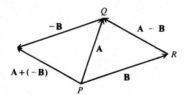

Figure 13-5

Vector sums and differences also have the following useful geometric interpretation. If A and B originate at the same point P, then $A + B$ and $A - B$ lie along the two diagonals of the parallelogram with adjacent sides A and B (Fig. 13-6).

Now let's turn our attention to an **operation** called scalar multiplication, denoted by cA.

Definition of Scalar Multiplication. The product cA (where c is a scalar) is a vector whose magnitude is $|c|$ times that of A. The direction of cA is the same as that of A if $c > 0$ and opposite to that of A if $c < 0$. (See Fig. 13-7).

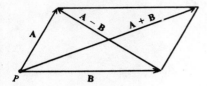

Figure 13-6

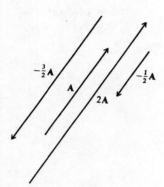

Figure 13-7

We have the following properties immediately from the definition:

$$(-1)\mathbf{A} = -\mathbf{A}$$

$$|c\mathbf{A}| = |c|\,|\mathbf{A}|$$

$$(c + d)\mathbf{A} = c\mathbf{A} + d\mathbf{A}$$

Scalar multiplication distributes over vector addition; that is,

$$c(\mathbf{A} + \mathbf{B}) = c\mathbf{A} + c\mathbf{B}$$

This property is illustrated in Fig. 13-8.

We previously stated that for two vectors to be equal they have to have the same direction and magnitude. Now let us look at what conditions must exist in a coordinate plane for

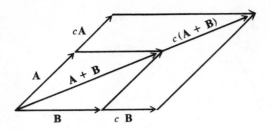

Figure 13-8

$$\vec{PQ} = \vec{RS}$$

If $P = (x_1, y_1)$, $Q = (x_2, y_2)$, $R = (x_1', y_1')$, and $S = (x_2', y_2')$, then $\vec{PQ} = \vec{RS}$ if $x_2' - x_1' = x_2 - x_1$, and $y_2' - y_1' = y_2 - y_1$ (Fig. 13-9).

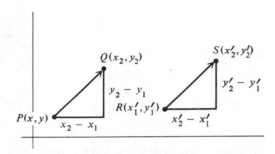

Figure 13-9

Example 1 Find the coordinates of Q so that $\vec{PQ} = \vec{RS}$ if $P = (3, 2)$, $R = (-2, 4)$, $S = (5, -1)$.

Solution. If $Q = (x, y)$ then the condition is

$$x - 3 = 5 - (-2) \quad \text{and} \quad y - 2 = -1 - 4,$$
$$x = 10 \quad\quad\quad\quad\quad y = -3$$

Example 2 Find a representative of $\vec{PQ} - \vec{RS}$ if $P = (1, 2)$, $Q = (5, 3)$, $R = (-1, 4)$, $S = (0, 1)$.

Solution. We first need to find the point T so that $\vec{PT} = \vec{RS}$. Then $\vec{PQ} - \vec{RS} = \vec{PQ} - \vec{PT} = \vec{TQ}$. If $T = (x, y)$, then $x - 1 = 0 - (-1)$, and $y - 2 = 1 - 4$. Hence, $T = (2, -1)$, and $\vec{TQ} = (2, -1)(5, 3)$.

Example 3 Let 0 be the origin, P and Q two given points, and R the midpoint of segment PQ. If $\mathbf{A} = \overrightarrow{OP}$, $\mathbf{B} = \overrightarrow{OQ}$, express $\mathbf{C} = \overrightarrow{OR}$ in terms of $\mathbf{A}$ and $\mathbf{B}$ (Fig. 13-10).

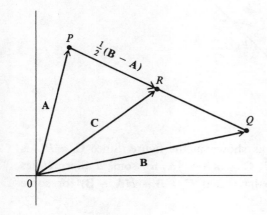

Figure 13-10

Solution. $\overrightarrow{PQ} = \mathbf{B} - \mathbf{A}$, so $\overrightarrow{PR} = \frac{1}{2}(\mathbf{B} - \mathbf{A})$. Therefore, $\overrightarrow{OR} = \overrightarrow{OP} + \overrightarrow{PR} = \mathbf{A} + \frac{1}{2}(\mathbf{B} - \mathbf{A}) = \frac{1}{2}\mathbf{A} + \frac{1}{2}\mathbf{B}$.

Let $\mathbf{A}$ and $\mathbf{B}$ be non-parallel vectors with $\mathbf{A} \neq \mathbf{0}$, $\mathbf{B} \neq \mathbf{0}$. It is easy to see geometrically that the only way you can have $a\mathbf{A} + b\mathbf{B} = \mathbf{0}$ is if $a = b = 0$. It follows that if $\mathbf{A}$ and $\mathbf{B}$ are non-parallel and $a_1 \mathbf{A} + b_1 \mathbf{B} = a_2 \mathbf{A} + b_2 \mathbf{B}$, then $(a_1 - a_2)\mathbf{A} + (b_1 - b_2)\mathbf{B} = \mathbf{0}$, and consequently $a_1 = a_2$, $b_1 = b_2$. We use this principle in the following example.

Example 4 Let $\mathbf{A} = \overrightarrow{PQ}$ and $\mathbf{B} = \overrightarrow{PS}$ be adjacent sides of the parallelogram $PQRS$ (Fig. 13-11). Let T be the midpoint of side RS. Write $\mathbf{C} = \overrightarrow{PU}$ in terms of $\mathbf{A}$ and $\mathbf{B}$.

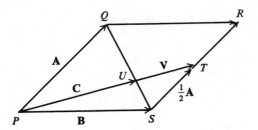

Figure 13-11

Solution. $\overrightarrow{PT} = \mathbf{B} + \frac{1}{2}\mathbf{A}$ as shown in the figure. Since $\mathbf{C} = \overrightarrow{PU}$ is some scalar multiple of $\overrightarrow{PT}$, $\mathbf{C} = a(\mathbf{B} + \frac{1}{2}\mathbf{A})$ for some a. Since U lies on $\overrightarrow{SQ} = \mathbf{A} - \mathbf{B}$, we also have that $\mathbf{C} = \mathbf{B} + b(\mathbf{A} - \mathbf{B})$ for some scalar b. Therefore,

$$\mathbf{C} = a(\mathbf{B} + \tfrac{1}{2}\mathbf{A}) = \mathbf{B} + b(\mathbf{A} - \mathbf{B}),$$
$$a\mathbf{B} + \tfrac{1}{2}a\mathbf{A} = (1 - b)\mathbf{B} + b\mathbf{A},$$

and, consequently, $a = 1 - b$, $\tfrac{1}{2}a = b$. It follows that

$$a = 1 - (\tfrac{1}{2})a, \quad a = 2/3,$$

so
$$\mathbf{C} = \tfrac{2}{3}(\mathbf{B} + \tfrac{1}{2}\mathbf{A}) = \tfrac{1}{3}\mathbf{A} + \tfrac{2}{3}\mathbf{B}$$

SECTION 13-1, REVIEW QUESTIONS

1. A quantity characterized by a magnitude and a direction is called a _____.

   vector

2. A vector has both magnitude and direction, but a scalar quantity has only _____.

   magnitude

3. Vectors are represented geometrically by _____ line segments.

   directed

4. Two directed segments represent the same vector if they have the same length and the same _____.

   direction

5. If $\vec{PQ}$ represents a vector, then the magnitude of the vector is the _____ of $\vec{PQ}$. **length**

6. The vector with zero length is denoted _____ or _____. **$\vec{0}$ 0**

7. The vector $\vec{0}$ is the only vector with no _____ specified. **direction**

8. The vector sum **A** + **B** represents the diagonal of a _____ with adjacent sides **A** and **B**. **parallelogram**

9. The vector −**A** has the same magnitude as **A**, but _____ direction. **opposite**

10. The magnitude of a vector **A** is denoted _____. **|A|**

11. The vector 2**A** is the vector with magnitude 2|**A**| and direction the same as _____. **A**

12. The following laws hold for scalar multiplication and vector addition:
$c(\mathbf{A} + \mathbf{B}) = $ _____, and $(c + d)\mathbf{A} = $ _____. **$c\overline{\mathbf{A}} + c\overline{\mathbf{B}}$** **$c\overline{\mathbf{A}} + d\overline{\mathbf{A}}$**

13. Vector addition is commutative (**A** + **B** = _____), and associative (**A** + (**B** + **C**) = _____). **$\overline{\mathbf{B}} + \overline{\mathbf{A}}$** **$(\overline{\mathbf{A}} + \overline{\mathbf{B}}) + \overline{\mathbf{C}}$**

SECTION 13-1, EXERCISES

In the parallelogram below, express each of the following as a single vector of the form $\vec{PP'}$.

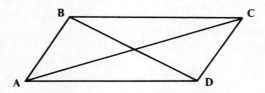

1. $\vec{BC} + \vec{CD}$
2. $\vec{AD} + \vec{CD}$
3. $\vec{DC} + \vec{DA}$
4. $\vec{AD} - \vec{AB}$
5. $\vec{AC} - \vec{BC}$
6. $\vec{BA} - \vec{CB}$
7. $\vec{DB} - \vec{CB}$
8. $\vec{AD} - \vec{BA}$

Find the coordinates of the point $Q = (x, y)$ such that $\vec{PQ} = \vec{RS}$ for each R and S below, if P is the origin $(0, 0)$.

9. $R = (1, 0), S = (3, 2)$
10. $R = (2, 1), S = (3, 4)$ **Example 1**
11. $R = (1, -2), S = (2, -1)$
12. $R = (-4, 2), S = (3, -7)$

Vectors and Scalars 489

13. $R = (0, 2), S = (0, 5)$ 14. $R = (1, 7), S = (1, 3)$

Find a representative of the form $\vec{PT}$ for $\vec{PQ} + \vec{RS}$ for the values of P, Q, R, S given.

Example 2
15. $P = (1, 0), Q = (2, 3), R = (3, 1), S = (5, 0)$
16. $P = (0, 3), Q = (1, 1), R = (5, 3), S = (1, 2)$
17. $P = (2, -1), Q = (1, 3), R = (-2, 0), S = (1, 4)$
18. $P = (3, 8), Q = (-3, 1), R = (2, 4), S = (0, 3)$

Let O be the origin, and P and Q be two given points. Express $\vec{OR}$ in terms of $\mathbf{A} = \vec{OP}, \mathbf{B} = \vec{OQ}$ for the following points R on the segment PQ.

Example 3
19. R is $1/3$ the way from P to Q
20. R is $2/3$ the way from P to Q
21. R is $1/4$ the way from P to Q
22. R is $7/8$ the way from P to Q

Let $\vec{PQ} = \mathbf{A}$ and $\vec{QR} = \mathbf{B}$ be two sides of a triangle PQR. Express the following in terms of $\mathbf{A}$ and $\mathbf{B}$.

Example 4
23. $\vec{RP}$ 24. $\vec{PQ} - \vec{PR}$
25. The median from P to the midpoint of QR.
26. The median from Q to the midpoint of RP.
27. The vector $\vec{PT}$, where T is the intersection of the medians of Exercises 25 and 26.
28. The vector $\vec{PS}$ where S is the intersection of the median from R to the midpoint of PQ.
29. Let $|\mathbf{A}| = |\mathbf{B}| = |\mathbf{C}|$ in the diagram below.
    (a) Express **D** in terms of **B**.
    (b) Express **C** in terms of **A** and **B**. (Hint: $\mathbf{A} + \mathbf{B} + \mathbf{C} = \mathbf{D}$.)

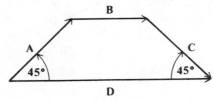

30. Let $\mathbf{A} = \vec{PQ}$ and $\mathbf{B} = \vec{QR}$ where $PQRS$ is a square. Let U, V be the respective midpoints of RS and SP. Express the following in terms of $\mathbf{A}$ and $\mathbf{B}$.
    (a) $\vec{PU}$ (b) $\vec{UV}$
    (c) $\vec{QS}$ (d) $\vec{UQ}$
31. Let R, S, T be the median vectors from the vertices of a triangle to the midpoints of the opposite sides. Show that $\mathbf{R} + \mathbf{S} + \mathbf{T} = 0$.

13-2 COMPONENTS

In this section we show how any vector can be written as a sum of two vectors parallel to the coordinate axes. Representing vectors as such a sum makes the relationships between vectors and coordinate geometry much clearer and simpler. We begin by defining what is meant by the angle between two vectors.

The angle between two vectors **A** and **B** is defined as the angle between any two representatives that originate at the same point (Fig. 13-12).

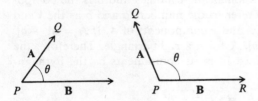

Figure 13-12

The *component* of vector **A** along another vector **B** is plus or minus the length of the projection of **A** on **B**. If the angle between **A** and **B** is acute, then the component of **A** along **B** is positive. If the angle between **A** and **B** is obtuse, the component of **A** along **B** is negative (Fig. 13-13). In either case, from trigonometry we see that the component of **A** along **B** is $|A| \cos \theta$ where θ is the angle between **A** and **B**.

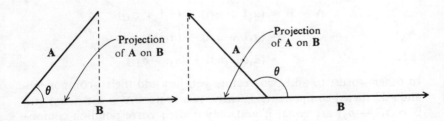

Figure 13-13

Components 491

Vectors, vector addition, scalar multiplication, etc., make sense in a plane without coordinates. However, vector ideas are also useful when used in conjunction with a cartesian coordinate system. In a given coordinate system, the vector of unit length in the direction of the positive x-axis is denoted **i**, and the unit vector in the direction of the positive y-axis is denoted **j**. Any vector, $\mathbf{A} = \overrightarrow{PR}$, in the plane is the sum of a horizontal vector $\overrightarrow{PQ}$ and a vertical vector $\overrightarrow{QR}$ (Fig. 13-14). The horizontal vector is some scalar multiple $a\mathbf{i}$ of **i** and the vertical vector is a multiple $b\mathbf{j}$ of **j**. Since **i** and **j** are unit vectors ($|\mathbf{i}| = |\mathbf{j}| = 1$), a is the component of **A** along **i** and b is the component of **A** along **j**. We will refer to the numbers a and b as the **i** and **j** components of **A**, or the x- and y-components of **A**. If $\mathbf{A} = a\mathbf{i} + b\mathbf{j}$, then the three vectors $a\mathbf{i}$, $b\mathbf{j}$, **A** form a right triangle. Therefore, the magnitude of **A** is determined from its components by the Pythagorean theorem:

$$|\mathbf{A}| = \sqrt{a^2 + b^2}$$

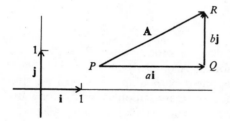

Figure 13-14

If $\mathbf{A} = a_1\mathbf{i} + a_2\mathbf{j}$ and $\mathbf{B} = b_1\mathbf{i} + b_2\mathbf{j}$, the rules for vector addition and scalar multiplication imply that

$$\mathbf{A} + \mathbf{B} = (a_1\mathbf{i} + a_2\mathbf{j}) + (b_1\mathbf{i} + b_2\mathbf{j})$$
$$= (a_1\mathbf{i} + b_1\mathbf{i}) + (a_2\mathbf{j} + b_2\mathbf{j})$$
$$= (a_1 + b_1)\mathbf{i} + (a_2 + b_2)\mathbf{j}$$

In other words, to add two vectors you just add their **i**-components and add their **j**-components. Clearly, two vectors $\mathbf{A} = a_1\mathbf{i} + a_2\mathbf{j}$ and $\mathbf{B} = b_1\mathbf{i} + b_2\mathbf{j}$ are equal if and only if their corresponding components are equal:

A = **B** if and only if $a_1 = b_1$ and $a_2 = b_2$

From this discussion we see that the vector $\overrightarrow{OP}$ from the origin 0 to a point $P = (x, y)$ can be written as $x\mathbf{i} + y\mathbf{j}$ (Fig. 13-15). Consequently, the vector $\overrightarrow{PQ}$, where $P = (x_1, y_1)$, $Q = (x_2, y_2)$ can be written:

$$\overrightarrow{PQ} = \overrightarrow{OQ} - \overrightarrow{OP}$$
$$= (x_2\mathbf{i} + y_2\mathbf{j}) - (x_1\mathbf{i} + y_1\mathbf{j})$$
$$= (x_2 - x_1)\mathbf{i} + (y_2 - y_1)\mathbf{j}$$

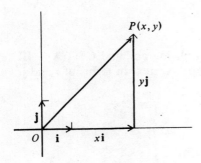

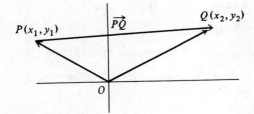

Figure 13-15

Example 1 Find the coordinates of the vertex opposite (1, 2) in the parallelogram with vertices at (3, 5), (1, 2), and (4, 1). (Fig. 13-16.)

Solution.
$$\mathbf{A} = \overrightarrow{(1, 2)(3, 5)}$$
$$\mathbf{B} = \overrightarrow{(1, 2)(4, 1)}$$
$$\mathbf{C} = \overrightarrow{(0, 0)(1, 2)}$$

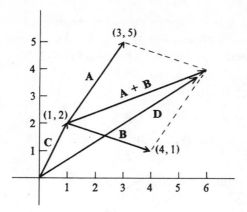

Figure 13-16

Then, **A** + **B** is the diagonal of the parallelogram, and **C** + (**A** + **B**) is the vector from the origin to the fourth vertex. In terms of **i** and **j**, we have

$$\mathbf{C} = \mathbf{i} + 2\mathbf{j}$$
$$\mathbf{A} = 2\mathbf{i} + 3\mathbf{j}$$
$$\mathbf{B} = 3\mathbf{i} - \mathbf{j}$$
$$\mathbf{A} + \mathbf{B} = 5\mathbf{i} + 2\mathbf{j}$$
$$\mathbf{C} + (\mathbf{A} + \mathbf{B}) = 6\mathbf{i} + 4\mathbf{j}$$

Therefore, (6, 4) are the coordinates of the vertex opposite (1, 2).

Example 2 Find the midpoint R of the segment joining $P = (-3, 2)$ and $Q = (2, 4)$. We have

$$\overrightarrow{OP} = -3\mathbf{i} + 2\mathbf{j},$$
$$\overrightarrow{OQ} = 2\mathbf{i} + 4\mathbf{j}.$$

Solution. From Example 3, in Section 13-1, we know that $\overrightarrow{OR} = \frac{1}{2}\overrightarrow{OP} + \frac{1}{2}\overrightarrow{OQ}$. Therefore,

$$\overrightarrow{OR} = \tfrac{1}{2}(-3\mathbf{i} + 2\mathbf{j}) + \tfrac{1}{2}(2\mathbf{i} + 4\mathbf{j})$$
$$= -\tfrac{1}{2}\mathbf{i} + 3\mathbf{j}$$

Hence, $R = (-1/2, 3)$.

Example 3 Find the coordinates of the point, R, 2/5 of the way from $P = (1, 3)$ to $Q = (4, -2)$.

Solution. We have $\overrightarrow{OR} = \overrightarrow{OP} + \frac{2}{5}\overrightarrow{PQ}$. Since

$$\overrightarrow{OP} = \mathbf{i} + 3\mathbf{j},$$

$$\overrightarrow{PQ} = 3\mathbf{i} - 5\mathbf{j},$$

$$\overrightarrow{OR} = (\mathbf{i} + 3\mathbf{j}) + \tfrac{2}{5}(3\mathbf{i} - 5\mathbf{j})$$

$$= \frac{11}{5}\mathbf{i} + \mathbf{j}.$$

Hence, $R = (11/5, 1)$.

Example 4 Find the unit vector $\mathbf{V}_0$ parallel to $7\mathbf{i} - 3\mathbf{j} = \mathbf{V}$.

Solution. The unit vector parallel to $\mathbf{V}$ is $\mathbf{V}/|\mathbf{V}|$. Since $|\mathbf{V}| = \sqrt{49 + 9} = \sqrt{58}$, the desired unit vector is

$$\mathbf{V}_0 = \frac{7}{\sqrt{58}}\mathbf{i} - \frac{3}{\sqrt{58}}\mathbf{j}$$

SECTION 13-2, REVIEW QUESTIONS

1. The angle between two vectors is the angle between two directed segments $\overrightarrow{PQ}$ and $\overrightarrow{PR}$ which _____ the vectors.   **represent**

2. The component of **A** along **B** is plus or minus the length of the _____ of **A** onto **B**.   **projection**

3. The component of **A** along **B** can be positive or negative but the projection of **A** onto **B** is always _____ .   **positive**

4. If the angle θ between **A** and **B** is acute, the component of **A** along **B** is positive, and if θ is obtuse, the component of **A** along **B** is _____ .   **negative**

5. If the angle between **A** and **B** is θ, the component c of **A** along **B** is given by the formula: $c = $ _____ . $|\overline{\mathbf{A}}|\cos\theta$

6. If θ is obtuse, $\cos\theta < 0$, so $|\mathbf{A}|\cos\theta$ is a _____ number.   **negative**

7. The components of a vector are not vectors, but are _____ .   **scalars**

8. The unit vector in the direction of the positive *x*-axis in a coordinate system is denoted _____.

î

ĵ

9. The unit vector along the positive *y*-axis is _____.

10. Every vector can be written **A** = *a***i** + *b***j** where *a* is the component of **A**

î component

along _____, and *b* is the _____ of **A** along **j**.

11. The *x*-component of **A** + **B** is the *x*-component of **A** plus the

x

_____ component of **B**.

SECTION 13-2, EXERCISES

Perform the indicated additions and subtractions.

1. $(3i + 4j) + (i - 2j)$
2. $(2i - j) + (-3i + 4j)$
3. $(i + 6j) + (2i - j) + (3i + 2j)$
4. $(3i - j) + (i + 4j) + (-3i - 2j)$
5. $(i + 7j) - (2i + j)$
6. $(3i - j) - (4i - 3j)$

Find the magnitude, |**A**|, of the given vector **A**.

7. $A = 2i - 3j$
8. $A = i + 6j$
9. $A = -2i + 2j$
10. $A = 5i - 2j$

Write the vectors $\overrightarrow{PQ}$ in the form *a***i** + *b***j** for the following points *P* and *Q*.

11. $P = (1, 3), Q = (2, 5)$
12. $P = (2, 1), Q = (4, 6)$
13. $P = (1, -2), Q = (-3, 2)$
14. $P = (-1, -3), Q = (2, -4)$
15. $P = (-2, 0), Q = (-3, 1)$
16. $P = (5, 7), Q = (1, -2)$

Find the coordinates of the fourth vertex of the parallelogram which has the three consecutive vertices given.

Example 1

17. $(2, 0), (1, 2), (4, 3)$
18. $(1, 4), (0, 0), (3, 0)$
19. $(2, -1), (3, 4), (-2, 1)$
20. $(-2, 0), (1, -2), (2, 2)$

Example 2,3

21. Find the coordinates of the point half way from (2, 5) to (4, 3).
22. Find the coordinates of the point half way from (1, −3) to (−5, 1).
23. Find the coordinates of the point 2/3 of the way from (4, 2) to (1, 8).
24. Find the coordinates of the point 1/4 of the way from (−3, 5) to (5, 1).

496 *Vectors*

Find a unit vector parallel to the given vector **V**.

25. $\mathbf{V} = 3\mathbf{i} + 4\mathbf{j}$
26. $\mathbf{V} = 5\mathbf{i} - 12\mathbf{j}$
27. $\mathbf{V} = \mathbf{i} - 2\mathbf{j}$
28. $\mathbf{V} = -3\mathbf{i} - 2\mathbf{j}$
29. $\mathbf{V} = \overrightarrow{(1, 4)(2, 1)}$
30. $\mathbf{V} = \overrightarrow{(-2, -1)(1, 2)}$

Example 4

13-3 DOT PRODUCT

Let us continue our discussion of vectors by defining the *dot product* of two vectors **A** and **B**.

Definition of Dot Product

$$\mathbf{A} \cdot \mathbf{B} = |\mathbf{A}| \, |\mathbf{B}| \cos \theta$$

where θ is the angle between the vectors.

The dot product of two vectors is a number, or scalar, and consequently $\mathbf{A} \cdot \mathbf{B}$ is also referred to as the scalar product of **A** and **B**. It is clear from the definition that

$$\mathbf{A} \cdot \mathbf{B} = \mathbf{B} \cdot \mathbf{A} \quad \text{and} \quad (c\mathbf{A}) \cdot \mathbf{B} = c(\mathbf{A} \cdot \mathbf{B}).$$

Since $|\mathbf{A}| \cos \theta$ is the component of **A** along **B**, the dot product of **A** and **B** is the product of $|\mathbf{B}|$ and the component of **A** along **B**. $\mathbf{A} \cdot \mathbf{B}$ is also the product of $|\mathbf{A}|$ and the component of **B** along **A**.

From Fig. 13-17 we see that the component of $\mathbf{B} + \mathbf{C}$ along a given

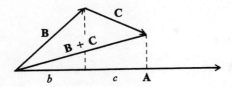

Figure 13-17

vector **A** is the component b of **B** along **A** plus the component c of **C** along **A**. Therefore $\mathbf{A} \cdot (\mathbf{B} + \mathbf{C})$, which is $|\mathbf{A}|(b + c)$ is the same as $|\mathbf{A}|a + |\mathbf{A}|b$, or $\mathbf{A} \cdot \mathbf{B} + \mathbf{A} \cdot \mathbf{C}$. Therefore, we have the distributive law for dot product over vector sums:

$$\mathbf{A} \cdot (\mathbf{B} + \mathbf{C}) = \mathbf{A} \cdot \mathbf{B} + \mathbf{A} \cdot \mathbf{C}.$$

Since dot product is commutative, we also have

$$(\mathbf{B} + \mathbf{C}) \cdot \mathbf{A} = \mathbf{B} \cdot \mathbf{A} + \mathbf{C} \cdot \mathbf{A}.$$

For any vector $\mathbf{A}$, $\mathbf{A} \cdot \mathbf{A} = |\mathbf{A}| |\mathbf{A}| \cos 0$, or

$$\mathbf{A} \cdot \mathbf{A} = |\mathbf{A}|^2.$$

If $\mathbf{A}$ and $\mathbf{B}$ are perpendicular, $\cos \theta = 0$, and $\mathbf{A} \cdot \mathbf{B} = 0$. Conversely, if

$$\mathbf{A} \cdot \mathbf{B} = |\mathbf{A}| |\mathbf{B}| \cos \theta = 0,$$

then $\mathbf{A} = \mathbf{0}$, or $\mathbf{B} = \mathbf{0}$, or $\mathbf{A} \perp \mathbf{B}$. Notice that there is no possible division associated with the dot product. From $\mathbf{A} \cdot \mathbf{B} = \mathbf{A} \cdot \mathbf{C}$ for a nonzero vector $\mathbf{A}$, we cannot conclude that $\mathbf{B} = \mathbf{C}$, but only that

$$\mathbf{A} \cdot (\mathbf{B} - \mathbf{C}) = 0,$$

which says that $\mathbf{B} = \mathbf{C}$ or $\mathbf{B} - \mathbf{C} \perp \mathbf{A}$.

One interesting application of the dot product is a simple derivation of the Law of Cosines. We let $\mathbf{A}$ and $\mathbf{B}$ be two adjacent sides of a triangle, so that $\mathbf{A} - \mathbf{B}$ represents the third side $\mathbf{C}$. Let $|\mathbf{A}| = a$, $|\mathbf{B}| = b$, and $|\mathbf{C}| = c$, and let θ be the angle between $\mathbf{A}$ and $\mathbf{B}$ (see Fig. 13-18). Notice that $\mathbf{A} \cdot \mathbf{A} = a^2$, $\mathbf{B} \cdot \mathbf{B} = b^2$, and $\mathbf{C} \cdot \mathbf{C} = c^2$. Hence,

$$\begin{aligned} c^2 &= \mathbf{C} \cdot \mathbf{C} \\ &= (\mathbf{A} - \mathbf{B}) \cdot (\mathbf{A} - \mathbf{B}) \\ &= \mathbf{A} \cdot \mathbf{A} + \mathbf{B} \cdot \mathbf{B} - 2\mathbf{A} \cdot \mathbf{B} \\ &= a^2 + b^2 - 2ab \cos \theta. \end{aligned}$$

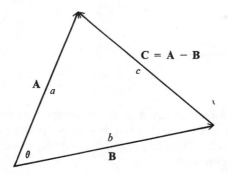

Figure 13-18

Finally, we should note that since i and j are unit vectors, which are perpendicular to each other, we have

$$\mathbf{i} \cdot \mathbf{i} = 1, \quad \mathbf{j} \cdot \mathbf{j} = 1, \quad \mathbf{i} \cdot \mathbf{j} = 0$$

These equations allow us to compute the dot product of vectors given in terms of their $\mathbf{i}$ and $\mathbf{j}$ components. Let $\mathbf{A} = a_1\mathbf{i} + a_2\mathbf{j}$ and $\mathbf{B} = b_1\mathbf{i} + b_2\mathbf{j}$; then

$$\begin{aligned}\mathbf{A} \cdot \mathbf{B} &= (a_1\mathbf{i} + a_2\mathbf{j}) \cdot (b_1\mathbf{i} + b_2\mathbf{j}) \\ &= a_1 b_1 \mathbf{i} \cdot \mathbf{i} + a_2 b_1 \mathbf{j} \cdot \mathbf{i} + a_1 b_2 \mathbf{i} \cdot \mathbf{j} + a_2 b_2 \mathbf{j} \cdot \mathbf{j} \\ &= a_1 b_1 + a_2 b_2 .\end{aligned}$$

Example 1 Show that $\mathbf{A} = 4\mathbf{i} - 2\mathbf{j}$ is perpendicular to $\mathbf{B} = 3\mathbf{i} + 6\mathbf{j}$.

Solution. $\mathbf{A} \cdot \mathbf{B} = (4)(3) + (-2)(6) = 12 - 12 = 0$. Therefore $\mathbf{A}$ is perpendicular to $\mathbf{B}$, since $\mathbf{A} \neq 0$ and $\mathbf{B} \neq 0$.

Example 2 Find the angle θ between PQ and PR where $P = (1, 1)$, $Q = (5, 3)$, and $R = (3, 5)$.

Solution. We use the formula

$$\overrightarrow{PQ} \cdot \overrightarrow{PR} = |\overrightarrow{PQ}| \, |\overrightarrow{PR}| \cos \theta .$$

Since $\overrightarrow{PQ} = 4\mathbf{i} + 2\mathbf{j}$ and $\overrightarrow{PR} = 2\mathbf{i} + 4\mathbf{j}$, we have $|\overrightarrow{PQ}| = \sqrt{16 + 4} = 2\sqrt{5}$, $|\overrightarrow{PR}| = \sqrt{4 + 16} = 2\sqrt{5}$, and

$$\overrightarrow{PQ} \cdot \overrightarrow{PR} = (4\mathbf{i} + 2\mathbf{j}) \cdot (2\mathbf{i} + 4\mathbf{j}) = 8 + 8 = 16 .$$

Therefore,

$$16 = (2\sqrt{5})(2\sqrt{5}) \cos \theta ,$$
$$\cos \theta = 16/20 = 0.8 ,$$
$$\theta = 36°52' .$$

Example 3 Find a unit vector perpendicular to $\mathbf{A} = -4\mathbf{i} + 2\mathbf{j}$.

Solution. It is easy to check with the dot product that $-b\mathbf{i} + a\mathbf{j}$ and $b\mathbf{i} - a\mathbf{j}$ are perpendicular to $a\mathbf{i} + b\mathbf{j}$. That is, if we switch the components of a vector, and then change the sign of one component, we get a vector perpendicular to the first vector. Hence $\mathbf{B} = 2\mathbf{i} + 4\mathbf{j}$ is perpendicular to $-4\mathbf{i} + 2\mathbf{j}$. A unit vector parallel to $\mathbf{B}$ is $\mathbf{B}_o = \mathbf{B}/|\mathbf{B}|$. Since $|\mathbf{B}| = \sqrt{4 + 16} = 2\sqrt{5}$, $\mathbf{B}_o = \dfrac{1}{2\sqrt{5}} (2\mathbf{i} + 4\mathbf{j}) =$

$\frac{1}{\sqrt{5}}\mathbf{i} + \frac{2}{\sqrt{5}}\mathbf{j}$. The other unit vector perpendicular to **A** is $-\mathbf{B}_o =$ $-\frac{1}{\sqrt{5}}\mathbf{i} - \frac{2}{\sqrt{5}}\mathbf{j}$.

Example 4 Show that PQR is a right triangle, and identify the right angle if

$$P = (-3, 4), \quad Q = (1, 2), \quad R = (-2, -4)$$

Solution. The vectors along the sides are

$$\overrightarrow{PQ} = 4\mathbf{i} - 2\mathbf{j}$$
$$\overrightarrow{PR} = \mathbf{i} - 8\mathbf{j}$$
$$\overrightarrow{QR} = -3\mathbf{i} - 6\mathbf{j}$$

$\overrightarrow{PQ} \cdot \overrightarrow{PR} = 4 + 16 = 20$, $\overrightarrow{PR} \cdot \overrightarrow{QR} = -3 + 48 = 45$, and $\overrightarrow{PQ} \cdot \overrightarrow{QR} = -12 + 12 = 0$. Since $\overrightarrow{PQ} \cdot \overrightarrow{QR} = 0$, the angle at Q is a right angle.

Example 5 If $P = (0, 4)$, find the point Q on the line $y = 2x - 1$ such that $\overrightarrow{PQ}$ is perpendicular to the vector $\mathbf{V} = 2\mathbf{i} + \mathbf{j}$.

Solution. Any point Q on the line has coordinates $(x, 2x - 1)$. The vector from $P = (0, 4)$ to the point $Q = (x, 2x - 1)$ is $(x - 0)\mathbf{i} + (2x - 1 - 4)\mathbf{j} = x\mathbf{i} + (2x - 5)\mathbf{j}$. $\overrightarrow{PQ}$ will be perpendicular to **V** if $\overrightarrow{PQ} \cdot \mathbf{V} = 0$:

$$[x\mathbf{i} + (2x - 5)\mathbf{j}] \cdot [2\mathbf{i} + \mathbf{j}] = 0,$$
$$2x + 2x - 5 = 0,$$
$$x = 5/4,$$
$$2x - 1 = 3/2.$$

Hence, $$Q = (5/4, 3/2).$$

SECTION 13-3, REVIEW QUESTIONS

|**A**| |**B**| cos θ 1. The dot product of **A** and **B** is defined as $\mathbf{A} \cdot \mathbf{B} =$ _____, where θ is the angle between **A** and **B**.

scalar   2. Since $\mathbf{A} \cdot \mathbf{B}$ is the scalar $|\mathbf{A}| |\mathbf{B}| \cos \theta$, the dot product is also called the _____ product of two vectors.

3. The dot product is commutative; that is, $\mathbf{A} \cdot \mathbf{B} =$ _____. $\overline{\mathbf{B}} \cdot \overline{\mathbf{A}}$

4. The dot product is distributive over vector addition; that is, $\mathbf{A} \cdot (\mathbf{B} + \mathbf{C}) =$
_____. $\overline{\mathbf{A}} \cdot \overline{\mathbf{B}} + \overline{\mathbf{A}} \cdot \overline{\mathbf{C}}$

5. If $\mathbf{A}$ and $\mathbf{B}$ are perpendicular, then $\mathbf{A} \cdot \mathbf{B} =$ _____. 0

6. Conversely, if $\mathbf{A}$ and $\mathbf{B}$ are nonzero vectors with $\mathbf{A} \cdot \mathbf{B} = 0$, then $\mathbf{A}$ and $\mathbf{B}$ are _____. perpendicular

7. The dot product of $a_1\mathbf{i} + a_2\mathbf{j}$ and $b_1\mathbf{i} + b_2\mathbf{j}$ is _____. $a_1b_1 + a_2b_2$

8. A vector perpendicular to $a_1\mathbf{i} + a_2\mathbf{j}$ is $a_2\mathbf{i} + ($_____$)\mathbf{j}$. $-a_1$

SECTION 13-3, EXERCISES

In Exercises 1-8 find $\mathbf{A} \cdot \mathbf{B}$, and then find $\cos \theta$, where θ is the angle between $\mathbf{A}$ and $\mathbf{B}$.

1. $\mathbf{A} = 2\mathbf{i} + \mathbf{j}, \mathbf{B} = 3\mathbf{i} - 2\mathbf{j}$ Example 1,2
2. $\mathbf{A} = \mathbf{i} - 3\mathbf{j}, \mathbf{B} = 4\mathbf{i} - 2\mathbf{j}$
3. $\mathbf{A} = 5\mathbf{i} - \mathbf{j}, \mathbf{B} = \mathbf{i} + 5\mathbf{j}$
4. $\mathbf{A} = -7\mathbf{i} - 2\mathbf{j}, \mathbf{B} = 2\mathbf{i} - 7\mathbf{j}$
5. $\mathbf{A} = \overrightarrow{(2, 4)(1, 5)}, \mathbf{B} = \overrightarrow{(1, -2)(3, 1)}$
6. $\mathbf{A} = \overrightarrow{(-1, 5)(3, 2)}, \mathbf{B} = \overrightarrow{(1, 4)(3, 3)}$
7. $\mathbf{A} = \overrightarrow{(0, 1)(2, 3)}, \mathbf{B} = \overrightarrow{(3, -1)(4, 2)}$
8. $\mathbf{A} = \overrightarrow{(-3, 1)(2, -1)}, \mathbf{B} = \overrightarrow{(4, 0), (0, 4)}$

Find a unit vector perpendicular to each of the following.

9. $3\mathbf{i} - \mathbf{j}$ 10. $-4\mathbf{i} + 3\mathbf{j}$ Example 3
11. $5/13\,\mathbf{i} + 12/13\,\mathbf{j}$ 12. $-1/\sqrt{2}\,\mathbf{i} + 1/\sqrt{2}\,\mathbf{j}$
13. $2\mathbf{i} + \mathbf{j}$ 14. $\mathbf{i} - 4\mathbf{j}$

Show that PQR is a right triangle. At which vertex is the right angle?

15. $P = (1, 3), Q = (4, 1), R = (3, 6)$ Example 4
16. $P = (0, -1), Q = (3, 1), R = (5, -2)$
17. $P = (-3, 2), Q = (5, 1), R = (6, 9)$
18. $P = (4, 2), Q = (-1, -1), R = (1, 4)$
19. $P = (-2, 5), Q = (3, 1), R = (2, 10)$
20. $P = (3, -4), Q = (4, 3), R = (0, 0)$
21. If $P = (2, 0)$, find the point Q on the line $y = x + 2$ so that $\overrightarrow{PQ}$ is perpendicular to $\mathbf{V} = 3\mathbf{i} + \mathbf{j}$.   Example 5
22. If $P = (5, 7)$, find the point Q on the line $y = 3x - 1$ so that $\overrightarrow{PQ}$ is perpendicular to $\mathbf{V} = 2\mathbf{i} - 3\mathbf{j}$.

23. If $P = (1, 3)$, find the point Q on the line $y = -x + 2$ so that $\overrightarrow{PQ}$ is perpendicular to $\mathbf{V} = \mathbf{i} - \mathbf{j}$.
24. If $P = (2, -5)$, find the point Q on the line $y = 2x - 3$ so that $\overrightarrow{PQ}$ is perpendicular to $\mathbf{V} = 4\mathbf{i} + \mathbf{j}$.
25. If $P = (4, 4)$ find the point Q on the line $y = 2x + 1$ such that $\overrightarrow{PQ}$ is perpendicular to the line. (Hint: $(0, 1)$ and $(1, 3)$ are the line, so $\mathbf{V} = \mathbf{i} + 2\mathbf{j}$ is parallel to the line and PQ must be perpendicular to $\mathbf{V}$.)
26. If $P = (5, 0)$ find the point Q on the line $y = -x + 1$ such that $\overrightarrow{PQ}$ is perpendicular to the line.
27. If $P = (1, 6)$ find the point Q on the line $y = -2x + 3$ such that $\overrightarrow{PQ}$ is perpendicular to the line.
28. If $P = (7, 5)$ find the point Q on the line $y = 5x - 4$ such that $\overrightarrow{PQ}$ is perpendicular to the line.

13-4 VECTORS AND STRAIGHT LINES

We already developed several forms of the equation of a straight line. These are:

$$y = mx + b \qquad \text{slope-intercept form}$$

$$y - y_1 = m(x - x_1) \qquad \text{point-slope form}$$

$$y - y_1 = \frac{y_2 - y_1}{x_2 - x_1}(x - x_1) \qquad \text{two-point form}$$

$$\frac{x}{a} + \frac{y}{b} = 1 \qquad \text{intercept form}$$

These equations allow us to write down the equation of a line directly when we know certain geometric facts. Here, we develop one more such form as an instructive application of vector methods.

Consider any line not through the origin, and let p be the distance of the line from the origin. Let φ be the inclination of the perpendicular from the origin to the given line. We want to write the equation of the line in terms of p and φ (Fig. 13-19).

Let $\mathbf{N} = \cos\varphi\,\mathbf{i} + \sin\varphi\,\mathbf{j}$. Then $\mathbf{N}$ is a unit vector perpendicular to the line. Let (x, y) be any point on the line, so that $\mathbf{V} = x\mathbf{i} + y\mathbf{j}$ is the vector from the origin to this point. For any such vector $\mathbf{V}$ to a point on the line, the component of $\mathbf{V}$ along $\mathbf{N}$ is the constant p. That is,

$$\mathbf{V} \cdot \mathbf{N} = p.$$

Writing this in terms of components, we get

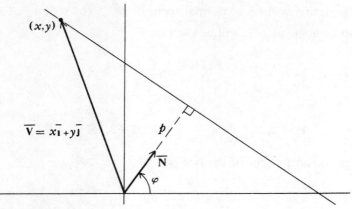

Figure 13-19

$$(x\mathbf{i} + y\mathbf{j}) \cdot (\cos \varphi \mathbf{i} + \sin \varphi \mathbf{j}) = p,$$

$$x \cos \varphi + y \sin \varphi = p.$$

This last equation is called the *normal form* of the equation of a line. If $p = 0$, the equation is that of the line through the origin and perpendicular to $\mathbf{N}$.

Example 1 Write the equation of the line three units from the origin and perpendicular to the vector $\mathbf{N} = 3\mathbf{i} - 4\mathbf{j}$.

Solution. The unit vector parallel to $\mathbf{N}$ is $\mathbf{N}_0 = \mathbf{N}/|\mathbf{N}|$. Since $|\mathbf{N}| = \sqrt{9 + 16} = 5$, $\mathbf{N}_0 = \frac{3}{5}\mathbf{i} - \frac{4}{5}\mathbf{j}$. Therefore, $3/5 = \cos \varphi$, $-4/5 = \sin \varphi$, where φ is the inclination of $\mathbf{N}$ or $\mathbf{N}_0$. The equation of the line is therefore

$$\tfrac{3}{5}x - \tfrac{4}{5}y = 3$$

Example 2 How far is the line $4x - 2y + 7 = 0$ from the origin?

Solution. We put the constant on the right, and multiply through by -1 so the constant is positive. This gives

$$-4x + 2y = 7.$$

Now divide through by a constant c to put the equation in normal form:

$$-\frac{4}{c}x + \frac{2}{c}y = \frac{7}{c}.$$

The equation will be in normal form if $-\dfrac{4}{c} = \cos\phi$ and $\dfrac{2}{c} = \sin\phi$ for some angle ϕ. This will be the case if

$$(-4/c)^2 + (2/c)^2 = 1.$$

$$16 + 4 = c^2.$$

$$c = \sqrt{20} = 2\sqrt{5}.$$

Hence the normal for of the line is

$$-\dfrac{4}{2\sqrt{5}} x + \dfrac{2}{2\sqrt{5}} y = \dfrac{7}{2\sqrt{5}}.$$

or

$$-\dfrac{2}{\sqrt{5}} x + \dfrac{1}{\sqrt{5}} y = \dfrac{7}{2\sqrt{5}},$$

and the line is $\dfrac{7}{2\sqrt{5}}$ units from the origin.

The same kind of technique used to find the normal form of the equation of a line can also be used to find the distance from any point to a line.

Example 3 Find the distance from (3, 4) to the line $2x + y = 2$.

Solution. Since (0, 2) and (1, 0) are points on the line, $\overline{\mathbf{W}} = -\mathbf{i} + 2\mathbf{j}$ is a vector along the line. The vector $\mathbf{N} = 2\mathbf{i} + \mathbf{j}$ is perpendicular to $\overline{\mathbf{W}}$. $\mathbf{N}_0 = (2/\sqrt{5})\mathbf{i} + (1/\sqrt{5})\mathbf{j}$ is a unit vector perpendicular to the line (Fig. 13-20). If $\mathbf{V}$ is the vector from (3, 4) to *any* point on the line, then $|\mathbf{V}||\cos\theta| = |\mathbf{V} \cdot \mathbf{N}_0|$ is the distance from (3, 4) to the line. We know that (1, 0) is on the line, so we let $\mathbf{V} = \overrightarrow{(3, 4)(1, 0)} = -2\mathbf{i} - 4\mathbf{j}$. Then,

$$\mathbf{V} \cdot \mathbf{N}_0 = (-2\mathbf{i} - 4\mathbf{j}) \cdot \left(\dfrac{2}{\sqrt{5}}\mathbf{i} + \dfrac{1}{\sqrt{5}}\mathbf{j} \right)$$

$$= -\dfrac{4}{\sqrt{5}} - \dfrac{4}{\sqrt{5}} = -\dfrac{8}{\sqrt{5}}.$$

Therefore, the distance is

$$|V \cdot N_0| = \left|-\frac{8}{\sqrt{5}}\right| = \frac{8}{\sqrt{5}}.$$

Example 4 Find the angle between the lines $x + 2y - 4 = 0$ and $2x + y - 6 = 0$.

Solution. We find a vector **A** along one line and a vector **B** along the other, and take the dot product. Since $(0, 2)$ and $(4, 0)$ are on the first line, $\mathbf{A} = \overrightarrow{(0, 2)(4, 0)} = 4\mathbf{i} - 2\mathbf{j}$ is a vector along the line. Similarly, $(0, 6)$ and $(3, 0)$ are points on the second line. Hence, $\mathbf{B} = \overrightarrow{(0, 6)(3, 0)} = 3\mathbf{i} - 6\mathbf{j}$ is a vector along the second line.

$$|\mathbf{A}|\,|\mathbf{B}| \cos \theta = \sqrt{16 + 4}\,\sqrt{9 + 36} \cos \theta$$
$$= 2\sqrt{5}\,3\sqrt{5} \cos \theta$$
$$= 30 \cos \theta$$

and

$$\mathbf{A} \cdot \mathbf{B} = (4\mathbf{i} - 2\mathbf{j}) \cdot (3\mathbf{i} - 6\mathbf{j}) = 12 + 12 = 24.$$

Therefore,

$$30 \cos \theta = 24,$$
$$\cos \theta = .8,$$
$$\theta = 36°52'.$$

Suppose we are given a vector $\mathbf{V} = a\mathbf{i} + b\mathbf{j}$ and a point $P = (x_0, y_0)$ and asked to find the line through P and parallel to $\mathbf{V}$. We can write the equation for this in vector form by expressing the

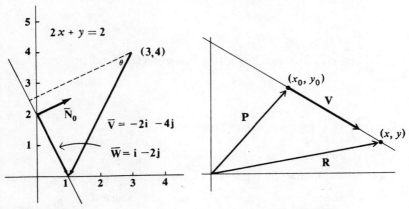

Figure 13-20   Figure 13-21

vector **R** from the origin to any point on the line. We let **P** be the vector from the origin to the given point (x_0, y_0), so that $\mathbf{P} = x_0\mathbf{i} + y_0\mathbf{j}$ (Fig. 13-21). Then any vector **R** to a point on the line can be written

$$\mathbf{R} = \mathbf{P} + t\mathbf{V}$$

for some scalar t. In terms of components, $\mathbf{R} = x\mathbf{i} + y\mathbf{j}$, and we have the vector equation:

$$x\mathbf{i} + y\mathbf{j} = (x_0\mathbf{i} + y_0\mathbf{j}) + t(a\mathbf{i} + b\mathbf{j}).$$

This is the same as the *two* scalar equations:

$$x = x_0 + ta,$$
$$y = y_0 + tb.$$

These two equations are the *parametric equations* of the line. The scalar t, which ranges over all real numbers, is called the parameter. The ordinary form of this equation can be obtained by eliminating the parameter. Since $(x - x_0)/a = t$ and $(y - y_0)/b = t$, we get the equation

$$\frac{x - x_0}{a} = \frac{y - y_0}{b}.$$

Example 5 Find the parametric equations of the line through (1, 3) and perpendicular to the line $L: 2x + 3y = 12$.

Solution. The points (0, 4) and (6, 0) are on L, so that $6\mathbf{i} - 4\mathbf{j}$ is a vector along L. Hence $4\mathbf{i} + 6\mathbf{j}$ is a vector perpendicular to L. Therefore,

$$x = 1 + 4t$$
$$y = 3 + 6t$$

are parametric equations for a line through (1, 3) and perpendicular to L.

SECTION 13-4, REVIEW QUESTIONS

inclination

1. The vector $\cos \varphi \mathbf{i} + \sin \varphi \mathbf{j}$ is a unit vector, and φ is its _____ .

2. The normal form of the equation of a line is given in terms of φ, which is the inclination of the perpendicular line, and p, which is the

distance

_____ from the origin.

3. The normal form of the equation of a line is $x \cos \varphi + y \sin \varphi = $ _____ . $\quad$ p

4. To put $Ax + By = C$ in normal form, you divide by $\sqrt{A^2 + B^2}$ if $C > 0$, and by _____ if $C < 0$. $\quad -\sqrt{A^2 + B^2}$

5. The line $\dfrac{A}{\sqrt{A^2 + B^2}} x + \dfrac{B}{\sqrt{A^2 + B^2}} y = \dfrac{C}{\sqrt{A^2 + B^2}}$, with $C > 0$, is
   _____ units from the origin. $\quad C/\sqrt{A^2 + B^2}$

6. To obtain a vector perpendicular to a given vector, switch the components and change the _____ of one of them. $\quad$ sign

7. To find the component of one vector along another, you use the
   _____ product. $\quad$ dot

8. The equations $x = x_0 + at$, $y = y_0 + bt$ are the parametric equations of the line through _____ and parallel to the vector _____ . $\quad (x_0, y_0)$ $\quad a\bar{i} + b\bar{j}$

9. The scalar t is called the _____ in the equations $x = x_0 + at$, $y = y_0 + bt$. $\quad$ parameter

10. If we eliminate the parameter, we obtain a single cartesian equation in _____ and _____ . $\quad$ x $\quad$ y

SECTION 13-4, EXERCISES

Write the equation of the line which is p units from the origin and perpendicular to the given vector N.

1. $p = 4$, $N = 3i + 4j$
2. $p = 7$, $N = 4i - 3j$ $\quad$ Example 1
3. $p = 1$, $N = i - 2j$
4. $p = 3$, $N = 3i + j$
5. $p = 5$, $N = i - j$
6. $p = 2$, $N = -2i + 3j$

Write the normal form of the following lines, and give the distance to the origin.

7. $x + y = 3$
8. $2x - y = 1$ $\quad$ Example 2
9. $-3x + y - 2 = 0$
10. $4x + y - 5 = 0$
11. $x - 5y + 3 = 0$
12. $2x + 2y + 5 = 0$
13. $3x - 5y + 7 = 0$
14. $-2x + 4y + 3 = 0$

Find the distance from the given point P to the given line.

15. $P = (1, 4)$; $x + y = 2$
16. $P = (2, 2)$; $x - 2y = 4$ $\quad$ Example 3
17. $P = (3, 0)$; $2x - y = 0$
18. $P = (0, 2)$; $x - 2y = 0$
19. $P = (1, -1)$; $y = x + 1$
20. $P = (3, 2)$; $y = -3x + 1$

Find the angle between each pair of lines.

Example 4
21. $x + 3y + 6 = 0; 2x + y - 4 = 0$
22. $x - 2y + 2 = 0; x - 3y + 9 = 0$
23. $x + 2y - 2 = 0; 2x + y + 4 = 0$
24. $2x + 4y + 3 = 0; -4x - 2y + 2 = 0$
25. $x + 4y + 8 = 0; 4x - y - 6 = 0$
26. $-3x + 2y + 5 = 0; 2x + 3y + 6 = 0$
27. $x + 2y - 2 = 0; 3x + y - 3 = 0$
28. $2x - y + 1 = 0; 4x + 2y + 5 = 0$

Find the parametric equations of the following lines.

Example 5
29. through $(1, 2)$, parallel to $5\mathbf{i} + \mathbf{j}$.
30. through $(5, -1)$, parallel to $3\mathbf{i} - 2\mathbf{j}$.
31. through $(4, -2)$, parallel to $\mathbf{i} - \mathbf{j}$.
32. through $(3, 2)$, parallel to $\mathbf{i} + 5\mathbf{j}$.
33. through $(1, 5)$, parallel to $\mathbf{i}$.
34. through $(-1, 3)$, parallel to $\mathbf{j}$.
35. through $(1, 1)$, perpendicular to $2x - y - 4 = 0$.
36. through $(-4, 0)$, perpendicular to $3x - y + 9 = 0$.
37. through $(2, -3)$, perpendicular to $x + 2y - 5 = 0$.
38. through $(1, 5)$, perpendicular to $2x + 3y + 12 = 0$.

13-5 PARAMETRIC EQUATIONS AND VECTOR FUNCTIONS

A vector-valued function on the real line is a rule which assigns a unique vector $\mathbf{r}(t)$ to each real number t. Since a vector is determined only when both components are determined, a vector-valued function is essentially the same thing as an ordered pair of real valued functions $(x(t), y(t))$:

$$\mathbf{r}(t) = x(t)\mathbf{i} + y(t)\mathbf{j}$$

We have already dealt with many different kinds of plane curves. These have been the graphs of equations of the form

$$y = f(x) \quad \text{or} \quad F(x, y) = 0$$

The most general definition of a plane curve is that it is a continuous function on the line into the plane. This means that we have a pair of coordinates $(x(t), y(t))$ for each t on the line, and $x(t), y(t)$ are continuous functions. We will call such a pair of real functions,

$$x = x(t), \quad y = y(t),$$

the *parametric equations* of the curve. We have already seen the form that such parametric equations take for a straight line.

We can of course interpret **r**(*t*) as the vector from the origin to the point (*x*(*t*), *y*(*t*)). Then, we have the picture of the curve being traced out by the point of the moving vector **r**(*t*). In any case, there is no real difference in the information given by a single vector function

$$\mathbf{r}(t) = x(t)\mathbf{i} + y(t)\mathbf{j}$$

or two parametric equations,

$$x = x(t), \quad y = y(t).$$

We frequently think of *t* as time, and (*x*(*t*), *y*(*t*)) as the point at time *t* (Fig. 13-22).

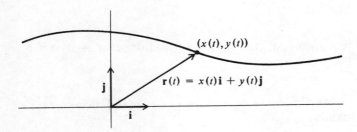

Figure 13-22

Parametric equations of a geometrically described curve are frequently easier to obtain than a direct relationship between the *x*- and *y*-coordinates of an arbitrary point on the curve. The parameter frequently also has some geometric significance, as the following examples illustrate.

Example 1 Show that $x = a \cos \theta$, $y = b \sin \theta$ are the parametric equations of the ellipse $x^2/a^2 + y^2/b^2 = 1$.

Solution. Since $x/a = \cos \theta$, $y/b = \sin \theta$, we have $x^2/a^2 + y^2/b^2 = \cos^2 \theta + \sin^2 \theta = 1$. The line through the origin with inclination θ intersects the circle of radius a at ($a \cos \theta$, $a \sin \theta$), and the circle of radius b at ($b \cos \theta$, $b \sin \theta$). (See Fig. 13-23.) Hence the point on the ellipse corresponding to angle θ has the *x*-coordinate of the point where the ray intersects the *a*-circle and the *y*-coordinate of the point where the ray intersects the *b*-circle. If $b = a$, we get the

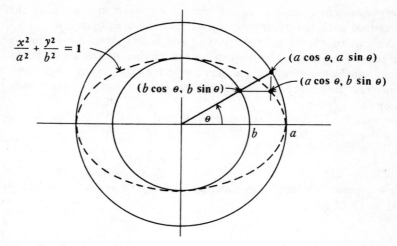

Figure 13-23

parametric equations of the circle of radius a: $x = a \cos \theta$, $y = a \sin \theta$.

Example 2 Plot the curve $x = t^2$, $y = t + 2$ and determine its cartesian equation.

Solution. We obtain sample points on the curve by assigning values to t, and computing the corresponding values for x and y. These points are shown in Fig. 13-24.

t	-3	-2	-1	0	1	2	3
x	9	4	1	0	1	4	9
y	-1	0	1	2	3	4	5

To find the cartesian equation of this curve, we eliminate the parameter t between the two equations. Thus $t = y - 2$, and

$$x = (y - 2)^2$$

Example 3 Graph the curve and find its cartesian equation if $x = \sin \varphi$, $y = \cos 2\varphi$.

Solution. We notice that $|x| \leq 1$, $|y| \leq 1$, so the entire curve lies inside the box $-1 \leq x \leq 1$, $-1 \leq y \leq 1$.

510 Vectors

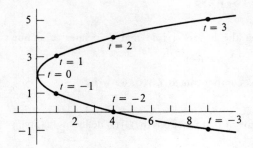

Figure 13-24

φ	$-\pi/2$	$-\pi/4$	0	$\pi/4$	$\pi/2$
x	-1	-0.7	0	0.7	1
2φ	$-\pi$	$-\pi/2$	0	$\pi/2$	π
y	-1	0	1	0	-1

The curve traces out the curve $ABCDE$ shown in Fig. 13-26 as φ goes from $-\pi/2$ to $\pi/2$. As φ then goes from $\pi/2$ to $3\pi/2$, the curve is traced out backwards: $EDCBA$. As φ increases, the point (x, y) oscillates back and forth across the curve shown. To eliminate φ and obtain the cartesian equation, we write

$$y = \cos 2\varphi = \cos^2\varphi - \sin^2\varphi$$
$$= 1 - 2\sin^2\varphi$$

Hence,

$$y = 1 - 2x^2$$

The graph of $y = 1 - 2x^2$ is the entire parabola shown (dotted) in Fig. 13-25, whereas the given parametric curve is just the arc $ABCDE$.

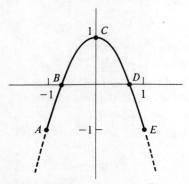

Figure 13-25

SECTION 13-5, REVIEW QUESTIONS

vector

1. A vector-valued function on the line is a rule that associates a unique _____ with each number.

2. If the vector function $\mathbf{r}(t)$ has components $x(t)$, $y(t)$, we write

x(t)ī + y(t)j̄

$\mathbf{r}(t) = $ _____.

3. A vector function $\mathbf{r}(t)$ or a pair of real functions $x(t)$, $y(t)$ defines a mapping

plane

of the line into the _____.

line

4. A curve is any continuous mapping of the _____ into the plane.

5. A pair of continuous functions $x = x(t)$, $y = y(t)$ defines a curve, and these

parametric

equations are called the _____ equations of the curve.

6. A vector function $\mathbf{r}(t) = x(t)\mathbf{i} + y(t)\mathbf{j}$ defines the curve traced out by the

point

_____ of the vector $\mathbf{r}(t)$ if its base is considered fixed at the origin.

7. The parameter sometimes has geometric significance, but sometimes we think of $(x(t), y(t))$ as giving the coordinates of a moving point at a given

time

_____.

origin

8. If $\mathbf{r}(t) = x(t)\mathbf{i} + y(t)\mathbf{j}$, then $\mathbf{r}(t)$ is the vector from the _____ to the moving point at time t.

9. To find the cartesian equation of a curve, $x = x(t)$, $y = y(t)$, we eliminate

t

_____ from the two equations.

SECTION 13-5, EXERCISES

Example 1

1. Give parametric equations of the ellipse $x^2/9 + y^2/4 = 1$.
2. Give parametric equations of the ellipse $x^2/4 + y^2/25 = 1$.

Use the construction of figure 13-23 to find three points in the first quadrant for the given ellipse, and sketch the part in the first quadrant.

3. $x^2/9 + y^2/4 = 1$ 4. $x^2/4 + y^2/25 = 1$

Write parametric equations of the following circles.

5. center (0, 0), radius 3
6. center (0, 0), radius 8
7. $x^2 + y^2 = 4$
8. $x^2 + y^2 = 5$

Plot the following curves and determine their cartesian equations

9. $x = (t+1)^2, y = t+1$
10. $x = t-1, y = t^2$ Example 2,3
11. $x = 2t+1, y = t-1$
12. $x = t+2, y = -3t$
13. $x = 2 + 1/t, y = 1 + t$
14. $x = t + 1/t, y = t$
15. $x = t^2, y = 1$
16. $x = 3, y = t^2 - 1$
17. $x = \cos\theta, y = \cos^2\theta$
18. $x = \cos^2\theta + 1, y = \sin^2\theta$
19. $x = \sin 2t, y = \cos t$
20. $x = \cos 2t, y = \sin t$
21. $x = 3 - \cos t, y = 2 + \sin t$
22. $x = 1 + 2\cos t, y = 2 + 2\sin t$
23. $x = 2 + 3\cos\theta, y = 3 - 2\sin\theta$
24. $x = -1 + 4\cos\theta, y = 2 - 2\sin\theta$

13-6 REVIEW FOR CHAPTER

1. If $\mathbf{A} = \vec{PQ}$ and $\mathbf{B} = \vec{PR}$, express $\vec{PS}$, where S is 3/4 of the way from Q to R, in terms of $\mathbf{A}$ and $\mathbf{B}$.
2. If $\mathbf{A} = \vec{PQ}$ and $\mathbf{B} = \vec{PR}$, express $\vec{PS}$, where S is 2/3 of the way from Q to R, in terms of $\mathbf{A}$ and $\mathbf{B}$.

If A, B, and C have coordinates $(2, 1)$, $(-4, 3)$, and $(-1, 2)$, find the coordinates of $X = (x, y)$.

3. $\vec{AB} = \vec{CX}$
4. $\vec{AX} = \vec{CB}$
5. $\vec{XA} = \vec{CB}$
6. $\vec{XA} = \vec{BC}$

If $\mathbf{A}$, $\mathbf{B}$, $\mathbf{C}$, $\mathbf{D}$ are the vectors from the origin to the points $(2, -1)$, $(3, 4)$, $(-4, -5)$, and $(-6, 1)$, then the following represent the vectors from the origin to what points?

7. $\mathbf{A} + \mathbf{B}$
8. $\mathbf{A} - \mathbf{B}$
9. $\mathbf{A} + \mathbf{C}$
10. $\mathbf{A} + \mathbf{D}$
11. $\mathbf{B} - \mathbf{C}$
12. $\mathbf{B} + \mathbf{D}$
13. $2(\mathbf{D} - \mathbf{C})$
14. $\mathbf{A} + \mathbf{B} + \mathbf{C}$

15. Find the coordinates of the vertex opposite $(-4, -5)$ in the parallelogram with three contiguous vertices at $(-1, 3)$, $(-4, -5)$, and $(2, -7)$.
16. Find the coordinates of the midpoint R of the segment joining $(-4, -5)$ and $(2, -7)$.
17. Find the unit vector $\mathbf{V}_0$ parallel to $\mathbf{V} = 5\mathbf{i} - 3\mathbf{j}$.
18. Find the unit vector $\mathbf{V}_0$ perpendicular to $\mathbf{V} = -2\mathbf{i} + \mathbf{j}$.

Find **A** ⋅ **B** if:

19. $\mathbf{A} = 5\mathbf{i} - 3\mathbf{j}$, $\mathbf{B} = -2\mathbf{i} + \mathbf{j}$
20. $\mathbf{A} = -4\mathbf{i} - 3\mathbf{j}$, $\mathbf{B} = 3\mathbf{i} + 4\mathbf{j}$
21. Find the angle between **A** and **B** in Exercises 19 and 20.
22. If $P = (0, 4)$, find the point Q on the line $y = 2x - 3$ such that $\overrightarrow{PQ}$ is perpendicular to the vector, $\mathbf{V} = 2\mathbf{i} - 3\mathbf{j}$.
23. If $P = (2, 3)$, find the point Q on the line $y = -3x + 5$ such that $\overrightarrow{PQ}$ is perpendicular to the vector, $\mathbf{V} = -4\mathbf{i} + 5\mathbf{j}$.
24. Write the equation of a line that passes through the origin and is perpendicular to $\mathbf{N} = 5\mathbf{i} + \mathbf{j}$.
25. Write the equation of a line six units from the origin and perpendicular to $\mathbf{N} = -3\mathbf{i} + \mathbf{j}$.
26. What is the perpendicular distance from $y = -\frac{2}{3}x + 4$ to the origin?
27. What is the perpendicular distance from $(4, -7)$ to $y = -\frac{2}{3}x + 4$?
28. What is the angle between $y = 3x - 1$ and $y = -\frac{1}{2}x + 3$?
29. Find the equation of the line through $(-2, 3)$ and parallel to $\mathbf{V} = \mathbf{i} - 3\mathbf{j}$.
30. Find the parametric equations of the line through $(2, -3)$ and perpendicular to $y = 4 - x$.

Plot the curve whose parametric equations are given below and find its cartesian equation.

31. $x = t$, $y = -3t$
32. $x = -4t$, $y = 2t$
33. $x = t^2 + 2t$, $y = t - 2$
34. $x = t - 1$, $y = t^2$
35. $x = \cos t$, $y = \cos^2 t$
36. $x = \cos \theta$, $y = 2 \sin \theta$
37. $x = 2 \cos \theta$, $y = 3 \sin \theta$
38. $x = -1 + \cos \theta$, $y = 2 + 2 \sin \theta$

Appendix A

ANSWERS TO ODD NUMBERED PROBLEMS

CHAPTER 1

SECTION 1-1, EXERCISES

1. $\{1, 2, 3, 4, 5, 6, 7, 8, 27\}$
3. $\{1, 2, 3, 4, 5, 6, 7, 8, 27, 64, 125\}$
7. (a) $\{3, 8, 27\}$
   (b) $\{3, 8, 27\}$
9. $\{a, b, c, d, e, f, g, h\}$
11. $\{a, b, d, e, f, g, h, i, j, k\}$
    (b) $\{a, b\}$
15. (a) $\{a, b\}$
17. S
19. $\emptyset$
21. $\emptyset$
23. (d)
25. (a) $\emptyset, \{x\}$
    (b) $\emptyset, \{x\}, \{y\}, \{x, y\}$
    (c) $\emptyset, \{x\}, \{y\}, \{z\}, \{x, y\}, \{x, z\}, \{y, z\}, \{x, y, z\}$ sixteen

SECTION 1-2, EXERCISES

1. $+$
3. $+, -, \times, \div$
5. Rational
7. Irrational
9. Integer
11. Integer
13. $>$
15. $<$
17. $>$
19. $>$
21. $<$
23. $>$

SECTION 1-3, EXERCISES

1. 3
3. -2
5. $x - 2$
7. $-3x + 5$
9. $-2 - x - y$
11. $-4 + u + v$
13. $-b$
15. $x - z$
17. $-b$
19. No

SECTION 1-4, EXERCISES

1. 3
3. $\dfrac{1}{2}$
5. $\dfrac{x}{2}$
7. $\dfrac{5}{3x}$
9. $\dfrac{1}{2xy}$
11. $\dfrac{1}{4}uv$
13. $\dfrac{3 \cdot 5}{4 \cdot 8}$; $(5')$; $(3 - 4) + (5 - 8) = (3 + 5) - (4 + 8)$
15. $\dfrac{2 \cdot 8}{3 \cdot 7}$; $(6')$; $(2 - 3) - (7 - 8) = (2 + 8) - (3 + 7)$
17. $\dfrac{1}{2 \cdot 5}$; $(3')$; $(-2) + (-5) = -(2 + 5)$
19. $\dfrac{1}{b}$
21. $\dfrac{1}{a}$
23. 1
25. No

SECTION 1-5, EXERCISES

1. $ax + ay$
3. $-2ur - 2us$
5. $6ux - 3vx$
7. x
9. $5ab - 3t$
11. $-xy + 4ab$
13. $-9ab + 11bc$
15. $\dfrac{3a - 2b}{x}$
17. $\dfrac{-3x}{2a}$
19. $\dfrac{x + y}{xy}$
21. $\dfrac{ax + bc}{bx}$
23. $\dfrac{a^2 - b^2}{ab}$

SECTION 1-6, EXERCISES

1. Positive
3. Negative
5. Positive
7. >
9. <
11. >
13. <
15. >
17. >
19. <

SECTION 1-7, EXERCISES

1. (a), (c), (e)
3. (a), (c), (e)
5. $(r + s\sqrt{3})^{-1} = \dfrac{r}{r^2 - 3s^2} - \dfrac{s}{r^2 - 3s^2}\sqrt{3}$

A-2   Appendix

7. (a) yes (b) yes; z
   (c) $-z = z, -u = t, -t = u$ (d) yes
   (e) yes (f) yes; u
   (g) $u^{-1} = u, t^{-1} = t$ (h) yes
   (i) yes

CHAPTER 2

SECTION 2-1, EXERCISES

1. $\dfrac{y^4}{x^3}$
3. x
5. $x^2 y^5$
7. $\dfrac{x^3 y^5}{81}$
9. $\dfrac{9}{4x^4 y^5}$
11. $\dfrac{1}{9x^2}$
13. $\dfrac{1}{7}$
15. $\dfrac{x^2}{y^2}$
17. $\dfrac{-64}{a^5 b^4}$
19. $\dfrac{1}{3}$
21. 49
23. 6^4
25. x^2
27. 1
29. $(xy)^{7n}$
31. $\dfrac{y + x}{xy}$
33. $\dfrac{1}{x + y}$
35. $\dfrac{ab}{b + a}$
37. $\dfrac{b^2 + a^2}{a^2 b^2}$
39. $\dfrac{x^6}{(x^4 + 1)^3}$

SECTION 2-2, EXERCISES

1. 2^2
3. $\dfrac{1}{5}$
5. $\dfrac{1}{2}$
7. 15
9. $\dfrac{27}{64}$
11. $8x^3$
13. $\dfrac{1}{a^4}$
15. $\dfrac{b^2}{4a}$
17. $\dfrac{1}{6} x^{1/3}$
19. $\dfrac{x^2}{9}$
21. $a^{1/2} b$
23. $\dfrac{x^{2/3}}{y^{1/4}}$
25. $x^{2/3} y^{1/3}$
27. $\dfrac{x + 1}{x^{1/2}}$
29. $\dfrac{a^{2/3} - 1}{a^{1/3}}$

SECTION 2-3, EXERCISES

1. $2\sqrt{2}$
3. $\dfrac{3\sqrt{3}}{2}$
5. 6
7. $2\sqrt[3]{2}$
9. $a\sqrt{a}$
11. $a^2 x^4 \sqrt{a}$

13. $x\sqrt[3]{x^2}$ 15. $\sqrt{x}$ 17. $x^2\sqrt{a}$
19. $a^3b^3\sqrt[3]{9b^2}$ 21. 3 23. $3ab^3\sqrt{a}$
25. $\dfrac{\sqrt{3a}}{a}$ 27. $\dfrac{2\sqrt{6ab}}{3b}$ 29. $\dfrac{2\sqrt{2x}}{x^3}$
31. $\dfrac{x^2\sqrt{xy}}{y^2}$ 33. $\dfrac{\sqrt{3xy}}{y}$ 35. $\dfrac{x\sqrt{2y}}{6a^2}$
37. $\dfrac{\sqrt[3]{18a^2b}}{6ab^2}$ 39. $\dfrac{3xy\sqrt{5xb}}{a^3b^3}$ 41. $\dfrac{3x\sqrt{5}}{y^2}$
43. $\dfrac{y\sqrt{x}}{ab}$

SECTION 2-4, EXERCISES

1. $x^3 + xy^2 - yx^2 - y^3$
3. $2x^4 + 2xy + yx^3 + y^2$
5. $ax + 3ay - 2bx - 6by$
7. $x^3 - xy^2 + yx^2 - y^3$
9. $2x^3 + 3x^2 + 4x + 3$
11. $-x^3 - 4x^2 + x + 12$
13. $5x^5 - 3x^4 + 5x^3 - 3x^2 - 10x + 6$
15. $x^4 + x^3 + 2x^2 + 5x + 3$
17. $2x^5 - x^4 + x^3 + x + 2$
19. $8x^4 - 14x^3 + 15x^2 - 8x + 4$
21. $a^2 - b^2$
23. $x^4 - 1$
25. $16x^2 - 9y^2$
27. $x^2 + 4xy + 4y^2$
29. $9a^2 + 6ab + b^2$
31. $x^4 - 2x^2y^2 + y^4$
33. $9a^4 - 18a^2b^2 + 9b^4$
35. $2x^2 + 5x + 3$
37. $3x^2 - 17x + 10$
39. $2x^2 - 3bx - 2b^2$
41. $1 - x^2$
43. $x^{2/3} - 1$
45. $a^{4/3} - b^{4/3}$
47. $x - 1$
49. $x - 2\sqrt{xy} + y$
51. $\dfrac{\sqrt{x} + 1}{x - 1}$
53. $\dfrac{\sqrt{xy} - 2}{xy - 4}$
55. $\dfrac{x\sqrt{3} - 5\sqrt{x}}{3x - 25}$

SECTION 2-5, EXERCISES

1. $3ax(x + 5a)$
3. $6a^2b^2(4a^2 - 3abc^3 + 2b^2)$
5. $7c(a + 5b - 3c)$
7. $(x - 2)(x - 3)$
9. $(x + 1)(x + 2)$
11. $(x - 2)(x + 8)$
13. $(x + 3)(x + 6)$
15. $(x + 2)(x - 15)$
17. $(2x + 1)(2x - 1)$
19. $(5x + 4y)(5x - 4y)$
21. $(x - 2)^2$
23. $(x + 4)^2$
25. $(x + 9)^2$
27. $(2x + 1)^2$
29. $(2x + 1)(x - 3)$
31. $(5x + 2)(2x - 3)$
33. $3(x + 2)(x + 3)$
35. $5(x - 2)^2$
37. $b(ax + 1)^2$
39. $(3 - a)(9 + 3a + a^2)$
41. $(2x - 3)(4x^2 + 6x + 9)$
43. $3xy(y + 3x)(y^2 - 3xy + 9x^2)$

SECTION 2-6, EXERCISES

1. $\dfrac{3}{4ab}$ 3. $\dfrac{x-3}{3(x-1)}$ 5. $x-y$ 7. $\dfrac{x-5}{x+5}$

9. $\dfrac{a-3}{2a-5}$ 11. $\dfrac{x^2+x+1}{x(x+5)}$ 13. $\dfrac{16a^3}{b}$ 15. $\dfrac{4}{a-3b}$

17. $\dfrac{1}{x+1}$ 19. $\dfrac{y+9}{y+1}$ 21. 2   23. $\dfrac{1+a}{1+a^2}$

25. $\dfrac{2(x^2+y^2)}{(3x-2y)(2x+3y)}$ 27. $\dfrac{2x+3y}{xy(x+y)}$

29. $\dfrac{2x^2-3x-2}{(x+2)(x-1)(x+1)}$ 31. $\dfrac{6x+9}{9-x^2}$ 33. $x+1$

35. $\dfrac{2x}{x-1}$ 37. $\dfrac{b+a}{b-a}$ 39. $\dfrac{4x^2-3x}{4x+1}$

SECTION 2-7, CHAPTER REVIEW

1. $\dfrac{1}{a^3x^2}$ 3. $\dfrac{-a^2b^3}{343}$ 5. $\dfrac{1}{x^2}+\dfrac{1}{y^2}$

7. $\dfrac{1}{x}+y$ 9. $\left(\dfrac{1}{x}+\dfrac{1}{y}\right)(x+y)=\dfrac{(x+y)^2}{xy}$

11. $\dfrac{-x}{y^2\sqrt[3]{2y}}=\dfrac{-x\sqrt[3]{4y^2}}{2y^3}$ 13. $\dfrac{2y}{x^2}$ 15. $\dfrac{y\sqrt{3xy}}{3x^2}$

17. $\dfrac{\sqrt{x^2-y^2}}{x-y}$ 19. $\dfrac{a}{3}\sqrt{15}$ 21. $2ab\sqrt[6]{ab^2}$

23. $\dfrac{(\sqrt{x}+\sqrt{y})^2}{x-y}$ 25. $8x^2-29x+15$ 27. x^3+8

29. $x^6-6x^4-3x^2+8$ 31. a^3+125 33. $-x^2-39$

35. x^4-1 37. $x^{4n}-2x^{2n}y^{2n}+y^{4n}$

39. $x^{2/3}-2x^{1/3}y^{1/3}+y^{2/3}$ 41. $(11x-7)(11x+7)$

43. $5x(3x^4-4x^2+9)$ 45. $-3x(x^2+9)$

47. $(3x-4y^2)(9x^2+12xy^2+16y^4)$ 49. $-(2+x)$

51. $2/x$ 53. $1/a$ 55. $\dfrac{x+12y}{3xy^2}$

57. $\dfrac{x+1}{x^2-9}$ 59. $\dfrac{21-5x+5y}{3(x^2-y^2)}$ 61. $\dfrac{2y-x}{2y+x}$

63. $\dfrac{4xy}{y^2+4x^2}$

CHAPTER 3

SECTION 3-1, EXERCISES

1. Identity; all x
3. Conditional
5. Identity; $a \neq b$
7. Identity; $x \geq 0$
9. Identity; all x
11. 11
13. -2
15. $\frac{9}{5}$
17. $-\frac{7}{3}$
19. $-\frac{21}{10}$
21. $-\frac{7}{2}$
23. -35
25. $\frac{3}{11}$
27. $\frac{20}{9}$
29. $\frac{30}{7}$
31. $\dfrac{3}{(1+a)}$
33. $\dfrac{(c+d)}{(b-a)}$
35. b
37. $\dfrac{Rr}{(R-r)}$
39. 20
41. 7
43. 5 ft.
45. 20
47. 23, 25
49. 10, 12, 14, 16
51. 15 minutes

SECTION 3-2, EXERCISES

1. $-1, 2$
3. $4, -\frac{1}{2}$
5. $0, 3$
7. $3, -3$
9. $2, -2$
11. $0, 3, -3$
13. $2, -1$
15. $2, -2$
17. 1
19. $3, 2$
21. $2, 4$
23. $0, -1$
25. $1, -1$
27. $1, -1$
29. No roots
31. $6a, -4a$
33. $-a, -b$
35. $3, -3$
37. $-7, 5$
39. $\frac{1}{2}, -3$
41. $-\frac{3}{2}, 1$
43. $\frac{1}{4}(-1 \pm \sqrt{41})$
45. $\frac{1}{2}(-3 \pm \sqrt{3})$
47. $\dfrac{1}{2c}(-b \pm \sqrt{b^2 - 4ac})$
51. 2 ft.
53. $5, -4$
55. 16
57. $2, -2$

SECTION 3-3, EXERCISES

1. $0, 5$
3. $0, \frac{3}{2}$
5. 7
7. No roots
9. 2
11. 7
13. -1
15. 1
17. 3
19. 4
21. $0, 3$
23. $\frac{1}{5}$
25. $\pm 1, \pm \sqrt{3}$
27. $\pm 2, \pm 3$
29. No roots
31. $1, 2, 4, -1$
33. $1, 3, -1$
35. $2, \frac{4}{3}$
37. 2
39. $5, -1$
41. -1
43. -3
45. $0, -6$
47. 225
49. 7

SECTION 3-4, EXERCISES

1. $x > -1$
3. $x < 2$
5. $x < -1$
7. $x \leq \frac{1}{2}$
9. $x \geq \frac{4}{3}$
11. $x < \frac{5}{2}$
13. $x > -13$
15. $-1 < x < 7$
17. $-1 < x < 4$
19. $-2 < x < 0$
21. No solution
23. $-\frac{3}{2} \leq x \leq 1$
25. $x \geq -\frac{1}{7}$
27. $x > 3$ or $x < 1$
29. $x > 2$ or $x < -1$
31. $x \geq 2$ or $x \leq \frac{6}{5}$

33. $x \geq -\frac{1}{7}$ or $x \leq -1$
35. $x > 6$ or $x < -14$
37. $1 < x < 2$
39. $x > 2$ or $x < -1$
41. $-4 < x < -2$
43. $x > 2$ or $x < -1$

SECTION 3-5, EXERCISES

1. $x = 1$
3. $1 \leq x \leq 2$
5. $x > 5$ or $x < 2$
7. $-4 < x < \frac{1}{2}$
9. $x \geq \frac{3}{2}$ or $x \leq -\frac{1}{5}$
11. $-\frac{1}{2} \leq x \leq 2$
13. $x > 1$
15. $x > 3$ or $1 < x < 2$
17. $x \leq -3$ or $-2 \leq x \leq 1$
19. $x > \frac{7}{3}$ or $-1 < x < \frac{3}{2}$
21. $x < -\frac{3}{2}$ or $-1 < x < \frac{1}{5}$
23. $x > 2$ or $x < 0$
25. $x = -1$ or $0 \leq x \leq 2$
27. $x < 0$
29. $x > 2$ or $x < 0$
31. $x < -4$ or $0 < x < 1$
33. $x > 1$
35. $-\sqrt{6} < x < -\sqrt{2}$ or $\sqrt{2} < x < \sqrt{6}$
37. $-2 < x < 2$
39. No solution
41. All x
43. $x < -2$ or $x > 2$
45. $-1 < x < 1$ or $x < -\sqrt{15}$ or $x > \sqrt{15}$

SECTION 3-6, CHAPTER REVIEW

1. 2
3. 33/5
5. 3
7. $\frac{11}{10}$
9. -6
11. 22
13. 5
15. $4P$
17. $\frac{a+b}{2}$
19. $\frac{bc+ac}{ab}$
21. $\frac{83}{9}$
23. 7
25. $0, -\frac{5}{3}$
27. $\frac{1}{3}, \frac{2}{5}$
29. $\pm \frac{5}{2}$
31. $-1 \pm \sqrt{10}$
33. $\emptyset$
35. $\frac{1 \pm \sqrt{73}}{12}$
37. $\frac{4 \pm \sqrt{2}}{2}$
39. $1, -4$
41. $-2, 1$
43. $0, 17$
45. $\emptyset$
47. $-\frac{1}{2}$
49. $-1, \frac{1}{3}$
51. ± 3
53. $4, -3, 2, -1$
55. $2, -8$
57. $0, \frac{4}{3}$
59. $\frac{4}{3}, 2$
61. $x < 2$
63. $x < \frac{8}{3}$
65. $x > \frac{8}{3}$
67. $x \leq -\frac{1}{3}$
69. $\emptyset$
71. $\emptyset$
73. $x < -\frac{2}{3}$ or $x > 3$
75. $-4 < x < -3$ or $x > 2$
77. $3 < x < 6$
79. 39 cm.
81. 15
83. 20/9 min.
85. 5 meters

CHAPTER 4

SECTION 4-1, EXERCISES

1. $\{x: |x| < 3\}, \{x: -3 < x < 3\}$
3. $(-6, 6), \{x: -6 < x < 6\}$
5. $[-\frac{3}{2}, \frac{3}{2}], \{x: |x| \leq \frac{3}{2}\}$
7. $(-1, 5)$
9. $(-5, 1)$

Answers to Odd Numbered Problems **A-7**

11. $[6, 8]$ 13. $[-3, 1]$ 15. $|x - 4| < 3$
17. $|x - 1| < 3$ 19. $|x - 8| \leq 3$ 21. $|x - 1| \leq 4$
23. $|x + \frac{15}{2}| < \frac{5}{2}$ 25. $|x + 3| \leq 2$ 27. $(-\infty, 0] \cup [4, \infty)$
29. $(-\infty, -1) \cup (3, \infty)$ 31. $(-\infty, -4] \cup [-2, \infty)$
33. $(-\infty, -7) \cup (-3, \infty)$ 35. $|x| > 5$
37. $|x - 2| \geq 1$ 39. $|x + 4| > 2$

SECTION 4-2, EXERCISES

1. 0 3. (a) I, IV
      (b) II, III
5. III 7. IV 9. I 11. II
13. 5 15. 5 17. $\sqrt{2}$ 19. $4\sqrt{2}$
21. $\sqrt{26}$ 27. (2, 4) 29. (2, 3) 31. $(-1, 2)$
33. $(3, -4)$ 35. $y = -2$
37. $2x - y = 0$ 39. $x^2 - 8y + 16 = 0$
41. $x^2 + y^2 - 10x + 21 = 0$

SECTION 4-3, EXERCISES

1. $x^2 + y^2 = 1$ 3. $x^2 + y^2 = 25$
5. $x^2 + y^2 - 2x - 4y + 4 = 0$ 7. $x^2 + y^2 + 6x - 4y - 12 = 0$
9. $(0, 0); 6$ 11. No graph 13. $(1, 3); 1$
15. $(-2, -1); 5$ 17. No graph 19. $(3, -4); 0$
21. $(x + 1)^2 + (y - 3)^2 = 10$ 23. $(x - \frac{1}{2})^2 + (y + \frac{1}{2})^2 = \frac{5}{2}$
25. $(x - 1)^2 + (y - 3)^2 = 10$ 27. $(x - 3)^2 + (y - 4)^2 = 8$
29. $(x - \frac{5}{4})^2 + y^2 = (\frac{5}{4})^2$ 31. $(x - \frac{5}{3})^2 + (y - 1)^2 = (\frac{5}{3})^2$
33. $(0, 0); 2\sqrt{2}$ 35. $(0, 0); 2\sqrt{5}$
37. $(-\frac{1}{2}, 0); \frac{3}{2}$ 39. (c) 15. Circle
      16. Circle
      17. No graph
      18. No graph
      19. Point
      20. Point

SECTION 4-4, EXERCISES

1. $y = 4x - 5$ 3. $y = -2x - 1$ 5. $y = \frac{3}{4}x - \frac{7}{2}$
7. $3; 1$ 9. $2; 5$ 11. $-3; 4$
13. $2; -\frac{5}{2}$ 15. $y = \frac{4}{3}x$; yes
17. $y = 2x - 1$; no 19. $y = -\frac{4}{3}x + \frac{1}{3}$; no
21. $y = \frac{3}{2}x - \frac{5}{4}; (0, -\frac{5}{4}); (\frac{5}{6}, 0)$ 23. $y = 4x - \frac{5}{2}; (0, -\frac{5}{2}); (\frac{5}{8}, 0)$
25. $y = -x + 3$ 27. $y = -\frac{1}{3}x + 3$ 29. $y = -2x + 12$
31. $y = 2$ 33. $y = 5x + 3$ 35. $y = 2x - 11$
37. $y = \frac{4}{5}x + 4$ 39. $y = 4x + 9$ 43. $4x + 3y - 25 = 0$

SECTION 4-5, EXERCISES

31. $y \geq \frac{3}{2}x; y \geq -\frac{1}{2}x + 4$
33. $y \leq \frac{1}{2}x; y \geq \frac{3}{2}x$
35. $y \geq -\frac{1}{2}x + 4; y \leq \frac{1}{2}x$

SECTION 4-6, EXERCISES

11. $x = 1; (1, 0)$
13. $x = -2; (-2, -3)$
15. $x = 2; (2, -3)$
17. $x = -3; (-3, 1)$
19. $x = 1; (1, 3)$
21. $y = -x^2$
23. $y = 4(x - 3)^2 + 1$
25. $y = -5(x - \frac{3}{2})^2 + 3$
27. $y = \frac{1}{2}x^2 + \frac{1}{2}x$
39. $x > y^2; x < 4$
41. $x > y^2 - 1; y > x - 1$
43. $y > \frac{1}{2}x^2 - 2; y < \frac{1}{2}x$
45. $x > y^2; x < 2 - y^2$
47. $y < \frac{1}{2}x^2 + 1; y > 2x^2$

SECTION 4-7, EXERCISES

9. $\dfrac{x^2}{4} + \dfrac{y^2}{1} = 1$
11. $\dfrac{x^2}{5} + \dfrac{y^2}{7} = 1$
13. $\dfrac{x^2}{4} + \dfrac{3y^2}{4} = 1$
15. $\dfrac{x^2}{5} + \dfrac{y^2}{3} = 1$
29. $\dfrac{x^2}{4} - \dfrac{y^2}{9} = 1$
31. $\dfrac{x^2}{9} - \dfrac{4y^2}{9} = 1$
33. $-x^2 + y^2 = 1$
35. $-\dfrac{4x^2}{25} + \dfrac{y^2}{25} = 1$
37. $(\pm\sqrt{7}, 0)$
39. $(\pm 2, 0)$
41. $(0, \pm\sqrt{5})$
43. $(0, \pm 5)$
45. $(\pm\sqrt{13}, 0)$
47. $(\pm 3, 0)$
49. $(0, \pm\sqrt{13})$
51. $(0, \pm\sqrt{8})$

SECTION 4-8, EXERCISES

1. $|x| < 5$
3. $|x - 3| < 2$
5. $(4, 6)$
7. $[-4, 2]$
9. $(-\infty, 3] \cup [3, \infty)$
11. $(-\infty, -7) \cup (3, \infty)$
13. 5
15. $\sqrt{2}$
17. Collinear
19. $y = x$
21. $(1, 3); 2$
23. $(2, -4); 3$
25. $(x + 1)^2 + (y + 4)^2 = 25$
27. $x^2 + y^2 = 4$
33. $y - 3 = 2(x - 1)$
35. $y = -3x + 5$
37. $y - 5 = \frac{1}{2}(x - 2)$
47. $x = -2; (-2, 0)$
49. $x = 2; (2, -1)$
51. $y = 2; (1, 2)$
53. $(\pm 1, 0)$
55. $(\pm 3, 0)$

CHAPTER 5

SECTION 5-1, EXERCISES

1. $(-\infty, \infty), [0, \infty)$
3. $(-\infty, \infty), [1, \infty)$

5. $(-\infty, 0], (-\infty, \infty)$
7. $x \neq 0, y \neq 1$
9. $(-\infty, \infty), (-\infty, \infty)$
11. $(-\infty, \infty), (-\infty, 1]$
13. $(-1, 1), (-1, 1)$
15. $[1, 3], [2, 4]$
17. $(-7, 1), (-3, 5)$
19. $(-\infty, \infty), (-\infty, \infty)$
21. $(-\infty, \infty), (-\infty, \infty)$
23. $(-\infty, \infty), (\infty, \infty)$
25. $(-\infty, \infty), (-\infty, 1]$
27. $(-\infty, \infty), (-\infty, \infty)$
29. $(-\infty, \infty), (-\infty, \infty)$
31. $[-2, 2], (-\infty, 2)$
33. $[0, \infty), [-3, 3]$
35. $[-2, 2], (-\infty, 3]$

SECTION 5-2, EXERCISES

1. Function; $F(1) = 2, F(-2) = 3, F(-3) = 4, F(4) = -5$.
3. Not a function
5. Function; $J(1) = 0, J(2) = 0, J(3) = 0$.
7. 1
9. 0
11. Function; $y = 2x + 3$
13. Function; $y = x^2 - 1$
15. Not a function
17. Function; $y = -\sqrt{1 - x^2}$
19. Not a function
21. Function; $y = \dfrac{1}{|x|}$
23. Not a function
25. Function; $y = -x$
27. 4
29. -1
31. 2, 0
33. -2
35. $1 + \sqrt{2}$

SECTION 5-3, EXERCISES

7. $x^2 + x; x^2 - x; x^3; x$
9. $2; 2x; 1 - x^2; \dfrac{1 + x}{1 - x}$
11. $2 + x - x^2; -x - x^2; 1 + x - x^2 - x^3; 1 - x$
13. $\sqrt{x - 1}; x \geq 1$
15. $x; x \geq 0$
17. $4x - 1$; all x
19. $\sqrt{x}; x \geq 0$
21. $x^2; -1 \leq x \leq 1$
23. $|x|; -3 \leq x \leq 3$
25. $t(r(x))$
27. $s(s(x))$
29. $t(m(x))$
31. $s(t(x))$
33. $m(s(x))$
35. $D(A(x))$
37. $A(D(x))$
39. $K(A(x))$
41. $S(A(x))$
43. $A(A(x))$
45. $D(D(x))$

SECTION 5-4, EXERCISES

1. $3x^3 + 2x + 1; 3x$
3. $x^2 - x + 2; 2x - 3$
5. $3x^2 + 10x + 10; 5$
7. $x(x + 4)(x + 2)$
9. $(x - 1)(x - 2)(x - 3)$
11. $(x - 1)(x^2 + 1)$

13. No real factors
15. $x(x^2 + 2x + 5)$
25. $+\infty; -\infty$
27. $-\infty; +\infty$
29. $-\infty; -\infty$

SECTION 5-5, EXERCISES

1. $y = 0, x = 0$
3. $y = 0, x = -2$
5. $y = 1, x = -1$
7. $y = \frac{3}{2}, x = \frac{1}{2}$
9. $y = -1, x = \frac{5}{2}$
11. $y = 0$
13. $y = 0, x = 1, x = -1$
15. $y = \frac{3}{2}x + \frac{1}{2}, x = 0$
17. $y = 2x + 1, x = 1$
19. $y = x + 1, x = 1$
21. $y = 1, x = 1, x = -1$
23. $y = \frac{2}{3}$
25. $y = \frac{1}{2}$
27. $y = \frac{3}{2}$

SECTION 5-6, EXERCISES

9. $y = \dfrac{1}{x}$
11. $y = 4/(x^2 + 2)$
13. $y = x^2/(x - 2)$
15. $y = \pm\sqrt{\dfrac{1 - x^2}{x}}$
17. $y = -1 \pm \sqrt{1 - x}$
19. $y = \pm \frac{3}{2}\sqrt{4 - x^2}$
21. $y = \pm \frac{3}{2}\sqrt{x^2 - 4}$
23. $y = 3x, y = -2x + 1$
25. $y = \pm x$
27. $y = 0, y = x - 1$
29. $y = 0, y = x^2$
31. $y = -1, y = -x^2 - 1$
33. $y = x, y = 2x$
35. $y^2 - x - 1 = 0$
37. $xy - x^2 - 1 = 0$
39. $xy^2 - x^2 - 1 = 0$
41. $y^3 - x^2 - x = 0$

SECTION 5-7, EXERCISES

1. $f^{-1} = \{(2, 0), (1, 1), (4, 2), (3, 3)\}$
3. Not one-to-one
5. $f^{-1} = \{(2, -1), (3, -2), (-2, 3), (-1, 0), (1, 2)\}$
7. $f^{-1}(x) = \frac{1}{2}x$
9. $f^{-1}(x) = \frac{1}{3}x + \frac{2}{3}$

11. No inverse
13. $f^{-1}(x) = \dfrac{1}{x} + 1$

15. No inverse
17. $f^{-1}(x) = x^2 - 2 (x \geq 0)$

19. $f^{-1} = f$
21. $y = \dfrac{2}{x + 1} (x \neq -1)$

23. $y = 3 + \sqrt{x}, (x \geq 0)$
25. $y = \dfrac{-1 + \sqrt{1 + 4x}}{2}, (x \geq 0)$

27. $y = \sqrt{9 - x^2} \, (0 \leq x \leq 3)$
29. $f^{-1}(x) = \frac{1}{4}x$

31. $f^{-1}(x) = x - 5$
33. $f^{-1}(x) = \dfrac{x - 1}{2}$

Answers to Odd Numbered Problems   A-11

35. $f^{-1}(x) = (x + 9)^{-1}$ 37. $f^{-1}(x) = \dfrac{1}{x} + 1$

39. $f^{-1}(x) = \sqrt[3]{x} - 2$ 41. $f^{-1}(x) = x^3 + 2$

43. (a) $f^{-1}(x) = x; f^{-1}(x) = -x; f^{-1}(x) = -x + 1;$
$f^{-1}(x) = -x - 8$
 (b) Lines perpendicular to $y = x$
 (c) $f(x) = -x + b$
 (d) $m^2 x + b(m + 1)$

REVIEW FOR CHAPTER 5

1. Function 3. Not a function 5. Not a function
7. Domain of A, range of A: $\{1, 2, 3\}$, domain of B: $\{-2, -3, -4, -5\}$, range of B: $\{5, 7\}$, domain of C: $\{3\}$, range of C, domain of D: $\{4, 5, 6, 7\}$, range of D: $\{3\}$, domain of E: $\{0\}$, range of E, domain of F, range of F: $\{-1, -2, -3\}$
9. $(f - g)(x) = \sqrt{25 - x^2} - 5 + x$, $(g - f)(x) = 5 - x - \sqrt{25 - x^2}$, domains: interval $[-5, 5]$
11. $(f/g)(x) = \sqrt{25 - x^2}/(5 - x)$, domain: interval $[-5, 5)$, $(g/f)(x) = (5 - x)/\sqrt{25 - x^2}$, domain: interval $(-5, 5)$
13. $f^{-1}(x) = \sqrt{25 - x^2}$ for $x \geq 0$, domain: interval $[0, 5]$
15. $x + 3\sqrt{x} - 1$ 17. $r \circ t(x)$ or $r(t(x))$
19. $t \circ f(x)$ or $t(f(x))$ 21. $r \circ f(x)$ or $r(f(x))$
23. $Q(x) = 2x^3 - 6x^2 + x - 3$, $R(x) = 5$
25. $Q(x) = x^2 + 7x + 12$, $R(x) = 0$
27. Similar in shape to the graph in Fig. 5-15, Section 5-4, a peak and valley for x in the intervals $(0, 1)$ and $(-1, 0)$, respectively, an intercept at $(0, -5)$, and an x-intercept in the interval $(-2, -1)$
29. Similar in shape to the parabola $y = -x^2$
31. Similar in shape to the parabola $y = x^2$ with vertex at $(0, -13)$, symmetry about the y-axis, and x-intercepts in the intervals $(1, 2)$ and $(-2, -1)$
33. ∞, ∞
35. Similar in shape to the graph in Fig. 5-16, Section 5-5, with asymptotes $x = 7$ and $y = 0$, and an intercept at $(0, -5/7)$
37. Similar in shape to the graph in Fig. 5-16, Section 5-5, reflected in the line $y = 1$, with asymptotes $x = -5$ and $y = 1$ and intercept at $(0, 0)$
39. Similar in shape to the graph in Fig. 5-16, Section 5-5, with asymptotes $x = 4$ and $y = 1$, and intercepts at $(0, -1/4)$ and $(-1, 0)$
41. Asymptotes: $x = -3$, $x = -2$, $y = 0$, for $-3 < x < -2$ similar in shape to the parabola $y = (x + 5/2)^2$, intercept at $(0, 0)$, negative for $x < -3$ and $-2 < x < 0$, and has a peak for x in the interval $(2, 3)$
43. Asymptotes: $x = 3$, $y = 0$, for $x > 3$ similar in shape to the right branch of the graph in Fig. 5-16, Section 5-5, intercepts at $(0, -4/27)$ and $(-4, 0)$, negative for $-4 < x < 3$, and has a peak for x in the interval $(-7, -6)$
45. $y^2 - 4y + 4 - x = 0$

47. $y^2 - 4y - 21 + x^2 = 0$
49. $xy^2 + 6xy + 9x - 1 = 0$
51. f is a line with intercepts $(-3, 0)$ and $(0, 3)$. $f^{-1}(x) = x - 3$ is a line with intercepts $(0, -3)$ and $(3, 0)$
53. f is the upper half of the circle, centered at $(0, 0)$ with radius 2, which is not in Quadrant II. f is its own inverse.

CHAPTER 6

SECTION 6-1, EXERCISES

1. 16
3. $\frac{1}{9}$
5. 1
7. 3
9. $\frac{1}{5}$
11. 4
13. 8
15. $\frac{1}{16}$
17. $\frac{1}{625}$
19. 8
21. $\frac{1}{9}$
23. $\frac{64}{27}$
25. $2x^5/y^2$
27. $\frac{1}{2}$
29. e^5
31. e^{3x}
33. e^{2x}
35. $1/e^{5x}$
37. e^{3x}

SECTION 6-2, EXERCISES

1. 2
3. -2
5. 4
7. -1
9. $2^4 = 16$
11. $10^3 = 1,000$
13. $\log_6 36 = 2$
15. $\log_8 2 = \frac{1}{3}$
17. $\log_{10}(.01) = -2$
19. $\log_9 3 = \frac{1}{2}$
21. $\log_a 9 + \log_a x + \log_a y$
23. $\log_a 3 + 2\log_a x + \log_a y$
25. $\log_a 2 + 1 + 2\log_a x$
27. $3\log_a x - \log_a y - 2\log_a z$
29. $\frac{1}{2}\log_a x + \frac{1}{2}\log_a y$
31. $-\frac{3}{5}\log_a x$
33. $\log_e 100 = 4.6052$
35. $\log_e 350 = 5.8580$
37. $\log_{10} 7.388 = 0.8686$
39. $\log_{10} 207 = 2.3160$

SECTION 6-3, EXERCISES

1. 2
3. -4
5. 2.4771
7. 1.1931
9. $-1 + 0.7007$
11. 0.4807
13. 0.3676
15. 1.3652
17. 2
19. 5,030
21. 7.515
23. 302.2
25. 0.009250
27. 0.06044

SECTION 6-4, EXERCISES

1. 39.75
3. 0.388
5. 3.662×10^{-9}
7. 3.125×10^{-5}
9. 1447000
11. 0.4197
13. 0.3360
15. 0.09544
*17. 198,600,000.0002685
*19. 5.424

SECTION 6-5, EXERCISES

1. 2.10
3. 1.21
5. 1.71

7. 2.32
9. -1.11
11. 1
13. $-\frac{5}{8}$
15. 9
17. No solution
19. $\frac{105}{90} = \frac{7}{6}$
21. 10
23. 3
25. $\frac{1}{2}, -3$
27. No solution
29. 0
31. 0, ln 2
33. $\ln(3 \pm 2\sqrt{2})$
35. $3206
37. Approximately 16 years
39. 3.8%
41. (a) 32,000
    (b) 512,000

REVIEW FOR CHAPTER 6

1. $2^{-3}, (2^{-2/9})^9, 2^{2/3}, (1/2)^{-4/3}, (4^{5/2})(8^{-1})$
3. 3
5. 5
7. (a) Graph of f is similar in shape to the graph of $y = 2^x$
   (b) f is symmetric about the line $y = x$ to the graph in part (a)
   (c) Yes
9. $10^y = 35$
11. $e^b = d$
13. $e^x = 42$
15. (a) Domain: real numbers, range $\{1\}$
    (b) No
    (c) No
17. 3
19. 4
21. -9.55
23. 100, 1000
25. 9.8

CHAPTER 7

SECTION 7-1, EXERCISES

1. 1, 0
3. 0, -1
5. $-1, 0$
7. 0, 1
9. II
11. I
13. $-\dfrac{\sqrt{2}}{2}$
15. $\dfrac{3\sqrt{11}}{10}$
17. $\dfrac{\sqrt{3}}{2}$
19. $\dfrac{\sqrt{7}}{4}$
21. $\sqrt{.708} = .84$
23. $\sqrt{.826} = .91$
25. $\sqrt{0.082} = 0.29$
27. -0.540
29. 0.540
31. 0.958
33. 0.990
35. 0.757
37. 0.757
39. $\cos s = -\cos s_1, \sin s = \sin s_1$
41. $\cos s = \cos s_1, \sin s = -\sin s_1$

SECTION 7-2, EXERCISES

3. $(\sqrt{3}/2, 1/2)$
5. $(-\sqrt{3}/2, 1/2)$

7. $(-1/2, -\sqrt{3}/2)$ 9. $(-1, 0)$
11. $(-\sqrt{2}/2, -\sqrt{2}/2)$

SECTION 7-3, EXERCISES

1. $2, 2\pi$ 3. $\frac{1}{2}, 2\pi$ 5. $1, 2\pi$
7. $1, \pi$ 9. $2, 4\pi$ 11. $\frac{3}{2}, \pi$
13. $\frac{1}{4}, \frac{4\pi}{3}$ 15. $2, 2\pi$ 17. $1, \pi$
19. $1, 2$

SECTION 7-4, EXERCISES

1. all s; $[-1, 1]$
3. $s \neq \pm\frac{\pi}{2}, \pm\frac{3\pi}{2}, \ldots ; (-\infty, \infty)$
5. $s \neq 0, \pm\pi, \pm 2\pi, \ldots ; (-\infty, -1] \cup [1, \infty)$
7. $\frac{1}{2}$ 9. $\sqrt{2}$
11. $-\sqrt{3}$ 13. -1
15. $-\dfrac{1}{\sqrt{3}}$ 17. Undefined
19. 2 21. Undefined
23. $\sin s = \frac{3}{5}, \cos s = \frac{4}{5}$ 25. $\sin s = \frac{12}{13}, \cos s = -\frac{5}{13}$
27. $\sin s = -\dfrac{2}{\sqrt{5}}, \cos s = -\dfrac{1}{\sqrt{5}}$
29. $\sec s = \sqrt{5}$ 31. $\sec s = \dfrac{\sqrt{10}}{3}$
33. $\sec s = \frac{5}{3}$ 35. $\csc s = \dfrac{\sqrt{10}}{3}$
37. $\csc s = -\dfrac{\sqrt{34}}{3}$ 39. $\sin s = \dfrac{1}{\sqrt{5}}$
41. $\sin s = -\dfrac{4}{\sqrt{17}}$ 43. $\tan s = \sqrt{8}$
45. $\tan s = -\frac{4}{3}$ 47. $\tan s = \sqrt{3}$

SECTION 7-5, EXERCISES

1. $\dfrac{1}{4}(\sqrt{6} - \sqrt{2})$ 3. $\dfrac{1}{4}(\sqrt{6} - \sqrt{2})$

13. $\dfrac{\sqrt{3}}{2}\sin s + \dfrac{1}{2}\cos s$ 15. $\dfrac{\sqrt{2}}{2}\cos s + \dfrac{\sqrt{2}}{2}\sin s$

17. $\dfrac{1}{2}\sin s - \dfrac{\sqrt{3}}{2}\cos s$ 19. $\sin s$

21. $-\sin s$ 23. $-\dfrac{1}{2}\cos s + \dfrac{\sqrt{3}}{2}\sin s$

25. $\sin\left(s - \dfrac{\pi}{3}\right)$ 27. $\sin\left(s + \dfrac{\pi}{3}\right)$ or $\cos\left(s - \dfrac{\pi}{6}\right)$

29. $\cos\left(s - \dfrac{\pi}{3}\right)$ or $\sin\left(s + \dfrac{\pi}{6}\right)$

31. $\dfrac{\tan s + 1}{1 - \tan s}$ 33. $\dfrac{\tan s + \sqrt{3}}{1 - \sqrt{3}\tan s}$

35. $\dfrac{\tan s_1 - \tan s_2}{1 + \tan s_1 \tan s_2}$ 37. $\dfrac{\cot s_1 \cot s_2 - 1}{\cot s_1 + \cot s_2}$

SECTION 7-6, EXERCISES

1. $\cos^3 s - 3\sin^2 s \cos s$ 3. $4\sin s \cos^3 s - 4\cos s \sin^3 s$

5. $\dfrac{1}{2}\sqrt{2 + \sqrt{2}}$ 7. $\dfrac{1}{2}\sqrt{2 + \sqrt{3}}$

9. $\dfrac{1}{2}\sqrt{2 + \sqrt{2}}$ 11. $\dfrac{1}{2}\sqrt{2 - \sqrt{3}}$

13. $\dfrac{(2 - \sqrt{3})}{(2 + \sqrt{3})}$ 15. $\dfrac{(2 + \sqrt{2})}{(2 - \sqrt{2})}$

17. $\dfrac{(2 + \sqrt{2})}{(2 - \sqrt{2})}$ 19. $\dfrac{2\tan s}{1 - \tan^2 s}$

21. $\tfrac{1}{2}(1 - \cos 4s)$ 23. $\tfrac{1}{8}(3 - 4\cos 2s + \cos 4s)$

SECTION 7-7, EXERCISES

1. Replace $\tan s$ with $\sin s/\cos s$ 3. Replace $\sec^2 s$ with $1 + \tan^2 s$
5. Multiply numerator and denominator by $(1 - \sin s)$. In the numerator and denominator replace $\sin s \cos s$ and $\cos^2 s$ with $(1/2)\sin 2s$ and $(1/2)(1 + \cos 2s)$, respectively
7. Subtract fractions. Replace the common denominator, $\sec s - \tan s)$ $(\sec s + \tan s) = \sec^2 s - \tan^2 s$, with 1
9. Replace $\cot^2 s$ with $\csc^2 s - 1$
11. Write $\csc 2s$ as $1/\sin 2s = (1/2)(1/\sin s)(1/\cos s)$. Use the identities $1/\sin s = \csc s$ and $1/\cos s = \sec s$
13. Replace $\tan s$ and $\sin 2s$ by $\sin s/\cos s$, and $2\sin s \cos s$, respectively

15. Replace cot s with $\cos s/\sin s$ and multiply. Use $\cos^2 s = 1 - \sin^2 s$ to write the result as $1/\sin s - \sin s$
17. Multiply numerator and denominator by $(1 - \sin s)$. Replace $(1 - \sin^2 s)$ by $\cos^2 s$
19. Write $\sin^2 s$ as $(1 - \cos^2 s) = (1 - \cos s)(1 + \cos s)$
21. Add fractions. Replace the common denominator, $(\sec s + 1)(\sec s - 1) = \sec^2 s - 1$, with $\tan^2 s$. Replace the numerator, $2 \sin s \sec s = 2 \sin s/\cos s$, with $2 \tan s$.
23. $s_1 = s_2 = 0$
25. $s = \pi/4$

SECTION 7-8, EXERCISES

1. Domain $[-1, 1]$, range $[-\pi, \pi]$ 3. Domain $[-\frac{1}{2}, \frac{1}{2}]$, range $[0, \pi]$
5. $\dfrac{\pi}{4}$ 7. $\dfrac{\pi}{6}$ 9. $-\dfrac{\pi}{3}$
11. $\dfrac{3\pi}{4}$ 13. $\dfrac{\pi}{4}$ 15. $\dfrac{2\pi}{3}$
17. $\dfrac{\pi}{3}$ 19. $\dfrac{3\pi}{4}$ $\dfrac{\sqrt{5}}{3}$
23. $\dfrac{\sqrt{15}}{4}$ 25. $\dfrac{1}{\sqrt{10}}, \dfrac{3}{\sqrt{10}}$ 27. $\dfrac{-2}{\sqrt{13}}, \dfrac{3}{\sqrt{13}}$
29. $\dfrac{\sqrt{2}}{2} \cdot \dfrac{3}{5} + \dfrac{\sqrt{2}}{2} \cdot \dfrac{4}{5}$ 31. 1

SECTION 7-9, EXERCISES

1. $\pi/3, 5\pi/3$
3. $0, \pi$
5. $\pi/6, 5\pi/6, 7\pi/6, 11\pi/6$
7. π
9. $\pi/6, 5\pi/6$
11. $4\pi/3, 5\pi/3$
13. $\pi/4, 3\pi/4, 5\pi/4, 7\pi/4$
15. $\pi/2, 3\pi/2, \pi/6, 5\pi/6$
17. 0
19. $\pi/4, 5\pi/4$

REVIEW FOR CHAPTER 7

1. (a) I, II   (g) I, IV
   (b) III, IV (h) II, III
   (c) I, IV   (i) I, III
   (d) II, III (j) II, IV
   (e) I, III   (k) I, II
   (f) II, IV   (l) III, IV

Answers to Odd Numbered Problems

3. $s = 2\pi - s_1$
5. $2, 2\pi$
7. $1, \pi$
9. $2, 4\pi$
11. $1, 2\pi$. The graph is the graph of $y = \cos x$ moved one unit downward
13. $2, 2\pi$. A cycle begins at $x = -\pi/2$
15. $0, -1, 0$
17. $-\sqrt{3}/2, 1/2, -\sqrt{3}$
19. $-\sqrt{2}/2$
21. $\sqrt{2} - \sqrt{3}/2 = \sqrt{2}(\sqrt{3} - 1)/4$
23. $2 + \sqrt{3}$
25. Let $s = \text{Tan}^{-1} 3$ and $t = \text{Tan}^{-1}(1/3)$. Then $\sin(s + t) = (3/\sqrt{10})(3/\sqrt{10}) + (1/\sqrt{10})(1/\sqrt{10}) = 1$. Since $0 < s + t < \pi$ this shows that $s + t = \text{Sin}^{-1} 1 = \pi/2$
27. Let $s_1 = \pi/2$ and $s_2 = s$ in the formula for $\cos(s_1 - s_2)$
29. Let $s_1 = s$ and $s_2 = 2s$ in the formula for $\cos(s_1 + s_2)$
31. Use $2\cos^2(s/2) = (1 + \cos s)$
33. Use $\sin^2 s + \cos^2 s = 1$ and $2\sin s \cos s = \sin 2s$ in the expansion of the binomial
35. Factor into linear factors the double angle formula for $\cos 2s$
37. Use $\tan s = \sin s/\cos s$, add the fractions, and use $\sin^2 s + \cos^2 s = 1$
39. Factor the left side as $(\sin^2 s + \cos^2 s)(\sin^2 s - 2\cos^2 s)$
41. Use $\tan s = \sin s/\cos s$ to show that $\tan s - \tan s_1 = (\cos s_1 \sin s - \sin s_1 \cos s)/(\cos s \cos s_1)$. Then apply the formulas for $\sin(s - s_1)$ and $1/\cos s$
43. $\cos A = \cos[\pi - (B + C)]$. Let $s = \pi$ and $s_1 = B + C$ in the formula for $\cos(s - s_1)$
45. $s_1 = s_2 = \pi/2$
47. $\pi/6, 5\pi/6$
49. $\pi/4, 5\pi/4$
51. $0, \pi$
53. $\text{Cos}^{-1}(1/3)$ or $2\pi - \text{Cos}^{-1}(1/3)$

CHAPTER 8

SECTION 8-1, EXERCISES

1. $60°$
3. $90°$
5. $120°$
7. $300°$
9. $225°$
11. $22.5°$
13. $18°$
15. $114.6°$
17. $28.6°$
19. $\dfrac{2\pi}{3}$
21. $\dfrac{\pi}{2}$
23. $\dfrac{3\pi}{2}$
25. 0.0350
27. 0.175
29. $-\dfrac{\pi}{2}, -90°$
31. $-\dfrac{11\pi}{6}, -330°$
33. $-\dfrac{\pi}{6}, -30°$
35. $-5.28, -302.7°$
37. $\pi, 180°$
39. $\dfrac{3\pi}{2}, 270°$
41. $\dfrac{\pi}{6}, 30°$
43. $\dfrac{7\pi}{4}, 315°$
45. $\dfrac{3\pi}{2}, 270°$
47. $\dfrac{\pi}{4}, 45°$

SECTION 8-2, EXERCISES

1. $70°$
3. $10°$
5. $20°$
7. $45°$
9. $70°$
11. $\sin 91° = \sin 89°$, $\cos 91° = -\cos 89°$
13. $\sin(-45°) = -\sin 45°$, $\cos(-45°) = \cos 45°$
15. $\sin 200° = -\sin 20°$, $\cos 200° = -\cos 20°$
17. $\sin(-195°) = \sin 15°$, $\cos(-195°) = -\cos 15°$
19. (a) $y = \pi/2 - x$
    (b) $y = \pi/2 - x$
21. -0.5807
23. -1.630
25. -1.630
27. -0.5911
29. 0.4752
31. -1.506
33. $43° \; 10'$
35. $34° \; 5'$
37. $16° \; 45'$

SECTION 8-3, EXERCISES

1. $\beta = 70°$, $a = 1.710$, $b = 4.698$
3. $\beta = 50°$, $a = 6.428$, $b = 7.660$
5. $\alpha = 65°$, $a = 2.719$, $b = 1.268$
7. $\alpha = 14° \; 2'$, $\beta = 75° \; 58'$, $c = 4.124$
9. $\alpha = 63° \; 26'$, $\beta = 26° \; 34'$, $c = 11.18$
11. $\alpha = 66°$, $\beta = 24°$, $c = 10{,}950$
13. $\cos \theta = \dfrac{\sqrt{21}}{5}$, $\tan \theta = \dfrac{2}{\sqrt{21}}$
15. $\sin \theta = \dfrac{2}{\sqrt{5}}$, $\cos \theta = \dfrac{2}{\sqrt{5}}$
17. $\cos \theta = \dfrac{\sqrt{15}}{4}$, $\tan \theta = \dfrac{1}{\sqrt{15}}$
19. $\sin \theta = -\dfrac{\sqrt{3}}{2}$, $\tan \theta = -\sqrt{3}$
21. $\sin \theta = -\dfrac{4}{5}$, $\cos \theta = -\dfrac{3}{5}$
23. $\cos \theta = -\dfrac{\sqrt{24}}{5}$, $\tan \theta = -\dfrac{1}{\sqrt{24}}$
25. 17.10 yards
27. 264.2 feet
29. 634.8 feet
31. $73° \; 44'$, and $106° \; 16'$

SECTION 8-4, EXERCISES

1. $\beta = 111°$, $c \doteq 17.2$, $a \doteq 11.67$
3. $\gamma = 24°$, $b = c \doteq 73.9$
5. $\gamma = 37°$, $a \doteq 85.82$, $b \doteq 57.56$
7. One solution
9. No solution
11. Two solutions
13. Problem 7, $\beta \doteq 48° \; 45'$, $\gamma \doteq 21° \; 15'$, $c \doteq 1.929$
    Problem 8, $\gamma \doteq 46° \; 13'$, $\alpha \doteq 73° \; 47'$, $a \doteq 13.30$
    Problem 10, $\beta \doteq 25° \; 57'$, $\gamma \doteq 124° \; 3'$, $c \doteq 13.27$
    Problem 11, $\beta \doteq 53° \; 50'$, $\gamma \doteq 81° \; 10'$, $c \doteq 19.56$,
    $\beta' \doteq 126° \; 10'$, $\gamma' \doteq 8° \; 50'$, $c' \doteq 3.041$
    Problem 12, $\beta \doteq 35° \; 16'$, $\gamma \doteq 24° \; 44'$, $c \doteq 5.798$

*15. Drop a perpendicular from the vertex of γ. The height of the triangle is $b \sin \alpha = a \sin \beta$. Apply this to the formula for the area of a triangle.

17. 179.1
19. 12.20, 17.85

21. $[200/(\tan 15° + \tan 20°)]$ yd $\doteq$ 316.5 yd

SECTION 8-5, EXERCISES

1. $a \doteq 8.9$
3. $\alpha \doteq 10° \, 30', \beta \doteq 114° \, 36', \gamma \doteq 54° \, 54'$
5. $\alpha \doteq 53° \, 58', \beta \doteq 59° \, 36', \gamma \doteq 66° \, 26'$
7. $b \doteq 41.74, \gamma \doteq 27° \, 31', \alpha \doteq 112° \, 29'$
9. $\alpha \doteq 22° \, 19', \gamma \doteq 45° \, 49', \beta \doteq 111° \, 52'$
11. 130.5 ft

SECTION 8-6, EXERCISES

1. $(-2, \pi), (-2, -\pi)$
3. $\left(4, -\dfrac{7\pi}{4}\right), \left(-4, \dfrac{5\pi}{4}\right), \left(-4, -\dfrac{3\pi}{4}\right)$
5. $\left(3, \dfrac{5\pi}{6}\right), \left(3, -\dfrac{7\pi}{6}\right), \left(-3, \dfrac{11\pi}{6}\right)$
7. $\left(2\sqrt{2}, \dfrac{\pi}{4}\right)$
9. $\left(2, \dfrac{\pi}{6}\right)$
11. $(1, \pi)$
13. $\left(-\dfrac{3}{2}, \dfrac{3\sqrt{3}}{2}\right)$
15. $\left(-\dfrac{7\sqrt{2}}{2}, \dfrac{7\sqrt{2}}{2}\right)$
17. $(0, 8)$
19. $r = 5/(\sin \theta - 3 \cos \theta)$
21. $r = \sin \theta / \cos^2 \theta$
23. $x^2 + y^2 - 3y = 0$
25. $3x^2 + 4y^2 - 4x - 4 = 0$

SECTION 8-7, EXERCISES

1. $\bar{x} = x - 3, \bar{y} = y - 2, \bar{x}^2/4 + \bar{y}^2 = 1$, ellipse
3. $\bar{x} = x + 3, \bar{y} = y + 1, \bar{x}^2/9 - \bar{y}^2/4 = 1$, hyperbola
5. $x = \sqrt{5}(\bar{x} - 2\bar{y})/5, y = \sqrt{5}(2\bar{x} + \bar{y})/5, \bar{x}^2/9 + \bar{y}^2/4 = 1$, ellipse
7. $x = \sqrt{10}(\bar{x} - 3\bar{y})/10, y = \sqrt{10}(3\bar{x} + \bar{y})/10,$
   $\bar{x} = 2\bar{y}$ or $\bar{x} = -2\bar{y}$, two lines
9. $x = \sqrt{5}(\bar{x} - 2\bar{y})/5, y = \sqrt{5}(2\bar{x} + \bar{y})/5,$
   $\bar{x} = 3\sqrt{5}/5$ or $\bar{x} = -3\sqrt{5}/5$, two line, each parallel to the $\bar{y}$-axis
11. $x = \sqrt{2}(\bar{x} - \bar{y})/2, y = \sqrt{2}(\bar{x} + \bar{y})/2,$
    $\bar{\bar{x}} = \bar{x}, \bar{\bar{y}} = \bar{y} + \sqrt{2}, 4\bar{\bar{x}}^2 + \bar{\bar{y}}^2 = 4$, ellipse
13. Choose 2α between $\pi/2$ and π. $x = \sqrt{5}(\bar{x} - 2\bar{y})/5, \bar{y} = \sqrt{5}(2\bar{x} + \bar{y})/5,$
    $\bar{y} = 8\bar{x}$, parabola
15. (a) Empty set
    (b) Point $(2, -1)$
    (c) Line $x = 2y$
    (d) Lines $x = y, x = -y$
17. $\bar{A} = A \cos^2 \alpha + B \cos \alpha \sin \alpha + C \sin^2 \alpha,$
    $\bar{B} = 2(C - A) \cos \alpha \sin \alpha + B(\cos^2 \alpha - \sin^2 \alpha),$

$$\bar{C} = A\sin^2\alpha - B\sin\alpha\cos\alpha + C\cos^2\alpha,$$
$$\bar{D} = D\cos\alpha + E\sin\alpha,$$
$$\bar{E} = D\sin\alpha + E\cos\alpha,$$
$$\bar{F} = F$$

REVIEW FOR CHAPTER 8

1. $\pi/2$
3. $8\pi/15$
5. $1/2$
7. $180°$
9. $1800°$
11. $(16200/\pi)°$
13. $4/5, 3/5, 4/3$
15. $\sin\theta = \sqrt{5}/5, \cos\theta = -2\sqrt{5}/5, \csc\theta = \sqrt{5}, \sec\theta = -\sqrt{5}/-2,$ $\cot\theta = -2$
17. $c \doteq 2.646, \alpha \doteq 79°\, 7', \beta \doteq 40°\, 53'$
19. $\alpha = 42°, a \doteq 132.3, b \doteq 196.7$
21. $c \doteq 152.2, \alpha \doteq 64°\, 46', \beta \doteq 71°\, 44'$
23. $\gamma = 75°, b \doteq 977.7, c \doteq 979.3$
25. $\alpha \doteq 19°\, 31', \beta \doteq 71°\, 29', \gamma \doteq 89°$
27. $a \doteq 959.8, \beta \doteq 23°\, 42', \gamma \doteq 31°\, 8'$
29. $\alpha \doteq 27°\, 57', \beta \doteq 33°\, 46', \gamma \doteq 118°\, 17'$
31. $r = \cos\theta + \sqrt{\cos^2\theta + 35}$ or $r = \cos\theta - \sqrt{\cos^2\theta + 35}$
33. $r^2 = 225/(9\cos^2\theta + 25\sin^2\theta)$
    $r = 15/\sqrt{(9 + 16\sin^2\theta)}$ or $r = -15/\sqrt{9 + 16\sin^2\theta}$
35. Circle centered at $(1, 0)$ and radius 6
37. Ellipse centered at $(0, 0)$ with intercepts $(5, 0), (5, \pi), (3, \pi/2),$ and $(3, -\pi/2)$
39. Four-leaved rose with peak points at $(1, \pi/4), (-1, 3\pi/4), (1, 5\pi/4),$ and $(-1, 7\pi/4)$
41. Similar in shape to a cardoid passing through the points $(7, 0), (5, \pi/2), (3, \pi),$ and $(5, -\pi/2)$, with only a slight indention at the point $(3, \pi)$
43. Propeller shaped curve passing through the points $(2, \pi/2), (-2, \pi/2),$ and $(0, 0)$
45. $(\sqrt{13} - 3)\bar{x}^2 - (\sqrt{13} + 3)\bar{y}^2 = 10$, hyperbola
47. $x = \sqrt{2}(\bar{x} - \bar{y})/2, y = \sqrt{2}(\bar{x} + \bar{y})/2, \bar{y}^2 - 3\bar{x}^2 = 32$, hyperbola
49. $x = \sqrt{2}(\bar{x} - \bar{y})/2, y = \sqrt{2}(\bar{x} + \bar{y})/2, 32(\bar{x} - \sqrt{2}/8)^2 - 48\bar{y}^2 = 9$, hyperbola

CHAPTER 9

SECTION 9-1, EXERCISES

1. $(0, 0), (\frac{1}{2}, \frac{1}{2})$
3. $(1, 3), (-3, -1)$
5. $(\sqrt{2}, 2), (-\sqrt{2}, 2)$
7. $(\pm 5, \pm 3)$
9. $(0, \pm 2)$
11. $(1, -3), (4, 0)$
13. $(4, 6), (-4, -6)$
15. $(0, 1), (\ln 2, 4)$
17. $(\ln 3, 18)$
19. No solutions
21. 9 by 12
23. 13, 9

SECTION 9-2, EXERCISES

1. $(1, 2)$
3. $(0, 2)$
5. Dependent, $(x, 2x - 3)$
7. Inconsistent
9. $(\frac{1}{2}, \frac{1}{3})$
11. Inconsistent
13. $(2, 1)$
15. $(-1, 3)$
17. $4\frac{1}{2}$ mph, $1\frac{1}{2}$ mph
19. Jeff's rate: $6\frac{7}{8}$ mph
    Jim's rate: $3\frac{1}{8}$ mph
21. $\frac{15}{2}, \frac{9}{2}$

SECTION 9-3, EXERCISES

1. $(3, 4, 5)$
3. $(2, -1, 1)$
5. $(\frac{7}{3}, \frac{16}{3}, \frac{11}{3})$
7. $(0, 2, 1)$
9. Inconsistent
11. $(4z + 26, z + 5, z)$
13. $(\frac{15}{4}(z - 1), \frac{3}{4}(z - 1), z)$
15. Inconsistent
17. $\left(-9 - w, \dfrac{4 - w}{2}, \dfrac{3w + 26}{2}, w\right)$
19. $(-3, -7, 5)$

SECTION 9-4, EXERCISES

1. $(3, -1)$
3. $(0, 2)$
5. $(5 + 2y, y)$
7. Inconsistent
9. $(3z + 10, -2z - 5, z)$
11. $(-1, 2, -2)$
13. $1, -2, 3$
15. $(1, 0, 1, 2)$
17. Inconsistent
19. $(z, 1, z, 0)$

SECTION 9-5, EXERCISES

1. $(4, -1)$
3. $\left(\dfrac{5}{2}, -\dfrac{1}{2}\right)$
5. $\left(\dfrac{6 + a}{4 + a}, \dfrac{-1}{4 + a}\right), a \neq -4$
7. $(2, 1, 4)$
9. $(-1, 3, 2)$
11. $(0, 2, 1)$
13. $\left(\dfrac{1}{3}, \dfrac{1}{2}, 1\right)$
    $\left(\dfrac{1}{x} = 3, \dfrac{1}{y} = 2, \dfrac{1}{z} = 1\right)$

SECTION 9-6, EXERCISES

1. $\frac{26}{19}$
3. $-\frac{18}{19}$
5. $(-3, -1, 2, 1)$
7. $(3, -2, 1, -1)$

9. $\begin{vmatrix} 1 & 2 & 1 & -1 \\ 0 & -3 & -1 & 5 \\ 0 & 1 & 3 & 2 \\ 0 & 0 & -2 & 7 \end{vmatrix}$ 11. $\begin{vmatrix} 1 & 0 & 0 & 0 \\ 2 & -3 & 2 & -5 \\ -1 & 4 & 3 & 5 \\ 1 & -1 & -1 & -1 \end{vmatrix}$

13. $\begin{vmatrix} 1 & 2 & -1 \\ 0 & 1 & 2 \\ 0 & 0 & 7 \end{vmatrix}$ 15. Factor 2 from 3rd row

17. Interchange rows and columns   19. Factor $\frac{1}{3}$ from 3rd column

SECTION 9-7, CHAPTER REVIEW

1. $\left(\frac{\pm 5\sqrt{2}}{2}, \frac{\pm 5\sqrt{2}}{2}\right)$ 3. $(\frac{5}{2}, -\frac{3}{2})$ 5. $(-12, 1), (-2, 6)$
7. $(\sqrt{10}, \sqrt{6}), (\sqrt{10}, -\sqrt{6}), (-\sqrt{10}, \sqrt{6}), (-\sqrt{10}, -\sqrt{6})$
9. $(4, \sqrt{5}), (4, -\sqrt{5}), (-4, \sqrt{5}), (-4, -\sqrt{5})$ 11. $(0, 1), (1, \frac{1}{2})$
13. $(\ln(1 + \sqrt{2}), 3 + 2\sqrt{2})$ 15. $(\frac{4}{9}, -\frac{1}{6})$ 17. Inconsistent
19. $(8, -12)$ 21. $\frac{y + 3z}{2}, y, z$ 23. $2z, \frac{z}{2}, z$
25. $(\frac{1}{3}, -\frac{2}{5}, \frac{1}{2})$ 27. $(\frac{3}{4}, \frac{1}{2}, 3)$ 29. $(-5, 3, 2, 1)$
31. 396   33. 39 dimes, 13 quarters   35. 4, 8

CHAPTER 10

SECTION 10-1, EXERCISES

1. $4 + 9i$ 3. $-1 + 7i$ 5. $-13 + 26i$
7. $6 - 36i$ 9. $7 - 24i$ 11. $x^2 + y^2$
13. $-5 + 4i$ 15. $-1 - i$ 17. $-a - bi$
19. $-i$ 21. $\frac{1 - i}{2}$ 23. $\frac{4 + 3i}{25}$
25. $\frac{-3i}{2}$ 27. $\frac{-3 + 46i}{25}$ 29. $\frac{23 - 14i}{29}$

SECTION 10-2, EXERCISES

1. Yes   3. No
5. $-\frac{1}{5} - \frac{2}{5}i$ 7. $\frac{12}{17} + \frac{31}{17}i$
9. 1   11. $5 - 10i$
13. $1 - 2i$ 15. -1
17. $\frac{7}{5} + \frac{4}{5}i$ 19. $10 + 5i$
21. $-5 + 14i$ 23. $(2 - \sqrt{15}) + (2\sqrt{5} - \sqrt{3})i$

25. $-1, 2$ 27. $1 \pm i$

29. $\dfrac{1 \pm \sqrt{3}i}{2}$ 31. $\dfrac{1 \pm \sqrt{7}i}{4}$

33. $\pm(1 + i)$ 35. $\pm(3 - i)$

37. $\pm(1 - 2i)$

SECTION 10-3, EXERCISES

1. 5 3. $\sqrt{26}$

5. $\sqrt{34}$ 7. $5\sqrt{2}$

9. $\sqrt{5}$ 11. $\sqrt{34}$

13. $1 + 2i$ 15. $3 + 2i$

17. $\tfrac{7}{29} - \tfrac{26}{29}i$ 19. $\overline{zw} = \bar{z}\,\bar{w} = 1 - 3i$

                                         $\overline{(z/w)} = \bar{z}/\bar{w} = \tfrac{3}{2} + \tfrac{1}{2}i$

21. $\overline{zw} = \bar{z}\,\bar{w} = -11 - 7i$ 23. $\dfrac{\sqrt{130}}{10}$

     $\overline{(z/w)} = \bar{z}/\bar{w} = -\tfrac{11}{17} - \tfrac{1}{17}i$

25. $\dfrac{\sqrt{26}}{2}$ 27. $\dfrac{\sqrt{170}}{2}$

SECTION 10-4, EXERCISES

1. $(x + 3i)(x - 3i)$ 3. $(x - (1 + 2i))(x - (1 - 2i))$

5. $(x - (2 + 2\sqrt{2}))(x - (2 - 2\sqrt{2}))$ 7. $2(x - (3 + i))(x - (3 - i))$

9. $x^2 - 6x + 10 = 0$ 11. $x^2 - 8x + 17 = 0$

13. $x^2 + 4x + 8 = 0$ 15. $x^2 + x + 1 = 0$

17. $(x - 1)(x^2 + 4)(x^2 - 2x + 2)$ 19. $(x + 2)(x - 7)(x^2 + 6x + 25)$

21. $(x^2 - 6x + 34)(x^2 + 4x + 20)$ 23. $1, i, -i$

25. $1 + i, 1 - i, i, -i$ 27. $1 - i, 1 + i, 3 + i, 3 - i$

SECTION 10-5, EXERCISES

1. $\operatorname{mod} z = 3\sqrt{2}$, $\arg z = -\dfrac{\pi}{4}$, $\operatorname{mod}\left(\dfrac{1}{z}\right) = \dfrac{\sqrt{2}}{6}$, $\arg\left(\dfrac{1}{z}\right) = \dfrac{\pi}{4}$

3. $\operatorname{mod} z = 6$, $\arg z = -\dfrac{\pi}{3}$, $\operatorname{mod}\left(\dfrac{1}{z}\right) = \dfrac{1}{6}$, $\arg\left(\dfrac{1}{z}\right) = -\dfrac{\pi}{3}$

5. $z = 3\sqrt{2}\left(\cos\dfrac{3\pi}{4} + i\sin\dfrac{3\pi}{4}\right)$ $w = 2\left(\cos\left(-\dfrac{\pi}{6}\right) + i\sin\left(-\dfrac{\pi}{6}\right)\right)$

    $zw = 6\sqrt{2}\left(\cos\dfrac{7\pi}{12} + i\sin\dfrac{7\pi}{12}\right)$, $z/w = \dfrac{3\sqrt{2}}{2}\left(\cos\dfrac{11\pi}{12} + i\sin\dfrac{11\pi}{12}\right)$

7. $z = 2\sqrt{2}\left(\cos\left(-\dfrac{\pi}{4}\right) + i\sin\left(-\dfrac{\pi}{4}\right)\right)$, $w = \dfrac{8\sqrt{3}}{3}\left(\cos\dfrac{5\pi}{6} + i\sin\dfrac{5\pi}{6}\right)$

    $zw = \dfrac{16\sqrt{6}}{3}\left(\cos\dfrac{7\pi}{12} + i\sin\dfrac{7\pi}{12}\right)$, $z/w = \dfrac{\sqrt{6}}{4}\left(\cos\left(-\dfrac{\pi}{12}\right) + i\sin\left(-\dfrac{\pi}{12}\right)\right)$

9. $\sqrt{26}$
11. $\sqrt{5}$
13. $\dfrac{10\sqrt{13}}{13}$
15. $-2, +2i$
17. $-\dfrac{1}{2} + \dfrac{\sqrt{3}}{2}$
19. -8
21. $i, -1, -i, 1$
23. $-\dfrac{3}{2} + \dfrac{3\sqrt{3}}{2}i, -\dfrac{3}{2} - \dfrac{3\sqrt{3}}{2}i, 3$
25. $-\sqrt{2} + \sqrt{2}i, \sqrt{2} - \sqrt{2}i$
27. $1 + \sqrt{3}i, -2, 1 - \sqrt{3}i$

SECTION 10-6, CHAPTER REVIEW

1. $z + w = (a + bi) + (c + di) = (a + c) + (b + d)i$ and if $z + w$ is real, $b + d = 0$ or $d = -b$.
3. $z + w = (a + c) + (b + d)i$, $zw = ac - bd + (cb + ad)i$.
   If $z + w$ and zw are real, we have the following system of equations.

   $\begin{cases} b + d = 0 \\ cb + ad = 0 \end{cases}$ $\begin{cases} b = -d \\ cb + ad = 0 \end{cases}$ $\begin{cases} b = -d \\ -cd + ad = 0 \end{cases}$

   $\begin{cases} b = -d \\ d(-c + a) = 0 \end{cases}$

   Solving $d(-c + a) = 0$, $d = 0$ or $c = a$, but $d \neq 0$, hence $c = a$ and $b = -d$. Therefore $z = a - di$ and $w = a + di$ which are conjugate complex numbers.

5. $-\dfrac{1}{5} + \dfrac{2}{5}i$
7. $\dfrac{-32 - 9i}{85}$
9. $-\dfrac{20}{29} - \dfrac{21}{29}i$
11. $\tfrac{1}{2}, -4$
13. $x = -5, y = 4$
15. $x = 5, y = 2$
17. $\tfrac{1}{2}\sqrt{10}$
19. $i, -i, 0; x(x - i)(x + i) = 0$
21. $1, 3 \pm 4i; (x - 1)(x - 3 + 4i)(x - 3 - 4i) = 0$
23. $\mod z = \sqrt{2}, \arg z = \dfrac{7\pi}{4}, z = \sqrt{2}\left(\cos\dfrac{7\pi}{4} + i\sin\dfrac{7\pi}{4}\right)$

   $\mod w = 2, \arg w = \dfrac{7\pi}{4}, w = 2\left(\cos\dfrac{7\pi}{4} + i\sin\dfrac{7\pi}{4}\right)$

   $\mod zw = 2\sqrt{2}, \arg zw = \dfrac{7\pi}{2}, zw = 2\sqrt{2}\left(\cos\dfrac{7\pi}{2} + i\sin\dfrac{7\pi}{2}\right)$

   $\mod \dfrac{z}{w} = \dfrac{\sqrt{2}}{2}, \arg zw = 0, \dfrac{z}{w} = \dfrac{\sqrt{2}}{2}(\cos 0 + i\sin 0) = \dfrac{\sqrt{2}}{2}$

25. $\mod z = 8$, $\arg z = \dfrac{5\pi}{3}$, $z = 8\left(\cos\dfrac{5\pi}{3} + i\sin\dfrac{5\pi}{3}\right)$

$\mod w = \sqrt{2}$, $\arg w = \dfrac{5\pi}{4}$, $w = \sqrt{2}\left(\cos\dfrac{5\pi}{4} + i\sin\dfrac{5\pi}{4}\right)$

$\mod zw = 8\sqrt{2}$, $\arg zw = \dfrac{11\pi}{12}$, $zw = 8\sqrt{2}\left(\cos\dfrac{11\pi}{12} + i\sin\dfrac{11\pi}{12}\right)$

$\mod \dfrac{z}{w} = 4\sqrt{2}$, $\arg \dfrac{z}{w} = \dfrac{5\pi}{12}$, $\dfrac{z}{w} = 4\sqrt{2}\left(\cos\dfrac{5\pi}{12} + i\sin\dfrac{5\pi}{12}\right)$

27. $-3i$ 29. $-1 + \sqrt{3}i$ 31. $\dfrac{1}{2} - \dfrac{\sqrt{3}}{2}i$

33. $\left(\dfrac{5}{2} + \dfrac{5\sqrt{3}}{2}i\right), -5, \left(\dfrac{5}{2} - \dfrac{5\sqrt{3}}{2}i\right)$ 35. $1, i, -1, -i$

CHAPTER II

SECTION 11-1, EXERCISES

1. 27	3. 6	5. 32, 1/32
7. 243	9. 50	11. 60
13. 24	15. 12	17. $24^2 \cdot 9^4$
19. 120	21. 24	23. 24
25. 120	27. 720	29. 144
31. 336	33. 2184	35. 861
37. 21	39. 1275	41. $(n + 2)(n + 1)$

43. $\dfrac{1}{n(n + 1)}$

SECTION 11-2, EXERCISES

1. 42	3. 5,040	5. 15
7. 45	9. 358,800	11. 24
13. 3,022	15. 6,840	17. 552
19. 126	21. 3,003	23. 70
25. 45	27. 15	29. 120
31. 10	33. 495	

SECTION 11-3, EXERCISES

1. 35   3. 210   5. 34,650
7. 6; 4488, 4848, 4884, 8484, 8448, 8844.   9. 10
11. 3; {{ab}, {c, d}}, {{ac}, {b, d}}, {{ad}, {bc}}.   13. 210
15. 2,520   17. 2,260   19. 56

SECTION 11-4, EXERCISES

1. 5/36
3. 1/12
5. 5/36
7. 1/2
9. 5/12
11. $48/C(52; 5) = 0.0000184$
13. $C(13; 5)/C(52; 5) = 0.0004951$
15. $C(8; 2)/C(52; 2) = 0.02112$
17. $C(6; 2)/C(52, 2) = 0.01131$
19. 2/9
21. 1/12
23. 1/4
25. 5/18
27. 2/3
29. 2/13
31. 17/52
33. 4/13
35. 1/54
37. 1/36
39. 11/36
41. 3/4
43. 7/8
45. 22,000/28,561

SECTION 11-5, CHAPTER REVIEW

1. $21 \cdot 31 \cdot 61$
3. $\dfrac{n(n-1)(n-2)}{3!}$
5. $\dfrac{n}{n+1}$
7. 4^4
9. $5!$
11. 336
13. $27 \cdot 26 \cdot 25 \cdot 24$
15. 165
17. 462
19. $5!$
21. 165
23. 210
25. $32x^5 + 160x^4y + 320x^3y^2 + 320x^2y^3 + 160xy^4 + 32y^5$
27. $-945x^4$
29. (a) $\dfrac{36}{270,725}$ (b) $\dfrac{1}{270,725}$
31. 63/170
33. 15/32

CHAPTER 12

SECTION 12-1, EXERCISES

1. $1^2 + 1$ is even. $(k+1)^2 + (k+1) = k^2 + 2k + 1 + k + 1 = (k^2 + k) + 2(k+1)$. If 2 divides $k^2 + k$, then 2 divides both terms and hence $(k+1)^2 + (k+1)$.
3. $1^3 + 2 \cdot 1 = 3$. $(k+1)^3 + 2(k+1) = k^3 + 3k^2 + 3k + 1 + 2k + 2 = (k^3 + 2k) + 3(k^2 + k + 1)$. If 3 divides $k^3 + 2k$, then 3 divides both terms and hence $(k+1)^3 + 2(k+1)$.
5. $_32^1 - 1 = 8$. $_32^{k+1} - 1 = (_32^k)^2 - 1 = (_32^k - 1)(_32^k + 1)$. If 4 divides $_32^k - 1$, then 4 divides the product and hence $_32^{k+1} - 1$.
7. $_23^1 + 1 = 9$. $_23^{k+1} + 1 = (_23^k)^3 + 1 = (_23^k + 1)\cdot[(_23^k)^2 - _23^k + 1]$. If 3 divides $(_23^k + 1)$, then 3 divides the product and hence $_32^{k+1} + 1$.
9. $3 \cdot 1 = 3 \geq 1 + 2$. If $3k \geq k + 2$, then $3(k+1) = (3k) + 3 \geq (k+2) + 3 = (k+1) + 2 + 2 > (k+1) + 2$.
11. $1^2 + 1 \geq 2 \cdot 1$. If $k^2 + 1 \geq 2k$, then $(k+1)^2 + 1 = k^2 + 2k + 1 + 1 = (k^2 + 1) + (2k + 1) \geq 2k + 2 = 2(k+1)$.
13. $2 \cdot 1 \leq 2^1$. If $2k \leq 2^k$, then $2(k+1) = 2k + 2 \leq 2^k + 2 \leq 2^k + 2^k = 2 \cdot 2^k = 2^{k+1}$.

15. $2^0 = 1 = 0!$. If $2^{k-1} \leq k!$, then $2^k = 2 \cdot 2^{k-1} \leq 2 \cdot k! \leq (k+1)k! = (k+1)!$.
17. $2 = 1(1+1)$. If $2 + \cdots + 2k = k(k+1)$, then $2 + \cdots + 2k + 2(k+1) = k(k+1) + 2(k+1) = (k+1)(k+2)$.
19. $1 = 1 \cdot (2 \cdot 1 - 1)$. If $1 + \cdots + (4k-3) = k(2k-1)$, then $1 + \cdots + (4k-3) + (4k+1) = k(2k-1) + (4k+1) = -2k^2 + 3k + 1 = (k+1)(2(k+1) - 1)$.
21. $1 = 2^1 - 1$. If $1 + \cdots + 2^{k-1} = 2^k - 1$, then $1 + \cdots + 2^{k-1} + 2^k = 2^k - 1 + 2^k = 2 \cdot 2^k - 1 = 2^{k+1} - 1$.
23. $1 = \dfrac{1 \cdot (1+1)(2 \cdot 1 + 1)}{6}$. If $1 + \cdots + k^2 = \tfrac{1}{6}k(k+1)(2k+1)$, then
$1 + \cdots + k^2 + (k+1)^2 = \tfrac{1}{6}k(k+1)(2k+1) + (k+1)^2$
$= \tfrac{1}{6}(k+1)[k(2k+1) + 6(k+1)] = \tfrac{1}{6}(k+1)(2k^2 + 7k + 6)$
$= \tfrac{1}{6}(k+1)(k+2)(2(k+1) + 1)$.
25. $2 \cdot 3 + 1 = 7 \leq 2^3 = 8$. If $2k + 1 \leq 2^k$, then $2(k+1) + 1 = (2k+1) + 2 \leq 2^k + 2 \leq 2^k + 2^k = 2 \cdot 2^k = 2^{k+1}$.
27. $4^2 = 16 \leq 2^4 = 16$. If $k^2 \leq 2^k$, then $(k+1)^2 = k^2 + (2k+1) \leq 2^k + 2^k = 2^{k+1}$.
29. $3^3 = 27 \leq 2^3$. If $k^3 \leq 3^k$ and $k \geq 3$, then $(k+1)^3 = k^3 + (3k^2 + 3k + 1) \leq 3^k + 2 \cdot 3^k = 3 \cdot 3^k = 3^{k+1}$.
31. $5^2 = 25 \leq 5! = 60$. If $k^2 \leq k!$ and $k \geq 5$, then $(k+1)^2 = k^2 + (2k+1) \leq k! + k! = 2k! \leq (k+1) \cdot k! = (k+1)!$

SECTION 12-2, EXERCISES

1. $1, 3, 9, 27 \ldots$ 7th term $= 3^6$
3. $2\tfrac{1}{3}, 2\tfrac{1}{9}, 2\tfrac{1}{27}, 2\tfrac{1}{81}$ 6th term $= 2\tfrac{1}{3^6}$
5. $4n$
7. $(-1)^n 2^n$
9. $n - 5$
11. $\dfrac{n}{n+1}$
13. $(-1)^{n+1}(4n - 1)$
15. $\dfrac{n}{2^{n-1}}$
17. $-1, 1, 3, 5, 7$
19. $2, -6, 18, -54, 162$
21. $2, 2^2, 2^4, 2^8, 2^{16}$

SECTION 12-3, EXERCISES

1. $28, 3 + (n-1)5$
3. $2, \tfrac{1}{2} + (n-1)\tfrac{1}{4}$
5. $-42, 8 + (n-1)(-10)$
7. $d = 5, a_n = 1 + (n-1)5, 1, 6, 11, 16, 21$
9. $d = -10\tfrac{1}{2}, a_n = 36 + (n-1)(-10\tfrac{1}{2}), 36, 25\tfrac{1}{2}, 15, 4\tfrac{1}{2}, -6, -16\tfrac{1}{2}$
11. $5 \cdot 2^7, 5 \cdot 2^{n-1}$
13. $-\tfrac{1}{32}, (-\tfrac{1}{2})^{n-1}$
15. $27\sqrt{2}, \sqrt{2}(\sqrt{3})^{n-1}$
17. $\$5.12, \$163.84$
19. $-3, -6, -9, -12, -15, -18, -21, -24, -27, -30, -33, -36$

21. $15, 5, \frac{5}{3}, \frac{5}{9}, \frac{5}{27}$ or $15, -5, \frac{5}{3}, -\frac{5}{9}, \frac{5}{27}$
23. $\sqrt{5}, \sqrt[3]{5^2}, \sqrt[6]{5^5}, 5$

SECTION 12-4, EXERCISES

1. $2, 6, 12, 20;\ S_n = n(n + 1)$
3. $1, 5, 12, 22;\ S_n = \frac{1}{2}n(3n - 1)$
5. $\frac{1}{2}, \frac{3}{4}, \frac{7}{8}, \frac{15}{16};\ S_n = 1 - \dfrac{1}{2^n}$
7. 50
9. 40
11. 61
13. 168
15. $\frac{1}{2}n(3n + 5)$
17. $\frac{1}{3}n(n + 1)(n + 2)$
19. $2^n - 1 + n(n + 1)$
21. $3(2^n - 1) - \frac{1}{2}n(n + 1)$
23. 275
25. 133
27. 2240
29. $4.65
31. 121
33. 2,860
35. $3,095
37. 619.02 feet

SECTION 12-5, EXERCISES

1. 1
3. $\frac{1}{2}$
5. 2
7. 0
9. 0
11. $\frac{1}{2}$
13. 1
15. 0
17. 0
19. $\frac{1}{2}$
21. $\frac{1}{4}$
23. $-\frac{100}{11}$
25. $\dfrac{1}{1 - x}$
27. $\frac{50}{9}$
29. $\frac{13}{99}$
31. $12\frac{7}{33}$
33. $S_n = 3 - \dfrac{3}{n + 1};\ S = 3$
35. $S_n = -1 + \dfrac{1}{n + 1};\ S = -1$
37. $S_n = \dfrac{2}{3} - \dfrac{2}{3n + 3};\ S = \dfrac{2}{3}$
39. $S_n = -1 + \dfrac{1}{(n + 1)^3};\ S = -1$

SECTION 12-6, CHAPTER REVIEW

7. $\frac{5}{2}, \frac{5}{6}, \frac{5}{12}, \frac{1}{4}, \frac{1}{6}; \frac{5}{72}$
9. $9, -27, 81, -243, 729;\ a_9 = (-1)^8\, 3^{10} = 3^{10}$
11. $(-1)^n$
13. $\dfrac{5}{n(n + 1)}$
15. 98, 14

17. $d = -2, a_n = -8 + (n-1)(-2) = -6 - 2n$
    $-8, -10, -12, -14, -16, -18$
    $S_{10} = -170$
19. $-\frac{11}{32}$
21. $\frac{1}{3}n(n+1)(n-1)$
23. 170
25. $5(x + 66)$
27. 1
29. $\frac{4}{5}$
31. $\dfrac{138,374}{999,000}$

CHAPTER 13

SECTION 13-1, EXERCISES

1. $\overrightarrow{BD}$
3. $\overrightarrow{DB}$
5. $\overrightarrow{DC}$ or $\overrightarrow{AB}$
7. $\overrightarrow{AB}$ or $\overrightarrow{DC}$
9. (2, 2)
11. (1, 1)
13. (0, 3)
15. $\overrightarrow{(1,0)(4,2)}$
17. $\overrightarrow{(2,-1)(4,7)}$
19. $\frac{2}{3}\bar{A} + \frac{1}{3}\bar{B}$
21. $\frac{3}{4}\bar{A} + \frac{1}{4}\bar{B}$
23. $-\bar{A} - \bar{B}$
25. $\bar{A} + \frac{1}{2}\bar{B}$
27. $\frac{2}{3}\bar{A} + \frac{1}{3}\bar{B}$
29. (a) $\bar{D} = (1 + \sqrt{2})\bar{B}$
    (b) $\bar{C} = \sqrt{2}\,\bar{B} - \bar{A}$

SECTION 13-2, EXERCISES

1. $4i + 2j$
3. $6i + 7j$
5. $-i + 6j$
7. $\sqrt{13}$
9. $2\sqrt{2}$
11. $\bar{i} + 2\bar{j}$
13. $-4\bar{i} + 4\bar{j}$
15. $-\bar{i} + \bar{j}$
17. (5, 1)
19. $(-3, -4)$
21. (3, 4)
23. (2, 6)
25. $\frac{3}{5}\bar{i} + \frac{4}{5}j$
27. $\dfrac{1}{\sqrt{5}}i - \dfrac{2}{\sqrt{5}}j$
29. $\dfrac{1}{\sqrt{10}}i - \dfrac{3}{\sqrt{10}}j$

SECTION 13-3, EXERCISES

1. $4, \dfrac{4}{\sqrt{65}}$
3. 0, 0
5. $1, \dfrac{1}{\sqrt{26}}$
7. $8, \dfrac{2}{\sqrt{5}}$
9. $\dfrac{1}{\sqrt{10}}\bar{i} + \dfrac{3}{\sqrt{10}}\bar{j}$
11. $\dfrac{12}{13}\bar{i} - \dfrac{5}{13}\bar{j}$
13. $\dfrac{1}{\sqrt{5}}\bar{i} - \dfrac{2}{\sqrt{5}}\bar{j}$
15. P
17. Q
19. P
21. (1, 3)
23. (0, 2)
25. (2, 5)
27. $(-1, 5)$

SECTION 13-4, EXERCISES

1. $\dfrac{3}{5}x + \dfrac{4}{5}y = 4$

3. $\dfrac{1}{\sqrt{5}}x - \dfrac{2}{\sqrt{5}}y = 1$

5. $\dfrac{1}{\sqrt{2}}x - \dfrac{1}{\sqrt{2}}y = 5$

7. $\dfrac{1}{\sqrt{2}}x + \dfrac{1}{\sqrt{2}}y = \dfrac{3}{\sqrt{2}}; \dfrac{3}{\sqrt{2}}$

9. $-\dfrac{3}{\sqrt{10}}x + \dfrac{1}{\sqrt{10}}y = \dfrac{2}{\sqrt{10}}; \dfrac{2}{\sqrt{10}}$

11. $-\dfrac{1}{\sqrt{26}}x + \dfrac{5}{\sqrt{26}}y = \dfrac{3}{\sqrt{26}}; \dfrac{3}{\sqrt{26}}$

13. $-\dfrac{3}{\sqrt{34}}x + \dfrac{5}{\sqrt{34}}y = \dfrac{7}{\sqrt{34}}; \dfrac{7}{\sqrt{34}}$

15. $\dfrac{3}{\sqrt{2}}$

17. $\dfrac{6}{\sqrt{5}}$

19. $\dfrac{3}{\sqrt{2}}$

21. $\dfrac{\pi}{4}$

23. $\cos\theta = .8; \theta = 36°\,52'$

25. $\dfrac{\pi}{2}$

27. $\dfrac{\pi}{4}$

29. $x = 1 + 5t; y = 2 + t$
31. $x = 4 + t; y = -2 - t$
33. $x = 1 + t; y = 5$
35. $x = 1 + 2t; y = 1 - t$
37. $x = 2 + t; y = -3 + 2t$

SECTION 13-5, EXERCISES

1. $x = 3\cos\theta, y = 2\sin\theta$
5. $x = 3\cos\theta, y = 3\sin\theta$
7. $x = 2\cos\theta, y = 2\sin\theta$
9. $x = y^2$

11. $x - 2y - 3 = 0$

13. $y = \dfrac{x-1}{x-2}$

15. $y = 1 (x \geq 0)$

17. $y = x^2 (-1 \leq x \leq 1)$
19. $4y^2 - x^2 = 4y^4 (-1 \leq x \leq 1)$

21. $(x-3)^2 + (y-2)^2 = 1$

23. $\dfrac{(x-2)^2}{9} + \dfrac{(y-3)^2}{4} = 1$

SECTION 13-6, CHAPTER REVIEW

1. $(1/4)\mathbf{A} + (3/4)\mathbf{B}$
3. $(-7, 4)$
5. $(5, 0)$
7. $5\mathbf{i} + 3\mathbf{j}$
9. $-2\mathbf{i} - 6\mathbf{j}$
11. $7\mathbf{i} + 9\mathbf{j}$
13. $-4\mathbf{i} + 12\mathbf{j}$
15. $(5, 1)$
17. $(5\sqrt{34}/34)\mathbf{i} - (3\sqrt{34}/34)\mathbf{j}$
19. -13
21. Arc cos $(-13\sqrt{170}/170)$
    Arc cos $(-24/25)$
23. $(18/19, 41/19)$
25. $y - 3x = 6\sqrt{10}$
27. $25\sqrt{13}/13$
29. $y + 3x = -3$
31. $y = -3x$, line
33. $x = (y + 3)^2 - 1$. The graph is a parabola, similar in shape to $x = y^2$ with vertex at $(-1, -3)$
35. $y = x^2$ for $|x| \le 1$ and $|y| \le 1$, part of a parabola
37. $9x^2 + 4y^2 = 36$, ellipse

Appendix B

TABLES

TABLE I. FOUR-PLACE COMMON LOGARITHMS OF NUMBERS

TABLE II. NATURAL LOGARITHMS

TABLE III. FOUR-PLACE VALUES OF TRIGONOMETRIC FUNCTIONS

TABLE IV. FOUR-PLACE LOGARITHMS OF TRIGONOMETRIC FUNCTIONS

TABLE I FOUR-PLACE COMMON LOGARITHMS OF NUMBERS

N	0	1	2	3	4	5	6	7	8	9
10	0000	0043	0086	0128	0170	0212	0253	0294	0334	0374
11	0414	0453	0492	0531	0569	0607	0645	0682	0719	0755
12	0792	0828	0864	0899	0934	0969	1004	1038	1072	1106
13	1139	1173	1206	1239	1271	1303	1335	1367	1399	1430
14	1461	1492	1523	1553	1584	1614	1644	1673	1703	1732
15	1761	1790	1818	1847	1875	1903	1931	1959	1987	2014
16	2041	2068	2095	2122	2148	2175	2201	2227	2253	2279
17	2304	2330	2355	2380	2405	2430	2455	2480	2504	2529
18	2553	2577	2601	2625	2648	2672	2695	2718	2742	2765
19	2788	2810	2833	2856	2878	2900	2923	2945	2967	2989
20	3010	3032	3054	3075	3096	3118	3139	3160	3181	3201
21	3222	3243	3263	3284	3304	3324	3345	3365	3385	3404
22	3424	3444	3464	3483	3502	3522	3541	3560	3579	3598
23	3617	3636	3655	3674	3692	3711	3729	3747	3766	3784
24	3802	3820	3838	3856	3874	3892	3909	3927	3945	3962
25	3979	3997	4014	4031	4048	4065	4082	4099	4116	4133
26	4150	4166	4183	4200	4216	4232	4249	4265	4281	4298
27	4314	4330	4346	4362	4378	4393	4409	4425	4440	4456
28	4472	4487	4502	4518	4533	4548	4564	4579	4594	4609
29	4624	4639	4654	4669	4683	4698	4713	4728	4742	4757
30	4771	4786	4800	4814	4829	4843	4857	4871	4886	4900
31	4914	4928	4942	4955	4969	4983	4997	5011	5024	5038
32	5051	5065	5079	5092	5105	5119	5132	5145	5159	5172
33	5185	5198	5211	5224	5237	5250	5263	5276	5289	5302
34	5315	5328	5340	5353	5366	5378	5391	5403	5416	5428
35	5441	5453	5465	5478	5490	5502	5514	5527	5539	5551
36	5563	5575	5587	5599	5611	5623	5635	5647	5658	5670
37	5682	5694	5705	5717	5729	5740	5752	5763	5775	5786
38	5798	5809	5821	5832	5843	5855	5866	5877	5888	5899
39	5911	5922	5933	5944	5955	5966	5977	5988	5999	6010
40	6021	6031	6042	6053	6064	6075	6085	6096	6107	6117
41	6128	6138	6149	6160	6170	6180	6191	6201	6212	6222
42	6232	6243	6253	6263	6274	6284	6294	6304	6314	6325
43	6335	6345	6355	6365	6375	6385	6395	6405	6415	6425
44	6435	6444	6454	6464	6474	6484	6493	6503	6513	6522
45	6532	6542	6551	6561	6571	6580	6590	6599	6609	6618
46	6628	6637	6646	6656	6665	6675	6684	6693	6702	6712
47	6721	6730	6739	6749	6758	6767	6776	6785	6794	6803
48	6812	6821	6830	6839	6848	6857	6866	6875	6884	6893
49	6902	6911	6920	6928	6937	6946	6955	6964	6972	6981
50	6990	6998	7007	7016	7024	7033	7042	7050	7059	7067
51	7076	7084	7093	7101	7110	7118	7126	7135	7143	7152
52	7160	7168	7177	7185	7193	7202	7210	7218	7226	7235
53	7243	7251	7259	7267	7275	7284	7292	7300	7308	7316
54	7324	7332	7340	7348	7356	7364	7372	7380	7388	7396

TABLE 1—continued

N	0	1	2	3	4	5	6	7	8	9
55	7404	7412	7419	7427	7435	7443	7451	7459	7466	7474
56	7482	7490	7497	7505	7513	7520	7528	7536	7543	7551
57	7559	7566	7574	7582	7589	7597	7604	7612	7619	7627
58	7634	7642	7649	7657	7664	7672	7679	7686	7694	7701
59	7709	7716	7723	7731	7738	7745	7752	7760	7767	7774
60	7782	7789	7796	7803	7810	7818	7825	7832	7839	7846
61	7853	7860	7868	7875	7882	7889	7896	7903	7910	7917
62	7924	7931	7938	7945	7952	7959	7966	7973	7980	7987
63	7993	8000	8007	8014	8021	8028	8035	8041	8048	8055
64	8062	8069	8075	8082	8089	8096	8102	8109	8116	8122
65	8129	8136	8142	8149	8156	8162	8169	8176	8182	8189
66	8195	8202	8209	8215	8222	8228	8235	8241	8248	8254
67	8261	8267	8274	8280	8287	8293	8299	8306	8312	8319
68	8325	8331	8338	8344	8351	8357	8363	8370	8376	8382
69	8388	8395	8401	8407	8414	8420	8426	8432	8439	8445
70	8451	8457	8463	8470	8476	8482	8488	8494	8500	8506
71	8513	8519	8525	8531	8537	8543	8549	8555	8561	8567
72	8573	8579	8585	8591	8597	8603	8609	8615	8621	8627
73	8633	8639	8645	8651	8657	8663	8669	8675	8681	8686
74	8692	8698	8704	8710	8716	8722	8727	8733	8739	8745
75	8751	8756	8762	8768	8774	8779	8785	8791	8797	8802
76	8808	8814	8820	8825	8831	8837	8842	8848	8854	8859
77	8865	8871	8876	8882	8887	8893	8899	8904	8910	8915
78	8921	8927	8932	8938	8943	8949	8954	8960	8965	8971
79	8976	8982	8987	8993	8998	9004	9009	9015	9020	9025
80	9031	9036	9042	9047	9053	9058	9063	9069	9074	9079
81	9085	9090	9096	9101	9106	9112	9117	9122	9128	9133
82	9138	9143	9149	9154	9159	9165	9170	9175	9180	9186
83	9191	9196	9201	9206	9212	9217	9222	9227	9232	9238
84	9243	9248	9253	9258	9263	9269	9274	9279	9284	9289
85	9294	9299	9304	9309	9315	9320	9325	9330	9335	9340
86	9345	9350	9355	9360	9365	9370	9375	9380	9385	9390
87	9395	9400	9405	9410	9415	9420	9425	9430	9435	9440
88	9445	9450	9455	9460	9465	9469	9474	9479	9484	9489
89	9494	9499	9504	9509	9513	9518	9523	9528	9533	9538
90	9542	9547	9552	9557	9562	9566	9571	9576	9581	9586
91	9590	9595	9600	9605	9609	9614	9619	9624	9628	9633
92	9638	9643	9647	9652	9657	9661	9666	9671	9675	9680
93	9685	9689	9694	9699	9703	9708	9713	9717	9722	9727
94	9731	9736	9741	9745	9750	9754	9759	9763	9768	9773
95	9777	9782	9786	9791	9795	9800	9805	9809	9814	9818
96	9823	9827	9832	9836	9841	9845	9850	9854	9859	9863
97	9868	9872	9877	9881	9886	9890	9894	9899	9903	9908
98	9912	9917	9921	9926	9930	9934	9939	9943	9948	9952
99	9956	9961	9965	9969	9974	9978	9983	9987	9991	9996

TABLE II NATURAL LOGARITHMS

This table contains logarithms of numbers from 1 to 10 to the base e. To obtain the natural logarithms of other numbers use the formulas:

$$\log_e (10^r N) = \log_e N + \log_e 10^r$$

$$\log_e \left(\frac{N}{10^r}\right) = \log_e N - \log_e 10^r$$

$\log_e 10 = 2.302585$ $\log_e 10^4 = 9.210340$
$\log_e 10^2 = 4.605170$ $\log_e 10^5 = 11.512925$
$\log_e 10^3 = 6.907755$ $\log_e 10^6 = 13.815511$

N	0	1	2	3	4	5	6	7	8	9
1.0	0.0 0000	0995	1980	2956	3922	4879	5827	6766	7696	8618
1.1	0.0 9531	*0436	*1333	*2222	*3103	*3976	*4842	*5700	*6551	*7395
1.2	0.1 8232	9062	9885	*0701	*1511	*2314	*3111	*3902	*4686	*5464
1.3	0.2 6236	7003	7763	8518	9267	*0010	*0748	*1481	*2208	*2930
1.4	0.3 3647	4359	5066	5767	6464	7156	7844	8526	9204	9878
1.5	0.4 0547	1211	1871	2527	3178	3825	4469	5108	5742	6373
1.6	0.4 7000	7623	8243	8858	9470	*0078	*0682	*1282	*1879	*2473
1.7	0.5 3063	3649	4232	4812	5389	5962	6531	7098	7661	8222
1.8	0.5 8779	9333	9884	*0432	*0977	*1519	*2078	*2594	*3127	*3658
1.9	0.6 4185	4710	5233	5752	6269	6783	7294	7803	8310	8813
2.0	0.6 9315	9813	*0310	*0804	*1295	*1784	*2271	*2755	*3237	*3716
2.1	0.7 4194	4669	5142	5612	6081	6547	7011	7473	7932	8390
2.2	0.7 8846	9299	9751	*0200	*0648	*1093	*1536	*1978	*2418	*2855
2.3	0.8 3291	3725	4157	4587	5015	5442	5866	6289	6710	7129
2.4	0.8 7547	7963	8377	8789	9200	9609	*0016	*0422	*0826	*1228
2.5	0.9 1629	2028	2426	2822	3216	3609	4001	4391	4779	5166
2.6	0.9 5551	5935	6317	6698	7078	7456	7833	8208	8582	8954
2.7	0.9 9325	9695	*0063	*0430	*0796	*1160	*1523	*1885	*2245	*2604
2.8	1.0 2962	3318	3674	4028	4380	4732	5082	5431	5779	6126
2.9	1.0 6471	6815	7158	7500	7841	8181	8519	8856	9192	9527
3.0	1.0 9861	*0194	*0526	*0856	*1186	*1514	*1841	*2168	*2493	*2817
3.1	1.1 3140	3462	3783	4103	4422	4740	5057	5373	5688	6002
3.2	1.1 6315	6627	6938	7248	7557	7865	8173	8479	8784	9089
3.3	1.1 9392	9695	9996	*0297	*0597	*0896	*1194	*1491	*1788	*2083
3.4	1.2 2378	2671	2964	3256	3547	3837	4127	4415	4703	4990
3.5	1.2 5276	5562	5846	6130	6413	6695	6976	7257	7536	7815
3.6	1.2 8093	8371	8647	8923	9198	9473	9746	*0019	*0291	*0563
3.7	1.3 0833	1103	1372	1641	1909	2176	2442	2708	2972	3237
3.8	1.3 3500	3763	4025	4286	4547	4807	5067	5325	5584	5841
3.9	1.3 6098	6354	6609	6864	7118	7372	7624	7877	8128	8379
4.0	1.3 8629	8879	9128	9377	9624	9872	*0118	*0364	*0610	*0854
4.1	1.4 1099	1342	1585	1828	2070	2311	2552	2792	3031	3270
4.2	1.4 3508	3746	3984	4220	4456	4692	4927	5161	5395	5629
4.3	1.4 5862	6094	6326	6557	6787	7018	7247	7476	7705	7933
4.4	1.4 8160	8387	8614	8840	9065	9290	9515	9739	9962	*0185
4.5	1.5 0408	0630	0851	1072	1293	1513	1732	1951	2170	2388
4.6	1.5 2606	2823	3039	3256	3471	3687	3902	4116	4330	4543
4.7	1.5 4756	4969	5181	5393	5604	5814	6025	6235	6444	6653
4.8	1.5 6862	7070	7277	7485	7691	7898	8104	8309	8515	8719
4.9	1.5 8924	9127	9331	9534	9737	9939	*0141	*0342	*0543	*0744
5.0	1.6 0944	1144	1343	1542	1741	1939	2137	2334	2531	2728
N	0	1	2	3	4	5	6	7	8	9

TABLE II—continued

N	0	1	2	3	4	5	6	7	8	9
5.0	1.6 0944	1144	1343	1542	1741	1939	2137	2334	2531	2728
5.1	1.6 2924	3120	3315	3511	3705	3900	4094	4287	4481	4673
5.2	1.6 4866	5058	5250	5441	5632	5823	6013	6203	6393	6582
5.3	1.6 6771	6959	7147	7335	7523	7710	7896	8083	8269	8455
5.4	1.6 8640	8825	9010	9194	9378	9562	9745	9928	*0111	*0293
5.5	1.7 0475	0656	0838	1019	1199	1380	1560	1740	1919	2098
5.6	1.7 2277	2455	2633	2811	2988	3166	3342	3519	3695	3871
5.7	1.7 4047	4222	4397	4572	4746	4920	5094	5267	5440	5613
5.8	1.7 5786	5958	6130	6302	6473	6644	6815	6985	7156	7326
5.9	1.7 7495	7665	7843	8002	8171	8339	8507	8675	8842	9009
6.0	1.7 9176	9342	9509	9675	9840	*0006	*0171	*0336	*0500	*0665
6.1	1.8 0829	0993	1156	1319	1482	1645	1808	1970	2132	2294
6.2	1.8 2455	2616	2777	2938	3098	3258	3418	3578	3737	3896
6.3	1.8 4055	4214	4372	4530	4688	4845	5003	5160	5317	5473
6.4	1.8 5630	5786	5942	6097	6253	6408	6563	6718	6872	7026
6.5	1.8 7180	7334	7487	7641	7794	7947	8099	8251	8403	8555
6.6	1.8 8707	8858	9010	9160	9311	9462	9612	9762	9912	*0061
6.7	1.9 0211	0360	0509	0658	0806	0954	1102	1250	1398	1545
6.8	1.9 1692	1839	1986	2132	2279	2425	2571	2716	2862	3007
6.9	1.9 3152	3297	3442	3586	3730	3874	4018	4162	4305	4448
7.0	1.9 4591	4734	4876	5019	5161	5303	5445	5586	5727	5869
7.1	1.9 6009	6150	6291	6431	6571	6711	6851	6991	7130	7269
7.2	1.9 7408	7547	7685	7824	7962	8100	8238	8376	8513	8650
7.3	1.9 8787	8924	9061	9198	9334	9470	9606	9742	9877	*0013
7.4	2.0 0148	0283	0418	0553	0687	0821	0956	1089	1223	1357
7.5	2.0 1490	1624	1757	1890	2022	2155	2287	2419	2551	2683
7.6	2.0 2815	2946	3078	3209	3340	3471	3601	3732	3862	3992
7.7	2.0 4122	4252	4381	4511	4640	4769	4898	5027	5156	5284
7.8	2.0 5412	5540	5668	5796	5924	6051	6179	6306	6433	6560
7.9	2.0 6686	6813	6939	7065	7191	7317	7443	7568	7694	7819
8.0	2.0 7944	8069	8194	8318	8443	8567	8691	8815	8939	9063
8.1	2.0 9186	9310	9433	9556	9679	9802	9924	*0047	*0169	*0291
8.2	2.1 0413	0535	0657	0779	0900	1021	1142	1263	1384	1505
8.3	2.1 1626	1746	1866	1986	2106	2226	2346	2465	2585	2704
8.4	2.1 2823	2942	3061	3180	3298	3417	3535	3653	3771	3889
8.5	2.1 4007	4124	4242	4359	4476	4593	4710	4827	4943	5060
8.6	2.1 5176	5292	5409	5524	5640	5756	5871	5987	6102	6217
8.7	2.1 6332	6447	6562	6677	6791	6905	7020	7134	7248	7361
8.8	2.1 7475	7589	7702	7816	7929	8042	8155	8267	8380	8493
8.9	2.1 8605	8717	8830	8942	9054	9165	9277	9389	9500	9611
9.0	2.1 9722	9834	9944	*0055	*0166	*0276	*0387	*0497	*0607	*0717
9.1	2.2 0827	0937	1047	1157	1266	1375	1485	1594	1703	1812
9.2	2.2 1920	2029	2138	2246	2354	2462	2570	2678	2786	2894
9.3	2.2 3001	3109	3216	3324	3431	3538	3645	3751	3858	3965
9.4	2.2 4071	4177	4284	4390	4496	4601	4707	4813	4918	5024
9.5	2.2 5129	5234	5339	5444	5549	5654	5759	5863	5968	6072
9.6	2.2 6176	6280	6384	6488	6592	6696	6799	6903	7006	7109
9.7	2.2 7213	7316	7419	7521	7624	7727	7829	7932	8034	8136
9.8	2.2 8238	8340	8442	8544	8646	8747	8849	8950	9051	9152
9.9	2.2 9253	9354	9455	9556	9657	9757	9858	9958	*0058	*0158
10.0	2.3 0259	0358	0458	0558	0658	0757	0857	0956	1055	1154
N	0	1	2	3	4	5	6	7	8	9

TABLE III FOUR-PLACE VALUES OF TRIGONOMETRIC FUNCTIONS

Angle θ		sin θ	csc θ	tan θ	cot θ	sec θ	cos θ		
Degrees	Radians								
0° 00′	.0000	.0000	No value	.0000	No value	1.000	1.0000	1.5708	90° 00′
10	029	029	343.8	029	343.8	000	000	679	50
20	058	058	171.9	058	171.9	000	000	650	40
30	.0087	.0087	114.6	.0087	114.6	1.000	1.0000	1.5621	30
40	116	116	85.95	116	85.94	000	.9999	592	20
50	145	145	68.76	145	68.75	000	999	563	10
1° 00′	.0175	.0175	57.30	.0175	57.29	1.000	.9998	1.5533	89° 00′
10	204	204	49.11	204	49.10	000	998	504	50
20	233	233	42.98	233	42.96	000	997	475	40
30	.0262	.0262	38.20	.0262	38.19	1.000	.9997	1.5446	30
40	291	291	34.38	291	34.37	000	996	417	20
50	320	320	31.26	320	31.24	001	995	388	10
2° 00′	.0349	.0349	28.65	.0349	28.64	1.001	.9994	1.5359	88° 00′
10	378	378	26.45	378	26.43	001	993	330	50
20	407	407	24.56	407	24.54	001	992	301	40
30	.0436	.0436	22.93	.0437	22.90	1.001	.9990	1.5272	30
40	465	465	21.49	466	21.47	001	989	243	20
50	495	494	20.23	495	20.21	001	988	213	10
3° 00′	.0524	.0523	19.11	.0524	19.08	1.001	.9986	1.5184	87° 00′
10	553	552	18.10	553	18.07	002	985	155	50
20	582	581	17.20	582	17.17	002	983	126	40
30	.0611	.0610	16.38	.0612	16.35	1.002	.9981	1.5097	30
40	640	640	15.64	641	15.60	002	980	068	20
50	669	669	14.96	670	14.92	002	978	039	10
4° 00′	.0698	.0698	14.34	.0699	14.30	1.002	.9976	1.5010	86° 00′
10	727	727	13.76	729	13.73	003	974	981	50
20	756	756	13.23	758	13.20	003	971	952	40
30	.0785	.0785	12.75	.0787	12.71	1.003	.9969	1.4923	30
40	814	814	12.29	816	12.25	003	967	893	20
50	844	843	11.87	846	11.83	004	964	864	10
5° 00′	.0873	.0872	11.47	.0875	11.43	1.004	.9962	1.4835	85° 00′
10	902	901	11.10	904	11.06	004	959	806	50
20	931	929	10.76	934	10.71	004	957	777	40
30	.0960	.0958	10.43	.0963	10.39	1.005	.9954	1.4748	30
40	989	987	10.13	992	10.08	005	951	719	20
50	.1018	.1016	9.839	.1022	9.788	005	948	690	10
6° 00′	.1047	.1045	9.567	.1051	9.514	1.006	.9945	1.4661	84° 00′
		cos θ	sec θ	cot θ	tan θ	csc θ	sin θ	Radians	Degrees
								Angle θ	

TABLE III—continued

Angle θ									
Degrees	Radians	sin θ	csc θ	tan θ	cot θ	sec θ	cos θ		
6° 00′	.1047	.1045	9.567	.1051	9.514	1.006	.9945	1.4661	84° 00′
10	076	074	9.309	080	9.255	006	942	632	50
20	105	103	9.065	110	9.010	006	939	603	40
30	.1134	.1132	8.834	.1139	8.777	1.006	.9936	1.4573	30
40	164	161	8.614	169	8.556	007	932	544	20
50	193	190	8.405	198	8.345	007	929	515	10
7° 00′	.1222	.1219	8.206	.1228	8.144	1.008	.9925	1.4486	83° 00′
10	251	248	8.016	257	7.953	008	922	457	50
20	280	276	7.834	287	7.770	008	918	428	40
30	.1309	.1305	7.661	.1317	7.596	1.009	.9914	1.4399	30
40	338	334	7.496	346	7.429	009	911	370	20
50	367	363	7.337	376	7.269	009	907	341	10
8° 00′	.1396	.1392	7.185	.1405	7.115	1.010	.9903	1.4312	82° 00′
10	425	421	7.040	435	6.968	010	899	283	50
20	454	449	6.900	465	6.827	011	894	254	40
30	.1484	.1478	6.765	.1495	6.691	1.011	.9890	1.4224	30
40	513	507	6.636	524	6.561	012	886	195	20
50	542	536	6.512	554	6.435	012	881	166	10
9° 00′	.1571	.1564	6.392	.1584	6.314	1.012	.9877	1.4137	81° 00′
10	600	593	277	614	197	013	872	108	50
20	629	622	166	644	084	013	868	079	40
30	.1658	.1650	6.059	.1673	5.976	1.014	.9863	1.4050	30
40	687	679	5.955	703	871	014	858	1.4021	20
50	716	708	855	733	769	015	853	992	10
10° 00′	.1745	.1736	5.759	.1763	5.671	1.015	.9848	1.3963	80° 00′
10	774	765	665	793	576	016	843	934	50
20	804	794	575	823	485	016	838	904	40
30	.1833	.1822	5.487	.1853	5.396	1.017	.9833	1.3875	30
40	862	851	403	883	309	018	827	846	20
50	891	880	320	914	226	018	822	817	10
11° 00′	.1920	.1908	5.241	.1944	5.145	1.019	.9816	1.3788	79° 00′
10	949	937	164	974	066	019	811	759	50
20	978	965	089	.2004	4.989	020	805	730	40
30	.2007	.1994	5.016	.2035	4.915	1.020	.9799	1.3701	30
40	036	.2022	4.945	065	843	021	793	672	20
50	065	051	876	095	773	022	787	643	10
12° 00′	.2094	.2079	4.810	.2126	4.705	1.022	.9781	1.3614	78° 00′
10	123	108	745	156	638	023	775	584	50
20	153	136	682	186	574	024	769	555	40
30	.2182	.2164	4.620	.2217	4.511	1.024	.9763	1.3526	30
40	211	193	560	247	449	025	757	497	20
50	240	221	502	278	390	026	750	468	10
13° 00′	.2269	.2250	4.445	.2309	4.331	1.026	.9744	1.3439	77° 00′
		cos θ	sec θ	cot θ	tan θ	csc θ	sin θ	Radians	Degrees
								Angle θ	

Tables B-7

TABLE III—continued

Angle θ									
Degrees	Radians	sin θ	csc θ	tan θ	cot θ	sec θ	cos θ		
13° 00′	.2269	.2250	4.445	.2309	4.331	1.026	.9744	1.3439	77° 00′
10	298	278	390	339	275	027	737	410	50
20	327	306	336	370	219	028	730	381	40
30	.2356	.2334	4.284	.2401	4.165	1.028	.9724	1.3352	30
40	385	363	232	432	113	029	717	323	20
50	414	391	182	462	061	030	710	294	10
14° 00′	.2443	.2419	4.134	.2493	4.011	1.031	.9703	1.3265	76° 00′
10	473	447	086	524	3.962	031	696	235	50
20	502	476	039	555	914	032	689	206	40
30	.2531	.2504	3.994	.2586	3.867	1.033	.9681	1.3177	30
40	560	532	950	617	821	034	674	148	20
50	589	560	906	648	776	034	667	119	10
15° 00′	.2618	.2588	3.864	.2679	3.732	1.035	.9659	1.3090	75° 00′
10	647	616	822	711	689	036	652	061	50
20	676	644	782	742	647	037	644	032	40
30	.2705	.2672	3.742	.2773	3.606	1.038	.9636	1.3003	30
40	734	700	703	805	566	039	628	974	20
50	763	728	665	836	526	039	621	945	10
16° 00′	.2793	.2756	3.628	.2867	3.487	1.040	.9613	1.2915	74° 00′
10	822	784	592	899	450	041	605	886	50
20	851	812	556	931	412	042	596	857	40
30	.2880	.2840	3.521	.2962	3.376	1.043	.9588	1.2828	30
40	909	868	487	994	340	044	580	799	20
50	938	896	453	.3026	305	045	572	770	10
17° 00′	.2967	.2924	3.420	.3057	3.271	1.046	.9563	1.2741	73° 00′
10	996	952	388	089	237	047	555	712	50
20	.3025	979	357	121	204	048	546	683	40
30	.3054	.3007	3.326	.3153	3.172	1.048	.9537	1.2654	30
40	083	035	295	185	140	049	528	625	20
50	113	062	265	217	108	050	520	595	10
18° 00′	.3142	.3090	3.236	.3249	3.078	1.051	.9511	1.2566	72° 00′
10	171	118	207	281	047	052	502	537	50
20	200	145	179	314	018	053	492	508	40
30	.3229	.3173	3.152	.3346	2.989	1.054	.9483	1.2479	30
40	258	201	124	378	960	056	474	450	20
50	287	228	098	411	932	057	465	421	10
19° 00′	.3316	.3256	3.072	.3443	2.904	1.058	.9455	1.2392	71° 00′
10	345	283	046	476	877	059	446	363	50
20	374	311	021	508	850	060	436	334	40
30	.3403	.3338	2.996	.3541	2.824	1.061	.9426	1.2305	30
40	432	365	971	574	798	062	417	275	20
50	462	393	947	607	773	063	407	246	10
20° 00′	.3491	.3420	2.924	.3640	2.747	1.064	.9397	1.2217	70° 00′
		cos θ	sec θ	cot θ	tan θ	csc θ	sin θ	Radians	Degrees
								Angle θ	

TABLE III—continued

Angle θ		sin θ	csc θ	tan θ	cot θ	sec θ	cos θ			
Degrees	Radians									
20° 00′	.3491	.3420	2.924	.3640	2.747	1.064	.9397	1.2217		70° 00′
10	520	448	901	673	723	065	387	188		50
20	549	475	878	706	699	066	377	159		40
30	.3578	.3502	2.855	.3739	2.675	1.068	.9367	1.2130		30
40	607	529	833	772	651	069	356	101		20
50	636	557	812	805	628	070	346	072		10
21° 00′	.3665	.3584	2.790	.3839	2.605	1.071	.9336	1.2043		69° 00′
10	694	611	769	872	583	072	325	1.2014		50
20	723	638	749	906	560	074	315	985		40
30	.3752	.3665	2.729	.3939	2.539	1.075	.9304	1.1956		30
40	782	692	709	973	517	076	293	926		20
50	811	719	689	.4006	496	077	283	897		10
22° 00′	.3840	.3746	2.669	.4040	2.475	1.079	.9272	1.1868		68° 00′
10	869	773	650	074	455	080	261	839		50
20	898	800	632	108	434	081	250	810		40
30	.3927	.3827	2.613	.4142	2.414	1.082	.9239	1.1781		30
40	956	854	595	176	394	084	228	752		20
50	985	881	577	210	375	085	216	723		10
23° 00′	.4014	.3907	2.559	.4245	2.356	1.086	.9205	1.1694		67° 00′
10	043	934	542	279	337	088	194	665		50
20	072	961	525	314	318	089	182	636		40
30	.4102	.3987	2.508	.4348	2.300	1.090	.9171	1.1606		30
40	131	.4014	491	383	282	092	159	577		20
50	160	041	475	417	264	093	147	548		10
24° 00′	.4189	.4067	2.459	.4452	2.246	1.095	.9135	1.1519		66° 00′
10	218	094	443	487	229	096	124	490		50
20	247	120	427	522	211	097	112	461		40
30	.4276	.4147	2.411	.4557	2.194	1.099	.9100	1.1432		30
40	305	173	396	592	177	100	088	403		20
50	334	200	381	628	161	102	075	374		10
25° 00′	.4363	.4226	2.366	.4663	2.145	1.103	.9063	1.1345		65° 00′
10	392	253	352	699	128	105	051	316		50
20	422	279	337	734	112	106	038	286		40
30	.4451	.4305	2.323	.4770	2.097	1.108	.9026	1.1257		30
40	480	331	309	806	081	109	013	228		20
50	509	358	295	841	066	111	001	199		10
26° 00′	.4538	.4384	2.281	.4877	2.050	1.113	.8988	1.1170		64° 00′
10	567	410	268	913	035	114	975	141		50
20	596	436	254	950	020	116	962	112		40
30	.4625	.4462	2.241	.4986	2.006	1.117	.8949	1.1083		30
40	654	488	228	.5022	1.991	119	936	054		20
50	683	514	215	059	977	121	923	1.1025		10
27° 00′	.4712	.4540	2.203	.5095	1.963	1.122	.8910	1.0996		63° 00′
		cos θ	sec θ	cot θ	tan θ	csc θ	sin θ	Radians	Degrees	
								Angle θ		

TABLE III—continued

Angle θ		sin θ	csc θ	tan θ	cot θ	sec θ	cos θ		
Degrees	Radians								
27° 00′	.4712	.4540	2.203	.5095	1.963	1.122	.8910	1.0996	63° 00′
10	741	566	190	132	949	124	897	966	50
20	771	592	178	169	935	126	884	937	40
30	.4800	.4617	2.166	.5206	1.921	1.127	.8870	1.0908	30
40	829	643	154	243	907	129	857	879	20
50	858	669	142	280	894	131	843	850	10
28° 00′	.4887	.4695	2.130	.5317	1.881	1.133	.8829	1.0821	62° 00′
10	916	720	118	354	868	134	816	792	50
20	945	746	107	392	855	136	802	763	40
30	.4974	.4772	2.096	.5430	1.842	1.138	.8788	1.0734	30
40	.5003	797	085	467	829	140	774	705	20
50	032	823	074	505	816	142	760	676	10
29° 00′	.5061	.4848	2.063	.5543	1.804	1.143	.8746	1.0647	61° 00′
10	091	874	052	581	792	145	732	617	50
20	120	899	041	619	780	147	718	588	40
30	.5149	.4924	2.031	.5658	1.767	1.149	.8704	1.0559	30
40	178	950	020	696	756	151	689	530	20
50	207	975	010	735	744	153	675	501	10
30° 00′	.5236	.5000	2.000	.5774	1.732	1.155	.8660	1.0472	60° 00′
10	265	025	1.990	812	720	157	646	443	50
20	294	050	980	851	709	159	631	414	40
30	.5323	.5075	1.970	.5890	1.698	1.161	.8616	1.0385	30
40	352	100	961	930	686	163	601	356	20
50	381	125	951	969	675	165	587	327	10
31° 00′	.5411	.5150	1.942	.6009	1.664	1.167	.8572	1.0297	59° 00′
10	440	175	932	048	653	169	557	268	50
20	469	200	923	088	643	171	542	239	40
30	.5498	.5225	1.914	.6128	1.632	1.173	.8526	1.0210	30
40	527	250	905	168	621	175	511	181	20
50	556	275	896	208	611	177	496	152	10
32° 00′	.5585	.5299	1.887	.6249	1.600	1.179	.8480	1.0123	58° 00′
10	614	324	878	289	590	181	465	094	50
20	643	348	870	330	580	184	450	065	40
30	.5672	.5373	1.861	.6371	1.570	1.186	.8434	1.0036	30
40	701	398	853	412	560	188	418	1.0007	20
50	730	422	844	453	550	190	403	977	10
33° 00′	.5760	.5446	1.836	.6494	1.540	1.192	.8387	.9948	57° 00′
10	789	471	828	536	530	195	371	919	50
20	818	495	820	577	520	197	355	890	40
30	.5847	.5519	1.812	.6619	1.511	1.199	.8339	.9861	30
40	876	544	804	661	501	202	323	832	20
50	905	568	796	703	1.492	204	307	803	10
34° 00′	.5934	.5592	1.788	.6745	1.483	1.206	.8290	.9774	56° 00′
		cos θ	sec θ	cot θ	tan θ	csc θ	sin θ	Radians	Degrees
								Angle θ	

TABLE III—*continued*

Angle θ		sin θ	csc θ	tan θ	cot θ	sec θ	cos θ		
Degrees	Radians								
34° 00'	.5934	.5592	1.788	.6745	1.483	1.206	.8290	.9774	56° 00'
10	963	616	781	787	473	209	274	745	50
20	992	640	773	830	464	211	258	716	40
30	.6021	.5664	1.766	.6873	1.455	1.213	.8241	.9687	30
40	050	688	758	916	446	216	225	657	20
50	080	712	751	959	437	218	208	628	10
35° 00'	.6109	.5736	1.743	.7002	1.428	1.221	.8192	.9599	55° 00'
10	138	760	736	046	419	223	175	570	50
20	167	783	729	089	411	226	158	541	40
30	.6196	.5807	1.722	.7133	1.402	1.228	.8141	.9512	30
40	225	831	715	177	393	231	124	483	20
50	254	854	708	221	385	233	107	454	10
36° 00'	.6283	.5878	1.701	.7265	1.376	1.236	.8090	.9425	54° 00'
10	312	901	695	310	368	239	073	396	50
20	341	925	688	355	360	241	056	367	40
30	.6370	.5948	1.681	.7400	1.351	1.244	.8039	.9338	30
40	400	972	675	445	343	247	021	308	20
50	429	995	668	490	335	249	004	279	10
37° 00'	.6458	.6018	1.662	.7536	1.327	1.252	.7986	.9250	53° 00'
10	487	041	655	581	319	255	969	221	50
20	516	065	649	627	311	258	951	192	40
30	.6545	.6088	1.643	.7673	1.303	1.260	.7934	.9163	30
40	574	111	636	720	295	263	916	134	20
50	603	134	630	766	288	266	898	105	10
38° 00'	.6632	.6157	1.624	.7813	1.280	1.269	.7880	.9076	52° 00'
10	661	180	618	860	272	272	862	047	50
20	690	202	612	907	265	275	844	.9018	40
30	.6720	.6225	1.606	.7954	1.257	1.278	.7826	.8988	30
40	749	248	601	.8002	250	281	808	959	20
50	778	271	595	050	242	284	790	930	10
39° 00'	.6807	.6293	1.589	.8098	1.235	1.287	.7771	.8901	51° 00'
10	836	316	583	146	228	290	753	872	50
20	865	338	578	195	220	293	735	843	40
30	.6894	.6361	1.572	.8243	1.213	1.296	.7716	.8814	30
40	923	383	567	292	206	299	698	785	20
50	952	406	561	342	199	302	679	756	10
40° 00'	.6981	.6428	1.556	.8391	1.192	1.305	.7660	.8727	50° 00'
10	.7010	450	550	441	185	309	642	698	50
20	039	472	545	491	178	312	623	668	40
30	.7069	.6494	1.540	.8541	1.171	1.315	.7604	.8639	30
40	098	517	535	591	164	318	585	610	20
50	127	539	529	642	157	322	566	581	10
41° 00'	.7156	.6561	1.524	.8693	1.150	1.325	.7547	.8552	49° 00'
		cos θ	sec θ	cot θ	tan θ	csc θ	sin θ	Radians	Degrees
								Angle θ	

TABLE III—*continued*

Angle θ		sin θ	csc θ	tan θ	cot θ	sec θ	cos θ		
Degrees	Radians								
41° 00′	.7156	.6561	1.524	.8693	1.150	1.325	.7547	.8552	49° 00′
10	185	583	519	744	144	328	528	523	50
20	214	604	514	796	137	332	509	494	40
30	.7243	.6626	1.509	.8847	1.130	1.335	.7490	.8465	30
40	272	648	504	899	124	339	470	436	20
50	301	670	499	952	117	342	451	407	10
42° 00′	.7330	.6691	1.494	.9004	1.111	1.346	.7431	.8378	48° 00′
10	359	713	490	057	104	349	412	348	50
20	389	734	485	110	098	353	392	319	40
30	.7418	.6756	1.480	.9163	1.091	1.356	.7373	.8290	30
40	447	777	476	217	085	360	353	261	20
50	476	799	471	271	079	364	333	232	10
43° 00′	.7505	.6820	1.466	.9325	1.072	1.367	.7314	.8203	47° 00′
10	534	841	462	380	066	371	294	174	50
20	563	862	457	435	060	375	274	145	40
30	.7592	.6884	1.453	.9490	1.054	1.379	.7254	.8116	30
40	621	905	448	545	048	382	234	087	20
50	650	926	444	601	042	386	214	058	10
44° 00′	.7679	.6947	1.440	.9657	1.036	1.390	.7193	.8029	46° 00′
10	709	967	435	713	030	394	173	.7999	50
20	738	988	431	770	024	398	153	970	40
30	.7767	.7009	1.427	.9827	1.018	1.402	.7133	.7941	30
40	796	030	423	884	012	406	112	912	20
50	825	050	418	942	006	410	092	883	10
45° 00′	.7854	.7071	1.414	1.000	1.000	1.414	.7071	.7854	45° 00′
		cos θ	sec θ	cot θ	tan θ	csc θ	sin θ	Radians	Degrees
								Angle θ	

B-12 Appendix

TABLE IV   FOUR-PLACE LOGARITHMS OF TRIGONOMETRIC FUNCTIONS

Angle θ in Degrees

Attach -10 to Logarithms Obtained from This Table

Angle θ	L sin θ	L csc θ	L tan θ	L cot θ	L sec θ	L cos θ	
0° 00′	No value	No value	No value	No value	10.0000	10.0000	90° 00′
10	7.4637	12.5363	7.4637	12.5363	.0000	.0000	50
20	.7648	.2352	.7648	.2352	.0000	.0000	40
30	7.9408	12.0592	7.9409	12.0591	.0000	.0000	30
40	8.0658	11.9342	8.0658	11.9342	.0000	.0000	20
50	.1627	.8373	.1627	.8373	.0000	10.0000	10
1° 00′	8.2419	11.7581	8.2419	11.7581	10.0001	9.9999	89° 00′
10	.3088	.6912	.3089	.6911	.0001	.9999	50
20	.3668	.6332	.3669	.6331	.0001	.9999	40
30	.4179	.5821	.4181	.5819	.0001	.9999	30
40	.4637	.5363	.4638	.5362	.0002	.9998	20
50	.5050	.4950	.5053	.4947	.0002	.9998	10
2° 00′	8.5428	11.4572	8.5431	11.4569	10.0003	9.9997	88° 00′
10	.5776	.4224	.5779	.4221	.0003	.9997	50
20	.6097	.3903	.6101	.3899	.0004	.9996	40
30	.6397	.3603	.6401	.3599	.0004	.9996	30
40	.6677	.3323	.6682	.3318	.0005	.9995	20
50	.6940	.3060	.6945	.3055	.0005	.9995	10
3° 00′	8.7188	11.2812	8.7194	11.2806	10.0006	9.9994	87° 00′
10	.7423	.2577	.7429	.2571	.0007	.9993	50
20	.7645	.2355	.7652	.2348	.0007	.9993	40
30	.7857	.2143	.7865	.2135	.0008	.9992	30
40	.8059	.1941	.8067	.1933	.0009	.9991	20
50	.8251	.1749	.8261	.1739	.0010	.9990	10
4° 00′	8.8436	11.1564	8.8446	11.1554	10.0011	9.9989	86° 00′
10	.8613	.1387	.8624	.1376	.0011	.9989	50
20	.8783	.1217	.8795	.1205	.0012	.9988	40
30	.8946	.1054	.8960	.1040	.0013	.9987	30
40	.9104	.0896	.9118	.0882	.0014	.9986	20
50	.9256	.0744	.9272	.0728	.0015	.9985	10
5° 00′	8.9403	11.0597	8.9420	11.0580	10.0017	9.9983	85° 00′
10	.9545	.0455	.9563	.0437	.0018	.9982	50
20	.9682	.0318	.9701	.0299	.0019	.9981	40
30	.9816	.0184	.9836	.0164	.0020	.9980	30
40	8.9945	11.0055	8.9966	11.0034	.0021	.9979	20
50	9.0070	10.9930	9.0093	10.9907	.0023	.9977	10
6° 00′	9.0192	10.9808	9.0216	10.9784	10.0024	9.9976	84° 00′
	L cos θ	L sec θ	L cot θ	L tan θ	L csc θ	L sin θ	Angle θ

TABLE IV—*continued*

Attach −10 to Logarithms Obtained from This Table

Angle θ	L sin θ	L csc θ	L tan θ	L cot θ	L sec θ	L cos θ	
6° 00′	9.0192	10.9808	9.0216	10.9784	10.0024	9.9976	84° 00′
10	.0311	.9689	.0336	.9664	.0025	.9975	50
20	.0426	.9574	.0453	.9547	.0027	.9973	40
30	.0539	.9461	.0567	.9433	.0028	.9972	30
40	.0648	.9352	.0678	.9322	.0029	.9971	20
50	.0755	.9245	.0786	.9214	.0031	.9969	10
7° 00′	9.0859	10.9141	9.0891	10.9109	10.0032	9.9968	83° 00′
10	.0961	.9039	.0995	.9005	.0034	.9966	50
20	.1060	.8940	.1096	.8904	.0036	.9964	40
30	.1157	.8843	.1194	.8806	.0037	.9963	30
40	.1252	.8748	.1291	.8709	.0039	.9961	20
50	.1345	.8655	.1385	.8615	.0041	.9959	10
8° 00′	9.1436	10.8564	9.1478	10.8522	10.0042	9.9958	82° 00′
10	.1525	.8475	.1569	.8431	.0044	.9956	50
20	.1612	.8388	.1658	.8342	.0046	.9954	40
30	.1697	.8303	.1745	.8255	.0048	.9952	30
40	.1781	.8219	.1831	.8169	.0050	.9950	20
50	.1863	.8137	.1915	.8085	.0052	.9948	10
9° 00′	9.1943	10.8057	9.1997	10.8003	10.0054	9.9946	81° 00′
10	.2022	.7978	.2078	.7922	.0056	.9944	50
20	.2100	.7900	.2158	.7842	.0058	.9942	40
30	.2176	.7824	.2236	.7764	.0060	.9940	30
40	.2251	.7749	.2313	.7687	.0062	.9938	20
50	.2324	.7676	.2389	.7611	.0064	.9936	10
10° 00′	9.2397	10.7603	9.2463	10.7537	10.0066	9.9934	80° 00′
10	.2468	.7532	.2536	.7464	.0069	.9931	50
20	.2538	.7462	.2609	.7391	.0071	.9929	40
30	.2606	.7394	.2680	.7320	.0073	.9927	30
40	.2674	.7326	.2750	.7250	.0076	.9924	20
50	.2740	.7260	.2819	.7181	.0078	.9922	10
11° 00′	9.2806	10.7194	9.2887	10.7113	10.0081	9.9919	79° 00′
10	.2870	.7130	.2953	.7047	.0083	.9917	50
20	.2934	.7066	.3020	.6980	.0086	.9914	40
30	.2997	.7003	.3085	.6915	.0088	.9912	30
40	.3058	.6942	.3149	.6851	.0091	.9909	20
50	.3119	.6881	.3212	.6788	.0093	.9907	10
12° 00′	9.3179	10.6821	9.3275	10.6725	10.0096	9.9904	78° 00′
10	.3238	.6762	.3336	.6664	.0099	.9901	50
20	.3296	.6704	.3397	.6603	.0101	.9899	40
30	.3353	.6647	.3458	.6542	.0104	.9896	30
40	.3410	.6590	.3517	.6483	.0107	.9893	20
50	.3466	.6534	.3576	.6424	.0110	.9890	10
13° 00′	9.3521	10.6479	9.3634	10.6366	10.0113	9.9887	77° 00′
	L cos θ	L sec θ	L cot θ	L tan θ	L csc θ	L sin θ	Angle θ

B-14 Appendix

TABLE IV—*continued*

Attach −10 to Logarithms Obtained from This Table

Angle θ	L sin θ	L csc θ	L tan θ	L cot θ	L sec θ	L cos θ	
13° 00′	9.3521	10.6479	9.3634	10.6366	10.0113	9.9887	77° 00′
10	.3575	.6425	.3691	.6309	.0116	.9884	50
20	.3629	.6371	.3748	.6252	.0119	.9881	40
30	.3682	.6318	.3804	.6196	.0122	.9878	30
40	.3734	.6266	.3859	.6141	.0125	.9875	20
50	.3786	.6214	.3914	.6086	.0128	.9872	10
14° 00′	9.3837	10.6163	9.3968	10.6032	10.0131	9.9869	76° 00′
10	.3887	.6113	.4021	.5979	.0134	.9866	50
20	.3937	.6063	.4074	.5926	.0137	.9863	40
30	.3986	.6014	.4127	.5873	.0141	.9859	30
40	.4035	.5965	.4178	.5822	.0144	.9856	20
50	.4083	.5917	.4230	.5770	.0147	.9853	10
15° 00′	9.4130	10.5870	9.4281	10.5719	10.0151	9.9849	75° 00′
10	.4177	.5823	.4331	.5669	.0154	.9846	50
20	.4223	.5777	.4381	.5619	.0157	.9843	40
30	.4269	.5731	.4430	.5570	.0161	.9839	30
40	.4314	.5686	.4479	.5521	.0164	.9836	20
50	.4359	.5641	.4527	.5473	.0168	.9832	10
16° 00′	9.4403	10.5597	9.4575	10.5425	10.0172	9.9828	74° 00′
10	.4447	.5553	.4622	.5378	.0175	.9825	50
20	.4491	.5509	.4669	.5331	.0179	.9821	40
30	.4533	.5467	.4716	.5284	.0183	.9817	30
40	.4576	.5424	.4762	.5238	.0186	.9814	20
50	.4618	.5382	.4808	.5192	.0190	.9810	10
17° 00′	9.4659	10.5341	9.4853	10.5147	10.0194	9.9806	73° 00′
10	.4700	.5300	.4898	.5102	.0198	.9802	50
20	.4741	.5259	.4943	.5057	.0202	.9798	40
30	.4781	.5219	.4987	.5013	.0206	.9794	30
40	.4821	.5179	.5031	.4969	.0210	.9790	20
50	.4861	.5139	.5075	.4925	.0214	.9786	10
18° 00′	9.4900	10.5100	9.5118	10.4882	10.0218	9.9782	72° 00′
10	.4939	.5061	.5161	.4839	.0222	.9778	50
20	.4977	.5023	.5203	.4797	.0226	.9774	40
30	.5015	.4985	.5245	.4755	.0230	.9770	30
40	.5052	.4948	.5287	.4713	.0235	.9765	20
50	.5090	.4910	.5329	.4671	.0239	.9761	10
19° 00′	9.5126	10.4874	9.5370	10.4630	10.0243	9.9757	71° 00′
10	.5163	.4837	.5411	.4589	.0248	.9752	50
20	.5199	.4801	.5451	.4549	.0252	.9748	40
30	.5235	.4765	.5491	.4509	.0257	.9743	30
40	.5270	.4730	.5531	.4469	.0261	.9739	20
50	.5306	.4694	.5571	.4429	.0266	.9734	10
20° 00′	9.5341	10.4659	9.5611	10.4389	10.0270	9.9730	70° 00′
	L cos θ	L sec θ	L cot θ	L tan θ	L csc θ	L sin θ	Angle θ

TABLE IV—*continued*

Attach −10 to Logarithms Obtained from This Table

Angle θ	L sin θ	L csc θ	L tan θ	L cot θ	L sec θ	L cos θ	
20° 00′	9.5341	10.4659	9.5611	10.4389	10.0270	9.9730	70° 00′
10	.5375	.4625	.5650	.4350	.0275	.9725	50
20	.5409	.4591	.5689	.4311	.0279	.9721	40
30	.5443	.4557	.5727	.4273	.0284	.9716	30
40	.5477	.4523	.5766	.4234	.0289	.9711	20
50	.5510	.4490	.5804	.4196	.0294	.9706	10
21° 00′	9.5543	10.4457	9.5842	10.4158	10.0298	9.9702	69° 00′
10	.5576	.4424	5879	.4121	.0303	.9797	50
20	.5609	.4391	.5917	.4083	.0308	.9692	40
30	.5641	.4359	.5954	.4046	.0313	.9687	30
40	.5673	.4327	.5991	.4009	.0318	.9682	20
50	.5704	.4296	.6028	.3972	.0323	.9677	10
22° 00′	9.5736	10.4264	9.6064	10.3936	10.0328	9.9672	68° 00′
10	.5767	.4233	.6100	.3900	.0333	.9667	50
20	.5798	.4202	.6136	.3864	.0339	.9661	40
30	.5828	.4172	.6172	.3828	.0344	.9656	30
40	.5859	.4141	.6208	.3792	.0349	.9651	20
50	.5889	.4111	.6243	.3757	.0354	.9646	10
23° 00′	9.5919	10.4081	9.6279	10.3721	10.0360	9.9640	67° 00′
10	.5984	.4052	.6314	.3686	.0365	.9635	50
20	.5978	.4022	.6348	.3652	.0371	.9629	40
30	.6007	.3993	.6383	.3617	.0376	.9624	30
40	.6036	.3964	.6417	.3583	.0382	.9618	20
50	.6065	.3935	.6452	.3548	.0387	.9613	10
24° 00′	9.6093	10.3907	9.6486	10.3514	10.0393	9.9607	66° 00′
10	.6121	.3879	.6520	.3480	.0398	.9602	50
20	.6149	.3851	.6553	.3447	.0404	.9596	40
30	.6177	.3823	.6587	.3413	.0410	.9590	30
40	.6205	.3795	.6620	.3380	.0416	.9584	20
50	.6232	.3768	.6654	.3346	.0421	.9579	10
25° 00′	9.6259	10.3741	9.6687	10.3313	10.0427	9.9573	65° 00′
10	.6286	.3714	.6720	.3280	.0433	.9567	50
20	.6313	.3687	.6752	.3248	.0439	.9561	40
30	.6340	.3660	.6785	.3215	.0445	.9555	30
40	.6366	.3634	.6817	.3183	.0451	.9549	20
50	.6392	.3608	.6850	.3150	.0457	.9543	10
26° 00′	9.6418	10.3582	9.6882	10.3118	10.0463	9.9537	64° 00′
10	.6444	.3556	.6914	.3086	.0470	.9530	50
20	.6470	.3530	.6946	.3054	.0476	.9524	40
30	.6495	.3505	.6977	.3023	.0482	.9518	30
40	.6521	.3479	.7009	.2991	.0488	.9512	20
50	.6546	.3454	.7040	.2960	.0495	.9505	10
27° 00′	9.6570	10.3430	9.7072	10.2928	10.0501	9.9499	63° 00′
	L cos θ	L sec θ	L cot θ	L tan θ	L csc θ	L sin θ	Angle θ

TABLE IV—*continued*

Attach −10 to Logarithms Obtained from This Table

Angle θ	L sin θ	L csc θ	L tan θ	L cot θ	L sec θ	L cos θ	
27° 00′	9.6570	10.3430	9.7072	10.2928	10.0501	9.9499	63° 00′
10	.6595	.3405	.7103	.2897	.0508	.9492	50
20	.6620	.3380	.7134	.2866	.0514	.9486	40
30	.6644	.3356	.7165	.2835	.0521	.9479	30
40	.6668	.3332	.7196	.2804	.0527	.9473	20
50	.6692	.3308	.7226	.2774	.0534	.9466	10
28° 00′	9.6716	10.3284	9.7257	10.2743	10.0541	9.9459	62° 00′
10	.6740	.3260	.7287	.2713	.0547	.9453	50
20	.6763	.3237	.7317	.2683	.0554	.9446	40
30	.6787	.3213	.7348	.2652	.0561	.9439	30
40	.6810	.3190	.7378	.2622	.0568	.9432	20
50	.6833	.3167	.7408	.2592	.0575	.9425	10
29° 00′	9.6856	10.3144	9.7438	10.2562	10.0582	9.9418	61° 00′
10	.6878	.3122	.7467	.2533	.0589	.9411	50
20	.6901	.3099	.7497	.2503	.0596	.9404	40
30	.6923	.3077	.7526	.2474	.0603	.9397	30
40	.6946	.3054	.7556	.2444	.0610	.9390	20
50	.6968	.3032	.7585	.2415	.0617	.9383	10
30° 00′	9.6990	10.3010	9.7614	10.2386	10.0625	9.9375	60° 00′
10	.7012	.2988	.7644	.2356	.0632	.9368	50
20	.7033	.2967	.7673	.2327	.0639	.9361	40
30	.7055	.2945	.7701	.2299	.0647	.9353	30
40	.7076	.2924	.7730	.2270	.0654	.9346	20
50	.7097	.2903	.7759	.2241	.0662	.9338	10
31° 00′	9.7118	10.2882	9.7788	10.2212	10.0669	9.9331	59° 00′
10	.7139	.2861	.7816	.2184	.0677	.9323	50
20	.7160	.2840	.7845	.2155	.0685	.9315	40
30	.7181	.2819	.7873	.2127	.0692	.9308	30
40	.7201	.2799	.7902	.2098	.0700	.9300	20
50	.7222	.2778	.7930	.2070	.0708	.9292	10
32° 00′	9.7242	10.2758	9.7958	10.2042	10.0716	9.9284	58° 00′
10	.7262	.2738	.7986	.2014	.0724	.9276	50
20	.7282	.2718	.8014	.1986	.0732	.9268	40
30	.7302	.2698	.8042	.1958	.0740	.9260	30
40	.7322	.2678	.8070	.1930	.0748	.9252	20
50	.7342	.2658	.8097	.1903	.0756	.9244	10
33° 00′	9.7361	10.2639	9.8125	10.1875	10.0764	9.9236	57° 00′
10	.7380	.2620	.8153	.1847	.0772	.9228	50
20	.7400	.2600	.8180	.1820	.0781	.9219	40
30	.7419	.2581	.8208	.1792	.0789	.9211	30
40	.7438	.2562	.8235	.1765	.0797	.9203	20
50	.7457	.2543	.8263	.1737	.0806	.9194	10
34° 00′	9.7476	10.2524	9.8290	10.1710	10.0814	9.9186	56° 00′
	L cos θ	L sec θ	L cot θ	L tan θ	L csc θ	L sin θ	Angle θ

Tables **B-17**

TABLE IV—*continued*

Attach −10 to Logarithms Obtained from This Table

Angle θ	L sin θ	L csc θ	L tan θ	L cot θ	L sec θ	L cos θ	
34° 00′	9.7476	10.2524	9.8290	10.1710	10.0814	9.9186	56° 00′
10	.7494	.2506	.8317	.1683	.0823	.9177	50
20	.7513	.2487	.8344	.1656	.0831	.9169	40
30	.7531	.2469	.8371	.1629	.0840	.9160	30
40	.7550	.2450	.8398	.1602	.0849	.9151	20
50	.7568	.2432	.8425	.1575	.0858	.9142	10
35° 00′	9.7586	10.2414	9.8452	10.1548	10.0866	9.9134	55° 00′
10	.7604	.2396	.8479	.1521	.0875	.9125	50
20	.7622	.2378	.8506	.1494	.0884	.9116	40
30	.7640	.2360	.8533	.1467	.0893	.9107	30
40	.7657	.2343	.8559	.1441	.0902	.9098	20
50	.7675	.2325	.8586	.1414	.0911	.9089	10
36° 00′	9.7692	10.2308	9.8613	10.1387	10.0920	9.9080	54° 00′
10	.7710	.2290	.8639	.1361	.0930	.9070	50
20	.7727	.2273	.8666	.1334	.0939	.9061	40
30	.7744	.2256	.8692	.1308	.0948	.9052	30
40	.7761	.2239	.8718	.1282	.0958	.9042	20
50	.7778	.2222	.8745	.1255	.0967	.9033	10
37° 00′	9.7795	10.2205	9.8771	10.1229	10.0977	9.9023	53° 00′
10	.7811	.2189	.8797	.1203	.0986	.9014	50
20	.7828	.2172	.8824	.1176	.0996	.9004	40
30	.7844	.2156	.8850	.1150	.1005	.8995	30
40	.7861	.2139	.8876	.1124	.1015	.8985	20
50	.7877	.2123	.8902	.1098	.1025	.8975	10
38° 00′	9.7893	10.2107	9.8928	10.1072	10.1035	9.8965	52° 00′
10	.7910	.2090	.8954	.1046	.1045	.8955	50
20	.7926	.2074	.8980	.1020	.1055	.8945	40
30	.7941	.2059	.9006	.0994	.1065	.8935	30
40	.7957	.2043	.9032	.0968	.1075	.8925	20
50	.7973	.2027	.9058	.0942	.1085	.8915	10
39° 00′	9.7989	10.2011	9.9084	10.0916	10.1095	9.8905	51° 00′
10	.8004	.1996	.9110	.0890	.1105	.8895	50
20	.8020	.1980	.9135	.0865	.1116	.8884	40
30	.8035	.1965	.9161	.0839	.1126	.8874	30
40	.8050	.1950	.9187	.0813	.1136	.8864	20
50	.8066	.1934	.9212	.0788	.1147	.8853	10
40° 00′	9.8081	10.1919	9.9238	10.0762	10.1157	9.8843	50° 00′
10	.8096	.1904	.9264	.0736	.1168	.8832	50
20	.8111	.1889	.9289	.0711	.1179	.8821	40
30	.8125	.1875	.9315	.0685	.1190	.8810	30
40	.8140	.1860	.9341	.0659	.1200	.8800	20
50	.8155	.1845	.9366	.0634	.1211	.8789	10
41° 00′	9.8169	10.1831	9.9392	10.0608	10.1222	9.8778	49° 00′
	L cos θ	L sec θ	L cot θ	L tan θ	L csc θ	L sin θ	Angle θ

TABLE IV—*continued*

Attach −10 to Logarithms Obtained from This Table

Angle θ	L sin θ	L csc θ	L tan θ	L cot θ	L sec θ	L cos θ	
41° 00′	9.8168	10.1831	9.9392	10.0608	10.1222	9.8778	49° 00′
10	.8184	.1816	.9417	.0583	.1233	.8767	50
20	.8198	.1802	.9443	.0557	.1244	.8756	40
30	.8213	.1787	.9468	.0532	.1255	.8745	30
40	.8227	.1773	.9494	.0506	.1267	.8733	20
50	.8241	.1759	.9519	.0481	.1278	.8722	10
42° 00′	9.8255	10.1745	9.9544	10.0456	10.1289	9.8711	48° 00′
10	.8269	.1731	.9570	.0430	.1301	.8699	50
20	.8283	.1717	.9595	.0405	.1312	.8688	40
30	.8297	.1703	.9621	.0379	.1324	.8676	30
40	.8311	.1689	.9646	.0354	.1335	.8665	20
50	.8324	.1676	.9671	.0329	.1347	.8653	10
43° 00′	9.8338	10.1662	9.9697	10.0303	10.1359	9.8641	47° 00′
10	.8351	.1649	.9722	.0278	.1371	.8629	50
20	.8365	.1635	.9747	.0253	.1382	.8618	40
30	.8378	.1622	.9772	.0228	.1394	.8606	30
40	.8391	.1609	.9798	.0202	.1406	.8594	20
50	.8405	.1595	.9823	.0177	.1418	.8582	10
44° 00′	9.8418	10.1582	9.9848	10.0152	10.1431	9.8569	46° 00′
10	.8431	.1569	.9874	.0126	.1443	.8557	50
20	.8444	.1556	.9899	.0101	.1455	.8545	40
30	.8457	.1543	.9924	.0076	.1468	.8532	30
40	.8469	.1531	.9949	.0051	.1480	.8520	20
50	.8482	.1518	9.9975	.0025	.1493	.8507	10
45° 00′	9.8495	10.1505	10.0000	10.0000	10.1505	9.8495	45° 00′
	L cos θ	L sec θ	L cot θ	L tan θ	L csc θ	L sin θ	Angle θ

Tables B-19

INDEX

a^x (exponential function), 205–231
Abscissa, 110
Absolute value, 7, 103
Addition of vectors, 482
Algebraic expression, multiplication, 50–54
  product, 59–66
  rational, 59
  simplify, 59
  sum, 59–66
Amplitude of a curve, 248
Angle(s), 293–299
  initial side of, 293
  in standard position, 293
  measurement of, 294
  negative, 297
  reference, 301
  right, 295, 307–310
  sides of, 293
  straight, 295
  sum of, 294–296
  terminal side of, 293
  trigonometric functions defined on, 300
  vertex of, 293
Antilogarithm, 219
Arccos, 282
Arcsin, 282
Arctan, 282
Argument of complex number, 389–394
ASA, 313
Associative law of addition, 9
Associative law of multiplication, 12
Asymptotes, 147, 182
Axes, coordinate, 108
Axioms of addition, 9, 10, 14, 22
Axioms for multiplication, 12–14

Binomial theorem, coefficients, 436
  expansion, 436
Bisector, perpendicular, 127

Cancellation law, of addition, 11
  multiplication, 16
Catenary, 208

Characteristic, 217
Circle(s), 115–120
   degenerate, 118
   equations of, 115–117
   graph of, 116–117
   unit, 233
Circular functions, 233
   graphs of, 247–260
   inverse, 280–285
Circular inequalities, 133
Closed half-plane, 132
Closed interval, 102
Closure, 28
Combinations, 433
Common logarithms, 216
Commutative group, 27
Commutative law, of addition, 9
   multiplication, 12
Completeness axiom, 27
Completing the square, 77, 142
Complex numbers, 389–394
   operations, 389–394
   polar form, 415
   properties, 389–394
Complex plane, 402
Components of vectors, 491
Composition of functions, 170
Compound fractions, 62–64
Compound interest, 228
Conditional equations, 70
Conic sections, 137, 335
Continuous function, 167
Convergence, 471
Coordinate axes, 109
Coordinate plane, 109
Coordinates, 110
Cosecant function, 233, 254
   graph of, 255
Cosine function, 233–236
   double-angle formula, 271
   graph of, 241–246
   half-angle formula, 271
   inverse, 282
   reduction formulas for, 264–266
   sum formula for, 266

Cotangent function, 115, 137, 233, 254
   graph of, 257
Cramer's rule, 369
Cycle of a curve, 248
Cycloid, 510

Degenerate circle, 118
Degree (angle measurement), 294
Degree, of polynomial, 182
De Moivre's theorem, 419
Determinants, 368–372
   properties, 375–382
Distance, in plane, 110
Distance formula, 115
Distributive laws, 17–20
Division theorem for polynomials, 176
Domain, 158
Dot products, 497–500
Double-angle formulas, 271

, 208
Ellipse, 137, 145, 146
   equation of, 145
   graph of, 145, 146, 335
Elliptic cone, 137
Equations, conditional, 70
   equivalent, 70
   of geometric figure, 115
   graph of, 115
   involving absolute value, 84
   involving radicals, 82, 83
   involving rational functions, 77–79
   linear, 69, 70
   polynomial, 76, 77
   quadratic, 77
   system of,
Exponent, 29–60, 343–382
   bases, 40
   laws of, 36, 37, 40–44
   negative, 35
   positive, 35
   rational, 40, 41
   zero, 36

Index

Exponential decay, 228
Exponential equations, 224–229
Exponential function, 205–206
    limit of, 206
Exponentiating, 225

Factoral notation, 429
Fields, 26
Function(s), 162–167
    algebraic, 192
    circular, 233
    composition of, 169
    continuous, 167, 206
    defined explicitly, 189
    defined implicitly, 189
    difference of, 170
    domain of, 162
    implicit, 188–192
    inverse, 194–200
    one-to-one, 194, 210
    periodic, 235
    polynomial, 175–180
    product of, 170
    quotient of, 170
    range of, 162
    rational, 181–186
    strictly decreasing, 194
    strictly increasing, 194
    sum of, 170
    transcendental, 205
    vector, 481

Geometric figure, 115
Graph, of equation, 115
Greater than ($>$), 6, 101
Group, 26

Half-angle formulas, 272
Half-line, 293
Half-open (or half-closed) interval, 103
Half-plane, 132
Horizontal asymptote, 182
Hyperbola, 137, 147–151
    asymptotes of, 147
    equations of, 147
    graph of, 147, 148, 335

Identity, additive, 9
    algebraic, 69
    multiplicative, 13
    trigonometric, 235–237, 258
    275–278

Identity element of vector addition, 483
Implicit function, 188–192
Induction, 451
Inequalities, 6, 23, 130–135, 141
    absolute, 87
    circular, 102–104, 133
    conditional, 87
    graphs of, 130, 132
    linear, 131
    on number line, 104
    simultaneous, 134
    solution of, 87–89, 92–94
Infinity (∞), 103
Initial side, 182, 293
Interval, 102
    closed, 102
    half-open or half-closed, 102
    open, 102
Inverse, additive, 9, 10
    multiplicative, 13
Inverse circular functions, 280–285
    principal values, 285
Inverse functions, 194–200

Law of cosines, 322
Law of sines, 314
Less than ($<$), 6, 101
Limit, 144, 145, 178, 179, 471
Line(s), equation of, 122, 123
    parallel, 126
    perpendicular, 127, 128
    shape of, 123
Linear equations, 122–128
    graph of, 123
    normal form, 503
    point-slope form, 123
    slope-intercept form, 124
Linear equations, two, 349–354
    dependent, 349–354
    inconsistent, 349–354
    independent, 349–354
Linear inequalities, 131
Linear interpolation, 219, 222
Logarithm(s), 210
    common, 216
    computation with, 221–223
    fundamental properties of,
        210–212, 221
    natural, 216
    table of common, 216–219

Index

Logarithm(s) (cont.)
  table of natural, 227
Logarithmic equations, 224–228
Log   210–214

Mantissa, 217
Matrix, 362
  augmented, 362
  columns, 364, 365
  main diagonal, 362
  operations, 363
  rows, 363
Mean, arithmetic, 461
  geometric, 461
Minus infinity ($-\infty$), 103
Minute (angle measurement), 297
Mixed axioms, 27
Multiplication axioms, 26

Natural logarithms, 216
Negative numbers, 103
Normal form of equation, 503
n-tuple, 358–360
Numbers, counting, 5
  integers, 5, 10
  irrational, 5
  natural, 5
  rational, 5
Number line, 6, 102–105
  intervals on, 102–105
  origin, 6

One-to-one correspondence, 102
One-to-one functions, 194
Open half-plane, 132
Open interval, 102
Order axioms, 22–24, 27
Ordered pair, 158
Order relations, 102
Ordinate, 110
Origin, 6, 109

Parabola, 137–143
  axis of, 138
  equation of, 137
  graph of, 138, 139, 334, 335
  symmetry of, 138
  vertex of, 138
Parallel lines, 126
Parameter, 357
Parametric equations, 508–511

Period (of a function), 235, 259
Periodicity, 235
Permutations, 427
  distinguishable, 438
Perpendicular bisector, 126
Perpendicular lines, 127
Plane(s)
  on number line, 101
  in plane, 110
Point-slope form, 123
Polar axis, 326
Polar coordinates, 326–332
  graphs of, 327
Polar equation, 328, 329
Pole, 326
Polynomial(s), factoring, 54–57
  multiplication, 50–52
Polynomial equations, complex roots, 409–414
Polynomial function, 175–180
  limit of, 178, 179
Positive numbers, 103
Principal values, 158
Probability, 442–446
  disjoint event, 444
  event,
  outcome, 443
Pythagorean theorem, 110, 321

Quadrant, 110
Quadratic formula, 77

Radian, 294
Radicals, index of, 46
  laws of, 46
Range, 158, 162
Rational exponent, 206
Rational function, 181–186
  limit of, 182
Ray, 293
Real numbers, 101–102
  on number line, 101–102
Reciprocal, 13
Reduction formulas, 264–266
Reference angle, 301
Relation(s), 158
  as set of ordered pairs, 158
  graph of, 158
Right angle, 295, 307, 310
Right triangle, 307
  solving, 308, 309

Index

Ring, 28
Root, 70
Rotation, 335

SAS, 313
Scalar(s), 481
Scalar multiplication, 484
Secant function, 233, 254
    graph of,
Second (angle measurement), 297
Sequence, 450
    arithmetic, 459
    geometric, 459
Series, 450, 464
    arithmetic, 467
    geometric, 467
    infinite, 471
Sets, element of, 2
    empty, 2
    equal, 3
    intersection, 2
    notation, 2
    subset, 2
    union, 2
Simultaneous inequalities, 134
Sine curve, 256
Sine function, 233–236
    double-angle formula, 271
    graph of, 241–246
    half-angle formula, 271
    inverse, 282
    reduction formulas for, 264–266
    sum formula for, 266
Sine wave, 256
Slope, 123
    of parallel lines, 126
    of perpendicular lines, 127, 128
Slope-intercept form 124
Solving right triangles, 308, 309
Solving triangles, 308
SSS, 313
Standard position, 291
Straight angle, 295
Sum formulas, 265
Summation notation, 465
Symmetry, 138
    about   or axis, 145
System of equations, 343–382
    echelon form, 356
    triangular form, 356

Table of common logarithms, 216–219
Table of natural logarithms, 227
Table of trigonometric functions, 176–177
Tangent function, 233, 254
    double-angle formulas, 271, 273
    graph of, 140, 256
    half-angle formula, 275
    inverse, 282
    sum formula for, 268–269
Tangent line, 208
Terminal side, 293
Transcendental functions, 205
Transitivity law, 22
Translation, 334
Trichotomy law, 22, 23
Trigonometric equations, 286–288
Trigonometric functions (see Circular functions)
Trigonometric identities, 235–237, 258, 275–278

Unit circle, 233
Unit vectors, 491
Upper bound, 27

Vector(s), 481
    addition of, 481–483
    associative law of addition, 484
    commutative law of addition, 482
    components, 491
    dot product of, 497–500
    function, 508
    identity element for addition, 483
    inverse, 483
    magnitude of, 482
    scalar multiplication, 484
    subtraction of, 484
    unit, 491
    zero, 481
Vertex, 291
Vertical asymptote, 182

  coordinate (abscissa), 110

  coordinate (ordinate), 110

Zero vector, 481